Handbook of Research on Innovations and Applications of AI, IoT, and Cognitive Technologies

Jingyuan Zhao
University of Toronto, Canada

V. Vinoth Kumar
MVJ College of Engineering, India

A volume in the Advances in Computational
Intelligence and Robotics (ACIR) Book Series

Published in the United States of America by
IGI Global
Engineering Science Reference (an imprint of IGI Global)
701 E. Chocolate Avenue
Hershey PA, USA 17033
Tel: 717-533-8845
Fax: 717-533-8661
E-mail: cust@igi-global.com
Web site: http://www.igi-global.com

Library of Congress Cataloging-in-Publication Data

Names: Zhao, Jingyuan, 1968- editor. | Kumar, V. Vinoth, 1988- editor.
Title: Handbook of research on innovations and Applications of AI, IoT, and
  Cognitive Technologies / Jingyuan Zhao, and V. Vinoth Kumar, editors.
Description: Hershey, PA : Engineering Science Reference, an imprint of IGI
  Global, [2021] | Includes bibliographical references and index. |
  Summary: "This book provides perspectives of academics and industrial
  practitioners offering the latest innovations and applications of
  artificial intelligence (AI), Internet-of-Things (IoT) and cognitive
  based smart systems to provide information for all those interested in
  AI, IoT and Cognitive Technologies in emerging and developed economies
  in terms of their respective development situation toward public
  policies, technologies and intellectual capital, innovation systems,
  competition and strategies, marketing and growth capability, governance
  and relegation model etc"-- Provided by publisher.
Identifiers: LCCN 2020046720 (print) | LCCN 2020046721 (ebook) | ISBN
  9781799868705 (hardcover) | ISBN 9781799868729 (ebook)
Subjects: LCSH: Internet of things--Industrial applications. | Artificial
  intelligence--Industrial applications.
Classification: LCC TK5105.8857 .I493 2021 (print) | LCC TK5105.8857
  (ebook) | DDC 004.67/8--dc23
LC record available at https://lccn.loc.gov/2020046720
LC ebook record available at https://lccn.loc.gov/2020046721

This book is published in the IGI Global book series Advances in Computational Intelligence and Robotics (ACIR) (ISSN: 2327-0411; eISSN: 2327-042X)

British Cataloguing in Publication Data
A Cataloguing in Publication record for this book is available from the British Library.

All work contributed to this book is new, previously-unpublished material. The views expressed in this book are those of the authors, but not necessarily of the publisher.

For electronic access to this publication, please contact: eresources@igi-global.com.

Advances in Computational Intelligence and Robotics (ACIR) Book Series

Ivan Giannoccaro
University of Salento, Italy

ISSN:2327-0411
EISSN:2327-042X

MISSION

While intelligence is traditionally a term applied to humans and human cognition, technology has progressed in such a way to allow for the development of intelligent systems able to simulate many human traits. With this new era of simulated and artificial intelligence, much research is needed in order to continue to advance the field and also to evaluate the ethical and societal concerns of the existence of artificial life and machine learning.

The **Advances in Computational Intelligence and Robotics (ACIR) Book Series** encourages scholarly discourse on all topics pertaining to evolutionary computing, artificial life, computational intelligence, machine learning, and robotics. ACIR presents the latest research being conducted on diverse topics in intelligence technologies with the goal of advancing knowledge and applications in this rapidly evolving field.

COVERAGE

- Artificial Intelligence
- Algorithmic Learning
- Computational Logic
- Brain Simulation
- Pattern Recognition
- Cognitive Informatics
- Adaptive and Complex Systems
- Synthetic Emotions
- Automated Reasoning
- Robotics

IGI Global is currently accepting manuscripts for publication within this series. To submit a proposal for a volume in this series, please contact our Acquisition Editors at Acquisitions@igi-global.com or visit: http://www.igi-global.com/publish/.

Titles in this Series

For a list of additional titles in this series, please visit:
http://www.igi-global.com/book-series/advances-computational-intelligence-robotics/73674

Genetic Algorithms and Applications for Stock Trading Optimization
Vivek Kapoor (Devi Ahilya University, Indore, India) and Shubhamoy Dey (Indian Institute of Management, Indore, India)
Engineering Science Reference • © 2021 • 262pp • H/C (ISBN: 9781799841050) • US $225.00

Decision Support Systems and Industrial IoT in Smart Grid, Factories, and Cities
Ismail Butun (Chalmers University of Technology, Sweden & Konya Food and Agriculture University, Turkey & Royal University of Technology, Sweden)
Engineering Science Reference • © 2021 • 285pp • H/C (ISBN: 9781799874683) • US $245.00

Deep Natural Language Processing and AI Applications for Industry 5.0
Poonam Tanwar (Manav Rachna International Institute of Research and Studies, India) Arti Saxena (Manav Rachna International Institute of Research and Studies, India) and C. Priya (Vels Institute of Science, Technology, and Advanced Studies, India)
Engineering Science Reference • © 2021 • 240pp • H/C (ISBN: 9781799877288) • US $245.00

AI Tools and Electronic Virtual Assistants for Improved Business Performance
Christian Graham (University of Maine, USA)
Business Science Reference • © 2021 • 300pp • H/C (ISBN: 9781799838418) • US $245.00

Transforming the Internet of Things for Next-Generation Smart Systems
Bhavya Alankar (Jamia Hamdard, India) Harleen Kaur (Hamdard University, India) and Ritu Chauhan (Amity University, India)
Engineering Science Reference • © 2021 • 173pp • H/C (ISBN: 9781799875413) • US $245.00

Handbook of Research on Machine Learning Techniques for Pattern Recognition and Information Security
Mohit Dua (National Institute of Technology, Kurukshetra, India) and Ankit Kumar Jain (National Institute of Technology, Kurukshetra, India)
Engineering Science Reference • © 2021 • 355pp • H/C (ISBN: 9781799832997) • US $295.00

Driving Innovation and Productivity Through Sustainable Automation
Ardavan Amini (EsseSystems, UK) Stephen Bushell (Bushell Investment Group, UK) and Arshad Mahmood (Birmingham City University, UK)
Engineering Science Reference • © 2021 • 275pp • H/C (ISBN: 9781799858799) • US $245.00

701 East Chocolate Avenue, Hershey, PA 17033, USA
Tel: 717-533-8845 x100 • Fax: 717-533-8661
E-Mail: cust@igi-global.com • www.igi-global.com

List of Contributors

Table of Contents

Section 2
Big Data and Cognitive Technologies: Current Applications and Future Challenges

Detailed Table of Contents

Section 1
AI and IoT: A Blend in Future Technologies and Systems

Section 1 introduces emerging AI and IoT technologies and systems, as well as their innovations and applications that are responsible for the digitization of all the sectors through automation and together rules the technological aspects of our lives. Section 1 is organized into 18 chapters.

Elmustafa Sayed Ali Ahmed, Sudan University of Science and Technology, Sudan & Red Sea University, Sudan
Zahraa Tagelsir Mohammed, Red Sea University, Sudan
Mona Bakri Hassan, Sudan University of Science and Technology, Sudan
Rashid A. Saeed, Sudan University of Science and Technology, Sudan & Taif University, Saudi Arabia

Internet of vehicles (IoV) has recently become an emerging promising field of research due to the increasing number of vehicles each day. It is a part of the internet of things (IoT) which deals with vehicle communications. As vehicular nodes are considered always in motion, they cause frequent changes in the network topology. These changes cause issues in IoV such as scalability, dynamic topology changes, and shortest path for routing. In this chapter, the authors will discuss different optimization algorithms (i.e., clustering algorithms, ant colony optimization, best interface selection [BIS] algorithm, mobility adaptive density connected clustering algorithm, meta-heuristics algorithms, and quality of service [QoS]-based optimization). These algorithms provide an important intelligent role to optimize the operation of IoV networks and promise to develop new intelligent IoV applications.

Rajeev Kumar, Central University of Karnataka, India
Ashraf Hossain, National Institute of Technology, Silchar, India

This chapter presents cooperative relaying networks that are helpful in Internet of Thing (IoT) applications for fifth-generation (5G) radio networks. It provides reliable connectivity as the wireless device is out of range from cellular network, high throughput gains and enhance the lifetime of wireless networks. These features can be achieved by designing the advanced protocols. The design of advanced protocols plays an important role to combat the effect of channel fading, data packet scheduling at the buffered relay, average delay, and traffic intensity. To achieve our goals, we consider two-way cooperative buffered relay networks and then investigate advanced protocols such as without channel state information (CSI) i.e., buffer state information (BSI) only and with partial transmit CSI i.e., BSI/CSI with the assistance of one dimensional Markov chain and transmission policies in fading environment. The outage probability of consecutive links and outage probability of multi-access and broadcast channels are provided in closed-form. Further, the buffered relay achieves maximum throughput gains in closed-form for all these protocols. The objective function of throughput of the buffered relay is evaluated in fractional programming that is transformed into linear program using standard CVX tool. Numerical results show that our proposed protocols performance better as compared to conventional method studied in the literature. Finally, this chapter provides possible future research directions.

Chapter 3

Supriya M. S., Ramaiah University of Applied Sciences, India
Kannika Manjunath, Ramaiah University of Applied Sciences, India
Kavana U. R., Ramaiah University of Applied Sciences, India

Uninvited disasters wreak havoc on society, both economically and psychologically. These losses can be minimized if events can be anticipated ahead of time. The majority of large cities in developing countries with increasing populations are highly vulnerable disaster areas around the world. This is due to a lack of situational information in their authorities in the event of a crisis, which is due to a scarcity of resources. Both natural and human-induced disasters need to be pre-planned and reactive to minimize the risk of causalities and environmental/infrastructural disruption. Disaster recovery systems must also effectively obtain relevant information. The developments in big data and the internet of things (IoT) have made a greater contribution to accuracy and timely decision-making in the disaster management system (DMS). The chapter explains why IoT and big data are needed to cope with disasters, as well as how these technologies work to solve the problem.

Chapter 4

Kishore Kumar K., Koneru Lakshmaiah Education Foundation, India

Smart farming is an evolving concept since IoT sensors are capable of providing agricultural field information and then acting on the basis of user feedback. The main factor in improving the yield of efficient crops is the control of environmental conditions. There is a small yard, farmland, or a plantation area for most of us. However, our busy timetable does not allow us to manage it well. But we can easily accomplish it with the use of technology. So, the authors make an IoT-based smart farming system that can control soil moisture. As data has become a critical component in modern agriculture to assist producers with critical decisions and make a decision with objective data obtained from sensors, significant advantages emerge. This chapter explores the current state of advanced farm management systems by revisiting each critical phase, from data collection in crop fields to variable rate applications, in order for growers to make informed decisions save money while also protecting the environment and transforming how food is grown to meet potential population growth.

Chapter 5

M. V. Ramana Rao, Osmania University, India

Thondepu Adilakshmi, Vasavi College of Engineering, India

M. Gokul Venkatesh, Sidhartha Medical College, India

*Jothikumar R, Department of Computer Science and Engineering, Shadan College of
    Engineering and Technology, India*

In a thickly populated nation like India, it is hard to forecast community transmission of COVID-19. Hence, a number of containment zones had been recognized all over the country separated into red, orange, and green zones, individually. People are restricted to move into these containment zones. This chapter focuses on informing the public about the containment zone when they are in travel and also sends an alert to the police when a person enters the containment zone without permission using the containment zone alert system. This chapter suggests a containment zone alert system by means of geo-fencing technology to identify the movement of public, deliver info about the danger to the public in travel and also send an alert to the police when there is an entry or exit detected in the containment zone by the use of location-based services (LBS). By creating a fence virtually called geo-fence at the containment zones established based on the government info, this system monitors public movements like entry and exit to fence.

Chapter 6

Senthil Murugan Nagarajan, VIT-AP University, India

Muthukumaran V., REVA University, India

Vinoth Kumar V., MVJ College of Engineering, India

Beschi I. S., St. Joseph's College, India

S. Magesh, Maruthi Technocrat Services, India

The workflow between business and manufacturing system level is changing leading to delay in exploring the context of innovative ideas and solutions. Smart manufacturing systems progress rapid growth in integrating the operational capabilities of networking functionality and communication services with cloud-based enterprise architectures through runtime environment. Fine tuning aims to process intelligent management, flexible monitoring, dynamic network services using internet of things (IoT)-based service oriented architecture (SOA) solutions in numerous enterprise systems. SOA is an architectural pattern for building software business systems based on loosely coupled enterprise infrastructure services and components. The IoT-based SOA enterprise systems incorporate data elicitation, integrating agile methodologies, orchestrate underlying black-box services by promoting growth in manufacturer enterprises workflow. This chapter proposes the integration of standard workflow model between business system level and manufacturing production level with an IoT-enabled SOA framework.

Vinoth Kumar, MVJ College of Engineering, India
V. R. Niveditha, Dr. M. G. R. Educational and Research Institute, Chennai, India
V. Muthukumaran, REVA University, India
S.Satheesh Kumar, REVA University, India
Samyukta D. Kumta, REVA University, India
Murugesan R., REVA University, India

Light fidelity (Li-Fi) is a technology that is used to design a wireless network for communication using light. Current technology based on wireless fidelity (Wi-Fi) has some drawbacks that include speed and bandwidth limit, security issues, and attacks by malicious users, which yield Wi-Fi as less reliable compared to LiFi. The conventional key generation techniques are vulnerable to the current technological improvement in terms of computing power, so the solution is to introduce physics laws based on quantum technology and particle nature of light. Here the authors give a methodology to make the BB84 algorithm, a quantum cryptographic algorithm to generate the secret keys which will be shared by polarizing photons and more secure by eliminating one of its limitations that deals with dependency on the classical channel. The result obtained is sequence of 0 and 1, which is the secret key. The authors make use of the generated shared secret key to encrypt data using a one-time pad technique and transmit the encrypted data using LiFi and removing the disadvantage of the existing one-time pad technique.

Arti Jain, Jaypee Institute of Information Technology, Noida, India
Rashmi Kushwah, Jaypee Institute of Information Technology, Noida, India
Abhishek Swaroop, Bhagwan Parshuram Institute of Technology, India
Arun Yadav, National Institute of Technology, Hamirpur, India

COVID-19 is caused by virus called SARS-CoV-2, which was declared by the WHO as global pandemic. Since the outbreak, there has been a rush to explore Artificial Intelligence (AI) and Internet of Things (IoT) for diagnosing, predicting, and treating infections. At present, individual technologies, AI and IoT, play important roles yet do not impact individually against the pandemic because of constraints like lack of historical data and the existence of biased, noisy, and outlier data. To overcome, balance among data privacy, public health, and human-AI-IoT interaction is must. Artificial Intelligence of Things (AIoT) appears to be a more efficient technological solution that can play a significant role to control COVID-19. IoT devices produce huge data which are gathered and mined for actionable effects in AI. AI converts data into useful results which are utilized by IoT devices. AIoT entails AI through machine learning and decision making to IoT and renovates IoT to add data exchange and analytics to AI. In this chapter, AIoT will serve as a potential analytical tool to fight against the pandemic.

The latest innovative technology products in the market are paving the way for a new growth in the medical field over medical wearable devices. Globally, the medical market is said to be segmented on the basis of global medical wearable report by its type, application level, regional level, and country level. In this medical advisory, these devices are classified as diagnostic, therapeutic, and respiratory. The regions covered include Europe, Asia-Pacific, and the rest of the world. These wearable devices are technically embedded with electronic devices which the users are able to adhere to their body parts. The main function of these wearable devices is said to be collecting users' personal health data (e.g., such devices include measurement on fitness of body, heartbeat measurement, ECG measurement, blood pressure monitoring, etc.).

Currently there are numerous portables in the retail which assist tracking the day-by-day actions of kids and furthermore help discover the kid utilizing Wi-Fi and Bluetooth directions available on the gadget. Bluetooth gives off an impression of being an untrustworthy mode of correspondence connecting the parent and kid. Along these lines, the pivotal motive of this chapter is to get an authorized corresponding medium that links the kid's portable and the parents. The genitor can send a book with explicit tags, for example, "location," "temperature," "UV," "SOS," "BUZZ," and so forth. The portable device will response with a book encompassing the continuous precise region of the kid, which after monitoring hand down headings to the youngster's region on Google Maps software will similarly provide the atmospheric temperature and UV ray emission file with a goal that genitors can pursue if temperature or UV emission isn't reasonable to the kid.

The integration of internet of things, artificial intelligence, and blockchain enabled the monitoring of structural health with unattended and automated means. Remote monitoring mandates intelligent automated decision-making capability, which is still absent in present solutions. The proposed solution in this chapter contemplates the architecture of smart sensors, customized for individual structures, to regulate the monitoring of structural health through stress, strain, and bolted joints looseness. Long range sensors are deployed for transmitting the messages a longer distance than existing techniques. From the simulated results, different sensors record the monitoring information and transmit to the blockchain platform in terms of pressure points, temperature, pre-tension force, and the architecture deems the criticality of transactions. Blockchain platform will also be responsible for storage and accessibility of information from a decentralized medium, automation, and security.

A modern wireless sensor and its development majorly depend on distributed condition maintenance protocol. The medium access and its computing have been handled by multi hope sensor mechanism. In this investigation, WSN networks maintenance is balanced through condition-based access (CBA) protocol. The CBA is most useful for real-time 4G and 5G communication to handle internet assistance devices. The following CBA mechanism is energy efficient to increase the battery lifetime. Due to sleep mode and backup mode mechanism, this protocol maintains its energy efficiency as well as network throughput. Finally, 76% of the energy consumption and 42.8% of the speed of operation have been attained using CBI WSN protocol.

Chapter 13

Image processing concepts are used in the biomedical domain. Brain tumors are among the dreadful diseases. The primary brain tumors start with errors which are the mutations that take place in the part of DNA. This mutation makes the cells breed in a huge manner and makes the healthier cells die. The mass of the unhealthier cells is called tumor, or the unwanted cell growth in the tissues of the brain are called brain tumors. In this chapter, images from the positron emission tomography (PET) scan and MRI scan are fused as a one image, and from that image, neural network concepts are applied to detect the tumor. The main intention of this proposed approach is to segment and identify brain tumors in an automatic manner using image fusion with neural network concepts. Segmentation of brain images is needed to segment properly from other brain tissues. Perfect detection of size and position of the brain tumor plays an essential role in the identification of the tumor.

Laboratories are essential facilities provided in professional institutes for scientific and technological work. The lab must ensure their accuracy to regulatory requirements and maintain their data records so that the laboratory environment can be monitored properly. The laboratory environment temperature (LET) is monitored to ensure proper regulation and maintenance of indoor conditions and also to correlate the collected samples with these conditions. The LET data collection must be stored to influence the quality of the results and to ensure the stability of the laboratory environment. Hence, an IoT solution is presented to supervise real-time temperature and is known as iRT. This method helps to monitor ambient object supervision in real time. It is composed of a hardware prototype to collect the temperature data and to use web application to provide the history of temperature evolution. The result gained from the study is promising, and it provides a significant contribution to IoT-based temperature monitoring systems.

Predictive modeling or predict analysis is the process of trying to predict the outcome from data using machine learning models. The quality of the output predominantly depends on the quality of the data that is provided to the model. The process of selecting the best choice of input to a machine learning model depends on a variety of criteria and is referred to as feature engineering. The work is conducted to classify the breast cancer patients into either the recurrence or non-recurrence category. A categorical breast cancer dataset is used in this work from which the best set of features is selected to make accurate predictions. Two feature selection techniques, namely the chi-squared technique and the mutual information technique, have been used. The selected features were then used by the logistic regression model to make the final prediction. It was identified that the mutual information technique proved to be more efficient and produced higher accuracy in the predictions.

Machine learning (ML) proven to be an emerging technology from small-scale to large-scale industries. One of the important industries is banking, where ML is being adapted all over the world by employing online banking. The online banking is using ML techniques in detecting fraudulent transactions like credit card fraud detection, etc. Hence, in this chapter, a Credit card Fraud Detection (CFD) system is devised using Luhn's algorithm and k-means clustering. Moreover, CFD system is also developed using Fuzzy C-Means (FCM) clustering instead of k-means clustering. Performance of CFD using both clustering techniques is compared using precision, recall and f-measure. The FCM gives better results in comparison to k-means clustering. Further, other evaluation metrics such as fraud catching rate, false alarm rate, balanced classification rate, and Mathews correlation coefficient are also calculated to show

how well the CFD system works in the presence of skewed data.

Chapter 17

  Bhavana D., Koneru Lakshmaiah Education Foundation, India
  Adada Neelothpala, Koneru Lakshmaiah Education Foundation, India
  Pamidimukkala Kalpana, Koneru Lakshmaiah Education Foundation, India

The orthogonal frequency division multiplexing (OFDM) is a multicarrier modulation scheme used for the transfer of multimedia data. Well-known systems like ADSL (asymmetric digital subscriber line) internet, wireless local area networks (LANs), long-term evolution (LTE), and 5G technologies use OFDM. The major limitation of OFDM is the high peak-to-average power ratio (PAPR). High PAPR lowers the power efficiency, thus impeding the implementation of OFDM. The PAPR problem is more significant in an uplink. A high peak-to-average power ratio (PAPR) occurs due to large envelope fluctuations in OFDM signal and requires a highly linear high-power amplifier (HPA). Power amplifiers with a large linear range are expensive, bulky, and difficult to manufacture. In order to reduce the PAPR, a hybrid technique is proposed in this chapter with repeated clipping and filtering (RCF) and precoding techniques. The proposed method is improving the PAPR as well as BER. Five types of pre-coding techniques are used and then compared with each other.

Chapter 18

  Jayashree R., College of Science and Humanities, SRM Institute of Science and Technology,
    Kattankulathur, India
  Vaithyasubramanian S., D. G. Vaishnav College, India

In this chapter, restricted Boltzmann machine-driven (RBM) algorithm is presented with an enhanced interactive estimation of distribution (IED) method for websites. Indian matrimonial websites are famous intermediates for finding marriage-partners. Matchmaking is one of the most pursued objectives in matrimonial websites. The complex evaluations and full of zip user preferences are the challenges. An interactive evolutionary algorithm with powerful evolutionary strategies is a good choice for matchmaking. Initially, an IED is generated as a probability model for the estimation of a user preference and then two RBM models, one for interested and the other for not-interested, is generated to endow with a set of appropriate matches simultaneously. In the proposed matchmaking method, the RBM model is combined with social group knowledge. Some benchmarks from the matrimonial internet site are pragmatic to empirically reveal the pre-eminence of the anticipated method.

Section 2
Big Data and Cognitive Technologies: Current Applications and Future Challenges

Section 2 provides insights on learning based on big data and cognitive computing technologies, including cutting edge topics (e.g., machine learning for Industrial IoT systems, deep learning, reinforced learning, decision trees for IoT systems, computational intelligence and cognitive systems, cognitive learning for IoT systems, cognitive-inspired computing systems)..Section 2 is organized into 17 chapters.

 Neeraja Koppula, MLR Institute of Technology, India
 K. Sarada, KLEF, India
 Ibrahim Patel, B. V. Raju Institute of Technology, India
 R. Aamani, Vignan's Institute of Information Technology, India
 K. Saikumar, Mallareddy University, India

This chapter explains the speech signal in moving objects depending on the recognition field by retrieving the name of individual voice speech and speaker personality. The adequacy of precisely distinguishing a speaker is centred exclusively on vocal features, as voice contact with machines is getting more pervasive in errands like phone, banking exchanges, and the change of information from discourse data sets. This audit shows the location of text-subordinate speakers, which distinguishes a solitary speaker from a known populace. The highlights are eliminated; the discourse signal is enrolled for six speakers. Extraction of the capacity is accomplished utilizing LPC coefficients, AMDF computation, and DFT. By adding certain highlights as information, the neural organization is prepared. For additional correlation, the attributes are put away in models. The qualities that should be characterized for the speakers were acquired and dissected utilizing back propagation algorithm to a format picture.

 Ebru Efeoglu, Istanbul Gedik University, Turkey
 Gurkan Tuna, Trakya University, Turkey

In this chapter, traditional and innovative approaches used in hazardous liquid detection are reviewed, and a novel approach for the detection of hazardous liquids is presented. The proposed system is based on electromagnetic response measurements of liquids in the microwave frequency band. Thanks to this technique, liquid classification can be made quickly without pouring the liquid from its bottle and without opening the lid of its bottle. The system can detect solutions with hazardous liquid concentrations of 70% or more, as well as pure hazardous liquids. Since it relies on machine learning methods and the success of all machine learning methods depends on provided data type and dataset, a performance evaluation study has been carried out to find the most suitable method. In the performance evaluation study naive Bayes and sequential minimal optimization has been evaluated, and the results have shown that naive Bayes is more suitable for liquid classification.

Sentiment evaluation alludes to separate the sentiments from the characteristic language and to perceive the mentality about the exact theme. Novel corona infection, a harmful malady ailment, is spreading out of the blue through the quarter, which thought processes respiratory tract diseases that can change from gentle to extraordinary levels. Because of its quick nature of spreading and no conceived cure, it ushered in a vibe of stress and pressure. In this chapter, a framework perusing principally based procedure is utilized to discover the musings of the tweets related to COVID and its effect lockdown. The chapter examines the tweets identified with the hash tags of crown infection and lockdown. The tweets were marked fabulous, negative, or fair, and a posting of classifiers has been utilized to investigate the precision and execution. The classifiers utilized have been under the four models which incorporate decision tree, regression, helpful asset vector framework, and naïve Bayes forms.

A privacy-preserving patient-centric clinical decision support system, called PPCD, is based on naive Bayesian classification to help the physician predict disease risks of patients in a privacy-preserving way. First, the authors propose a secure PPCD, which allows the service providers to diagnose a patient's disease without leaking any patient medical data. In PPCD, the past patient's historical medical data can be used by a service provider to train the naive Bayesian classifier. Then, the service provider can use the trained classifier to diagnose a patient's diseases according to his symptoms in a privacy-preserving way. Finally, patients can retrieve the diagnosed results according to their own preference privately without compromising the service provider's privacy.

The use of AI algorithms in the IoT enhances the ability to analyse big data and various platforms for a number of IoT applications, including industrial applications. AI provides unique solutions in support of managing each of the different types of data for the IoT in terms of identification, classification, and decision making. In industrial IoT (IIoT), sensors, and other intelligence can be added to new or existing plants in order to monitor exterior parameters like energy consumption and other industrial parameters levels. In addition, smart devices designed as factory robots, specialized decision-making systems, and other online auxiliary systems are used in the industries IoT. Industrial IoT systems need smart operations management methods. The use of machine learning achieves methods that analyse big data developed for decision-making purposes. Machine learning drives efficient and effective decision making, particularly in the field of data flow and real-time analytics associated with advanced industrial computing networks.

Komali Dammalapati, Koneru Lakshmaiah Education Foundation, India
B. Sankara Babu, Gokaraju Rangaraju Institute of Engineering and Technology, India
P. Gopala Krishna, Gokaraju Rangaraju Institute of Engineering and Technology, India
V. Subba Ramaiah, Mahatma Gandhi Institute of Technology, India

With rapid growth of various well-known methods implemented by the engineers in the software field in order to create a development in automated tasks for manufacturers and researchers working worldwide, the researchers in the field of software engineering (SE) root for concepts of machine learning (ML), a subfield that utilizes deep learning (DL) for the development of such SE tasks. In essence, these systems would highly cope with the featured automation with inbuilt capabilities in engineering to develop the software simulation models. Nevertheless, it is very tough to condense the present scenario in research of situations that necessitate failures, successes, and openings in DL for software-based technology. The survey works for renowned technology of SE and DL held for the latest journals and conferences leading to the span of 85 issued papers throughout 23 distinctive tasks for SE.

Dhilip Kumar, Vel Tech Rangarajan Dr. Sagunthala R&D Institute of Science and Technology, India
Swathi P., Vel Tech Rangarajan Dr. Sagunthala R&D Institute of Science and Technology, India
Ayesha Jahangir, Vel Tech Rangarajan Dr. Sagunthala R&D Institute of Science and Technology, India
Nitesh Kumar Sah, Vel Tech Rangarajan Dr. Sagunthala R&D Institute of Science and Technology, India
Vinothkumar V., MVJ College of Engineering, India

With recent advances in the field of data, there are many advantages of speedy growth of internet and mobile phones in the society, and people are taking full advantage of them. On the other hand, there are a lot of fraudulent happenings everyday by stealing the personal information/credentials through spam calls. Unknowingly, we provide such confidential information to the untrusted callers. Existing applications for detecting such calls give alert as spam to all the unsaved numbers. But all calls might not be spam. To detect and identify such spam calls and telecommunication frauds, the authors developed the application for suspicious call identification using intelligent speech processing. When an incoming call is answered, the application will dynamically analyze the contents of the call in order to identify frauds. This system alerts such suspicious calls to the user by detecting the keywords from the speech by comparing the words from the pre-defined data set provided to the software by using intelligent algorithms and natural language processing.

Chapter 26

Bhavana D., Koneru Lakshmaiah Education Foundation, India
K. Chaitanya Krishna, Koneru Lakshmaiah Education Foundation, India
Tejaswini K., Koneru Lakshmaiah Education Foundation, India
N. Venkata Vikas, Koneru Lakshmaiah Education Foundation, India
A. N. V. Sahithya, Koneru Lakshmaiah Education Foundation, India

The task of image caption generator is mainly about extracting the features and ongoings of an image and generating human-readable captions that translate the features of the objects in the image. The contents of an image can be described by having knowledge about natural language processing and computer vision. The features can be extracted using convolution neural networks which makes use of transfer learning to implement the exception model. It stands for extreme inception, which has a feature extraction base with 36 convolution layers. This shows accurate results when compared with the other CNNs. Recurrent neural networks are used for describing the image and to generate accurate sentences. The feature vector that is extracted by using the CNN is fed to the LSTM. The Flicker 8k dataset is used to train the network in which the data is labeled properly. The model will be able to generate accurate captions that nearly describe the activities carried in the image when an input image is given to it. Further, the authors use the BLEU scores to validate the model.

Chapter 27

Koppula Srinivas Rao, MLR Institute of Technology, India
Saravanan S., B. V. Raju Institute of Technology, India
Pattem Sampath Kumar, Malla Reddy Institute of Engineering and Technology, India
Rajesh V., Department of Electronics and Communication Engineering, Koneru Lakshmaiah
    Educational Foundation, India
K. Raghu, Mahatma Gandhi Institute of Technology, India

The benefits of data analytics and Hadoop in application areas where vast volumes of data move in and out are examined and exposed in this report. Developing countries with large populations, such as India, face several challenges in the field of healthcare, including rising costs, addressing the needs of economically disadvantaged people, gaining access to hospitals, and conducting medical research, especially during epidemics. This chapter discusses the role of big data analytics and Hadoop, as well as their effect on providing healthcare services to all at the lowest possible cost.

Chapter 28

Koppula Srinivas Rao, MLR Institute of Technology, India
B. Veerasekhar Reddy, MLR Institute of Technology, India
Sarada K., Koneru Lakshmaiah Education Foundation, India
Saikumar K., Malla Reddy Institute of Technology, India

In this survey, the issue of flaw zone identification of separation transferring in "FACTS-based transmission lines" is investigated. Presence of FACTS gadgets on transmission line (TL), while they have been remembered for issue zone, from separation transfer perspective, causes various issues in deciding the specific area of the flaw by varying impedance perceived by hand-off. The degree of these progressions relies upon boundaries that are set in FACTS gadgets. To tackle issues related with these, two instruments for partition and examination of three-line flows, from the hand-off perspective to blame occasion, have been used. In addition, to examine the impacts of TCSC area on deficiency zone recognition of separation hand-off, two spots, one of every 50% of line length and the other in 75% of line length, are deliberated as two situations for affirmation of suggested strategy. Reproductions show that this strategy is powerful in security of FACTS-based TLs.

Chapter 29

Abdullah Saleh Alqahtani, CFY Deanship King Saud University, Saudi Arabia

In endemic and pandemic situations, governments will implement social distancing to control the virus spread and control the affected and death rate of the country until the vaccine is introduced. For social distancing, people may forget in some public places for necessary needs. To avoid this situation, the authors develop the Android mobile application to notify the social distancing alert to people to avoid increased levels of the endemic and pandemic spread. This application works in both online and offline mode in the smart phones. This application helps the people to obey the government rule to overcome the endemic and pandemic situations. This application uses k-means clustering algorithms to cluster the data and form the more safety clusters for social distancing. It uses artificial intelligence to track the living location by the mobile camera without the internet facilities. It helps the user to follow social distancing even with no internet with user knowledge.

Chapter 30

D. Rajalakshmi, Sri Sairam Institute of Technology, India
Meena K., Vel Tech Rangarajan Dr. Sagunthala R&D Institute of Science and Technology,
India

A MANET (mobile ad hoc network) is a self-organized wireless network. This network is more vulnerable to security failure due to dynamic topology, infrastructure-less environment, and energy consumption. Based on this security issue, routing in MANET is very difficult in real time. In these kinds of networks, the mobility and resource constraints could lead to divide the networks and minimize the performance of the entire network. In real time it is not possible because some selfish nodes interacts with other nodes partially or may not share the data entirely. These kind of malicious or selfish nodes degrade the network performance. In this chapter, the authors proposed and implemented the effect of malicious activities in a MANETs using self-centered friendship tree routing. It's a novel replica model motivated by the social relationship. Using this technique, it detects the malicious nodes and prevents hacking issues in routing protocol in future routes.

Chapter 31

Sirasani Srinivasa Rao, Mahatma Gandhi Institute of Technology, India
K. Butchi Raju, GRIET, India
Sunanda Nalajala, Koneru Lakshmaiah Education Foundation, India
Ramesh Vatambeti, CHRIST University (Deemed), India

Wireless sensor networks (WSNs) have as of late been created as a stage for various significant observation and control applications. WSNs are continuously utilized in different applications, for example, therapeutic, military, and mechanical segments. Since the WSN is helpless against assaults, refined security administrations are required for verifying the information correspondence between hubs. Because of the asset limitations, the symmetric key foundation is considered as the ideal worldview for verifying the key trade in WSN. The sensor hubs in the WSN course gathered data to the base station. Despite the fact that the specially appointed system is adaptable with the variable foundation, they are exposed to different security dangers. Grouping is a successful way to deal with vitality productivity in the system. In bunching, information accumulation is utilized to diminish the measure of information that streams in the system.

Chapter 32

Sasank V. V. S., Koneru Lakshmaiah Education Foundation, India
Kranthi Kumar Singamaneni, Gokaraju Rangaraju Institute of Engineering and Technology, India
A. Sampath Dakshina Murthy, Vignan's Institute of Information Technology, India
S. K. Hasane Ahammad, Koneru Lakshmaiah Education Foundation, India

Various estimating mechanisms are present for evaluating the regional agony, neck torment, neurologic deficiencies of the sphincters at the stage midlevel of cervical spondylosis. It is necessary for the cervical spondylosis that the survey necessitates wide range of learning skills about the systemized life, experience, and ability of the expertise for learning the capability, life system, and experience. Doctors check the analysis of situation through MRI and CT scan, but additional interesting facts have been discovered in the physical test. For this, a programming approach is not available. The authors thereby propose a novel framework that accordingly inspects and investigates the cervical spondylosis employing computation of CNN-LSTM. Machine learning methods such as long short-term memory (LSTM) in fusion with convolution neural networks (CNNs), a kind of neural network (NN), are applied to this strategy to evaluate for making the systematization in various applications.

Chapter 33

Swapna B., Dr. M. G. R. Educational and Research Institute, India
S. Manivannan, Dr. M. G. R. Educational and Research Institute, India
M. Kamalahasan, Dr. M. G. R. Educational and Research Institute, India

The multivariate data analysis technique is used to determine the highly impacted data in soil and crop growth. The importance and relationship between soil variables were factored by using the regression analysis technique. The correlation matrix technique was used for comparing several variables to correlate positive and negative signs. From the soil testing procedure and understanding of results, it shows that soil nutrients and pH level have a powerful effect on variation in the usage of fertilizers, crop selection, and high crop yield. pH determination can be used to indicate whether the soil is suitable for the plant's growth or in need of adjustment to produce optimum plant growth. Based upon the predictive analysis results, nitrogen and potassium content are naturally high compared to other soil nutrients of this region and suggested fertilizers required for crop growth. To produce healthy crop yield, farmers should select the crops as per soil types, nutrients level, and pH level.

Preface

Recently, artificial intelligence (AI), Internet of Things (IoT) and cognitive technologies have successfully been applied to various research domains, including computer vision, natural language processing, voice recognition, etc. In addition, AI with IoT has made a significant breakthrough and technical direction in achieving high efficiency and adaptability in a variety of new applications, such as smart wearable devices in healthcare, smart automotive industry, recommender systems, and financial analysis. On the other hand, network design and optimization for AI applications addresses a complementing topic, namely the support of AI-based systems through novel networking techniques, including new architectures as well as performance models for IoT systems. IoT has paved the way to a plethora of new application domains, at the same time posing several challenges as a multitude of devices, protocols, communication channels, architectures and middleware exist. In particular, we are witnessing an incremental development of interconnections between devices (e.g., smartphones, tablets, smart watches, fitness trackers, and wearable devices in general, smart TVs, home appliances and much more like this), people, processes, and data. Big data generated by these devices calls for advanced learning and data mining techniques to effectively understand, learn, and reason with this volume of information, such as cognitive technologies.

Cognitive technologies play a major role in developing successful cognitive systems which mimic "cognitive" functions associated with human intelligence, such as "learning" and "problem solving". Thus, there is a continuing demand for recent research in these two linked fields. Nowadays, this cognitive technologies like Neural Networks, Deep learning, Reinforcement learning, Fuzzy Systems, Evolutionary Computation, Bio-inspired computing paradigms, Quantum-inspired Evolutionary Algorithm, Cognitive-inspired computing systems, Brain analysis for cognitive computing, Internet cognitive of things, Cognitive agents are combined with AI and IoT for many innovative applications and system developments.

This *Handbook of Research of Innovations and Applications of AI, IoT, and Cognitive Technologies* discusses the latest innovations and applications of AI, IoT, and cognitive-based smart systems. The chapters cover the intersection of these three fields in emerging and developed economies in terms of their respective development situation, public policies, technologies and intellectual capital, innovation systems, competition and strategies, marketing and growth capability, and governance and relegation models. These applications span areas such as healthcare, security and privacy, industrial systems, multidisciplinary sciences, and more. This book is ideal for technologists, IT specialists, policymakers, government officials, academics, students, and practitioners interested in the experiences of innovations and applications of AI, IoT, and cognitive technologies.

ORGANIZATION OF THE BOOK

This book has been divided into two sections:

Section 1, "AI and IoT: A Blend in Future Technologies and Systems," introduces emerging AI and IoT technologies and systems, as well as their innovations and applications that are responsible for the digitization of all the sectors through automation and together rules the technological aspects of our lives. Section 1 is organized into 18 chapters. A synopsis of each chapter is given below.

Chapter 1 (*Algorithms Optimization for Intelligent IoV Applications*) introduces different optimization algorithms i.e. clustering algorithms, Ant colony optimization, Best Interface Selection (BIS) Algorithm, Mobility adaptive density connected clustering algorithm, Meta-Heuristics Algorithms and Quality of Service (QoS) based optimization. These algorithms provide an important intelligent role to optimize the operation of IoV networks and promise to develop new intelligent IoV applications.

Chapter 2 (*Cooperative Relaying Communication in IoT Applications for 5G Radio Networks*) presents cooperative relaying networks that are helpful in Internet of Thing (IoT) applications for fifth generation (5G) radio networks. The design of advanced protocols plays an important role to combat the effect of channel fading, data packet scheduling at the buffered relay, average delay, and traffic intensity.

Chapter 3 (*The Rise of IoT and Big Data Analytics for Disaster Management System*) explains the reason why IoT and Big Data are needed to cope with disasters, as well as how these technologies work to solve the problem. The developments in Big Data and the Internet of Things (IoT) have made a greater contribution to accuracy and timely decision-making in the Disaster Management System (DMS).

Chapter 4 (*IoT-Based Smart Agriculture*) explores the current state of advanced farm management systems by revisiting each critical phase, from data collection in crop fields to variable rate applications, in order for growers to make informed decisions save money while also protecting the environment and transforming how food is grown to meet potential population growth.

Chapter 5 (*Mobile Geo-Fencing Triggers for Alerting Entries Into COVID-19 Containment Zones Using IoT*) discusses a Containment Zone alert system by means of Geo-fencing technology to identify the movement of public, deliver info about the danger to the public in travel and also send an alert to the police when there is an entry or exit detected in the containment zone by the use of Location-Based Services (LBS).

Chapter 6 (*Fine Tuning Smart Manufacturing Enterprise System: A Perspective of Internet of Things-Based Service-Oriented Architecture*) proposes the integration of standard workflow model between business system level and manufacturing production level with IoT-enabled SOA framework.

Chapter 7 (*A Quantum Technology-Based LiFi Security Using Quantum Key Distribution*) discusses Wireless Fidelity (Wi-Fi) and focuses on some drawbacks that include speed and bandwidth limit, security issues, and attacks by malicious users, which yield Wi-Fi as less reliable as compared to LiFi.

Chapter 8 (*Role of Artificial Intelligence of Things [AIoT] to Combat Pandemic COVID-19*) studies on AIoT entails AI through machine learning and decision making to IoT; and renovates IoT to add data exchange and analytics to AI. Through this chapter, AIoT will serve as potential analytical tool to fight against the pandemic.

Chapter 9 (*Recent Trends in Wearable Device Technology for Health State Monitoring*) proposes that the main function of wearable devices is said to be collecting the user personal health data. Eg: such devices include measurement on fitness of body, heart beat measurement, ECG Measurement, blood pressure monitoring etc.

Chapter 10 (*IoT-Based Delineate and Evolution of Kid's Safety Portable Device*) presents those numerous portables in the retail which assist track the day by day action of kids and furthermore help discover the kid utilizing Wi-Fi and Bluetooth directions available on the gadget.

Chapter 11 (*Blockchain-Enabled Internet of Things for Unsupervised Structural Health Monitoring in Potential Building Structures*) proposes a solution that contemplates the architecture of smart sensors, customized for individual structures, to regulate the monitoring of structural health through stress, strain and bolted joints looseness.

Chapter 12 (*An Advanced Wireless Sensor Networks Design for Energy Efficient Applications Using Condition Based Access Protocol*) states that, in this investigation WSN networks maintenance is balanced through Condition Based Access (CBA) protocol. The CBA most useful for real time 4G and 5G communication to handle internet assistance devices.

Chapter 13 (*Effective Image Fusion of PET-MRI Brain Images Using Wavelet Transforms and Neural Networks*) proposed approach is segment and identify brain tumor in automatic manner using image fusion with neural network concepts. Segmentation of brain images is needed to segment properly from other brain tissues.

Chapter 14 (*IoT-Enabled Non-Contact-Based Infrared Thermometer for Temperature Recording of a Person*) states that the LET data collection must be stored to influence the quality of the results and to ensure the stability of the laboratory environment. Hence an IoT solution is presented to supervise real time temperature and is known as iRT.

Chapter 15 (*Predictive Modeling for Classification of Breast Cancer Dataset Using Feature Selection Techniques*) conducts to classify the breast cancer patients into either the recurrence or non recurrence category. It is identified that the Mutual Information technique proved to be more efficient and produced higher accuracy in the predictions.

Chapter 16 (*Credit Card Fraud Detection Using K-Means and Fuzzy C-Means*) states that the CFD system is developed using Fuzzy C-Means (FCM) clustering instead of K-means clustering. Performance of CFD using both clustering techniques is compared using precision, recall and F-measure. The FCM gives better results in comparison to K-means clustering.

Chapter 17 (*A Hybrid Method to Reduce PAPR of OFDM to Support 5G Technology*) discusses Well-known systems like ADSL (asymmetric digital subscriber line) internet, wireless local area networks (LANs), Long term Evolution (LTE), and 5G technologies use OFDM.

Chapter 18 (*Restricted Boltzmann Machine-Driven Matchmaking Algorithm With Interactive Estimation of Distribution for Websites*) presents that an interactive evolutionary algorithm with powerful evolutionary strategies is a good choice for matchmaking, in which the RBM model is combined with social group knowledge. Some benchmarks from the matrimonial internet site are pragmatic to empirically reveal the pre-eminence of the anticipated method.

Section 2, "Big Data and Cognitive Technologies: Current Applications and Future Challenges" provides insights on learning based on big data and cognitive computing technologies, including cutting edge topics (e.g., machine learning for industrial IoT systems, deep learning, reinforced learning and decision trees for IoT systems, computational intelligence and cognitive systems, cognitive learning for IoT systems, cognitive-inspired computing systems). Section 2 is organized into 17 chapters. A synopsis of each chapter is given below.

Chapter 19 (*Identification and Recognition of Speaker Voice Using Neural Network-Based Algorithm: Deep Learning*) explains about the speech signal in moving objects depending on the recognition field by retrieving the name of individual voice speech and speaker personality.

Chapter 20 (*Traditional and Innovative Approaches for Detecting Hazardous Liquids*) presents traditional and innovative approaches used in hazardous liquid detection are reviewed and a novel approach for the detection of hazardous liquids. The proposed innovative system is based on electromagnetic response measurements of liquids in the microwave frequency band.

Chapter 21 (*COVID-19 Analysis of Sentiment Tweets on Pandemic Using Machine Learning Techniques*) gathers the tweets identified with the hash tags of crown infection and lock down, pre-framework them and was examined. The tweets had been marked fabulous, negative or fair and a posting of classifiers had been utilized to investigate the precision and execution. The classifiers utilized had been under the four models which incorporates decision Tree, Regression, helpful asset Vector framework and Naïve Bayes form.

Chapter 22 (*Intelligence Medical Data Analytics Using Classifiers and Clusters in Machine Learning*) proposes a secure and PPCD, which allows the service providers to diagnose a patient's disease without leaking any patient's medical data. Patients can retrieve the diagnosed results according to their own preference privately without compromising the service provider's privacy.

Chapter 23 (*Machine Learning for Industrial IoT Systems*) states that the new model of industrial IoT (IIoT), sensors and other intelligence are added to new or existing plants to monitor exterior parameters such like energy consumption and other industrial parameters levels. In addition, smart devices designed as factory robots, specialized decision-making systems, and other online auxiliary systems are used in the industries IoT.

Chapter 24 (*A Research on Software Engineering Research in Machine Learning: Machine Learning*) proposes that the researchers in the field of software engineering (SE) root for concepts of machine learning (ML), a sub field in which utilizing the deep learning (DL)for the development of such SE tasks.

Chapter 25 (*Intelligent Speech Processing Technique for Suspicious Voice Call Identification Using Adaptive Machine Learning Approach*) develops the application for suspicious call identification using intelligent speech processing. When an incoming call is answered, the application will dynamically analyze the contents of the call in order to identify frauds.

Chapter 26 (*Image Captioning Using Deep Learning*) designs a model that is able to generate accurate captions that nearly describe the activities carried in the image when an input image is given to it. Further, the BLEU scores can be used to validate the model.

Chapter 27 (*Leveraging Big Data Analytics and Hadoop in Developing India's Healthcare Services*) discusses the role of Big Data Analytics and Hadoop, as well as their effect on providing healthcare services to all at the lowest possible cost.

Chapter 28 (*A Sequential Data Mining Technique for Identification of Fault Zone Using Facts-Based Transmission*) investigates the issue of flaw zone identification of separation transferring in "FACTS-based transmission lines".

Chapter 29 (*K-Means Clustering Machine Learning Concept to Enable Social Distancing in Public Places*) studies on utilizing artificial intelligence to track the living location by the mobile camera without the internet facilities and it helps users to follow the social distancing.

Chapter 30 (*An Efficient Selfishness Control Mechanism for Mobile Ad Hoc Networks*) introduces a novel replica model, motivated by the social relationship. Using this technique, it detects the malicious nodes and prevent the hacking issues in routing protocol in future routes.

Chapter 31 (*Estimation of Secured Wireless Sensor Networks and Its Significant Observation for Improving Energy Efficiency Using Cross-Learning Algorithm*) presents Wireless Sensor Networks (WSNs) are helpless against assaults, refined security administrations that are required for verifying the information correspondence between hubs. Grouping is a successful way to deal with accomplishes vitality productivity in the system. In bunching, information accumulation is utilized to diminish the measure of information that streams in the system.

Chapter 32 (*Executing CNN-LSTM Algorithm for Recognizable Proof of Cervical Spondylosis Infection on Spinal Cord MRI Image: Machine Learning Image*) proposes that machine learning methods such as Long Short-Term Memory (LSTM) in fusion with Convolution Neural Networks (CNNs), a kind of neural network (NN) are applied to this strategy to evaluate for making the systematization in various applications.

Chapter 33 (*Soil Nutrients and Ph Level Testing Using Multivariate Statistical Techniques for Crop Selection*) explains that the importance and relationship between soil variables are factorized by using the Regression analysis technique.

Jingyuan Zhao
University of Toronto, Canada

V. Vinoth Kumar
MVJ College of Engineering, India
2021 June

Section 1
AI and IoT: A Blend in Future Technologies and Systems

Section 1 introduces emerging AI and IoT technologies and systems, as well as their innovations and applications that are responsible for the digitization of all the sectors through automation and together rules the technological aspects of our lives. Section 1 is organized into 18 chapters.

Chapter 1
Algorithms Optimization for Intelligent IoV Applications

Elmustafa Sayed Ali Ahmed
ⓘⓓ https://orcid.org/0000-0003-4738-3216
Sudan University of Science and Technology, Sudan & Red Sea University, Sudan

Zahraa Tagelsir Mohammed
Red Sea University, Sudan

Mona Bakri Hassan
Sudan University of Science and Technology, Sudan

Rashid A. Saeed
ⓘⓓ https://orcid.org/0000-0002-9872-081X
Sudan University of Science and Technology, Sudan & Taif University, Saudi Arabia

ABSTRACT

Internet of vehicles (IoV) has recently become an emerging promising field of research due to the increasing number of vehicles each day. It is a part of the internet of things (IoT) which deals with vehicle communications. As vehicular nodes are considered always in motion, they cause frequent changes in the network topology. These changes cause issues in IoV such as scalability, dynamic topology changes, and shortest path for routing. In this chapter, the authors will discuss different optimization algorithms (i.e., clustering algorithms, ant colony optimization, best interface selection [BIS] algorithm, mobility adaptive density connected clustering algorithm, meta-heuristics algorithms, and quality of service [QoS]-based optimization). These algorithms provide an important intelligent role to optimize the operation of IoV networks and promise to develop new intelligent IoV applications.

DOI: 10.4018/978-1-7998-6870-5.ch001

INTRODUCTION

Recently, Internet of vehicles (IoV) has become an emerging promising field of research due to the increasing number of vehicles each day. It is a branch of the internet of things (IoT) which deal with communication among vehicles (Mahmoud et al, 2012). IoV enables vehicles to send information among vehicles, road infrastructures, passengers, drivers, sensors and electric actuators through different communication media including IEEE 802.11p, vehicular cooperative media access control (VC-MAC), dynamic source routing (DSR), Ad hoc on demand distance vector (AODV), directional medium access control (DMAC) and general packet radio services (GPRS) (Amal et al, 2016). As know, that vehicular nodes are always in motion that causes the frequent changes in the network topology (Mahmoud et al,2014). These changes cause issues in IoV as scalability, dynamic topology changes and shortest path for routing.

The design of an effective application is a major challenge that should not be neglected in IoV, considering their special features and characteristics, such as high vehicle mobility and quick topology changes, which make the design and implementation of effective solutions for such networks a difficult task (Mayada et al,2018) (Nahla et al,2021). First, high mobility is the main factor distinguishing IoV from other networks. Vehicle speed varies according to road conditions and may be low or medium in urban areas and large on highways (Amal et al,2018). This speed variation has a direct impact on network stability and results in a dynamic network topology. Secondly, node density is not uniform but exhibits spatiotemporal variation. Typically, the density in urban areas is higher than in rural areas and depends on the time of day. Finally, network fragmentation generally occurs when vehicle density is low and irregular. Then, the vehicles move in disconnected isolated clusters, and therefore, end-to-end communication becomes difficult.

Generally, the focus on optimization in IoV applications is related to the importance of traffic accuracy and protection of information and entertainment network, as well as ensuring road safety when deploying the IoVs applications. According to what has been mentioned, IoV applications face a number of problems related to the reliability and consistency in the exchange of information between the vehicles and the appropriate decision-making to improve safety on the road (Mayada et al,2017). Artificial intelligence (AI) techniques provide appropriate and effective solutions to a number of the aforementioned problems, especially those related to decision-making in IoV systems. Recently, many studies have been presented that rely on the capabilities of AI and the use of its algorithms in various processes related to improving quality in IoV applications and services. Table 1 shows the summary of different optimization studies in IoV applications.

(Farhan et al, 2018), presented a method for using the metaheuristic dragonfly-based clustering algorithm (CAVDO) to optimize cluster-based packet route to create a stable IoV topology in a dynamic environment. The study is presented on the basis that the mobility aware dynamic transmission range algorithm (MA-DTR) algorithm is used with CAVDO to adapt the transmission range based on traffic density. A comparison of the proposed algorithm was also made with other algorithm such as the Ant Colony Improvement (ACO) and learning particle swarm optimization (LPSO). Through the analysis, it was found that the proposed algorithm has much better performance than ACO and LPSO as it provides the minimum number of clusters according to the current channel conditions, as it improves network availability and integrates the functions of the network infrastructure.

The study presented by (Ali et al, 2019), reviews two algorithms known as Energy Perceived QoE Optimization (PQO) and Store Perceived QoE Improvement (BQO) as their performance was compared, in addition to proposing a multimedia communication mechanism based on the two algorithms. Also, the researchers demonstrated a framework that improves QoE during IoV multimedia communication through mobile devices. Through the analysis, it was found that the proposed algorithms give a significant improvement in QoE, as they greatly assist in enabling applications during multimedia communication.

In the study presented by (Hongjing et al,2020) the potential of both MEC and AI in IoV applications and the possibility of combining them in a specific structure and topology are reviewed. The researchers analysed the application of MEC and AI in IoV and compared it with current approaches. The analysis shows that the MEC with AI will gives higher location awareness in addition to higher scalability. (Xiantao et al, 2020) studied the analysis of video applications on the IoV using multiple access computing (MEC) technology and integrating it with the blockchain i to improve transaction throughput as well as reduce MEC system latency. Researchers relied on the Markov decision process (MDP) and asynchronous critic-critic (A3C) algorithm to model the improvement problem. Through the analysis, the results showed that the proposed approach could rapidly converge and significantly improve the performance of the blockchain-powered IoV.

(Quoc et al, 2020), presented an overview of swarm intelligence(SI) technologies and their impact in the optimization process. The researchers covered the use of SI in spectrum management, resource allocation, wireless buffering, advanced computing and network security. in addition, they review the future challenges facing the optimization in many IoV aspects such as plectrum management and resource allocation for IoV based on 5G. In the study presented by (Ghassan et al,2020) a mechanism known as Whale Optimization Algorithm is used to improve the select optimum cluster head for vehicle communication according to two criteria: intelligence and capacity. Through the analysis, it was found that the proposed algorithm gave better results in finding optimum cluster compared to other traditional algorithms, as it gave an improvement in cluster optimization by 46%.

(Kayarga et al,2021), review the challenges facing the IoV deployment in addition to the reliability and consistently of messages between vehicles that enables drivers to make appropriate decisions to improve road safety. Where the authors presented a review of the methods proposed from a number of previous studies that present a new approach to determine the mechanism used to identify and route the vehicular node for the IoV environment. (elmustafa et al,2021), presented theories and general foundations for the use of machine learning and models of some algorithms in IoV applications, where they conducted a critical review with an analytical model of mobile edge computing decisions based on machine learning and a deep augmented learning (DRL) approaches. The study adopted the IoV secure edge computing unloading model with different data processing and traffic flow by taking an analytical model of Markov decision process (MDP) and ML in unpacking the decision process of the different task streams of the IoV network control cycle. The researchers also analysed the mechanisms of buffer and perceived energy in improving the quality of the experiment (QoE) enabled by ML.

Considering all issues disused in the previous studies, as new techniques and mechanisms have been studied to ensure a stable network topology structure as well as effective data routing and dissemination. This chapter provides a brief concept about algorithms optimization in mentioned issues related to the IoV applications. The chapter describes the background of Internet of Vehicles, in addition to the use of Artificial Intelligence in IoV. The clustering algorithms in IoV are reviewed in this chapter by explaining the concept of clustering and how it is used to optimize IoV networks, it also describes some of the most important clustering algorithms used for IoV; Dragonfly-based clustering algorithm for IoV, Moth Flame Clustering Algorithm and Mobility adaptive density connected clustering algorithm.

The chapter also focuses in three types of Swarm-Intelligence (SI) Algorithms which is inspired from the life of some swarms such as ants, bees and Cookoo Bird. In addition, the concept of Meta-Heuristics

Table 1. Summary of Algorithms Optimization Studies on IoV Applications

Approach	Feature	Advantages	Citations
Metaheuristic dragonfly-based clustering algorithm	Optimize cluster based packet route in IoV dynamic environment	Minimum number of clusters and improving in network availability	Farhan et al, 2018
Energy perceived QoE optimization & Store perceived QoE improvement algorithms	QoS for multimedia communication mechanism in IoV	Improved QoS multimedia communication through IoV mobile devices	Ali et al, 2019
MEC based AI algorithm	Combination of MEC and AI in IoV applications	Higher location awareness in addition to higher scalability	Hongjing et al,2020
Markov decision process (MDP) and asynchronous critic-critic (A3C) algorithm	Video services on IoV using multiple access computing (MEC) with blockchain	Reduce MEC system latency and improve the performance of the blockchain-powered IoV	Xiantao et al, 2020
Swarm intelligence (SI) technology	Impact of SI in IoV spectrum management, resource allocation, and network security	Possible improvements in management and resource allocation for IoV based on 5G	Quoc et al, 2020
Whale Optimization Algorithm	Optimal Cluster head selection in IoVs communication	Improvement in cluster optimization by 46%. Compared with traditional algorithms	Ghassan et al,2020
Intelligent decision making for IoV application	Road safety and route optimization in IoV	Improved methods to identify and route the vehicular node for the IoV environment	Kayarga et al,2021
Mobile edge computing decisions based on machine learning	Secure edge computing unloading model with different data processing and traffic flow in IoV	improving the quality of the experiment (QoE) enabled by ML	elmustafa et al,2021

algorithms and the Best Interface Selection (BIS) algorithm with their classifications in IoV are also discussed. Moreover, the chapter explains the Distributed Multichannel and Mobility-Aware Cluster-Based MAC Protocol. Chapter also describes the importance of considering Quality of Service with the Optimization Algorithms in IoV System, and the optimization problem in IoV clustering. Finally, the chapter presented the future research directions that should be followed to enhance the process of optimizing IoV networks, and summarizes the optimization and IoV.

IOV APPLICATIONS BACKRGOUND

The new era of the IoT is driving the development of Vehicle Ad-hoc Networks into the IoV. VANET is recognized from mobile ad hoc network (MANET). VANET technology is interfaced in vehicle-to-vehicle (V2V) and vehicle to roadside (V2R) communications technology to send data among vehicles. IoV is a new version of IoT used to enable communication among vehicles, things and environments to send information among IoV networks (yang et al, 2014) (Amal et al,2013). In other words, IoV is a combination of inter-vehicle network, an intra-vehicle network and vehicular mobile internet in urban environment. It improves VANET features and reduces traffic issues in urban traffic environment.

IoV is technology that concerned on travel plan and network access for passengers, drivers and individuals who are working in the traffic management department. The major applications of vehicles and road network (IoV) including smart traffic monitoring, self-drive cars and smart parking system (Mayada et al,2020). IoV allows managing traffic, developing intelligent dynamic information communication among vehicles, pollution and environmental protection, road safety management and energy management as well as road accident prevention (yang et al, 2014) (Priyan et al, 2019).

As shown in figure 1, IoV connects humans within and around vehicles, intelligent systems on board vehicles and various cyber-physical systems in urban environments, by integrating sensors, actuator, vehicles and smart phones into a wide area network (WAN) which allows varied services to vehicles and humans on board and around vehicles. Several researchers defined vehicle as a manned computer with four wheels or a manned large phone in IoV (yang et al, 2014). Thus, in contrast to other networks, existing multi-user, multi-vehicle, multi-thing and multi-network systems need multi-level collaboration in IoV.

Figure 1. Internet-of-Vehicles (IoV)

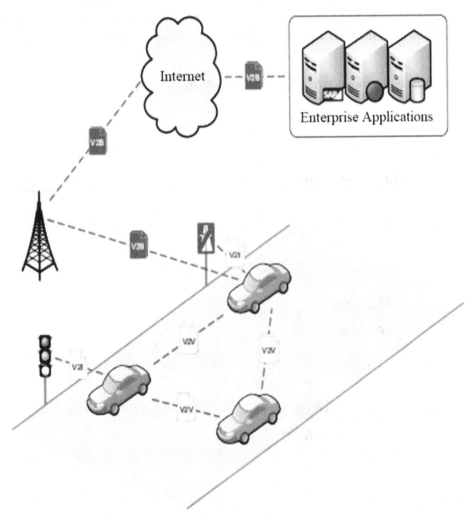

RESEARCH METHODOLOGY

The general aim of AI that has an ability machines integrate with human intelligence. ML is a method that using algorithms of AI techniques to analyse data, learn from data, and make a decisions and predictions for real-world events. In the next section research methodologies are discussed.

Artificial Intelligence in IoV

For realizing ML, it enables many applications and expands the scope of AI. Reinforcement learning (RL), also known as evaluation learning. It is a technique of ML. Deep reinforcement learning (DRL) is the combination of DL and RL techniques. It uses the benefits of deep neural networks (DNNs) to train the learning process, therefore the learning speed and performance of the RL algorithm and overcoming the unsuitability of RL for large-scale networks. Figure 2 shows relationship between ML, DL, RL and AI (Hongjing et al, 2020).

AI integrated with IoT in automated vehicles (AV) to provide high-performance of embedded systems that used to allow more dynamic and robust control systems. While the main software components of AVs are traditionally host by cloud computing systems, new edge computing paradigm has addressed some technical challenges includes bandwidth, latency and security of networks (Mona et al,2021).

Figure 2. The relationship between AI, ML, RL, DL and DRL

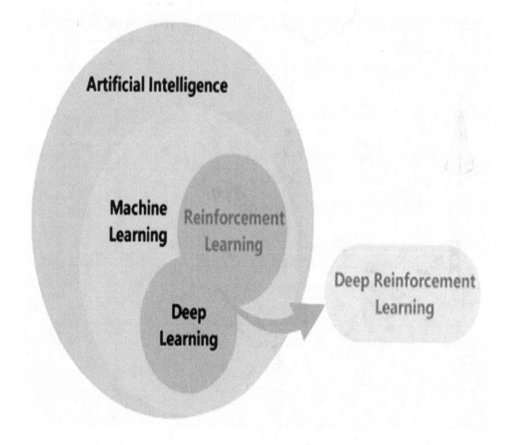

The amount of data is increase in AV which is used for advanced driver assistance dystems (ADAS) and entertainment (Hongjing et al, 2020). Hence, software and hardware have been required. Sensors, actuators devices and software are used to complete the purpose similar to the superhuman brain as aimed through AI.

The AV brings new approach to industrial manufacturers and dealerships, enabling companies to implement AI to increase value for their customers. The most efficient approach for AI to process this data is to use ML algorithms. The ML algorithms help a specific driver profiles and vehicle owners exactly what they require in vehicle and through their mobile phones via a corresponding application. They accomplish this by remembering their behaviour and analysing their driving history and the situation on the road (Hongjing et al, 2020).

Even though AI can handle large data from vehicles with some of the extra data conditions including traffic, pedestrians, and experiences, will need to be collected via IoT networks as LAN, PAN, WSN and wide area network (WAN). This large amount data required some equipment such as sensors, vehicles, embedded electronics devices, software and network connectivity for allowing to gathering and sharing data among vehicles. These IoT-allowed AVs to integrate equipment to provide more safety reduce the fuel consumption and security.

Figure 3. Architecture for AI based on IoV

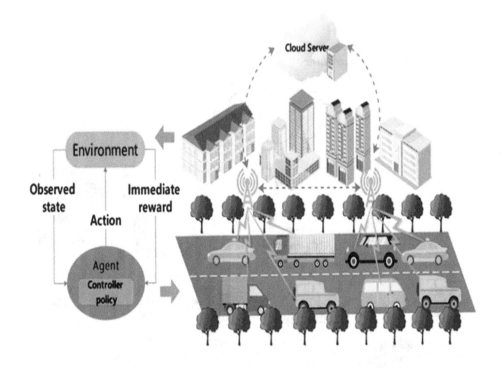

AI has ability to enhance the cognition and IoV networks and thus assist in optimally allocating resources for problems with diverse, time-variant and complex features (Hongjing et al, 2020). As shown in figure3, the architecture of AI in IoV, where the agent observed its current environmental state, takes action and receives its immediate reward with the new state. The observed information includes immediate rewards and the new status which used to adjust the agent's strategy and this process is repeated until the agent's strategy approaches the optimal strategy.

Clustering Algorithms in IoV

As mentioned before, IoV is a new version of IoT which focus on communication among vehicles. Vehicular nodes are usually in motion with time it will lead frequent changes in the topology. These changes cause issues in IoV network include dynamic topology changes, shortest path for routing and scalability (Rashid et al,2010). Clustering is among one of the popular topologies to solve issues in IoV network. Logical grouping of nodes like in UAV's, sensors, automobiles, ships, and etc. have a common topographical zone termed as clustering. Nodes can vary according to the applied problem; participating nodes are cluster members (CM).

Node has two fundamental topology types they are Centralized topology is known as a central governing node cluster head (CH) which responsible for intra-cluster and inter-cluster activities. And second type is De-centralized topology is considered as each group of the population consists of similar nodes having distributed control (Farhan et al, 2018). If control is centralized, then a cluster contains central governing node CH. Nodes can be connected with other by wireless networks (3G, 4G, and 5G). Nodes

Figure 4. 5G-enabled VANET architecture

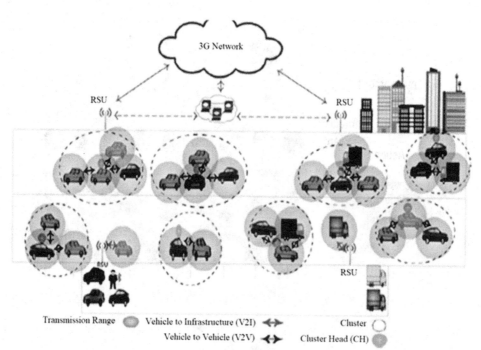

that connected with 5G network are more desirable CHs but not compulsory for enhanced management. Better CH selection/election is difficult in IoVs.

V2X communication in heterogeneous networks is considered the architecture that clarifies 5G allows using IoVs as well as it supports more benefits in term of bandwidth and availability of network resource (see figure 4). Also, 5G will enable new application scenarios such as parked cars and pedestrians. IoV issues including mobility, information broadcast and shortest path routing can be addressed more efficiently in 5G-based scenarios (Farhan et al, 2018). The future scenario that CAVDO can be interfaced with other future networks technologies such as Li-Fi and beyond network technologies.

Dragonfly-Based Clustering Algorithm for IoV

Clustering is one of the solution issues for IoV due to motivation such as scalability, dynamic topology changes and shortest path for routing. The Meta heuristic dragonfly-based clustering algorithm CAVDO is used for cluster-based packet route optimization for stability of IoV topology in a dynamic environment and mobility aware dynamic transmission range algorithm (MA-DTR) is used with CAVDO for transmission range adaptation on the basis of traffic density. Minimum number of clusters represents the shortest path in cluster-based routing (Farhan et al, 2018). To avoid broadcast overhead, CHs only are responsible to forward and exchange information between CMs and to other clusters. For efficient CHs selection/election, the unique swarming technique of dragonflies enable CAVDO and MA-DTR used to construct ideal clustering solution for IoV routing. In swarm-based techniques, a swarm is a set of solutions and each node defined as a single solution. A dragonfly algorithm represents a complete route containing CH IDs through which packet is travelled. Swarm-based approaches proved their working efficiency on both problemdomains: in discrete value problems and in continuous problems as well. Although in comparison, the implementation complexity is high, mostly applied to exhaustive search situations to collect best possible solution, this nature of algorithm makes it best suited for cluster-based routing in IoV. CAVDO with dynamic weights is first attempt to construct efficient clustering solution in IoVs. The algorithm is initialized by finding CHs, and then finds neighbouring nodes of CHs (Rofida et al,2017). For the application of swarm-based techniques in IoV, following are the requirements of problem encoding. First, a complete solution set formation by combination of multiple different subsets.

Second, the evaluation methodology can be applied to solution fitness measures.

And third, an associated heuristic measure for solution subsets which is not compulsory, but desirable.

Moth Flame Clustering Algorithm

This technique is inspired from Moths, Moths have a mechanism called transverse orientation, it is based on tracking path at night by following moon light by making use of the same angle towards the moon, this procedure assures that the moth will fly in a straight line as the moon is immensely distant from the moth, this is an efficient procedure for drifting in lengthy expanses in a strait trail (Muhammad et al, 2019). It is observed that moth fly spirally near artificial lights, moths try to retain an analogous angle with artificial light and try to move in a straight track; but as these light sources are near, preserving a same angle triggers a lethal spiral trail for moths. This is shown in figure 5.

Moths can fly in 1- Dimensional, 2- Dimensional, 3- Dimensional or hyper dimensional area with altering their positions. The sets of moths can be shown in the following matrix form.

Figure 5. Spiral flying path

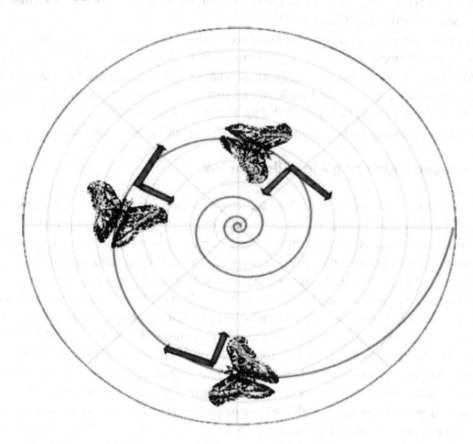

$$M = \begin{bmatrix} M1,1 & M1,2 & \text{...........} & M1,d \\ M2,1 & M2,2 & \text{...........} & M2,d \\ \vdots & \vdots & \vdots & \vdots \\ Mn,1 & Mn,2 & \text{...........} & Mn,d \end{bmatrix}. \tag{1}$$

Where, n: represents the number of moths. d: represents the number of dimensions.

The network of autonomous vehicles is created by randomly initializing their position within a certain region called as grid size. Next, the speed and direction of vehicles are also allocated arbitrary. The vehicle ID's are assigned for the identification so in the mesh topology of the network (Yasir et al, 2018). Afterward, Euclidian distance is measured between all the nodes to form a complete distance matrix of whole network.

An ideal number of clusters in IoV network mark the network more stable as the resources of network are effectively employed. Moth flame clustering algorithm optimizes the sum of clusters in the network. This is due to the evolutionary proficiency of MFO which empowers it to select the optimum number of solutions as it is efficient and appropriate for discrete and continuous variable problems (Yasir et al, 2018). Though the implementation of such algorithms is relatively challenging, these procedures are computationally inexpensive, specifically when competed with an extensive search to recognize the finest solution. As a result, these characteristics signify that MFO based procedures are successful for clustering in IoV networks.

Mobility Adaptive Density Connected Clustering Algorithm

Clustering is one of the popular approaches for topology management that can positively influence in the performance of networks. It plays important role in VANETs. VANET having highly mobile nodes that causes change in topology with a time and it is very difficult to construct stable clusters. More homogeneous environment creates more stable clusters. Homogeneous neighbourhood for a vehicle is strongly driven by density and standard deviation of average relative velocity of vehicles in its communication range. Mobility adaptive density connected clustering algorithm (MADCCA) is a density-based clustering algorithm.

This approach used standard deviation of relative velocity and density matrices in their neighbourhood for selecting the cluster head for a cluster as well as direction and location of movement in the clustering process (Anant Ram et al, 2017). The CHs will select with respect to standard deviation of average relative velocity and density matrices in their neighbourhood. Vehicle that has a more homogenous environment will be selected as CHs. The standard deviation of the average relative velocity will become low only when all the neighbour vehicles of a vehicle are homogeneous with respect to the average relative velocity. While the standard deviation of the average relative velocity of a vehicle will be very high, when average relative velocity difference of a vehicle is high with each of its neighbourhood. Standard deviation of average relative velocity parameter enhanced the intra-cluster homogeneity; the target of density variation parameter is to improve inter-cluster homogeneity (Anant Ram et al, 2017).

Swarm Intelligence (SI) Algorithms

Swarm Intelligence (SI) is a new branch of AI based on analysis the action of individuals in various decentralized system. SI provides the possibility of SI behavior through collaboration in individuals that have limited or no intelligence. Its potential parallelism and distribution characteristics used to enable the possibility of solving complex nonlinear problems with advanced capabilities in term of robustness, self-adaptability and search ability (Hazem et al, 2012). Since there are many SI optimization algorithms including classical particle swarm optimization (PSO) and ant colony optimization (ACO). Recently, many improvement algorithms appear such as artificial bee colony (ABC), bacterial foraging algorithm (BFO) and butterfly optimization algorithm (BOA). SI algorithms explore optimized solution based on heuristic information.

SI implemented in some area as routing, scheduling, medical, military, telecommunication networks and for process optimization problems (Hazem et al, 2012) (Priyan et al.2017). SI algorithm can be used on robots which is one of the most rapid developments today. There are many developments of SI that was utilized on Unmanned Aerial Vehicle (UAV) for industries because this algorithm has a several benefits such as flexibility, easy marketing and implementation, robustness and derivative free optimization (Priyan et al.2017). Moreover, SI has been developed and installed on autonomous surface vehicle (ASV). ASV implements the global positioning system (GPS) to determine the location, relative orientation to other ASV and ability to arrive the desire direction.

Ant Colony Optimization (ACO) in IoV

In ACO, each ant releases pheromones on the path. The whole ant colony can perceive pheromones. The ants in the ant colony will choose the path with higher pheromones and ants passing through the path will release pheromones so that after a while, the whole ant colony can follow the shortest path to reach the food. The advantages of ACO include strong global optimal ability and flexibility in implementation. It is suitable for integrating with other algorithms. Moreover, this algorithm uses forward and backward ant procedures to know the optimal route to reach the destination. The number of vehicles currently moving on a road and the road length are collected simultaneously with the help of IoV technology (Priyan et al.2017). The calculation of maximum number of vehicles $\text{Max}_{NV_{ij}}$ .can be on the road as follows:

$$\text{Max}_{NV_{ij}} = \frac{LL_{ij}}{L_v + \Delta L} \times NL_{ij} .$$
(2)

where, LL_{ij} . Represents road length. NL_{ij} . represents the number of roads in a street between two nodes (i and j). ΔL .: represents the average distance between two vehicles. L_v . represents the average length of vehicles. Density of vehicles (D_{ij} . is calculate by

$$D_{ij} = \frac{NV_{ij}}{\text{Max}_{NV_{ij}}} .$$
(3)

In ant colony optimization, forward ants used to find the optimal and shortest path to reach the desired destination. The forward ants could calculate the movement of new position as in the following equation.

$$p_{ij}^k(t) = \begin{cases} \dfrac{a\left(\partial_{ij}\right) + b\left(1 - \cdot_{ij}\right)}{\sum_{h \notin tabu_k} a\left(\partial_{ij}\right) + b\left(1 - \cdot_{ij}\right)} \times \left(\dfrac{1}{1 + \dfrac{1}{N_j}}\right) & ,\text{if } j \notin tabu_k \\ \\ 0 \; otherwise \end{cases}$$
(4)

Where: ∂_{ij} . describe the pheromone value of an ant in node i to move to node j. $\cdot_{ij}$. describe instantaneous state of the fuzzy value on the link from i to j and calculated by vehicle as ant. a: describe the weight for the importance of ∂_{ij} .b: describe the weight for the importance of $\cdot_{ij}$. $tabu_k$. describe a group of nodes connected to node i that an ant k has not visited until now. N_j. describe a number of neighbours for node j. If a forward ant reaches desired destination, then the forward ant changes as a backward ant. The memory of forward ant is known from the backward ant to find optimal route.

Ant Colony Optimization algorithm is deployed for efficiency in the CH selection/election to create desirable solutions of clustering for IoV. In methods, which are swarm-based, each node is considered to be a single solution, and the swarm is a group of solutions. An ant, in a given instance, depicts all CH IDs of the entire route. The initial attempt is to create optimum solutions in IoV clustering. The Ant Colony Optimization algorithm by using intelligent first-node selection and dynamic evaporation strategy makes this aim possible. The initialization of the algorithm is made by finding the CHs regarding node speed, distance, direction, and local traffic density and then locating the neighbouring nodes of the CHs (Mohd Nadhir et al,2015). Every vehicle is considered to have interfaces of 5G and 802.11p. The way the vehicular networks are organized is made up of clusters that are managed by the eNodeBs. The cluster sizes are of varying ranges at 802.11p based on the vehicular local traffic density.

Bee Colony Optimization in IoV

This algorithm is inspired by the intelligent behaviour of real honey bees in finding food sources, known as nectar, and the sharing of information about that food source among other bees in the nest. The bee colony divides the bee swarm into three types, the employed bee, the onlooker bee, and the scout bee. Each of these bees has different tasks assigned to them in order to complete the algorithm's process. The employed bees focus on a food source and retain the locality of that food source in their memories. The number of employed bees is equal to the number of food sources since each employed bee is associated with one and only one food source (Anuja et al, 2017). The onlooker bee receives the information of the food source from the employed bee in the hive. After that, one of the food sources is selected to gather the nectar. The scout bee is in charge of finding new food sources and the new nectar. Figure 6 shows the flow chart of Bee colony algorithm.

Within bee colony optimization met-heuristic (BCO), the agents that call artificial bees collaborate in order to solve difficult combinatorial optimization problem. All artificial bees are located in the hive at the beginning of the search process. During the search process, artificial bees communicate directly. Each artificial bee makes a series of local moves, and in this way incrementally constructs a solution of the problem. Bees are adding solution components to the current partial solution until they create one or more feasible solutions.

The search process is composed of iterations. The first iteration is finished when bees create for the first time one or more feasible solutions. The best discovered solution during the first iteration is saved, and then the second iteration begins. Within the second iteration, bees again incrementally construct solutions of the problem, etc. There are one or more partial solutions at the end of iterations. The analyst-decision maker prescribes the total number of iterations (Anuja et al, 2017) (José et al,2018). In the application of vehicle routing problem, the bees construct paths in different ways depending on their roles. Leaders only retrace the last path of the iterative procedure without changing the situation. The scouts and followers choose the next nectar source as follows:

Figure 6. The Bee Colony algorithm

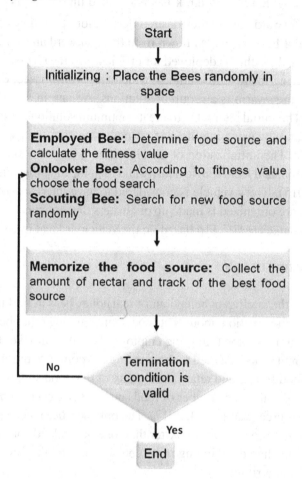

$$p_{ij}^{k}\left(t\right)=\begin{cases} \dfrac{[S_{ij}\left(t\right)]^{\bullet}[f_{y}\left(t\right)]^{\beta}}{\sum_{Allowed,k\,to\,i}[S_{ij}\left(t\right)]^{\bullet}[f_{y}\left(t\right)]^{\beta}},j\notin A_{i}^{k} \\ 0, Otherwise \end{cases}. \tag{5}$$

Where, $p_{ij}^{k}(t)$ represents the probability that bee k travels from node i to j in iteration t. $A_{i}^{k}=\Phi-Tabu_{k}$ denotes the available nodes for bee k, *i.e.*, the points that satisfy all constraints but have not been previously visited by the bee). Tabu k is the tabu list of bee k, which prevents the nodes already visited by bee k from being repeatedly visited. In some applications (*i.e.*, router optimization), it is also convenient that the bee can return through the original path. $sij(t)$ and $fij(t)$ represent the bee colony and heuristic information, respectively, while α and β are the weight coefficients (Ajay et al,2019).

Cookoo Bird Optimization in IoV

Cuckoo Optimization algorithm is based on a life of a bird called Cuckoo. The basic of this novel Optimization algorithm is specific breeding and egg lying of this bird, adult cuckoos and eggs used in this modelling. The adult cuckoos lay eggs in other bird's habitat. Those eggs grow and become a mature cuckoo if are not finds and not removed by host birds. The immigration of groups of cuckoos and environmental specifications hopefully lead them to converge and reach the best place for reproduction and breeding. The objective function is in this best place (Anuja et al, 2017). Figure 7, shows the flowchart of cuckoo optimization algorithm. The cuckoo search algorithm is described using the following idealized rules:

- Each cuckoo lays one egg at a time and dumps it in a randomly chosen nest.
- The best nests with high-quality eggs will be carried over to the next generations.
- The number of available host nests is fixed, and the egg laid by a cuckoo is discovered by the host bird with a probability pa ε [0,1]. In this case, the host bird can either get rid of the egg or simply abandon the nest and build a completely new nest (José et al, 2018).

Figure 7. Cuckoo Optimization Algorithm

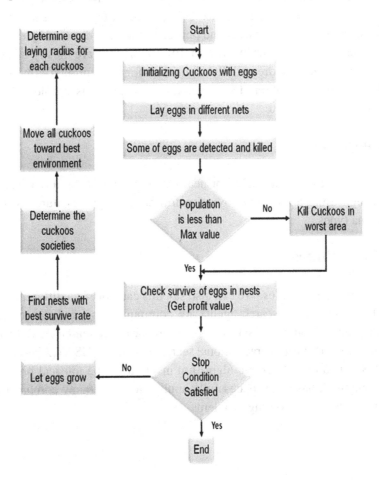

Optimization of the IoV network is formulated by cuckoo-based optimization algorithm. Nodes are deployed randomly and organized as static clusters by cuckoo search (CS). After the cluster heads are selected, the information is collected, aggregated and forwarded to the base station using generalized particle approach algorithm. The generalized particle model algorithm (GPMA) transforms the network energy consumption problem into dynamics and kinematics of numerous particles in a force-field.

The proposed approach can significantly lengthen the network lifetime when compared to traditional methods. The aspect of formulating this technique is incorporation of constraints that helps in energy efficient data collection or fusion, by eliminating data redundancy and minimization of energy consumption. Cuckoo Search is performed to form sub-optimal data fusion chains. The collected information is transmitted to the base station via cluster heads through GPM algorithm. In the GPM algorithm energy constraints are incorporated and the information is routed in shortest path (Shaik et al, 2019). The proposed technique shows better performance in optimum number of clusters, total energy consumption and prolonging of network lifetime.

Pigeon-Inspired Optimization (PIO) Algorithm

Pigeon-inspired optimization (PIO) algorithm, which was initially introduced by (Qiao and Duan, 2014). In this algorithm, each pigeon in the swarm has its own location, a speed, and a best historic location, conferring to which it travels in the space of search. The pigeon search procedure contains of two phases. In the first phase, pigeons search in search space being directed by the best location initiated by the swarm and the experience of flying. In the second phase, pigeons are directed by the positions centre of residual pigeons that successfully follow the swarm. PIO algorithms are mostly utilised for continuous issues, as well-known, papers have addressed PIO algorithms for conventional combinatory optimization issues, such as salesmen travel problem (TSP), quadratics assignments problem, knapsacks problem, and job-shops scheduling problem, etc.

Meta-Heuristic Algorithms

Meta-heuristic algorithms aim to find global or near-optimal solutions at a reasonable computational cost. Meta-heuristic algorithms are inspired by animal/insect behaviour, nature, evolutionary concepts and natural phenomena types.

Physics Based Algorithms

It is typically inspired by physical rules. We can also say this category is a different form of the swarm-based algorithms. Undefined usual search agents are communicating and stirring from one point to another due to physical rules. Some popular algorithms are harmony search (HS), black hole algorithm (BHO), simulated annealing (SA), and gravitational search algorithm (GSA) (Ajay et al,2019). Harmony search algorithm (HSA) is a meta-heuristic algorithm inspired by the musical process to solve different optimization problems. It has been successfully applied to solve many combinatorial optimization problems such as standard vehicle routing problems.

Evolutionary Algorithms

It is based on natural selection process concepts. The population tries to continue building on the fitness measurement in their environment and applied the best effort to get the best optimal solution in the search spaces (Ajay et al, 2019). Some evolutionary algorithms are genetic algorithms, evolution strategies, differential evolution, genetic programming, and evolutionary programming. Genetic Algorithm (GA) that handles the optimization of the static vehicle routing problem like-instances that correspond to the whole dynamic optimization problem. The GA is launched at any fixed duration and must run within an efficient amount of time. The fitness evaluation involves the vehicle routes obtained after the translation of the chromosome representation. It returns the total travel distance /cost of the routes. The best-cost route crossover (BCRC) is used as the crossover operator and the inversion operator is used as the mutation operator (Chin et al, 2017).

Differential evolution (DE) is a parallel direct search method and searches for a global optimum in a N dimensional parameter in IoV space. Initially, a randomly initiated population of NP-N dimensional real valued parameters, where NP is the number of vehicles populations, is created and should cover the entire search space. Parameters in DE are usually represented by vectors (Ali et al, 2019). DE optimises a problem by maintaining a population of candidate solution and creates new candidate solutions by combining existing ones through its formulas and then keeping the candidate solutions with the best score or fitness on the optimisation problem.

Distributed Multi-channel and Mobility Aware Cluster Based MAC Protocol

As Vehicular safety applications needs periodic advertise of status and urgent messages, contention-based medium-access-control (MAC) protocols like IEEE 802.11p have issues in low throughput, latency, Predictable, fairness and high collision rate, especially in overload networks due to that distributed multi-channel and mobility-aware cluster-based MAC (DMMAC) protocol is developed. In channel scheduling and an adaptive learning mechanism integrated within the fuzzy-logic inference system (FIS), vehicles ordered themselves into more stable and non-overlapped clusters. In DMMAC, each cluster uses any various sub-channel from its neighbours in a distributed way to overcome on hidden terminal issue. DMMAC provides Increase system reality; reduce delay time for vehicular safety application and efficiency clustering vehicles in highly dynamic and dense networks in a distributed way which considered as the main contributions method in MAC protocol.

Short-range communication opened the door to utilizing these technologies in using advanced vehicular safety applications. Especially, the new dedicated short-range communications (DSRC) or IEEE 802.11p allows a new class of vehicular safety applications and they will improve network performance such as safety, reliability and efficiency of the current transportation system (Sawsan et al,2018). In ITSs field, this technology provides a wide spectrum to avoid or to reduce the road accidents. Safety applications in VANETs allow vehicles to periodically exchange their event and emergency messages to surrounding environment for preventing accidents before they occur. Without efficient and reliable medium-access control (MAC) protocol it will cause serious degradation in the performance of VANETs' safety and non-safety applications (Khalid et al, 2013). The proposed MAC protocol will reduce transmission collisions between vehicles and emergency messages will send in a real-time. Furthermore, the wireless channel has to be reliability, high throughput, and limited delay time concurrently and fairly shared among vehicles.

In IEEE 80211p integrates OFDMA with the contention-based distributed coordination function (DCF) algorithm. CHs are elected depends on their minimal overhead and stability on the road so clustering information is embedded in vehicles' periodic status messages. The MAC protocol is acceptable to drivers' behaviour and has ability to learn mechanism for predicting the position and speeds of all cluster members applied the fuzzy-logic inference system (FIS). This makes the MAC protocol more efficient, increases the battery life of CH elected and its members and reliability. Moreover, OFDMA subcarriers of the IEEE 802.11p control channel (CCH) are sub-divided into four groups. Each cluster can use only one group that is various from its neighbours to remove the hidden terminal issues [14].

Best Interface Selection (BIS) Algorithm

The idea of using BIS algorithms that it permits the vehicular users to switch among interfaces belongs to various technologies. However, the standard of select the interface for connectivity depends on several QoS parameters like delay, throughput cost-effectiveness or other user preference. Therefore, the use of multiple wireless interfaces will ensure services and the best-connected user interface at all the times. Additionally, the multiple wireless interfaces such as WAVE, long-range Wi-Fi, and 4G/LTE could be considered as a back up to each other. There may be a different application running by vehicular users. This algorithm chooses the wireless interface randomly from the available options to access network.

The choose occurs according to checks if QoS requirements in terms of bandwidth and/or delay are successfully met by the chosen interface or it required to switch over to some new interface (Hafiz et al, 2015). Moreover, Cost is another user-defined preference. If the QoS parameters are satisfied, the interface with lowest cost would be select. Furthermore, the algorithm serves the purpose to manage load sharing among various interfaces.

In IoV applications, the best interface selection algorithm is depending on application requirements i.e., bandwidth, cost, delay, and network utilization (Hafiz et al, 2015). The procedure of selecting interface performed according to the mentioned requirements. In BIS algorithm to switch required application access, first the algorithm sets an interface by selecting random network ID for initialization. Then switching processes take places depending on decision made by comparing and fulfil the following case questions, is the current network interface meets application bandwidth requirements? Is the list of networks to iterate and sort in the increasing cost order? And the current network interface meets the delay requirements? Due to this decision questions, the algorithm must meet the requirements of highest bandwidth interface, network interface with least cost, and network interface with least delay.

SOLUTIONS AND RECOMMENDATIONS

QoS parameters optimization plays an important role in IoV network. It defines as analysis the behaviour of the network so as to transport data with minimum packet loss, delay and maximum bandwidth system. It is necessary to provide QoS in an environment like IoV (Mohd et al, 2015). The next section presents some of the solutions and recommendations.

Quality of Service (QoS) Optimization Algorithms in IoV System

In IoV system, many devices are deployed for sensing, processing, integrating and communicating with another vehicle via various open wireless technologies this will cause a massive load on the network so there is required to provide an optimal End-to-End delivery with QoS parameters like Jitter, latency, delay and bandwidth. Some of the applications that cause a heavy load on the network are voice, video and multimedia applications so the data does not ensure that it delivered over mediums (Zeinab et al,2018). Moreover, QoS in IoV scenario provides optimal delay services. Hence, there is a need to provide QoS Mechanisms to deal with some challenges in an environment that have billions of devices for exchanging information (Shaik et al, 2019).

Figure 8. Traffic types vs. Bandwidth

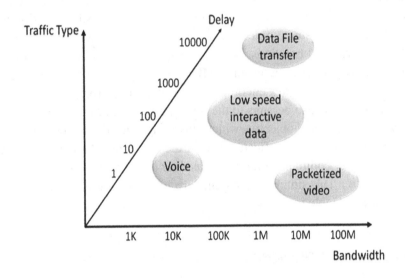

QoS provides control in delay in sensitive service, Jitter, dedicated bandwidths and improved packet loss characteristics. In IoV networks, there is need the best effort services to provide throughput, Reliability, availability, bandwidth, latency, and delay and jitters when the network receives the data. QoS parameters could be more emphasized as handling of Network traffic. Moreover, due to vehicles are mobile and dynamic and moving with varying speed it will lead the network resource is unpredictable then, there is a need of developing Highly efficient QOS schemes. Figure 8 shows types of multimedia traffic application characteristics relatively to bandwidth usage and delays. It is obvious that QOS becomes more challenging as the traffic varies significantly with a wide range of parameters (Shaik et al, 2019) (Ali et al, 2018).

Since the development of network technology, the multimedia applications have become increasingly popular. However, it is hard to achieve QoS and efficiency in multimedia application through IoV networks due to high mobility characteristic Therefore, a buffer-aware approach is proposed to allow client in the case of handover between different eNodeBs to play multimedia applications through vehicular 5G networks for realizing minimal delay and achieve better QoS. The use of buffer-aware QoS streaming approach through IoV network technology provides different priority levels of streaming service (Chin et al, 2017).

Buffer-aware QoS Optimization Algorithm is based on the buffer allocation mechanism to optimize the QoE during multimedia communication in the IoV system. The algorithm optimizes the QoE by controlling high peak variable rate of multimedia by allocating proper buffer size in IoV (Ali et al, 2019). Power-aware QoS Optimization Algorithm is based on the channel features in association with the IoV as an individual and realistic network.

Optimization Problem in IoV Clustering

Optimization challenges possess excellent significance within scientific engineering model along with decision making applications. Optimization is the term for discovering several remedies of an issue, which will correspond to extreme values connected with more than one objective. When an optimization problem possesses just one objective, the task of choosing the best possible solution is named single-objective problem. Typically, in a single-objective problem, the focus is on obtaining just a single solution with the exception of multimodal functions. If the optimization problem comes with several objective functions, the optimization problem is referred as MOP (Hadded et al, 2015). The majority of the real-world problems belong to MOPs, as they encompass a variety of objectives which have to be optimized concurrently. Clustering in IoV also belongs to the set of problems in MOPs.

Cluster stability is a fundamental objective which clustering algorithms endeavour to accomplish and is also regarded as a way of measuring effectiveness of the clustering algorithm. Stability is important for the upper as well as lower communication layers as their performances will rise apparently by using clusters (Abbasi et al, 2007). This simplifies routing, permits spatial reuse of resources and helps makes the network turn up further stable in the view of every cluster node.

FUTURE RESEARCH DIRECTIONS

In IoVs, effective communication and network management are challenging tasks. The optimization algorithms provide better network management. Clustering techniques provides minimum number of clusters with respect to dynamic transmission range, where transmission range is a high the number of clusters are minimum and vice versa. These are the major contributions for ITS and IoVs which can be employed toward different research directions. For the future work, proposed techniques can be enhanced for mobility awareness. Many algorithms aim to address general single-objective optimization problems. However, other types of optimization problems still deserve further attention from the community, unfolding vast opportunities for the development of the optimization algorithms (Daniel et al, 2020):

- Multi- and many-objective optimization, wherein the goal is to simultaneously optimize several conflicting objectives.
- Multimodal optimization, in which there may be several global optima to be found and retained during the search.
- Large-Scale Global Optimization, in which the number of vehicles (dimensionality of the search space) is huge.

- Mimetic Algorithms, where the algorithmic proposal is combined with other search techniques to further improve its performance.
- Parameter tuning, which refers to the search for the values of the parameters of the optimization algorithm itself that lead to its best search performance.
- Parameter adaptation, i.e., the design of adaptive parameters that allow solver to adapt themselves to the problem during the search.

CONCLUSION

With the rapid development of Internet and communication technologies, vehicles that often quickly move in cities or suburbs have strong computation and communication abilities. IoV is emerging as an important part of the smart or intelligent cities being proposed and developed around the world. IoV is a complex integrated network system that interconnects people within and around vehicles, intelligent systems on board vehicles, and various cyber-physical systems in urban environments. IoV goes beyond telematics, vehicle ad hoc networks, and intelligent transportation by integrating vehicles, sensors, and mobile devices into a global network to enable various services to be delivered to vehicular and transportation systems and to people on board and around vehicles.

With network optimization as one of the main challenges that IoV would face in the forthcoming years. This chapter depicts comprehensive survey on the most important aspects through some of the novel approaches related to network optimization for IoV communication is presented. Various algorithm types are presented in order to improve network utilization. Background knowledge regarding the clustering process, a brief historical note, definitions, cluster structure is presented.

REFERENCES

Aadil, F., Ahsan, W., Rehman, Z. U., Shah, P. A., Rho, S., & Mehmood, I. (2018). Clustering algorithm for internet of vehicles (IoV) based on dragonfly optimizer (CAVDO). Springer. *The Journal of Supercomputing*, 74(9), 4542–4567. doi:10.100711227-018-2305-x

Abbasi, A. A., & Younis, M. (2007). A survey on clustering algorithms for wireless sensor networks. *Computer Communications*, 30(14-15), 2826–2841. doi:10.1016/j.comcom.2007.05.024

Abdel Hafeez, K., Zhao, L., Shen, X., & Niu, N. (2013). Distributed Multichannel and Mobility-Aware Cluster-Based MAC Protocol for Vehicular Ad Hoc Networks. IEEE Transactions on Vehicular Technology, 62(8).

Abdelgadir, M., Saeed, R. A., & AbuAgla, B. (2017). Mobility Routing Model for Vehicular Ad-hoc Networks (VANETs), Smart City Scenarios. *Vehicular Communications*, 9, 154–161. doi:10.1016/j.vehcom.2017.04.003

Abdelgadir, Saeed, & Babiker. (2018). Cross Layer Design Approach for Efficient Data Delivery Based on IEEE 802.11P in Vehicular Ad-Hoc Networks (VANETs) for City Scenarios. *International Journal on Ad Hoc Networking Systems*, 8(4).

Abdelgadir & Saeed. (2020). Evaluation of Performance Enhancement of OFDM Based on Cross Layer Design (CLD) IEEE 802.11p Standard for Vehicular Ad-hoc Networks (VANETs), City Scenario. *International Journal of Signal Processing Systems, 8*(1).

Ahmed, H. R. (2012). *Swarm Intelligence: Concepts, Models and Applications*. Queen's University.

Ahmed, Z., Saeed, R. A., & Mukherjee, A. (2018). Vehicular Cloud Computing models, architectures, applications, challenges and opportunities. In Vehicular Cloud Computing for Traffic Management and Systems. IGI Global. Doi:10.4018/978-5225-3981-0

Alawi, Saeed, Hassan, & Alsaqour. (2014). Simplified gateway selection scheme for multi-hop relay vehicular ad hoc network. *International Journal of Communication Systems, 27*(12), 3855–3873. doi:10.1002/dac.2581

Alawi, M., Saeed, R., Hassan, A., & Khalifa, O. (2012). Internet Access Challenges and Solutions for Vehicular Ad-Hoc Network Environment. *IEEE International Conference on Computer & Communication Engineering (ICCCE2012)*.

Ali, E. S., Hasan, M. K., Hassan, R., Saeed, R. A., Hassan, M. B., Islam, S., Nafi, N. S., & Bevinakoppa, S. (2021). Machine Learning Technologies for Secure Vehicular Communication in Internet of Vehicles: Recent Advances and Applications. *Journal of Security and Communication Networks*.

Anuja, S. (2017). Cuckoo Search Optimization- A Review. *Materials Today: Proceedings, 4*(8), 7262–7269. doi:10.1016/j.matpr.2017.07.055

Dirar, Saeed, Hasan, & Mahmud. (2017). Persistent Overload Control for Backlogged Machine to Machine Communications in Long Term Evolution Advanced Networks. *Journal of Telecommunication, Electronic and Computer Engineering, 9*(3).

Eltahir, A. A., & Saeed, R. A. (2018). V2V Communication Protocols in Cloud Assisted Vehicular Networks. In Vehicular Cloud Computing for Traffic Management and Systems. IGI Global. doi:10.4018/978-1-5225-3981-0.ch006

Eltahir, A. A., Saeed, R. A., Mukherjee, A., & Hasan, M. K. (2016). Evaluation and Analysis of an Enhanced Hybrid Wireless Mesh Protocol for Vehicular Ad-hoc Network. *EURASIP Journal on Wireless Communications and Networking, 2016*(1), 1–11. doi:10.118613638-016-0666-5

Eltahir, S., & Alawi. (2013). An enhanced hybrid wireless mesh protocol (E-HWMP) protocol for multi-hop vehicular communications. *2013 International Conference on Computing, Electrical and Electronics Engineering (ICCEEE)*, 1 - 8.

García, J., Altimiras, F., Peña, A., Astorga, G., & Peredo, O. (2018). *A Binary Cuckoo Search Big Data Algorithm Applied to Large-Scale Crew Scheduling Problems*. Computational Intelligence in Modeling Complex Systems and Solving Complex Problems. doi:10.1155/2018/8395193

Hadded, M., Zagrouba, R., Laouiti, A., Muhlethaler, P., & Saidane, L. A. (2015). A multi-objective genetic algorithm-based adaptive weighted clustering protocol in VANET. *Proc. of Evolutionary Computation (CEC), IEEE Congress on*, 994-1002.

Hafiz, H. R. S., Khan, Z. A., Iqbal, R., Rizwan, S., Imran, M. A., & Awan, K. (2019). *A Heterogeneous IoV Architecture for Data Forwarding in Vehicle to Infrastructure Communication*. Hindawi Mobile Information Systems.

Hassan, M. B., Alsharif, S., Alhumyani, H., Ali, E. S., Mokhtar, R. A., & Saeed, R. A. (2021). An Enhanced Cooperative Communication Scheme for Physical Uplink Shared Channel in NB-IoT. *Wireless Personal Communications, 116*(2).

Husnain & Anwar. (2020). An intelligent cluster optimization algorithm based on whale optimization algorithm for VANETs(WOACNET). *PLoSONE, 16*(4). doi:10.1371/journal.pone.0250271

Hussain, S. A. (2019). A Review of Quality-of-Service Issues in Internet of Vehicles (IoV). *IEEE Access: Practical Innovations, Open Solutions*.

Ibrahim, S., Saeed, R. A., & Mukherjee, A. (2018). Resource management in Vehicular Cloud Computing. In Vehicular Cloud Computing for Traffic Management and Systems. IGI Global. doi:10.4018/978-1-5225-3981-0.ch004

Ji, H., Alfarraj, O., & Tolba, A. (2020). Artificial Intelligence-Empowered Edge of Vehicles: Architecture, Enabling Technologies, and Applications. *IEEE Access: Practical Innovations, Open Solutions, 8*, 61020–61034. doi:10.1109/ACCESS.2020.2983609

Jiang, X., Ma, Z., Yu, F. R., Song, T., & Boukerche, A. (2020). Edge Computing for Video Analytics in the Internet of Vehicles with Blockchain. In *Proceedings of the 10th ACM Symposium on Design and Analysis of Intelligent Vehicular Networks and Applications (DIVANet '20)*. Association for Computing Machinery. 10.1145/3416014.3424582

Kayarga, T., & Kumar, S.A. (2021). A Study on Various Technologies to Solve the Routing Problem in Internet of Vehicles (IoV). *Wireless Personal Communications*.

Khan, M. F. (2019). *Moth Flame Clustering Algorithm for Internet of Vehicle (MFCA-IoV)*. IEEE., doi:10.1109/ACCESS.2018.2886420

Kumar & Bawa. (2019). *A comparative review of meta-heuristic approaches to optimize the SLA violation costs for dynamic execution of cloud service*. Springer-Verlag GmbH Germany.

Kumar. (2017). Ant Colony Optimization algorithm with internet of vehicles for intelligent Traffic Control System. *Computer Networks Journal*.

Lai, Chang, Chao, Hossain, & Ghoneim. (2017). A Buffer-Aware QoS Streaming Approach for SDN-Enabled 5G Vehicular Networks. *IEEE Communications Magazine*.

Mohd, N. A. W. (2015). *A Comprehensive Review of Swarm Optimization Algorithms*. ResearchGate. doi:10.1371/journal.pone.0122827

Molina, Poyatos, Del Ser, García, Hussain, & Herrera. (2020). Comprehensive Taxonomies of Nature- and Bio-inspired Optimization: Inspiration versus Algorithmic Behavior, Critical Analysis and Recommendations. *arXiv journal*.

Nurelmadina, Hasan, Mamon, Saeed, Akram, Ariffin, Ali, Mokhtar, Islam, Hossain, & Hassan. (2021). *A Systematic Review on Cognitive Radio in Low Power Wide Area Network for Industrial IoT Applications*. MDPI, Sustainability.

Pham, Q.-V. (2020). *Swarm Intelligence for Next-Generation WirelessNetworks: Recent Advances and Applications*. arXiv:2007.15221v1 [cs.NI]

Priyan & Usha Devi. (2019). A survey on internet of vehicles: applications, technologies, challenges and opportunities. *Int. J. Advanced Intelligence Paradigms, 12*.

Ram, A. (2017). Mobility adaptive density connected clustering approach in vehicular ad hoc networks. *International Journal of Communication Networks and Information Security*.

Saeed, R. A., Amran, B. H. N., & Aris, A. B. (2010). Design and evaluation of lightweight IEEE 802.11p-based TDMA MAC method for road side-to-vehicle communications. *IEEE, ICACT'10 Proceedings of the 12th international conference on Advanced communication technology*, 1483 – 1488.

Shah, Y. A. (2018). *CAMONET: Moth-Flame Optimization (MFO) Based Clustering Algorithm for VANETs*. IEEE Access. doi:10.1109/ACCESS.2018.2868118

Sodhro, A. H., Luo, Z., Sodhro, G. H., Muzamal, M., Rodrigues, J. J. P. C., & Victor, H. C. (2019). Artificial Intelligence based QoS optimization for multimedia communication in IoV systems. *Future Generation Computer Systems*, *95*, 667–680. doi:10.1016/j.future.2018.12.008

Yang, F. (2014). An Overview of Internet of Vehicles. *China Communications Journal*. doi:10.1109/CC.2014.6969789

KEY TERMS AND DEFINITIONS

Advanced Driver Assistance Systems (ADAS): ADAS are electronic *systems* that *assist drivers* in *driving* and parking functions. Through a safe human-machine interface, *ADAS* increase car and road safety. It uses automated technology, such as sensors and cameras, to detect nearby obstacles or driver errors, and respond accordingly.

Bacterial foraging Algorithm (BFO): Is an optimization algorithm which is based on a lifecycle model that simulates some typical behaviours of E. coli bacteria during their whole lifecycle, including chemo-taxis, communication, elimination, reproduction, and migration

Butterfly Optimization Algorithm (BOA): Is a newcomer in the category of nature inspired meta-heuristic algorithms, inspired from food foraging behaviour of the butterflies. Its encounters two probable problems, entrapment in local optima and slow convergence speed such like any other meta-heuristic algorithms.

CAVDO: It's an algorithm known as Clustering algorithm for internet of vehicles (IoV) based on dragonfly optimizer. It used for cluster-based packet route optimization to make stable topology, and mobility aware optimization for reduced network overhead.

Method of Procedure (MOP): Greatly reduce risks and improve efficiency in the management of a network. Without proper change control, your enterprise can suffer irreparable losses. The method is an ordered arrangement of steps to achieve a specific task. In other words, given a specific set of circumstances, take these specific actions.

Chapter 2

Cooperative Relaying Communication in IoT Applications for 5G Radio Networks

Rajeev Kumar

https://orcid.org/0000-0001-6465-4247

Central University of Karnataka, India

Ashraf Hossain

National Institute of Technology, Silchar, India

ABSTRACT

This chapter presents cooperative relaying networks that are helpful in Internet of Thing (IoT) applications for fifth-generation (5G) radio networks. It provides reliable connectivity as the wireless device is out of range from cellular network, high throughput gains and enhance the lifetime of wireless networks. These features can be achieved by designing the advanced protocols. The design of advanced protocols plays an important role to combat the effect of channel fading, data packet scheduling at the buffered relay, average delay, and traffic intensity. To achieve our goals, we consider two-way cooperative buffered relay networks and then investigate advanced protocols such as without channel state information (CSI) i.e., buffer state information (BSI) only and with partial transmit CSI i.e., BSI/CSI with the assistance of one dimensional Markov chain and transmission policies in fading environment. The outage probability of consecutive links and outage probability of multi-access and broadcast channels are provided in closed-form. Further, the buffered relay achieves maximum throughput gains in closed-form for all these protocols. The objective function of throughput of the buffered relay is evaluated in fractional programming that is transformed into linear program using standard CVX tool. Numerical results show that our proposed protocols performance better as compared to conventional method studied in the literature. Finally, this chapter provides possible future research directions.

DOI: 10.4018/978-1-7998-6870-5.ch002

INTRODUCTION

The cooperative relaying communication (CRC) is one of most promising technology in Internet of Things (IoT) applications for fifth-generation (5G) networks (BenMimoune & Kadoch; Liu & Ansari, 2017). In IoT, CRC exploits the features of mutual cooperation along with relaying technologies to improve the spectral efficiency of cellular networks, to enhance the coverage area as the wireless portable devices are out of range from cellular networks, and to reduce the battery depletion problem. It can support tremendous access in cellular network for indoor environment and outdoor environment in real time.

Figure 1. Cooperative Relaying Technology for 5G radio networks and IoT applications: (a) the multihop relay network; (b) the heterogeneous relay scenario; (c) the relay connects to donor base station (BS); (d) the relay-assisted D2D communication; and (e) Public service by smart bus (SB) and the relay-assisted high-speed smart train (HST).

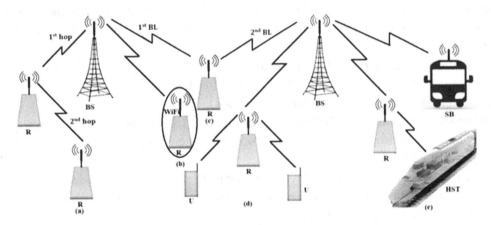

The relay in IoT applications (IoTAs) is the state-of-art for 5G radio networks that accommodates various physical things such as sensor nodes, smart mobile phones, home appliances, healthcare gadgets, machines, and intelligent furniture. The communications road maps of such autonomous things are essential for bonding them together to form the IoT. All these things are linked through a network which is referred to as IoT devices (IoTDs). The deployment strategies of the relay in the IoT applications for 5G network are shown in Fig.1 (BenMimoune, & Kadoch, 2017). The relay deployment scenarios in the IoTAs are summarized as follows: (a) the multi-hop relay can be used to extend the radio links between base station (BS) and user into more than two-hop, and performances of each of hop are expected to be better than direct link in cellular networks. (b) Heterogeneous relay is attractive technique to provide smart coverage to local area via WiFi technology. (c) the relay connects to several base stations (BSs) to achieve high throughput gains and balance the network loads across the BSs. (d) performance of device-to-device (D2D) communication is limited by excessive interference and poor propagation channel. These challenges can be minimized by providing relay-assisted links. The relay can efficiently improve the performance of D2D communication. (e) mobile relay stations are mounted on public buses and high speed smart train (HST) vehicles to provide better connectivity and compensate the vehicular penetration losses. In modern era, instruments in factory utilize the features of IoT to improve the performance and capability of factory and farm operations (Liu & Ansari, 2017). The relay in IoTAs also, provides wireless access links to the persons travelling in high speed trains to achieve highest possible transmission rate.

BACKGROUND

The CRC plays an important role in IoTAs for 5G networks by providing low-powered relay to increase the performance gains of wireless networks in real time applications. It includes various objective parameters for analysing the performance gains of relay networks such as transmission power, energy efficiency, throughput, average delay, low interference to prime networks, and distanced from network coverage. In IoTAs, the relay exploits several relaying protocols to receive the incoming information from the source and after some electronic processing, it sends the information to destination via available wireless links. The signalling protocols are summarized as follows.

1. **Amplify-and-forward (AF) Relaying Protocol**: In the AF relaying scheme, the relay amplifies the incoming noisy signal from the source and retransmits it to the destination (Beaulieu & Hu, 2006).
2. **Decode-and-forward (DF) Relaying Protocol:** In the DF method, the signal received from the source by the relay is decoded in the first phase, after that the relay retransmits the re-encoded information to the destination (Laneman et al., 2004).
3. **Compress-and-forward (CF) Relaying Protocol:** In the CF scheme, the relay allows to compress the incoming signal from the source and encode it into new packet that is sent to the destination (Akhbari et al., 2009).
4. **Randomize-and-forward (RF) Relaying Protocol:** In the RF method, the relay has freedom to receive random data packet from the source and then it uses different codebooks to forward the data packets to destination (Shafie et al., 2016).

The amplification of noise and propagation of error are occurred respectively in the AF and the DF relaying methods that limits the performance of the relay networks. Further, Li & Vucetic (2008) have investigated an adaptive relaying protocol for signalling protocol to diminish the limitation of AF and DF relaying techniques.

In IoTAs, the relay can operate on either half-duplex relaying (HDR) or full-duplex relaying (FDR) that depends on how it controls the simultaneity of forwarding and receiving of message signals. The authors of several article (Laneman et al., Beaulieu & Nabar, 2004; Hu, 2006; Li & Vucetic, 2008; Akhbari et al., 2009) have investigated time-division multi-access (TDMA) protocols for two-hop buffer-less HDR to enhance performance of wireless networks. Here, the HDR protocol causes multiplexing loss. Thus, the successive relaying protocol (Fan et al., 2007), the multiple relay selection protocol (Tannious & Nosratinia, 2008) and selective DF protocols (Alves & Souza, 2011) have been adopted to reduce the multiplexing loss. Further, authors have studied three-node FDR to investigate high spectral efficiency of wireless networks (Xia et al., 2008; Riihonen et al.; 2011; Zlatanov & Schober, 2013; Yu et al., 2015). Riihonen et al. (2011) have proposed hybrid FDR/HDR that opportunistically switches either HDR or FDR by exploring the concept of either instantaneous or statistical CSI. They have developed instantaneous and long-term transmit power adaptation protocol to reduce the residual self-interference (RSI). The optimization problems for power-location and joint power-location have been studied by exploiting full duplex (FD) DF relaying protocol to investigate outage performance of wireless networks (Yu et al., 2015). In addition, the buffered relay is introduced to exploit the characteristic of cooperation to achieve the high performance gains of wireless networks.

In modern era, the buffered relay networks provide flexibility of time-slot to sojourn the data packet without falling at the relay. The characteristics of buffer at the relay have been employed to receive data packets for fixed time-slots before forwarding the incoming data packets from source to destination (Xia et al., 2008). However, the fixed time-slots for relay-receiving and relay-transmitting of packets limit the flexibility of the relay networks. Further, Zlatanov and Schober (2013), have studied buffered relay networks that provides freedom to relay node to decide in which time slot to transmit and in which time slot to receive the data packets using adaptive nature of CSI of the assisted links. Moreover, the multi-hop relay has been employed (Jamali et al., 2015) to reduce the large path loss caused by long distance receiving node. Shafie et al. (2014) have obtained high throughput with known buffer state as well as the channel outage states. Furthermore, the buffer state and channel quality dependent on relay selection (RS) scheme have been studied (Luo and The, 2015) to derive closed-form expressions of outage probability and average delay of wireless networks. Manoj et al. (2016) have presented multi-hop relay networks to evaluate outage probability and average delay using max-link RS scheme and a state-clustering based method. The D2D is a promising technique that supports the 5G radio networks. Yang et al. (2017) have studied D2D communication by introducing the concept of joint resource block job and power allocation (PA) to achieve high performance in an FD cellular system. Moreover, the joint RS and PA have been exploited by (Ma et al., 2018) to lead high performance gains of FD relay-assisted D2D communication using in mm wave technology.

The two-way HDR/FDR networks establish the connection between two terminals/users via either one relay or more than relay that provides great interest of research in cellular system. First, we assume that traditional two-way relay networks (TWRNs) where, two terminals exchange information via relay in four succeeding transmission phases. They are affected by pre-log factor one-half and experienced loss in spectral efficiency. To overcome these challenges, time-division broadcast (TDBC) (Wu et al., 2005), and multi-access broadcast (MABC) (Popovski & Yomo, 2007) protocols have been studied to enhance the spectral efficiency. Moreover, the pre-log factor one-half from the channel capacity is removed by FDR technique that supports the simultaneous transmission and reception of information at the relay over same channel (Ju et al., 2009). Li et al. (2017) have proposed PA scheme and optimal relay placement strategies to investigate outage probability for two-way FDR networks. As the dedicated wireless channels suffer from heavy attenuation or large path losses, the buffer-less relay has no space to store incoming traffic flow from the source terminals. The two-way buffered HDR (TBHDR) has been studied by exploring the concept of buffer-aware network coding (BNC) with Markov-chain (Chen et al., 2012). The selection of adaptive transmission scheme in each time slot depends on qualities of associated links that supports maximum sum throughput for delay-unconstrained protocol and delay-constrained protocol (Jamali et al., 2015). The performance gains of TBHDR networks have been limited due to signaling overhead that is observed with feedback of CSI and scheduling of packets. Further, Shi et al. (2016) have studied the practical BNC to improve the capacity of wireless system by exploiting the signaling overhead, lossy links and finite buffered relay.

The buffered FDR plays an important role in IoTAs to achieve high spectral efficiency of 5G radio networks. Two-hop buffered FDR (THBFDR) has been presented by (Qiao, 2016) by introducing the relay selection scheme under statistical delay to achieve high capacity. Nomikos et al. (2018) have studied buffered relay cluster networks (BRCNs) to achieve high throughput gains subject to minimum

Figure 2. Two-way relay networks consisting of two users: (a) the relay with single buffer (B), (b) the relay with two buffers, B1 and B2 and their assisted links.

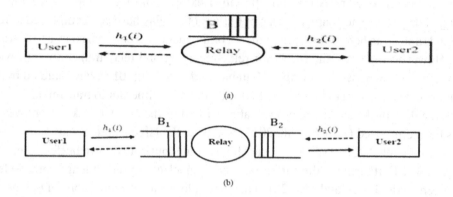

power consumption. A brief survey on HDR and FDR and their performance with advanced protocols for cooperative communications have been addressed by (Kumar & Hossain, 2019).

Motivated by performance gains of the buffered relaying networks, the main goals of this chapter is to explore the advanced new protocols design to boost up the throughput gains of TWRNs, where both users exchange information via buffered relay node without direct link. The investigative model such as two-way buffered relay networks is described with significant parameters. We obtain outage probabilities for multi-access and broadcasting in such a way to avoid data packets collisions at the buffered relay. For scheduling of data packets at the relay, the advanced protocols such as the buffer sate information (BSI) and buffer with the knowledge of channel state information (BSI/CSI) are developed using buffer size with the features of one dimension Markov chain. Moreover, the closed-form expressions of throughput for BSI/CSI protocol is derived in terms of fractional programming. For clarification of the objective problem, we adopt the standard CVX numerical tool to convert the fractional programme into linear program. Finally, our proposed approach shows that the numerical results outperforms the benchmarks scheme in the existing literature.

The remainder of the chapter is organized as follows. Next Section presents the TWRNs with different scenarios and their preliminaries. Then, advanced protocols such as BSI only and BSI/CSI are proposed to compute the throughput of relay node. Numerical results show that various insights on the variation of different parameters to achieve high throughput gains as compared to existing schemes in the literature. At the end, we conclude this chapter and provide possible future research directions.

Performance of Buffered Relay-Assisted Cooperative Communications

In this Section, we consider two-way buffered relay networks of two different scenarios to achieve high throughput gains by exploiting the advanced protocols.

System Model and Preliminaries

The TWRNs consist of two users, buffered relay and no direct wireless links present between the users that is depicted in Fig. 2. The half-duplex decode-and-forward (DF) relaying protocol is adopted for processing signal at the relay. The finite size buffer, B .at the relay can be stored up to N data packets in Fig. 2(a) while in the Fig. 2(b), the finite size buffers, B_1 .and B_2 .at the relay, r .can be stored up to N .and M .data packets from user, u_1 .and user, u_2 .respectively. In the $i-$ th specified time slot, the complex fading coefficients $h_1(i)$ .and $h_2(i)$ .are available between $u_1 - r$. and $u_2 - r$ .links respectively. In addition, the fading gain of channel, $\left|h_j(i)\right|^2$ .is ergodic and stationary random process with means $©_j = E\left\{\left|h_j(i)\right|^2\right\}$. $j \in \{1,2\}$ .where $E\{.\}$ .denotes expectation. Each node of system model consumes power, P .watts for transmitting a data packet from one node to another node and the noise variance is σ_n^2 .at the receiver. Moreover, signal-to-noise ratio (SNR) of links, $u_j - r$ .is defined as $\gamma_j = \gamma\left|h_j(i)\right|^2$. $j \in \{1,2\}$ .where $\gamma = P/\sigma_n^2$ .is SNR of each transmitting nodes. If the size of data packet is K .its and transmission rate, $\mathcal{R}_{jk} = \mathcal{R} = \dfrac{K}{T_s}$ .bits/sec/Hz, then outage probability of $j-k$ .link for Rayleigh fading channel is given by,

$$\mathcal{P}_{jk} = 1 - \exp\left(-\frac{2^R - 1}{©_j \gamma}\right). \tag{1}$$

Figure 3. The finite size birth-death Markov chain at the buffered relay for two-way relay networks.

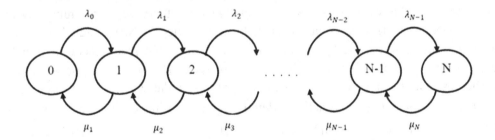

For reciprocal channels i.e., $\mathcal{P}_{jr} = \mathcal{P}_{rj}$. $j \in \{1,2\}$. the outage probability for multiple-access channel is formulated as,

$$\mathcal{P}_{MA} = 1 - \frac{©_1}{©_1 - ©_2} \exp\left(-\frac{\gamma_T^s}{©_1 \gamma} - \frac{\gamma_0}{©_2 \gamma}\right) - \frac{©_2}{©_2 - ©_1} \exp\left(-\frac{\gamma_T^s}{©_2 \gamma} - \frac{\gamma_0}{©_1 \gamma}\right). \tag{2}$$

where, $\gamma_T^s = (\gamma_0^s - \gamma_0) \cdot \gamma_0 = 2^{\mathcal{R}} - 1$ .and $\gamma_0^s = 2^{2\mathcal{R}} - 1$. If the relay sends incoming data packet from u_2 .and u_1 .respectively to u_1 .and u_2 .uccessfully via relay, then the outage probability for broadcast channel is given by,

$$\mathcal{P}_{BC} = 1 - \exp\left(-\left(1/\copyright_1 + 1/\copyright_2\right)\gamma_0/\gamma\right). \tag{3}$$

The proof of (2) and (3) is existing in (Kumar & Hossain, 2019).

Throughput Formulation of Two-Way Relay Networks

In this Section, we investigate the BSI only and BSI/CSI protocols that can be used to derive the relay throughput gains of the two-way buffered relay networks.

The BSI Protocol for Two-Way Relay Networks

The BSI protocol is examined with the queueing feature of one-dimensional birth-death Markov chain (BDMC) at the buffered relay. The BDMC is depicted in Fig. 3, where state n .denotes the relaying queue contains n .packets. The channel from $u_1 - \text{to} - r$ .is accessed with probability α_n .when the relay queue consists of n .packets. However, the $r - \text{to} - u_2$ .channel operates with probability $\overline{\alpha_n}$ .as the $u_1 - \text{to} - r$. channel is inactive, where $\overline{\alpha_n} = 1 - \alpha_n$. Again, this approach is same for $u_2 - \text{to} - r$ .and $r - \text{to} - u_1$ .channels. When the buffered relay is empty, the channels is accessed by both users with probability 1 i.e., $\alpha_0 = 1$. If the queue of the buffered relay is full, then both users access channels with probability, $\alpha_N = 0$.

The scheduling and identification of data packets at the buffered relay is state-of-art in cooperative relay networks. When the relay receives the data packet in specified time slot successfully from the users then it sends an acknowledgement (ACK) signal to users; otherwise, it transmits back to users a negative-acknowledgement (NACK) signal. Further, the users notice the NACK signal received from the relay and then retransmit the data packet in next time slot after observing received control signal from the relay. Our relaying system model works on half-duplex relaying (HDR) protocol for data processing at the relay. Thereby, at a time one user can be attempted to transmit data packet in the specified time slot.

Let $x(i)$ .be the number of packets at the buffered relay in the initial of time slot i . From the feature of the BDMC, the state transitions occur between the neighboring states i.e., $i \rightarrow (i+1)$ .or $i \rightarrow (i-1)$. Thus, the transition probabilities of states are expressed as,

$$\lambda_n = \Pr\{x(i) = (n+1) | x(i-1) = n\} = \alpha_n \left(\overline{\mathcal{P}_{1r}} + \overline{\mathcal{P}_{2r}}\right); \ 0 \le n \le (N-1) \tag{4}$$

$$\mu_n = \Pr\{x(i) = (n-1) | x(i-1) = n\} = \overline{\alpha_n} \left(\overline{\mathcal{P}_{r1}\mathcal{P}_{r2}}\right); \ 1 \le n \le N$$

where $\mathcal{P}_{1r}, \mathcal{P}_{2r}, \mathcal{P}_{r1}$ and $\mathcal{P}_{r2}$ denote the outage probabilities of the links $u_1 - r$, $u_2 - r$, $r - u_1$, and $r - u_2$ respectively. This chapter follows the notation $\overline{\mathcal{F}} = 1 - \mathcal{F}$, the terms, $\overline{\mathcal{F}_1 \mathcal{F}_2} = 1 - \mathcal{F}_1 \mathcal{F}_2$ and $\overline{\mathcal{F}_1 \mathcal{F}_2} \neq \overline{\mathcal{F}_1} \overline{\mathcal{F}_2}$. Thus, throughput of the relay is defined as,

$$\tau_r^B = \overline{\mathcal{P}_{r1} \mathcal{P}_{r2}} \sum_{n=1}^{N} \xi_n \overline{\alpha_n} \tag{5}$$

where $\xi_n = \xi_0 \prod_{m=0}^{n-1} \dfrac{\lambda_m}{\mu_{m+1}}$ with $\xi_0 = \left(1 + \sum_{n=1}^{N} \prod_{m=0}^{n-1} \dfrac{\lambda_m}{\mu_{m+1}}\right)^{-1}$ is steady state probability of the BDMC of state n (Gross et al., 2008). Notice that the links $u_1 - r$ and $u_2 - r$ are reciprocal to $r - u_1$ and $r - u_2$ links respectively. Thus, $\mathcal{P}_{1r} = \mathcal{P}_{r1} = \mathcal{P}_1$ and $\mathcal{P}_{2r} = \mathcal{P}_{r2} = \mathcal{P}_2$. After some mathematical operations, throughput of the buffered relay for BSI only is obtained as,

$$\tau_r^B = \dfrac{\overline{\mathcal{P}_1 \mathcal{P}_2} \left(\overline{\mathcal{P}_1} + \overline{\mathcal{P}_2}\right)}{\overline{\mathcal{P}_1} + \overline{\mathcal{P}_2} + \overline{\mathcal{P}_1 \mathcal{P}_2}} \tag{6}$$

The Buffer State with Partial Transmit Channel State Information (BSI/CSI) Protocol for Two-Way Relay Networks

For development of this protocol, the outage of channels is known to associated nodes before transmission of data packets. In general, the possible transmission states of the relay networks in each time slot are transmit, receive and silent states that form possible transmission modes as $3^3 = 27$. But useful transmission modes are only seven. If the $u_1 - r$ link is in outage stage and the buffer at the relay is not full, then two possibility will be occurred: first, the u_2 sends data packets to the relay; otherwise, the relay delivers data packet to u_1 when it has possibly one or more packet. Similar phenomena for transmission/reception is followed by the u_1 and the relay if the $u_2 - r$ link is in outage. For non-outage links $u_1 - r$ and $u_2 - r$, the empty buffer at the relay receives data packets either from the u_1 or, the u_2. Furthermore, when the $r - u_1$ link and the $r - u_2$ link are not in outage and buffer at the relay has one or more packet, then the relay node can send data packets to either the u_1 or, the u_2 or, simultaneously to both users.

When the links $u_1 - r$ and $u_2 - r$ are non-outage and the buffered relay contains n packets, then both users can send the data packet with probability, β_n while the relay can transmit with probability, $\overline{\beta_n}$. If buffer at the relay has zero packet, either the u_1 or u_2 sends data packets with probability, $\beta_0 = 1$. When buffer at the relay is full i.e., queue of buffer at state, N then transmission probability, β_N of both users are zero. Thus, the transition probabilities of the BDMC for increment and decrement are defined as,

$$\lambda_n = \overline{\mathcal{P}_{1r}}\left(\mathcal{P}_{2r} + \overline{\mathcal{P}_{r2}}\beta_n\right) + \overline{\mathcal{P}_{2r}}\left(\mathcal{P}_{1r} + \overline{\mathcal{P}_{r1}}\beta_n\right); \ 0 \le n \le (N-1)$$

$$\mu_n = \overline{\mathcal{P}_{r1}}\left(\mathcal{P}_{2} + \overline{\mathcal{P}_{2r}}\ \overline{\beta_n}\right) + \overline{\mathcal{P}_{r2}}\left(\mathcal{P}_{1} + \overline{\mathcal{P}_{1r}}\ \overline{\beta_n}\right); \ 1 \le n \le N \tag{7}$$

Throughput of the buffered relay in presence of partial transmit CSI is formulated as,

$$\tau_r^C = \left(\overline{\mathcal{P}_{r1}}\ \mathcal{P}_{r2} + \mathcal{P}_{r1}\overline{\mathcal{P}_{r2}} + 2\overline{\mathcal{P}_{r1}}\ \overline{\mathcal{P}_{r2}}\right)\sum_{n=1}^{N}\overline{\beta_n}\xi_n \tag{8}$$

After some mathematical operations, the throughput of relay for BSI/CSI can be obtained as,

$$\tau_r^C = \left(2 - \mathcal{P}_1 - \mathcal{P}_2\right) \rangle_r \tag{9}$$

where, set of variable $\left\{ \ddot{}_n \right\}_{n=1}^{N} = \prod_{m=0}^{n-1}\dfrac{\lambda_m}{\mu_{m+1}}$ and the most important term $\rangle_r$ for the optimization point of is expressed as,

$$\rangle_r = \frac{\sum_{n=1}^{N}\overline{\beta_n}\,\ddot{}_n}{1 + \sum_{n=1}^{N}\ddot{}_n} \tag{10}$$

Finally, we can achieve maximum throughput of the relay by maximizing the $\rangle_r$. Thus, objective function of the relay throughput is formulated in fractional program as,

$$Maximize, \Lambda_r = \frac{C^T\Psi + b}{1^T\Psi + 1}$$

$$subjected\ to\ 0 \le \frac{d_n^T\Psi + e_n}{1_n^T\Psi} \le 1$$

$$d_N^T\Psi + e_N = 0 \tag{11}$$

where

$$\Psi = \lceil \Psi_{\cdot\cdot}, \Psi_{\cdot\cdot}, \ldots \Psi_{\circ}, \Psi_{\cdot}\rceil^T\ ;\ C = \lceil -a, (1+a), -a, (1+a), \ldots\rceil^T$$

$$b = \begin{cases} (1+a) & when\ N = odd \\ 0 & when\ N = even \end{cases} \quad d_n = \left[\underset{N-n\,zeros}{0,...,0},(1+a),-(1+2a),(1+2a),-(1+2a),... \right]^T ;$$

$$e_n = (-1)^n (1+a);$$

$$1_n = \left[0,...,0,1,0,...,0 \right]^T \quad \&\ a = \frac{P_1\overline{P_2} + \overline{P_1}P_2}{2\overline{P_1}\,\overline{P_2}} \tag{12}$$

Please, notice that the proof has been shown in appendix (Kumar & Hossain, 2017). In Fig.2 (b), our system model is reformed to allow the data packets for multi-access and broadcast to avoid the collision at the buffered relay that has been shown in (Kumar & Hossain, 2019).

Figure 4. Throughput of the relay versus buffer size for Rayleigh fading channel where, SNR $\gamma_{u_1 r} = 20\,dB$; $\gamma_{u_2 r} = 5\,dB$ and R = 1 bits/sec/Hz.

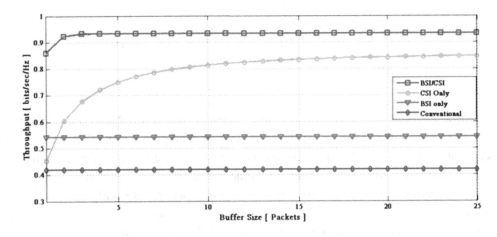

NUMERICAL RESULTS AND DISCUSSIONS

This Section evaluates throughput of the two-way relay networks for Rayleigh fading by providing significant parameters. The proposed numerical results are matched with conventional MABC protocol as benchmark (Popovski & Yomo, 2007). The objective function of the relay throughput for BSI/CSI protocol is obtained in the fractional program that is transformed into standard linear program (Grant & Boyd, 2014). For throughput analysis, we have considered results (1), (6), (11) and (12) and then the performances of the TWRNs are evaluated that are summarized as follows.

a)   *Variation of throughput with buffer-size:* For plotting the throughput with buffer-size result, we assume that the SNR of links, $u_1 - r$ and $u_2 - r$ are 20 dB and 5 dB respectively i.e., $u_2 - r$ link

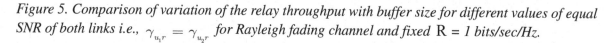

Figure 5. Comparison of variation of the relay throughput with buffer size for different values of equal SNR of both links i.e., $\gamma_{u_1 r} = \gamma_{u_2 r}$ for Rayleigh fading channel and fixed R = 1 bits/sec/Hz.

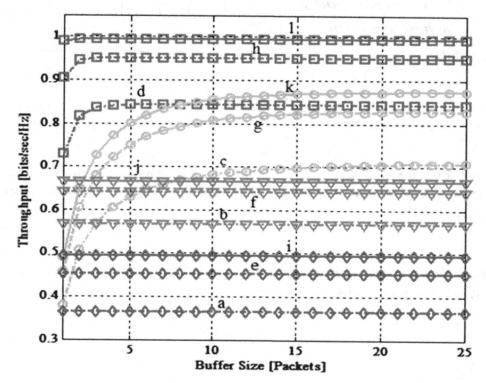

is more faded than another link, $u_1 - r$. The throughput of the buffered relay is achieved with buffer-size for fixed transmission rate, $R = 1$ bits/sec/Hz which is depicted in Fig.4. From observation, it is clear that the BSI/CSI protocol performs better as compared to BSI only, CSI only as well as conventional scheme, where conventional scheme is considered as benchmark scheme. The BSI protocol is superior to the CSI only as well as benchmark scheme at small size of buffer while the CSI protocol leads better as compared to BSI protocol as the size of buffer is more than two.

b) *The plot of throughput with buffer-size for various SNR value of links:* For comparison purpose, we have plotted various the throughput of relay with buffer size for miscellaneous SNR of links with fixed rate $R = 1$ bits/sec/Hz that is depicted in Fig.5, where links SNR $\gamma_{u_1 r} = \gamma_{u_2 r} = 5$ dB, 10 dB, 20 dB, for Conventional (a, e & i), BSI Only (b, f, & j), CSI Only (c, g, & k) and BSI/CSI (d, h,, & l). Notice that the spectral efficiency for all protocols achieves high at high SNR value of both links. Since, the outage probability of link decreases with transmit SNR for fixed transmission rate. Further, we observe that the BSI/CSI protocol is performed significantly better as compared to conventional, BSI only, and CSI only. In addition, the throughput of relay networks is saturated at large size of buffer for all protocols.

c) *The throughput versus transmit SNR:* The Fig. 6 shows variation of throughput gains of the relay with transmit SNR. For plotting the results, we assume that SNR of both links are equal i.e., $\gamma_{u_1 r} = \gamma_{u_2 r}$ i.e., both links are symmetric to each other and also, the buffer size, N is fixed to 5 packets. When the range of SNRs of both links is -10 dB $\leq \gamma_{u_1 r} \leq -4$ dB, the throughput reads

Figure 6. Plot the throughput with SNR for packet, $N = 5$ and rate, $R = 1$ bits/sec/Hz where, SNR of both links are equal i.e., $\gamma_{u_1 r} = \gamma_{u_2 r}$.

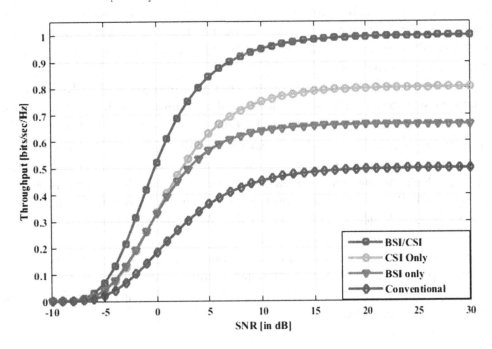

Figure 7. Variation of throughput with transmission rate, R for $N=5$ packets and SNR $\gamma_{u_1 r} = \gamma_{u_2 r} = 10\,dB$.

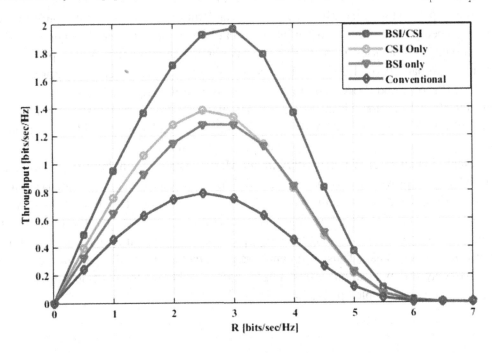

by the relay is very poor for all protocols. Moreover, the throughput for BSI/CSI protocol is achieved high as compared to all the cases if the SNRs of both links are greater than $-5\,\text{dB}$. In addition, the throughput for CSI and BSI protocols is almost approach to 0.4 bits/sec/Hz at $1\,\text{dB}$ while both CSI and BSI protocols are superior to conventional protocol. Throughput is reached to maximum value at high SNR. Since, the outage probability of the links almost approaches to zero at high SNR for fixed transmission rate.

d) *Variation of throughput with transmission rate:* In Fig. 7, throughput of buffered relay increases with transmission rate R. The specification of the parameters such as, $N = 5$ packets and $\gamma_{u_1 r} = \gamma_{u_2 r} = 10\,dB$ are used to plot the results. From the plot, it is found that the BSI/CSI protocol performs better as compared to others protocols over diverse range of transmission rate. The throughput of the relay for CSI protocol increases with the transmission rate and superior to BSI protocol. The high transmission rate of all nodes does not indicate that the throughput of the buffered relay is high. Because, the outage probability of both links is very high at high rate of transmission for fixed SNR values and fixed buffer size. Whenever the queue of buffered relay is full, we can mitigate the burden on available buffer-size with high transmission rate. In this way, we can improve the performance of relay networks. From observation of plot, throughput of the relay approaches to maximum values for all states as the transmission rate leads in the range $2.5\,\text{bits}/\sec/\text{Hz} \leq R \leq 3\,\text{bits}/\sec/\text{Hz}$. In addition, one can attain high buffer occupancy over this range of transmission rate.

GENERAL CONCLUSIONS AND POSSIBLE FUTURE DIRECTIONS

In this chapter, we have studied CRC in IoTAs for 5G radio networks. Particularly, we present the cooperative relaying networks model such as two-way HDR networks to achive high throughput gains with different possible parameters namely buffer-size, SNR, and transmission rate. For this purpose, advanced protocols have been proposed by exploring the concept of cooperative buffered relaying infrastructure, channel, and their preliminaries. Considering various tranmsission policies, outage probability of the links and outage probability of MABC are obtained in closed-from. The closed-form expressions of relay throughput for BSI only and BSI/CSI protocols are derived for Rayleigh fading channel. The objective function of the BSI/CSI protocol is evaluated in fractional program that is translated in linear program using standard CVX numerical tool. And then, the numerical results are plotted for all ptotocols and notic that our proposed schemes outperform the benchmark schemes.

Some potential challenges of CRC with/without Markov chains are helpful in IoT to enhance the spectral efficiency by providing low latency in real time applications that are summarized as follows. Most of the existing works focus on the half-duplex buffered relaying. The room is also, available for buffered FDR relay networks to achive high performance gains. Aditionally, the QoS constraints in FDR are engrossed a lot of interest to realize high performance gains in terms of secrecy throughput. The size of IoTDs is very small. Thus, an antenna isolation method for FDR operations are not suitable in coopeartive relay communications. Thus, single antenna technique for FDR operation is needed for small-size IoTDs that is possible open research issues.

ACKNOWLEDGMENT

The authors would like to thank the anonymous reviewers, and the Editor for their valuable suggestions to improve the quality of chapter.

REFERENCES

Akhbari, B., Mirmohseni, M., & Aref, M. (2009, June – July). Compress-and-forward strategy for the relay channel with non-causal state information. *Proc. IEEE Int. Symp. on Inf. Theory*, 1169 – 1173.

Alves, H., & Souza, R. D. (2011). Selective decode-and-forward using fixed relays and packet accumulation. *IEEE Communications Letters*, *15*(7), 707–709.

Beaulieu, N. C., & Hu, J. (2006). A noise reduction amplify-and-forward relay protocol for distributed spatial diversity. *IEEE Communications Letters*, *10*(11), 787–789. doi:10.1109/LCOMM.2006.060849

BenMimoune, A., & Kadoch, M. (2017). *Relay technology for 5G networks and IoT applications. In Internet of Things: Novel Advances and Envisioned Applications*. Springer.

Chen, W., Letaief, K. B., & Cao, Z. (2012). Buffer-aware network coding for wireless networks. *IEEE/ACM Transactions on Networking*, *20*(5), 1389–1401.

Fan, Y., Wang, C., Thompson, J., & Poor, H. V. (2007). Recovering multiplexing loss through successive relaying using repetition coding. *IEEE Transactions on Wireless Communications*, *6*(12), 4484–4493.

Grant, M., & Boyd, S. (2014). *CVX: MATLAB software for disciplined convex programming, version 2.1*. http://cvxr.com/cvx

Gross, D., Shortle, J. F., Thompson, J. M., & Harris, C. M. (2008). *Fundamentals of Queueing Theory* (4th ed.). John Wiley & Sons, Inc.

Jamali, V., Zlatanov, N., & Schober, R. (2015). Bidirectional buffer-aided relay networks with fixed rate transmission—Part I: Delay-unconstrained case. *IEEE Transactions on Wireless Communications*, *14*(3), 1323–1338.

Jamali, V., Zlatanov, N., & Schober, R. (2015). Bidirectional buffer-aided relay networks with fixed rate transmission—Part II: Delay-constrained case. *IEEE Transactions on Wireless Communications*, *14*(3), 1339–1355.

Jamali, V., Zlatanov, N., Shoukry, H., & Schober, R. (2015). Achievable rate of the half-duplex multi-hop buffer-aided relay channel with block fading. *IEEE Transactions on Wireless Communications*, *14*(11), 6240–6256.

Ju, H., Oh, E., & Hong, D. (2009). Catching resource-devouring worms in next-generation wireless relay systems: Two-way relay and full-duplex relay. *IEEE Communications Magazine*, *47*(9), 58–65.

Kumar, R., & Hossain, A. (2017). Optimisation of throughput of two-way buffer-aided relaying networks with wireless assisted links. *IET Communications*, *11*(10), 1626–1632.

Kumar, R., & Hossain, A. (2019). Survey on half- and full- duplex relay based cooperative communications and its potential challenges and open issues using Markov chains. *IET Communications*, *13*(11), 1537–1550.

Kumar, R., & Hossain, A. (2019). Performance of random access Markov modelling for two-way buffer-aided relaying networks with wireless assisted links. Wireless Personal Comm., 108(3), 1995 – 2015.

Laneman, J., Tse, D., & Wornell, G. (2004). Cooperative diversity in wireless networks: Efficient protocols and outage behavior. *IEEE Transactions on Information Theory*, *50*(12), 3062–3080. doi:10.1109/TIT.2004.838089

Li, C., Chen, Z., Wang, Y., Yao, Y., & Xia, B. (2017). Outage analysis of the full-duplex decode-and-forward two-way relay system. *IEEE Transactions on Vehicular Technology*, *66*(5), 4073–4086.

Li, Y., & Vucetic, B. (2008). On the performance of a simple adaptive relaying protocol for wireless relay networks. In *Proc. IEEE Semiannual Veh. Technol. Conf.* (pp. 2400–2405). Singapore: IEEE.

Liu, X., & Ansari, N. (2017). Green relay assisted D2D communications with dual batteries in heterogeneous cellular networks for IoT. *IEEE Internet of Things Journal*, *4*(5), 1707–1715. doi:10.1109/JIOT.2017.2717853

Luo, S., & The, K. C. (2015). Buffer state based relay selection for buffer-aided cooperative relaying systems. *IEEE Transactions on Wireless Communications*, *14*(10), 5430–5439.

Ma, B., Mansouri, H. S., & Wong, V. W. S. (2018). Full-duplex relaying for D2D communication in mm wave based 5G networks. *IEEE Transactions on Wireless Communications*, *17*(7), 4417–4431.

Manoj, B. R., Mallik, R. K., & Bhatnagar, M. R. (2016). Buffer-aided multi-hop DF cooperative networks: A state-clustering based approach. *IEEE Transactions on Communications*, *64*(12), 4997–5010.

Nabar, R. U., Bölcskei, H., & Kneubühler, F. W. (2004). Fading relay channels: Performance limits and space-time signal design. *IEEE Journal on Selected Areas in Communications*, *22*(6), 1099–1109.

Nomikos, N., Charalambous, T., Vouyioukas, D., Wichman, R., & Karagiannidis, G. K. (2018). Power adaptation in buffer-aided full-duplex relay networks with statistical CSI. *IEEE Transactions on Vehicular Technology*, *67*(8), 7846–7850.

Popovski, P., & Yomo, H. (2007). Wireless network coding by amplify-and-forward for bi-directional traffic flows. *IEEE Communications Letters*, *11*(1), 16–18.

Qiao, D. (2016). Effective capacity of buffer-aided full-duplex relay systems with selection relaying. *IEEE Transactions on Communications*, *64*(1), 117–129.

Riihonen, T., Werner, S., & Wichman, R. (2011). Hybrid full-duplex/half-duplex relaying with transmit power adaptation. *IEEE Transactions on Wireless Communications*, *10*(9), 3074–3085.

Shafie, A. E., Khafagy, M. G., & Sultan, A. (2014). Optimization of a relay-assisted link with buffer state information at the source. *IEEE Communications Letters*, *18*(12), 2149–2152.

Shafie, A. E., Sultan, A., & Dhahir, N. A. (2016). Physical-layer security of a buffer-aided full-duplex relaying system. *IEEE Communications Letters*, *20*(9), 1856–1859.

Shi, S., Li, S., & Tian, J. (2016). Markov modeling for two-way relay with finite buffer. *IEEE Communications Letters*, *20*(4), 768–771.

Shim, Y., & Park, H. (2014). A closed-form expression of optimal time for two-way relay using DF MABC protocol. *IEEE Communications Letters*, *18*(5), 721–724.

Tannious, R., & Nosratinia, A. (2008). Spectrally-Efficient Relay Selection with Limited Feedback. *IEEE Journal on Selected Areas in Communications*, *26*(8), 1419–1428.

Wu, Y., Chou, P. A., & Kung, S. Y. (2005, March). Information exchange in wireless network coding and physical-layer broadcast. In *Proc. 39th Annu. Conf. Sci. syst.*, (pp. 1 – 6). Academic Press.

Xia, B., Fan, Y., Thompson, J., & Poor, H. V. (2008). Buffering in a three-node relay network. *IEEE Transactions on Wireless Communications*, *7*(11), 4492–4496.

Yang, T., Zhang, R., & Cheng, X. (2017). Graph coloring based resource sharing (GCRS) scheme for D2D communications underlaying full-duplex cellular networks. *IEEE Transactions on Vehicular Technology*, *66*(8), 7506–7517.

Yu, B., Yang, L., Cheng, X., & Cao, R. (2015). Power and location optimization for full-duplex decode-and-forward relaying. *IEEE Transactions on Communications*, *63*(12), 4743–4753.

Zhou, B., Cui, Y., & Tao, M. (2015). Stochastic throughput optimization for two-hop systems with finite relay buffers. *IEEE Transactions on Signal Processing*, *63*(20), 5546–5560.

Zlatanov, N., & Schober, R. (2013). Buffer-aided relaying with adaptive link selection—Fixed and mixed rate transmission. *IEEE Transactions on Information Theory*, *59*(5), 2816–2840.

Chapter 3
The Rise of IoT and Big Data Analytics for Disaster Management Systems

Supriya M. S.

https://orcid.org/0000-0003-3465-6879

Ramaiah University of Applied Sciences, India

Kannika Manjunath

Ramaiah University of Applied Sciences, India

Kavana U. R.

Ramaiah University of Applied Sciences, India

ABSTRACT

Uninvited disasters wreak havoc on society, both economically and psychologically. These losses can be minimized if events can be anticipated ahead of time. The majority of large cities in developing countries with increasing populations are highly vulnerable disaster areas around the world. This is due to a lack of situational information in their authorities in the event of a crisis, which is due to a scarcity of resources. Both natural and human-induced disasters need to be pre-planned and reactive to minimize the risk of causalities and environmental/infrastructural disruption. Disaster recovery systems must also effectively obtain relevant information. The developments in big data and the internet of things (IoT) have made a greater contribution to accuracy and timely decision-making in the disaster management system (DMS). The chapter explains why IoT and big data are needed to cope with disasters, as well as how these technologies work to solve the problem.

DOI: 10.4018/978-1-7998-6870-5.ch003

INTRODUCTION AND BACKGROUND

Disasters are uninvited and cause huge destruction to society- economically and mentally such as human life, the networks, and the environment at anytime, anywhere. The majority of the large urban areas in developed countries with an ascending population are highly vulnerable responsive disaster regions worldwide. IFRC, World Catastrophe Report 2018, has recorded 3,751 calamities like as landslides, flooding, tsunamis, earthquakes and others in the last 10 years. The commercial and business damages linked with the calamities are estimated up to 1,658 billion USD (Fisher et al., 2018). The impact of a catastrophe is on the native economies, such as infrastructural damages, agriculture, homes, whereas lack of jobs, unemployment and industry destabilization are indirect impacts (Noji, 1997). Despite various attempts by security practitioners and regime firms, and the deaths caused due to disaster in recent years has remained high. Ergo, disaster management systems also need to effectively obtain affirmative cognizance, track and analyse the ground condition, facilitate voidances and soothsay of disasters. Government authorities, academics and professionals relevant to disaster management have attempted to strengthen disaster management processes by considering various research gatherings, such as information technology, health sciences, cartography, and environmental sciences. The collection and retrieval of the vast of disaster data volumes are actually most immensely colossal problems that is encountered by defence, fire departments, health ministry, police department, and others managing disasters order to respond and organise effectively, it is extremely important for these organisations to acquire collected Real-time emergency data. Their ultimate aim is to strengthen the compilation, managing, monitoring, analysis and analysis of data phases in emergency management systems for timely and precise decision-making. Timely dispatch of mitigation materials to hospitals from distribution centres in compliance with the plan of distribution centres is withal a major task in disaster management. Therefore, it is a necessity that adequate planning must be carried involving multiple stakeholders for successful and productive in a country-specific context.

Due to the advent of the new data analytics approaches, accommodation as well as connectivity technology like cloud computing, Internet of Things (IoT), Big Data Analytics (BDA) and others have come out to be very useful in managing disasters. With numerous incipient supporting data, cost-effective implements of analysis and process of data that is used in making decisions which involves rescuing people, mitigating them and preparing to face disaster damages. Reasonable and appropriate decision-making on the basis of reliability and knowledge dictates the efficacy of a crisis management system, during the course of any crisis (Hristidis et al., 2010).

Actually, emergency management focused on BDA and IoT is an under-investigated field of study that entails many exciting possibilities and challenges. With the capability of IoT to have an omnipresent network system with interlinked smart devices and sensors (Gubbi et al., 2013), Internet of Things has the capacity to be integrated into Disaster Management System (DMS) and will have a good effect on all emergency response phases (Yang, Yang, & Plotnick, 2013). BDA can be considered as to enable real time analysis of IoT and other relevant data's (Rathore et al., 2016), also it can provide meaningful outcomes to understand the scenario that remain in the affected areas; hence the resources has to be deployed is effective, reliable based on analytical findings (Wang et al., 2016(. In addition, Big Data created in the IoT areas is used to perform mapping, forecasting's and generate event warnings (Ahmed et al., 2017).

Thus, it can be said that, combined use of IoT and BDA technologies contributes for developing disaster management environment which is innovative, effective and highly-needed.

EXISTING PROBLEM

A major hurdle for India is the communication infrastructure which makes IoT a reality. When compared to China and other developed countries, India is lagging in the telecom industry. This is because private players dominate the Telecom Industry in India, whereas China's is undertaken by the regime. There are few examples of India's slow progress towards IoT such as National Optical Fiber Network and sluggish 4G momentum. India struggles because of challenging demographics and lack of favourable policies. Majority of the developing countries are experimenting with 5G technology, whereas India is not even starting the conversation regarding it. The main reason behind that is the cost of the spectrum. There are other challenges such as Right of Way (RoW), which needs to be addressed urgently. Another big concern is the cybersecurity which needs to be addressed as soon as possible. The development of India is totally dependent on 5G technology, which is the backbone of IoT.

APPLICATIONS OF IOT AND BIG DATA ANALYTICS IN DISASTER MANAGEMENT SYSTEM

Disaster Management is also called as emergency management lifecycle and is not an easy task to be accomplished. It is a step-by-step and long process which includes identification of jeopardy, preparedness for the disaster, resource allocation, emergency response, disaster recovery. IoT and BDA are a boon for us in identifying the natural disaster in every stage. Many traditional DMS miss the assistance of several novel data provenance and technologies for real-s time big data analytics which could avail decision makers to produce fast and reliable outcomes. With its planned integrated parameters, the reference model can provide guidelines for harvesting, transmitting, managing and analysing disaster information from numerous Sources for the provision of revised, reliable disaster management information.

In the 21st century, we are in the era of another industrial revolution, as technologies are transforming manufacturing into a highly intelligent, automated, and efficient industry. In which things such as global warming, climate change, potential disasters caused by our rapid change of technology and revolution could continue to cause harm to earth; thus, drastically increase the chances of having a collapse global community. Therefore, it is important to utilize IoT and BDA related sufficient technologies to provide proper solutions for disaster management. Professionals in the disaster management use vast capacities of data's from various sources in order to make apprised decisions to utilise the resources efficiently. Many technologies can be leveraged to strengthen this method and promote the same. The consistency, availability, and responsiveness of data must be protected and facilitated by a collaborative and immersive atmosphere during time-critical circumstances in any form of disaster scenario. Timely and efficient rescue responses can be the difference between life and death during disasters.

There is diversity of big data which is available for every level in disaster management system. The significant difficulty is that how multiple databases can be related to specific forms of disasters. Datasets of different types of will be kept circumscribed and big data technology's ability has not been completely surveyed. Telco Systems offers purpose-built, service-aware, intelligent networks and carrier grade networking solutions under the robust BiNOX operating system that supports a wide variety of technologies including Ethernet, MPLS, and circuit emulation services (CES). A financial arrangement with Telco is necessary to access the comprehensive call history of mobile subscribers. In comparison, some of the satellite data such as Landsat, SRTM is free and others such as Lidar data, GeoEye and many others needs to be purchased. GeoEye is a Very high-resolution data which has enhanced the use of remote sensed data in disaster management. From these data, we can compare the image acquired before the occurrence of event and after the occurrence of event, and predict the level of damage caused by the disaster. Figure1 shows the before and after image of tsunami which took place at the South-west suburb of Banda Aceh, Sumatra (Van Western, Soeters and Buchroithner, 1996).

Furthermore, financial data sets that is credit cards, debit cards constitute financial transactions data that cane used to keep a track on the movements of people and behavioural reactions before-and-after the catastrophes. In addition, transport datasets collected from conveyances fitted with Global positioning systems could also be utilized for the detecting the catastrophe damages.

Applications of IoT

Even with the help now technologies like IoT, disasters cannot be stopped, yet these technologies could be extremely helpful in preparing for disaster, like system of detection, Early warning systems etc.

Minimization and prevention of risk: surveillance of hazard possibilities by satellite communication and the Geographic Information System (GIS); development of early warning systems; use of social media to build awareness.

Emergency response: Coordination using real-time analysis for the moment reprieve and emergency response.

Recovery from Disaster: Searching for people lost through online services and fund management.

Some of the present day IoT applications:

One example that arises is the earthquake detection and warning effort of the US Geological Society. The system they are working on is called the Earthquake Early Warning system, and there are such systems in Mexico, Taiwan and Japan. In California, they have piloted the EEW prototype, called Shake Alert, since 2012. Recently, shake Alert managed to detect and admonish Pasadena residents about a 4.4 magnitude earthquake.

Another example of IoT is forest fire monitoring: Insertion of sensors on the trees will measure that and shows when tree is on fire or when there's high danger, such as temperature, humidity, CO2 and CO levels. Early warning systems notify the local community and request availability if there is a crucial mixture of these parameters. The fire-fighters when arrive, will have accurate details regarding location, intensity of wildfire. Applications of IoT is also seen in managing earthquakes where microwave sensors are used to monitor earth movements, and infrared sensors are used to track flooding as well as people's movements.

Applications of Big-data

BDA generated new chances to control of natural disasters, largely because of the due to different ways it presents in analysing, visualizing and presaging the natural calamities. Different way in which we human population implement the prevention techniques to minimise misery and damages have been dramatically transmuted.

In response to the increasing prevalence of rigorous forest fires, the National Centre for Atmospheric Research developed a computer model that simulates weather and fire interactions. Not only does the weather affect fires, the fires we see now engender their own weather. Scientists use satellite data to issue new estimates every 12 hours.

Along with computing paradigms such as the Hadoop ecosystem, the seamless integration of multiple data sources will enable data storage and process for efficient preparedness of disasters. It is also possible to acclimatise multi-sourced social media data to track the hurricanes.

In addition, the wireless sensor which are deployed and RFID (radio identification) devices helps in producing and monitoring public's environmental data. Cellular channel transparency in cellular mobile communication, however, provides persons with possibilities for wireless channel malevolent behaviours like insertion, modification, deletion of data to mislead the disaster mechanism. GIS, GPS and cloud-based tracking tools are capable of forecasting events like earthquakes and snowmelt floods. GIS data plus the data from the conveyance network can help to explain patterns of human migration during disasters situations. The data gathered from social-media such as Twitter, Instagram provides autonomous distribution of disaster vigilance and can provide proximate to authentic data about the happening of disasters close to real time.

NEED FOR IOT AND BIG DATA IN DMS

Disaster management is not a single step process. At either point of the disaster management period, effective intervention assures improved preparedness, enhanced and accurate early notice, decreased risk or catastrophe impact reduction during the resulting recurrence of incidents. The whole emergency recovery period involves the development of national policy and plans that can alleviate the effects of disasters or their effect on persons and infrastructure. The diverse complexity of the requirements and environment during a relief operation highlights the need to make quick and successful decisions in the shortest possible period. IoT and BDA technology, which can transmit rapid updates of information, can be a key player in creating complex adaptations to the workflow.

The timely dispatch of emergency supplies from distribution centres to hospitals in keeping with the scheduling of the medical teams is also a critical role in disaster recovery. Actually, disaster management focused on BDA and IoT is an under-investigated field of study that entails many exciting possibilities and challenges. IoT technology has the capacity to be implemented into crisis management with the ability to have an omnipresent network system of linked sensors and smart devices (Kitsuregawa, 2013), and will have a positive impact on each stage of emergency response (Yang and Plotnick, 2013). On the other hand, BDA is considered to assist IoT and other related data sources in real-time analysis (Rathore et al., 2016), and is able to provide useful findings to explain the situations that remain in the disaster-affected region. Big data gathered in Internet of Things environments can also be utilised for conducting irregular events for data collection, tracking, forecasting and warnings (Buribayeva et al.,2015). Thus, we may therefore infer that the combined use of BDA techniques and IoT technology may therefore infer that the combined use of BDA and IoT technologies would give rise to developing a mature, efficient and highly essential disaster management environment.

Benefits: In the sense of emergency recovery, the BDA and IoT-based disaster management environments has a variety of advantages.

The following sections discuss some of the main advantages.

Connectivity

Connectivity is needed to allow immense data volumes to be aggregated from data stores to high-performance tote up infrastructures and to foster share the information with the disaster management force concerned. Primary benefits of the disaster management environments hinge on IoT and BDA is to provide safe connectivity due to the presence of multiple communal populations. Connectivity between the DMS and interlinked data nodes serves as foundation for efficient operations allowance. With the availability of various communication technologies, the architecture as a whole must be scalable in order to cope with various protocols of communication, including remote and local (Lloret et al., 2016). In addition, including distraction from other developed contact networks, is given with the evolution of post-disaster communication networks.

Data Storage

The handling of vast amount of diverse data in real time can be difficult in traditional data management systems. With BDA technology such as Hadoop, enormous-sized structured or unstructured databases can be easily processed on low-cost stock hardware. Real-time environments with scalding storing capabilities for the IoT devices and related data sources can increase the performance of the overall data system and can provide the specified applications with a range of advantages (Cai et al., 2016). In addition, BDA technology will allow low latency, fast processing for data analytics although retaining the stockpile of the large unstructured datasets.

Real Time Analytics

Owing to the complex and challenging nature of disaster recovery, real time interaction is one of the principal requirements for proven crisis management systems. Connectivity with different data bases leads to huge high-speed data generation that can generate obstacles to real-time analytics output. To cope with such huge and high-speed data, a dedicated technical infrastructure with the software solution capabilities to implement real time encoding, cascade and an in memory computing is necessary (Storey and Song, 2017). BDA's ability to conduct fast analyses for real-time questions is important in order to help decision-makers achieve the requisite data for a successful emergency response.

Cost-Effectiveness

In contrast to purchasing specialised data processing software tools for emergency relief activities, BDA resources are largely open-sourced and provide a significant cost savings potential. To simulate and model the huge volume of geospatially enabled data collected from sources like remote sensing, satellite, digital camera, sensor networks, radar, LIDAR etc., it requires high performance computing or cloud computing environment.

Apache Hadoop is one such open-source framework which is actually written in JAVA language to perform computing and for distributed storage of very large dataset. The big data is broken into smaller parts in Hadoop and hence it is able to achieve distributed storage. (Bello-Orgaz et al., 2016)

Hadoop has its own processing and storage area known as Hadoop Distributed File System (HDFS) and Map Reduce.

Figure 1. Before and after tsunami images of Banda Aceh 2004. (Source: www.cnes.fr)

Spark is another tool which is written in Scala to analyse and work on Big Data. Apache Spark is almost 100-times faster than Hadoop and it provides real time processing and batch processing. Spark performs In-Memory computation. An interesting feature of Spark is that Machine-learning algorithms are best suited in this framework. Along with its fast batch processing ability it also does interactive data analysis, Real-time processing, which helps in taking fast decisions at the time of disaster.

In developing countries, cost-effectiveness is a crucial concern for disaster related agencies, wherein emergency recovery initiatives are not enforced because of a shortage of affluence. Map reduction is an optimal method for cost efficient data management and aims to reduce the total device operational expenditures.

Multiple Data Sources

A wide range of data sources can be combined to obtain advanced and usable knowledge and intelligence, taking into account the proliferation of IoT ecosystems with various data resources, such as smartphones, sensors, cameras etc and BDA techniques that enable data handling. Multiple data resources offer different ways of dealing with concerns that involve multidimensional data representations to derive universal patterns for a solution that is unavailable from a single source.

By using the high-resolution digital elevation models i.e. DEM like lnSAR and LIDAR with some on-site local data, imaging spectroscopy and the data collected from sensors like ASTER helps in monitoring earthquake and landslides. (Tralli et al., 2005)

Along with the High-resolution data provided by GeoEye etc, other geospatial data portals like Microsoft Virtual Earth, Google earth also provides access to the data.

From these data, we can compare the image acquired before the occurrence of event and after the occurrence of event, and predict the level of damage caused by the disaster. Figure 1 shows the before and after image of tsunami which took place at the South-west suburb of Banda Aceh, Sumatra.

BDA for Disaster Management

BDA offers a vast options to come upon secret trends and understand situations on the field on large multi sourced datasets obtained from the disaster region so that the relief operations can be performed out efficiently and logistic systems can be maintained optimally. A key benefit of BDA is to help data scientists to analyse large quantities of data that could not be gathered with standard methods, requiring multiple data sources (Mikalef et al., 2018). For analytical processes, BDA rely on various technology and resources in order to perform massive amounts of structured, semi-structured and unstructured data. BDA research developments for disaster management rely on both the material or the text and geographical point of view of the data for centuries of study and results (Wang et al., 2016). Some recent research such as Hadoop, Kafka, spark indicates that a number of data sources that are being in use currently for numerous open source BDA resources not beyond the framework of disaster management, regardless the restricted broadcasting on BDA for disaster management. For example, the authors in (Assis et al., 2018) have developed an interoperable framework to manage flood risk while integrating heterogeneous sensors which allow almost real-time access and filtering of data using Spark.

The method used for their analysis provides a means to improve the adjoining real time implementations using heterogeneous types of statistics which is crowd source and sensor data, a big data crisis visualisation framework was built in another study (Avvenuti et al., 2018) which is able to capture and analyse Twitter data using Spark and Kafka. The framework gathers the info related to the catastrophe from the collected geo tagged tweets by the implementing annotators and classification techniques. This info is therefore visualised on a web based platform for plight services in order to achieve better situational alertness in the early stages of the tragedy.

The authors in (Lin et al., 2018) have stated that the Spark based computation offers better results for the simulation to determine the viability of typhoon risk assessment on large sets of historical data. Similarly, the scientists used multiple regression algorithms using the Spark in another study (Cortés et al., 2018) to examine a broad catalogue of earthquake events. Concerning the estimation of earthquake frequency in the state of California, they showed very encouraging results. In (Liu et al., 2018), in order to generate an accurate heat map of the population impacted in the earthquake, the authors recommended a real time aggregation and classification algorithm for cell phone location data through stream processing environments such as Kafka and Spark. In (Wang et al., 2017), an advanced disaster response model built by the Hadoop and Spark combination was introduced. Their proposed framework tackles territorial and problems in large-scale datasets and includes prognostic risk modelling for the planning and management of fire response services. Their proposed framework tackles spatial and temporal problems in large-scale datasets and includes prognostic risk modelling to plan and manage fire response services.

Planning for evacuations. A study (Huang, Cervone & Zhang, 2017) was undertaken to illustrate a mechanism that synthesises multi-source data such as social media, remote sensing and Wikipedia in order to create a scalable approach that offers spatial data mining and text mining with previous and potential crisis research concerning Hadoop.

From WSN to IoT for Disaster Management

The WSN's are one of the Autonomous low powered sensor nodes that concentrate on particular areas with the potential to quantify and record environmental conditions (i.e. dust, vibration, temperature, locations).Wireless Sensor Networks have been used long for disaster audit study, such as event monitoring in plight response monitoring (Erd et al., 2016), natural disaster tracking (Chen et al., 2013), and multi agent system based studies (Sardouk et al., 2013). Unassisted WSN's, however, lack numerous factors such as societal, technological and economic expectations for comprehensive and productive implementation of disaster management (Ray, Mukherjee and Shu, 2017). Wireless Sensor Networks plays a vital role in the field of IoT and is useful for data management, interpretation and decision-making considerations for important interpretations and sensor sensed data. There has been a major change in emergency response

Figure 2. Big Data in Disaster Management

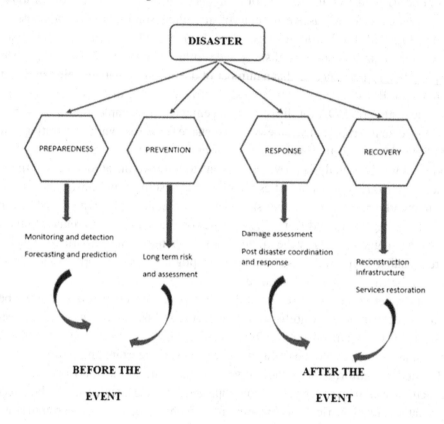

science over the last three years, with IoT on the front line. By the end of 2020 there will be major shift in the usage of IoT, as predicted there will be more than 50 billion new links that will be interconnected (Collier, 2016). IoT provides a resourceful forum, supported by coordination between various physical and virtual institutions, consisting of several tools and technologies to interpret, interact and process data. IoT has been generally regarded in recent studies on emergency response to provide multi-source and multi-dimensional information for effective decision-making.

To predict emergency incidents, IoT is efficient. IoT System proposes smart collection, incorporation, and interpretation of multi-dimensional and multi-sourced data that are the key steps for efficient decision-making in situational awareness. In a study (Greco, et al., 2018), the authors showed how IoT can be successfully developed for earthquake-related event prediction with semantic web technologies. For successful earthquake event prediction, the proposed framework was able to semantically annotate streams which were collected from web servers collecting IoT-based sensor data. Planning for evacuations. A study (Huang, Cervone and Zhang, 2017) was undertaken to illustrate a mechanism that synthesises multi-source data such as social media, remote sensing and Wikipedia in order to create a scalable approach that offers spatial data mining and text mining with previous and potential crisis research concerning Hadoop.

In disaster-stricken nations, IoT is the best option for data gathering, since it offers up an alternate means of connectivity on low powered battery and IoT enabled wireless smartphones. IoT has been generally regarded in recent studies on emergency response to provide multisource and multidimensional

Figure 3. IoT in Disaster Management

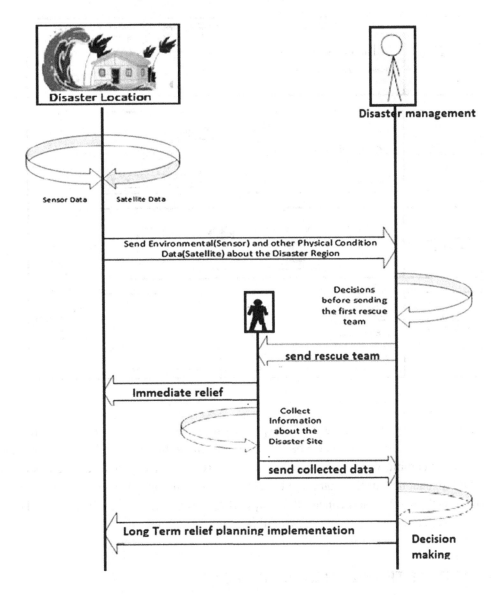

information for effective decision making. IoT System offers incorporation, smart collection, and interpretation of multidimensional and multisource data that are the key pace for efficient decision-making in situational awareness. In a study (Greco et al., 2018), the authors exhibit how IoT can be successfully put into action for earthquake-related occurrence prediction with the help of semantic web technologies. For successful earthquake event prediction, the proposed framework was able to semantically annotate streams obtained from web servers collecting IoT-based sensor data.

Added IoT based system (Xu et al., 2018) focused on the hasty and comprehensive vacation after disasters of huge numbers of people. The CLOTHO - Crowd lives-oriented track and help optimization system intent to reduce the loss of life by introducing a distributed cloud computing platform based on IoT. The data collection component of the system includes IoT-supported mobile phones, while the

Figure 4. IoT and Big Data based Disaster Management System Architecture

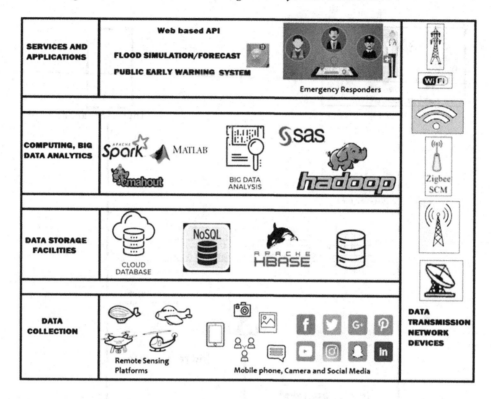

cloud-backed system is part of the storage and data processing component. An emergency and emergency response services monitored by the cloud-based IoT platform (Arbia et al., 2017). The framework is well recognised as Wearable Wireless Sensor Networks (CROW2) Critical and Rescue Operations and incorporates heterogeneous wireless equipment such as smartphones and sensors with diverse networking technologies such as Wi-Fi and Bluetooth to support end-to-end network connectivity. This device allows plight rescuers to be wired to any network that runs or to the internet.

BIG DATA AND IOT TECHNIQUES USED IN DMS

Big Data and IoT application in DMS are actually new, this is a modern opportunity with maximum benefits due to improvements in capacity level, management level, and research level. Big data is available from sensor networks like seismograph networks, remote sensing sensors and other technologies like GPS and GIS which are the components of IoT (Pu and Kitsuregawa, 2013). In each step of Disaster management one or the other technique of IoT or Big Data have been used. Disasters can be detected early using Big data technology which uses data from diverse sources (Roglá & Chalmeta, 2016) The characteristics of Big data are data-veracity, data-variety, data-velocity, data- volume and data-value. Usefulness of available data is data value and its access and enhancement are the chief ambitions of big data (Chen et al., 2014). Through Big Data it is easy to detect region prone to Disaster, and these cases later helps in responding effectively. (Pu and Kitsuregawa, 2013)

Figure 5. Data Collected from Various Sources for Disaster Management

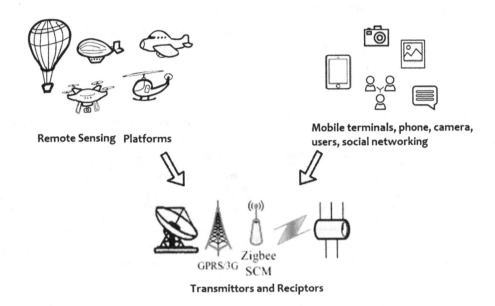

Disaster management using Big Data and IoT technique involves following process

- Collecting the data,
- Storing the collected data (Data base management system)
- Processing (on cloud)

Presentation (visualizing) (Chen and Zhang, 2014). This is depicted in Figure 4 and each layer is explained in brief below. DMS uses sensors and crowdsourcing techniques to gather information, sensed data is then processed and communicated through transmitters or Wi-Fi and finally would be stored in some cloud-based server.

Collecting the Data

Information gathered by social media, GIS and GPS are very useful and provides precise information about affected area and the damages caused by disasters.(Emmanouil and Nikolaos, 2015) Big data gathered by remote sensing platforms, stations, webcam, mobile phones, Single-Chip Microcomputer (SCM), Personal computers and GIS contributes largely to Early Warning Systems. The process of data collection is depicted in Figure 5.

These sensors, Geographical Information and Positioning Systems (GPS, GIS) helps in predicting earthquakes (Buribayeva et al., 2015) and floods caused by melting of snow (Fang et al., 2015). Other than these, data is also collected from public and users through social media (e.g. Twitter, Facebook, Instagram) which are External API (Ext API). It offers awareness and real time information about the eventuality of disasters (Cen et al., 2011). Not only these, even weather forecasting plays a major role and these data will be stored in information warehouse. Such well-timed data helps a lot in planning suitable process for managing disaster. Big data obtained from social media and mobile phones helps in effective response to disaster (Qadir et al., 2016).

Remote sensing is also one of the best source available, it is very useful in all the phases of volcanic eruptions disaster management and earth quakes, land slides disasters these are specified in Table1 (Van Western, Soeters and Buchroithner, 1996)

Table 1. Remote sensing data for disaster management (Van Western, Soeters and Buchroithner, 1996)

Disaster type	Disaster prevention	Disaster preparedness	Disaster relief
Earthquakes	Remote sensing data is useful	Remote sensing data is not useful	Remote sensing data has limited use
Volcanic eruptions	Remote sensing data is Very useful	Remote sensing data is Very useful	Remote sensing data is Very useful
Landslides	Remote sensing data has limited useful	Remote sensing data is useful	Remote sensing data is useful

Big data aids in identifying the areas affected and helps in providing assistance to the people. GIS can be used in assigning necessary resources to the needy (Pu & Kitsuregawa,2013). Zigbee, Wi-Fi and Bluetooth are used for short-range communication. The huge amount of gathered information has to be processed and transformed into useful data (Gibson et al., 2014). To solve this problem, Big data methods like Statistical analysis, ML (Machine learning), Data mining and Analytical tools are used. These methods are useful in detecting hidden patterns and trends.

Statistical analysis is used in collecting, analysing, interpreting, presenting and organizing the gathered data. XLSTAT, SPSS, MATLAB software, R studio and others are used in performing statistics. From Data mining, huge amount of raw data could be explored to find meaningful and useful patterns E.g. Framework called Apache Mahout is used to build Data mining algorithms. Artificial Intelligence (AI) led to Machine learning, which is programmed explicitly to make keen decisions. ML Algorithms are divided into supervised, unsupervised and semi-supervised learning algorithms. Spark is used for machine learning. Analytical tools unveil hidden patterns by evaluating the raw data to, correlations and insights. E.g. Spark.

Storing and Processing the Collected Data

In Big Data and IoT applications, the Data collected from variety of different sources has to be stored and processed. Big Data has techniques and tools for storage, processing and to provide security to the data (Chen and Zhang, 2014). This massive amount of Data needs large storage facilities.

Unstructured data is stored and managed using a data storage type called NoSQL Database Management system (DBMS). It is non-relational (Han et al., 2011), scalable, has distributed storage and offers quick arrangement and analysis of data. NoSQL databases is divided into three types. They are column stores, key-value, and document Stores (Grolinger et al., 2013). Any data can be stored without any previously stored structure in NoSQL. MongoDB is an example for document Stores and Hadoop HBase data store for Column stores. This data is used for disaster data presentation (Ishida and Ohyanagi, 2019).

Main-Memory DBMS store data in main memory. It provides speedy processing of incoming data which is useful for IoT Applications. It is also called as in-memory data base. DBMS can be combined with Hadoop Distributed File System (HDFS) to enhance the performance.

Hadoop supports distributed storage, analysis and process of large datasets. (Dittrich and Ruiz, 2012). Data warehouses can also store huge amount of data. Apache Tajo is one of the examples.

Cloud based warehouse is a union of database tools and databases which more storage capacity up to PB and EB (Petabyte, Exabyte).

Visual Representation

Visualization may be in the form of maps, graphs, tables or 3D models and it is the most important part of DMS. It has strong influence on people's data understanding as well as the system's usefulness (Fang et al., 2015). It provides glimpse about a particular objective and is easily understandable. Plotting geo-spatial maps could lead to the development of successful approach for disaster management. For example, Google graphs, Tableau software etc. Big data and Statistical analysis are used in forecasting with the help of atmospheric data and satellite (Jongman et al., 2015)

Wireless communication not only helps in pre-disaster management by laying out necessary data but also helps during post-disaster situations. Communication networks like towers and others gets damaged in large extent due to disaster. During such situation Radio frequency identification (RFID) and Wireless networks helps to track people. Device to Device (D2D) communications is also a promising technology in network, it allows UE's (User equipment's) to communicate with itself without the network. (Ali et al., 2016). The applications of big data and IoT technologies for catastrophe administration, their diverse datasets, technologies adopted are specified in Table 1.

Table 2. IoT, BDA Techniques used in Different Disaster Cases

SI no	Use Case	Type of Disaster	Data obtained from	IoT /BDA techniques used
1.	Emergency Warning system [13]	Fire, Earthquake, Snow melt floods	Sensors	Arduino, GPS, cloud server, and Mysql databases
2.	Real-time monitoring for disaster management [49]	Flood	Social media and crowdsourcing	GPS and Databases
3.	Disaster surveillance [37]	Fire, flood	Sensors	Zigbee and Hadoop
4.	Statical analysis using multisource data [50]	Typhoon	GIS and social media data	GIS, Apache Hive, Hadoop and Mahout
5.	Disaster collection and analysis system [51]	Flood, land slide	Crowd sourcing, mobile	Hadoop, HTML5

Huge information created from sensors and geological mappings contributes to early caution frameworks for disaster. (GIS)Geological Data Frameworks, Global Positioning Frameworks (GPS) and natural monitoring sensors with cloud administrations has a capability to foresee catastrophes like snowmelt surges (Ali et al., 2016) as well as seismic tremors (Buribayeva et al., 2015). Geo informatics data beside transportation data helps in understanding transportability of affected people in the area of disasters (Song et al., 2015) though social medias (e.g. Instagram, Facebook etc.) provides autonomously dissemination of calamity mindfulness (Grolinger et al., 2015 and Choi & Bae, 2015) and can gives close to genuine time data of the happening of disasters (Cen et al., 2011).Consistent combination of distinctive data streams, along with Hadoop like handling models helps in effective disaster readiness. It also helps in tracking typhoons.

CHALLENGES / ISSUES OF IoT AND BIG DATA IN DISASTER MANAGEMENT SYSTEM

There are certain challenges while working with the traditional ways of managing data for IoT and BDA based applications namely, Sematic and engineering challenges. The challenges faced while gathering meaningful and valuable data from the available datasets is considered as sematic, whereas the challenges regarding the query and storage is considered to be engineering (Bizer et al., 2012). By using new technologies like in-memory databases and Hadoop these problems can be resolved. (Thakur and Mann, 2014) Some problems faced during the processes are mentioned.

Resistance Against System Failure During Disaster

During a Disaster, the hardware as well as software components used in DBMS many get damaged and the communication networks will also be affected. The system should not be damaged by any ways due to calamity, so it is important to make system architecture resistant to disaster, so that, data can be recovered by itself and retrieve without any problems. The system must be available even during such scenario. To overcome this problem, alternative communication networks must be established, data must be managed effectively by storing data in cloud-based storage system (Ali et al., 2016) along with power backup. Over the last few years, use of cloud computing technology has also increased, but the incorporation of hybrid environments and the implementation of integrated information management systems (IIMS) has been remained as the major issues that needs explored. Storing data in multiple formats will help in reducing the loss of data.

Data Gathered from Multiple Sources

In spite the benefits of Big Data gathered from various sources, it is a big challenge to process this huge volume of heterogeneous data. This dataset must be filtered from unwanted data to improve data quality. To overcome this challenge, techniques like filtering and sampling must be used. Aggregation and combination of related data was done by Data Aggregator to resolve the issue of data duplication and minimise the use of resources. However, with the rising number of data sources, data aggregation challenges are rising, requiring more storage and power usage.

Quality of the Collected Disaster Data

The collected data may contain error prone, noisy, useless data which would waste precious time. This problem has to be solved before starting analysis. Quality of the data decides the accuracy of analysis.

Data cleansing is a process where data quality parameters like reliability, accuracy, consistency, completeness and others are checked.

To Prevent Misuse of Personal Information

Both in IoT and BDA, personal information's have been utilized which is arising the question of privacy concern. Misuse may lead to different kind of problems like separation, profiling, tracking, theft and many others (Tene and Polonetsky, 2011). Security level must be increased by isolating data for various requirements, since big data might contain confidential data about government or information like people's location, financial status, health which should not be misused. Open source tools like Hadoop usually provide poor security (Kim, Trimi and Chung, 2014) hence, sharing and segregating the data in cloud-based and other platforms are challenging.

Quick Response and Decision-Making During Emergency

Applications of IoT gathers large information from various sources, fully using this massive data during crisis is also a challenge. It needs fast data integration, aggregation, analysis, normalization and visualizing to make rapid decisions during crisis. Extracting quality data at minimum time is also important. Even though we use newer data analytical tools, data processing consumes a lot of time. Since quick response can save a lot of life it is important to obtain required results in less time. (Emmanouil and Nikolaos, 2015)

Metadata

Disaster management system is time sensitive and requires quality data, metadata plays a significant part in identification and maintenance of data. During a crisis, it is really a big challenge to collect and manage the metadata. This is because of the data from multiple sources, and its formats. The data will be of variety of quality which is a problem and issues related to data combination could be solved by metadata. Metadata assists in processing the variety of data. (Fertier et al.,2016)

CONCLUSION

Managing disaster will be a big challenge worldwide. The combination of BDA (Big Data Analytics) and IoT technologies is innovative, effective for catastrophe administration. Big data analytics along with the enabling technologies have improved effective decision-making during crisis. Huge amount of data is gathered from IoT applications. And these information gathered from huge variety of sources during disaster is extremely prone to unwanted useless information. So, it is necessary to implement data processing techniques like filtration, sampling and sorting to get rid of data inconsistencies. Most of the disaster forecasting systems depends upon BDA. Disaster management using BDA and IoT involves following steps, initially the disaster data will be collected from various sources which will be of different formats like image/ video etc. This data has to processed, stored, analysed and visualized. All these involves IoT techniques and BDA tools. Using these technologies, effective disaster management could be achieved. During the transfer of data and storage, it is important to constantly examine these processes to guarantee the realness of these information and also not to harm the privacy of the people, since these data can contain personal information which are prone to be misused. So, it is important to adopt security tools to maintain secrecy of data. Big data management also involves certain issues like data consistency, accuracy and completeness to make better decisions. We examined and compared a few crucial case studies to demonstrate the part played by big data and IoT in several management stages of disaster.

Further, the technologies have to be developed and implemented in such a way that the benefits of IoT and BDA is fully acquired. With the advancements in the field of Internet of Things and big data, their applications are increasing, which results in saving lakhs of lives during disaster. Future works has to be done by expanding the usage of applications like prediction of disaster, monitoring, quick evacuation of people from the affected region.

REFERENCES

Ahmed, Y., & Hashem, K. (2017). The role of big data analytics in Internet of Things. *Computer Networks*, *129*, 459–471. doi:10.1016/j.comnet.2017.06.013

Ali, N., & Shah, V. & Bhuvanasundaram. (2016, April). Architecture for public safety network using D2D communication. In *2016 IEEE Wireless Communications and Networking Conference Workshops (WCNCW)* (pp. 206-211). IEEE. 10.1109/WCNCW.2016.7552700

Asencio–Cortés, M.-E., Morales-Esteban, A., Shang, X., & Martínez-Álvarez, F. (2018). Earthquake prediction in California using regression algorithms and cloud-based big data infrastructure. *Computers & Geosciences*, *115*, 198–210. doi:10.1016/j.cageo.2017.10.011

Assis, Horita, & Freitas, Ueyama & De Albuquerque. (2018). A service-oriented middleware for integrated management of crowdsourced and sensor data streams in disaster management. *Sensors (Basel)*, *18*(6), 1689. doi:10.339018061689 PMID:29794979

Avvenuti, C., & Vigna, F. (2018). CrisMap: A big data crisis mapping system based on damage detection and geoparsing. *Information Systems Frontiers*, *20*(5), 993–1011. doi:10.100710796-018-9833-z

Basanta-Val, A., & Wellings, G. (2016). Architecting time-critical big-data systems. *IEEE Transactions on Big Data*, *2*(4), 310–324. doi:10.1109/TBDATA.2016.2622719

Bello-Orgaz, Gema, Jung, & Camacho. (2016). Social big data:Recent achievements and new challenges. *Information Fusion*, *28*, 45–59.

Ben Arbia, Alam, & Kadri, Hamida, & Attia. (2017). Enhanced IoT-based end-to-end emergency and disaster relief system. *Journal of Sensor and Actuator Networks*, *6*(3), 19.

Bizer, B., Boncz, P., Brodie, M. L., & Erling, O. (2012). The meaningful use of big data: Four perspectives—four challenges. *SIGMOD Record*, *40*(4), 56–60. doi:10.1145/2094114.2094129

Buribayeva, M., Miyachi, T., Yeshmukhametov, A., & Mikami, Y. (2015). An autonomous emergency warning system based on Cloud Servers and SNS. *Procedia Computer Science*, *60*, 722–729. doi:10.1016/j.procs.2015.08.225

Cai, X., Xu, B., Jiang, L., & Vasilakos, A. V. (2016). IoT-based big data storage systems in cloud computing: Perspectives and challenges. *IEEE Internet of Things Journal*, *4*(1), 75–87. doi:10.1109/JIOT.2016.2619369

Cen, J., Yu, T., Li, Z., Jin, S., & Liu, S. (2011). *Developing a disaster surveillance system based on wireless sensor network and cloud platform*. Academic Press.

Chen, L., Liu, Z., Wang, L., Dou, M., Chen, J., & Li, H. (2013). Natural disaster monitoring with wireless sensor networks: A case study of data-intensive applications upon low-cost scalable systems. *Mobile Networks and Applications*, *18*(5), 651–663. doi:10.100711036-013-0456-9

Chen, M., Mao, S., & Liu, Y. (2014). Big data: A survey. *Mobile Networks and Applications*, *19*(2), 171–209. doi:10.100711036-013-0489-0

Choi & Bae. (2015). The real-time monitoring system of social big data for disaster management. In *Computer science and its applications* (pp. 809–815). Springer.

Collier, S. E. (2016). The emerging enernet: Convergence of the smart grid with the internet of things. *IEEE Industry Applications Magazine*, *23*(2), 12–16. doi:10.1109/MIAS.2016.2600737

Dittrich & Quiané-Ruiz. (2014-2015). Efficient big data processing in Hadoop MapReduce. *Proceedings of the VLDB Endowment International Conference on Very Large Data Bases*, *5*(12).

Emmanouil & Nikolaos. (2015). Big data analytics in prevention, preparedness, response and recovery in crisis and disaster management. In *The 18th International Conference on Circuits, Systems, Communications and Computers (CSCC 2015), Recent Advances in Computer Engineering Series* (Vol. 32, pp. 476-482). Academic Press.

Erd, S., Schaeffer, F., Kostic, M., & Reindl, L. M. (2016). Event monitoring in emergency scenarios using energy efficient wireless sensor nodes for the disaster information management. *International Journal of Disaster Risk Reduction*, *16*, 33–42. doi:10.1016/j.ijdrr.2016.01.001

Fang, S., Xu, L., Zhu, Y., Liu, Y., Liu, Z., Pei, H., Yan, J., & Zhang, H. (2015). An integrated information system for snowmelt flood early-warning based on internet of things. *Information Systems Frontiers*, *17*(2), 321–335. doi:10.100710796-013-9466-1

Fertier, M., & Barthe-Delanoë, T., & Bénaben (2016, May). Adoption of big data in crisis management toward a better support in decision-making. *Proceedings of Conference on Information System for Crisis Response And Management (ISCRAM 16)*.

Fisher, Hagon, Lattimer, Callaghan, Swithern, & Walmsley. (2018). *Executive summary world disasters report: Leaving no one behind*. Academic Press.

Gibson, Andrews, Domdouzis, Hirsch, & Akhgar. (2014, December). Combining big social media data and FCA for crisis response. In *2014 IEEE/ACM 7th International Conference on Utility and Cloud Computing* (pp. 690-695). IEEE.

Greco, R., Ritrovato, P., Tiropanis, T., & Xhafa, F. (2018). IoT and semantic Web technologies for event detection in natural disasters. *Concurrency and Computation*, *30*(21), e4789. doi:10.1002/cpe.4789

Grolinger, Capretz, Mezghani, & Exposito. (2013, June). Knowledge as a service framework for disaster data management. In *2013 Workshops on Enabling Technologies: Infrastructure for Collaborative Enterprises* (pp. 313-318). IEEE.

Grolinger, Higashino, Tiwari, & Capretz. (2013). Data management in cloud environments: NoSQL and NewSQL data stores. *Journal of Cloud Computing: Advances, Systems and Applications, 2*(1), 22.

Gubbi, B., Buyya, R., Marusic, S., & Palaniswami, M. (2013). Internet of Things (IoT): A vision, architectural elements, and future directions. *Future Generation Computer Systems, 29*(7), 1645–1660. doi:10.1016/j.future.2013.01.010

Han, H., Le, G., & Du. (2011, October). Survey on NoSQL database. In *2011 6th international conference on pervasive computing and applications* (pp. 363-366). IEEE.

Hristidis, C., Chen, S.-C., Li, T., Luis, S., & Deng, Y. (2010). Survey of data management and analysis in disaster situations. *Journal of Systems and Software, 83*(10), 1701–1714. doi:10.1016/j.jss.2010.04.065

Huang, C. Jing & Chang (2015, November). DisasterMapper: A CyberGIS framework for disaster management using social media data. In *Proceedings of the 4th International ACM SIGSPATIAL Workshop on Analytics for Big Geospatial Data* (pp. 1-6). ACM.

Huang, C., Cervone, G., & Zhang, G. (2017). A cloud-enabled automatic disaster analysis system of multi-sourced data streams: An example synthesizing social media, remote sensing and Wikipedia data. *Computers, Environment and Urban Systems, 66*, 23–37. doi:10.1016/j.compenvurbsys.2017.06.004

Ishida & Ohyanagi. (2019). Implementation and evaluation of a visualization and analysis system for historical disaster records. *Journal of Ambient Intelligence and Humanized Computing*, 1–16.

Jongman, W., Wagemaker, J., Romero, B., & de Perez, E. (2015). Early flood detection for rapid humanitarian response: Harnessing near real-time satellite and Twitter signals. *ISPRS International Journal of Geo-Information, 4*(4), 2246–2266. doi:10.3390/ijgi4042246

Kim, T., Trimi, S., & Chung, J.-H. (2014). Big-data applications in the government sector. *Communications of the ACM, 57*(3), 78–85. doi:10.1145/2500873

Li, J., & Liu, H. (2017). Challenges of feature selection for big data analytics. *IEEE Intelligent Systems, 32*(2), 9–15. doi:10.1109/MIS.2017.38

Lin, F., & Wu, C. (2018). A spark-based high performance computational approach for simulating typhoon wind fields. *IEEE Access: Practical Innovations, Open Solutions, 6*, 39072–39085. doi:10.1109/ACCESS.2018.2850768

Liu, X., Li, X., Chen, X., Liu, Z., & Li, S. (2018). Location correction technique based on mobile communication base station for earthquake population heat map. *Geodesy and Geodynamics, 9*(5), 388–397. doi:10.1016/j.geog.2018.01.003

Lloret, T., Tomas, J., Canovas, A., & Parra, L. (2016). An integrated IoT architecture for smart metering. *IEEE Communications Magazine, 54*(12), 50–57. doi:10.1109/MCOM.2016.1600647CM

Mikalef, P., Pappas, I. O., Krogstie, J., & Giannakos, M. (2018). Big data analytics capabilities: A systematic literature review and research agenda. *Information Systems and e-Business Management, 16*(3), 547–578. doi:10.100710257-017-0362-y

Mo, L., Kim, Y. H. & Lee, J. K. (2015, June). Design of Disaster Collection and Analysis System Using Crowd Sensing and Beacon Based on Hadoop Framework. In *International Conference on Computational Science and Its Applications* (pp. 106-116). Springer. 10.1007/978-3-319-21410-8_8

Noji, E. K. (Ed.). (1997). *The public health consequences of disasters.* Oxford University Press.

Philip Chen, C. L., & Zhang, C.-Y. (2014). Data-intensive applications, challenges, techniques and technologies: A survey on Big Data. *Information Sciences, 275,* 314–347. doi:10.1016/j.ins.2014.01.015

Pu & Kitsuregawa. (2013). *Big Data and disaster management: a report from the JST/NSF Joint Workshop.* Georgia Institute of Technology, CERCS.

Qadir, A., & Rasool, Z. (2016). Crisis analytics: Big data-driven crisis response. *Journal of International Humanitarian Action, 1*(1), 1–21. doi:10.118641018-016-0013-9

Rathore, A., Ahmad, A., Paul, A., & Rho, S. (2016). Urban planning and building smart cities based on the internet of things using big data analytics. *Computer Networks, 101,* 63–80. doi:10.1016/j.comnet.2015.12.023

Ray, M., Mukherjee, M., & Shu, L. (2017). Internet of things for disaster management: State-of-the-art and prospects. *IEEE Access: Practical Innovations, Open Solutions, 5,* 18818–18835. doi:10.1109/ACCESS.2017.2752174

Roglá, O. (2016). Social customer relationship management: Taking advantage of Web 2.0 and Big Data technologies. *SpringerPlus, 5*(1), 1462. doi:10.118640064-016-3128-y PMID:27652037

Sardouk, M., & Merghem-Boulahia, G. (2013). Crisis management using MAS-based wireless sensor networks. *Computer Networks, 57*(1), 29–45. doi:10.1016/j.comnet.2012.08.010

Song, Z., & Sekimoto, S. (2015, February). A simulator of human emergency mobility following disasters: Knowledge transfer from big disaster data. *Twenty-Ninth AAAI Conference on Artificial Intelligence.*

Storey, V. C., & Song, I.-Y. (2017). Big data technologies and management: What conceptual modeling can do. *Data & Knowledge Engineering, 108,* 50–67. doi:10.1016/j.datak.2017.01.001

Tene & Polonetsky. (2011). Privacy in the age of big data: A time for big decisions. *Stan. L. Rev. Online, 64,* 63.

Thakur & Mann. (2014). Data mining for big data: A review. *International Journal of Advanced Research in Computer Science and Software Engineering, 4*(5), 469–473.

Tralli, B., & Zlotnicki, D. (2005). Satellite remote sensing of earthquake, volcano,flood, landslide and coatal industrial hazards. *ISPRS Journal of Photogrammetry and Remote Sensing, 59*(4), 185–198. doi:10.1016/j.isprsjprs.2005.02.002

Van Western, Soeters, & Buchroithner. (1996). Potential and limitations of satellite remote sensing for geo-disaster reduction. *International Archieves of Photogrammetry and Remote Sensing, 31*(B6).

Wang, V., & Salehi, R. (2017, May). A large-scale spatio-temporal data analytics system for wildfire risk management. In *Proceedings of the Fourth International ACM Workshop on Managing and Mining Enriched Geo-Spatial Data* (pp. 1-6). ACM.

Wang, W., Wu, Y., Yen, N., Guo, S., & Cheng, Z. (2016). Big data analytics for emergency communication networks: A survey. *IEEE Communications Surveys and Tutorials, 18*(3), 1758–1778. doi:10.1109/COMST.2016.2540004

Xu, Z., & Sotiriadis, A. (2018). CLOTHO: A large-scale Internet of Things-based crowd evacuation planning system for disaster management. *IEEE Internet of Things Journal, 5*(5), 3559–3568. doi:10.1109/JIOT.2018.2818885

Yang, S. H., & Plotnick. (2013). How the internet of things technology enhances emergency response operations. *Technological Forecasting and Social Change, 80*(9), 1854-1867.

Chapter 4
IoT–Based Smart Agriculture

Kishore Kumar K.

Koneru Lakshmaiah Education Foundation, India

ABSTRACT

Smart farming is an evolving concept since IoT sensors are capable of providing agricultural field information and then acting on the basis of user feedback. The main factor in improving the yield of efficient crops is the control of environmental conditions. There is a small yard, farmland, or a plantation area for most of us. However, our busy timetable does not allow us to manage it well. But we can easily accomplish it with the use of technology. So, the authors make an IoT-based smart farming system that can control soil moisture. As data has become a critical component in modern agriculture to assist producers with critical decisions and make a decision with objective data obtained from sensors, significant advantages emerge. This chapter explores the current state of advanced farm management systems by revisiting each critical phase, from data collection in crop fields to variable rate applications, in order for growers to make informed decisions save money while also protecting the environment and transforming how food is grown to meet potential population growth.

1. INTRODUCTION

In each region, farming is carried out from age. The science and specialization of growing plants is agriculture. The critical advance in the growth of passive human progress was horticulture. Agriculture is that the most vital sector for grouping to survive their existence. Agriculture is the primary occupation for most of the people in India. It enhances an enormous concern to manage food for folks everywhere the globe. Most of the farmers follow terribly ancient ways to cultivate their crops. They accustomed to be present physically on their farm to watch crops. Use of technology will create this job easier and time efficient. In smart agribusiness, IOT takes on a critical role. IOT sensors are ideal for supplying information on the fields of agriculture. We have suggested an IOT framework using computerization and a genius farming framework. Internet of things may be a technology which may send or receive any knowledge to server exploitation the net. Exploitation this technology, farmers can monitor the particular condition of the crops while not being gift in their field. In this project, we have projected a system to watch the farming field with the assistance of iot technology. In order to ensure the continu-

DOI: 10.4018/978-1-7998-6870-5.ch004

ity of this economic sector, the agricultural sector is one of the most significant economic resources in these countries, contributing to the importance of good management of the available water resources. In India, 10% of the area of the country is covered by rice plantations. Agriculture is undergoing a transformation driven by technological advances, which appears to be very exciting because it will enable this primary sector to achieve success of farm productivity and profitability. Precision agriculture, which entails applying inputs (what is required) when and where they are needed, has become the third wave of the modern agricultural revolution (the first was mechanization, and the second was the green revolution with its genetic modification), and it is now being intensified by the availability of larger amounts of data. In October 2016, the United States Department of Agriculture (USDA) released a survey. Precision Agriculture technologies boosted net returns and operating income, according to the report. In addition, when it comes to the climate, new technologies are constantly being used in farms to ensure the long-term viability of farm production. The implementation of these innovations, however, is fraught with risk and trade-offs. According to a market study, better farmer education and training, knowledge sharing, easy access to financial services and rising customer demand for organic food are all factors that would facilitate the adoption of sustainable farming technologies. When using these new technologies to retrieve data from crops, the challenge is to deliver something coherent and meaningful, since data is nothing other than numbers or photographs. Agriculture is India's main source of revenue. Agriculture employs 58 percent of Indians living in rural areas, according to the IBEF (India Brand Equity Foundation). Agriculture's contribution to India's Gross Value Addition is projected to be about 8%, according to the Central Statistics Office's second advised report, which is a substantial contribution. Agriculture would use a huge amount of water, especially fresh water supplies, in such a scenario, according to current market studies, agriculture uses 85 percent of available freshwater resources worldwide, and this number will continue to rise due to population growth and increased food demand. This necessitates the creation of plans and strategies to use water wisely while taking advantage of scientific and technological advances. There are a number of water-saving systems available for different crops, ranging from the most simple to the most technologically advanced. Thermal imaging is used in one of the existing systems to track plant water status and irrigation scheduling. Automation of irrigation systems is also possible by calculating the water level in the soil and controlling actuators to irrigate as and when required rather than on a predetermined schedule, thereby saving and using water more wisely. When the volumetric water content of the suction pipe exceeds a certain level, an irrigation controller opens a solenoid valve and waters the bedding plants.

2. LITERATURE SURVEY

The existing methodology and one amongst the oldest ways in which in agriculture is that the manual methodology of checking the parameters. During this methodology the farmers they themselves verify all the parameters and calculate the readings. (Gutiérrez et al., 2013)It focuses on developing devices and tools to manage, show and alert the users victimization the benefits of a wireless detector network system. (Ram Prasad et al., 2019)It aims at creating agriculture good victimization automation and IoT technologies. The highlight options square measure good GPS primarily based remote controlled mechanism to perform tasks like weeding, spraying, wetness sensing, human detection and keeping vigilance. (Venkatarao & Anup Kumar, 2019)The cloud computing devices which will produce a full automatic {data processing system | ADP, system |ADPS |system} from sensors to tools that observe data from agricultural field pictures and from human actors on the bottom and accurately feed the information into the repositories in conjunction with the situation as GPS coordinates.(Prasad et al., 2020)This plan proposes a unique methodology for good farming by linking {a good la sensible la wise} sensing system and smart irrigator system through wireless communication technology.(Kumar & Srinath, 2020)It proposes an occasional price and economical wireless detector network technique to accumulate the soil wetness and temperature from numerous location of farm and as per the requirement of crop controller to require the choice whether or not the irrigation is enabled or not.(Anupkumar et al., 2019)It proposes an inspiration concerning however machine-driven irrigation system was developed to optimize water use for agricultural crops. Additionally, a entry unit handles detector data.(Murali et al., 2020)The region conditions square measure monitored and controlled on-line by victimization LAN IEEE 802.3.The partial root zone drying method are often enforced to a most extent.(Chaitanya et al., 2020)It is intended for IoT primarily based watching system to research crop setting and also the methodology to boost the potency of higher cognitive process by analyzing harvest statistics.(Kekan & Kumar, 2019)In this paper image process is employed as a tool to observe the diseases on fruits throughout farming, right from plantation to harvest home. The variations square measure seen in color, texture and morphology. (Rao et al., 2018)In this paper, greenhouse may be a building during which plants square measure adult in closed setting. It's accustomed maintain the optimum conditions of the setting, greenhouse management and knowledge acquisition. Since IoT sensors can provide information about agriculture fields and then act on it based on user feedback, smart agriculture is a new concept. The aim of this paper is to create a smart agriculture system that takes advantage of cutting-edge technology like Arduino, IoT, and wireless sensor networks. The paper aims to make use of emerging technology, such as the Internet of Things (IoT) and smart agriculture by automation (Bhavana et al., 2021; Bhavana et al., 2020; Bhavana et al., 2016; Sepaskhah & Ahmadi, 2012; Vishnu et al., 2018). The ability to track environmental conditions is a crucial factor in increasing the yield of productive crops. The aim of this paper is to develop a system that can track temperature, humidity, moisture, and even the movement of animals that may kill crops in agricultural fields using sensors and, in the event of a discrepancy, send an SMS notification as well as a notification on the app developed for the same to the farmer's smart phone via Wi-Fi/3G/4G.The device uses a duplex communication connection based on a cellular-Internet interface, allowing data inspection and irrigation scheduling to be programmed via an android application. The device has the potential to be useful in water-scarce, geographically remote areas due to its energy independence and low cost (Howell et al., 2012; Nandurkar et al., 2014).

The results of the project demonstrate the real-time monitoring and statistical analysis of soil moisture of the soil automatically water the plants from every place shown in Fig.1. This framework can be extended by using PIR sensor and we can detect if any animals or humans enter into the farm and can produce an alarm (Jyothi et al., 2020; Jyothi & Sriadibhatla, 2019; Jyothi & Sridevi, 2018).

3. PROPOSED METHODOLOGY

In this methodology we are going to check the moisture level of the soil and water the plants if necessary through phone via app designed through kodular creator. The app displays the soil moisture values of the soil and waters the plants if necessary. In this project we also check the humidity and temperature of the of the atmosphere and take necessary actions for the growth of plants and the conditions suitable for the plants and soil. We have also placed a PIR sensor that detects the motion of any living thing that moves

Figure 1. Block diagram of Arduino

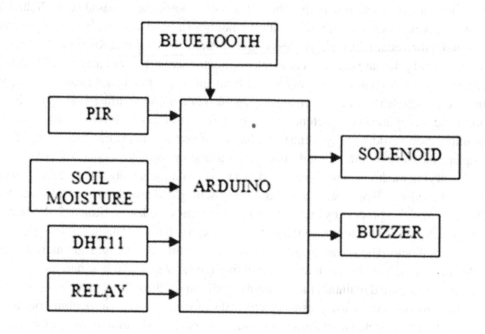

near the surrounding area and if it detects any living thing it gives an alarm sound though buzzer. This method of using PIR sensor helps in protecting the plant from surrounding animals and outside threats. In this project we are going to monitor the moisture of the soil, temperature, humidity and detect the moment of animals around the field .We are going to do this with the help of PIR sensor and DHT11 sensor. PIR sensor is used to detect the entry of any animals or human beings around the field or into the field and gives an alarm through buzzer. The DHT11 sensor is used to check the humidity and temperature of the atmosphere. This is very simple yet efficient kit. This kit is cost effective and we can maintain easily. We can monitor the field's condition from the comfort of our homes effortlessly, and we can water the plants if necessary. We followed the methodology as follows, Firstly we created an app through kodular creator shown in Fig.2, In this kodular app we are using few blocks some of them are IF THEN blocks, In this block if the condition is TRUE, then the block gets implemented, if the condition is false it ignores the block and goes to the another block. Various devices are used in this work are as follows.

3.1 Arduino Uno

Arduino is an online platform that is based on hardware that is simple to use. It is an open source software. (Gutiérrez et al., 2013) Arduino UNO is one of the most easily available low-cost Arduino board. The Arduino is an embedded system. The Arduino Uno is an open-source microcontroller board created by Arduino, based on the Microchip ATmega328P microcontroller. The board has 14 digital I/O pins, 6 analog I/O pins, and is Programmable with the Arduino IDE (Integrated Development Environment), it can be operated by a USB cable or an external 9-volt battery, although voltages ranging from 7 to 20 volts are approved in Fig.3.

3.2 Soil Moisture Sensor

Figure 2. Arduino Uno board

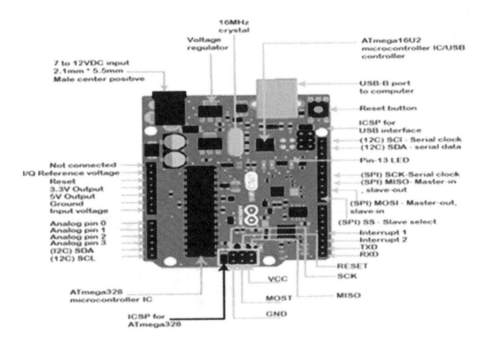

The majority of commercial irrigation controllers use soil moisture sensors that are based on dielectric permittivity, which is commonly calculated for a given soil, it is regarded as a constant. In the real world, however, it varies somewhat depending on the form of soil and/or the amount of water in the soil. Salt content (metals in the soil that are salty) soil texture, and the amount of nutrients dissolved in the soil solution Density in bulk. Permittivity of soil is made up of many variables. Permittivity values for each of the material's subcomponents a specific type of soil .The bulk permittivity of soil is measured. An electromagnetic wave's velocity was measured in units. It is analysed via the soil. The wave would be slowed enough to detect a difference in a substance with a higher dielectric constant, or Kab, such as water. Water has a much greater Kab than soil minerals or air, because it is the main material that is being detected.

This sensor primarily uses capacitance to measure the soil's water content. By inserting this sensor into the ground, the work of this sensor can be done and the status of the water content in the soil can be recorded as a percentage shown in Fig.4.

Characteristics

This sensor's specification contains the following.

For working, the necessary voltage is 5V

1. The required working current is <20mAA
2. The device style is analog.

4.   The required working temperature of this sensor is 10°C~30°C.

4.1 HC Bluetooth

Figure 3. Sensor

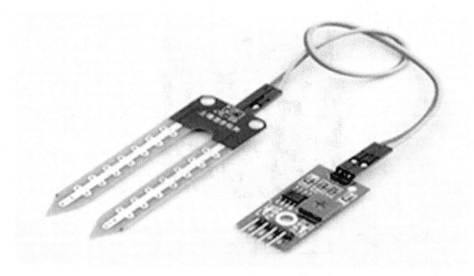

The HC-05 module, designed for transparent wireless serial communication setup, is an easy-to-use Bluetooth SPP (Serial Port Protocol) module. It is possible to use the HC-05 Bluetooth Module in a Master or Slave configuration, making it a perfect wireless communication solution. This Bluetooth serial port module is completely qualified with Bluetooth V2.0+EDR (Enhanced Data Rate) 3Mbps modulation with maximum 2.4GHz radiation. It is used for Communication with the app and our circuit (with Arduino). The HC-05 Bluetooth module pre - existing knowledge between the Application and the Sensing Node, i.e., sharing of information between the sensors and the Application is shown. The information received by both sensors will be received by the HC-05 module, which is connected to the android that contains the APPLICATION that we use. Now we can know the progress of the farm by the tests and the data. The HC-05 Bluetooth module allows you to switch between master and slave mode, which means you, could use it for both receiving and transmitting data. The HC-05 has a red LED that shows whether the Bluetooth is attached or not. This red LED blinks continuously in a periodic pattern before connecting to the HC-05 module. Its blinking slows down to two seconds when it connects to another Bluetooth system. This module operates at 3.3 volts shown in fig.5.

Specifications:

1.   Works with Serial communication (USART) and TTL compatible
2.   Follows IEEE 802.15.1 standardized protocol

3. Can operate in Master, Slave or Master/Slave mode
4. Supported baud rate: 9600,19200,38400,57600,115200,230400,460800.
5. Serial Bluetooth module for Arduino and also other microcontrollers
6. Operating Voltage: 4V to 6V (Typically +5V)
7. Uses Frequency-Hopping Spread spectrum (FHSS)
8. Operating Current: 30mA
9. It can be easily interfaced with Mobile phones or laptops with Bluetooth
10. Range: <100m

4.2 Relay:

Figure 4. HC Bluetooth

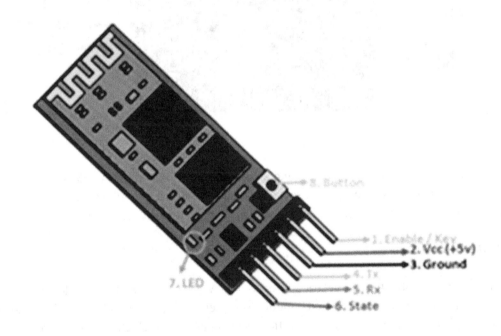

A relay is an electrically operated switch that can be switched on or off and can be regulated with low voltages, such as the 5V given by the Arduino pins, letting the current go through or not. It is as easy as manipulating any other output to power a relay module with the Arduino, as we can see later on. The benefits of a relay lie in its lower moving inertia, stability, long-term reliability and low volume. It is commonly used in power safety, automation technology, sports, remote control, reconnaissance and communication devices, as well as in electromechanical and electronic power devices. In general, a relay includes an induction component that can represent input variables such as current, voltage, strength, resistance, frequency, temperature, pressure, speed, light, etc. It also includes an actuator module (output) that can energize the managed circuit link or de-energize it. An electromagnet operates a power relay module, which is an electrical switch. A separate low-power signal from a microcontroller activates the electromagnet. The electromagnet pulls to open or close an electrical circuit when triggered. A simple

Figure 5. Relay board

relay is made up of a wire coil wrapped around a soft iron heart, known as A solenoid, an iron yoke, a movable iron armature, and one or more sets of contacts provide a low reluctance path for magnetic flux. The movable armature is hinged to the yoke and is attached to one or more sets of moving contacts.

4.3 Solenoid Valve

A solenoid valve is a valve that is electrically operated. A solenoid, which is an electric coil with a movable ferromagnetic core (plunger) in its middle, is present in the valve. In the remaining position, a small orifice seals off the plunger. A magnetic field is generated by an electric current through the coil. The magnetic field on the plunger that opens the orifice exerts an upward force. The basic theory used to open and close solenoid valves in fig.6.

4.4 Kodular Creator

First, we create an application layout. In the form of a circular ring, this app utilizes an extension plug in that visualizes the soil data. You can download the attached ".aia" file that includes all the features pre-made for that. Import it now to your maker, Kodular. In this app, we have added the following components that have been used to build a Smart Agriculture Device app layout.

Components

Circular bar extension

- Bluetooth Client
- List picker
- 4 Buttons
- 2 Text View

4.5 PIR Sensor

Figure 6. PIR Sensor

An electronic sensor that detects infrared light emitted by objects in its field of view is known as a passive infrared sensor. They're most commonly found in motion detectors with a PIR sensor. PIR sensors are widely used in security alarm systems and automatic lighting systems. In terms of technology, PIR is made up of a piezoelectric sensor that can detect various levels of infrared radiation. All, for example, emits varying amounts of radiation, with the dose of energy increasing as the object's temperature increases. PIR (Passive Infrared Detector) sensors are also known as PID (Passive Infrared Detectors). As a result, the PIR sensor is capable of detecting infrared radiation emitted by particles. In general, PIR sensors can detect animal/human movement within a specified range, which is determined by the sensor's spec. The detector does not emit energy; instead, it passively absorbs it and detects infrared radiation from the atmosphere. As compared to the other sensors, the PIR Sensor is very difficult. They have two slots, and the slots are made of a delicate material. The Fresnel lens is used to ensure that the PIR's two slots can see beyond a certain distance. When the sensor is turned off, both slots detect the same amount of infrared. The ambient amount comes from the outside, the walls, the bed, and other sources. As a consequence, the two bisects have a positive differential shift. The sensor, however, generates a negative differential shift between the two bisects when the body leaves the sensing field. The detection distances for indoor passive infrared range from 25 cm to 20 m. Indoor curtain-type detectors have a detection range of 25 cm to 20 m. The detection distance for outdoor passive infrared ranges are from 10 metres to 150 metres. Passive infrared curtain detector for outdoor use is with a range of 10 to 150 metres.

4.6 DHT 11 Sensor

Figure 7. DHT 11 Sensor

The DHT11 is a low-cost wireless temperature and humidity sensor. This sensor can easily be connected to any micro - controller (Arduino, Raspberry Pi, etc.) to calculate humidity and temperature in real time. The DHT11 humidity and temperature sensor comes in two versions: a sensor and a module. The pull-up resistor and a power-on LED distinguish this sensor from the module. A relative humidity sensor is the DHT11. This sensor uses a thermistor and a capacitive humidity sensor to test the ambient air. A capacitive humidity sensing module and a thermistor for temperature sensing make up the DHT11 sensor. A moisture-holding substrate serves as a dielectric between the two electrodes of the humidity sensing capacitor. With changes in humidity levels, the capacitance value changes. The IC measures, processes, and transforms the resistance values into digital form. This sensor uses a Negative Temperature Coefficient Thermistor to measure temperature, which causes the resistance value to decrease as the temperature rises. This sensor is normally made of semiconductor ceramics or polymers to get a higher resistance value even for the slightest change in temperature. DHT11 has a temperature range of 0 to 50 degrees Celsius and a 2-degree precision in Fig.7.

5. RESULTS AND DISCUSSION

In this paper we have successfully installed a kit which helps us to monitor the fields or gardens through mobile phone using Kodular app via Bluetooth shown in Fig8, Fig.9. . We have used soil moisture sensor, PIR sensor and DHT 11 sensor in order to monitor the soil moisture, detect the threat caused by animals to the field and monitor the temperature and humidity of the atmosphere respectively in order to produce better irrigation for the field and shown in Fig.10 and Fig.11.

Figure 8. The circuitry

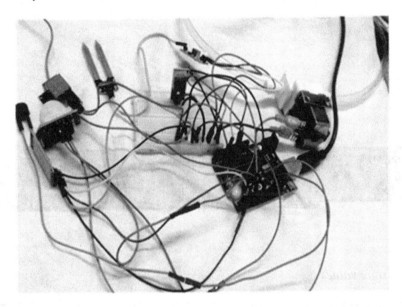

CONCLUSION

Smart farming is an evolving concept since IOT sensors are capable of providing agricultural field information and then acting on the basis of user feedback. The main factor in improving the yield of efficient crops is the control of environmental conditions. The feature of this paper includes the development of a device that can track the temperature, humidity, humidity and even the movement of animals that can kill crops in the agricultural field through sensors using the Arduino board and, in case of any discrepancy, send an SMS notification as well as a notification to the farmer's smart phone using Wi-Fi/3G/4G on the application developed for the same reason. In every country, agriculture is carried out from era. The science and art of growing plants is agriculture. The main development in the emergence of sedentary human civilization was agriculture. Agriculture is carried out manually from early on. IOT also plays a very important role in smart farming, as the world is moving towards new technology and implementations, it is an essential objective to trend with agriculture.IOT sensors are capable of providing agricultural field knowledge. We have suggested a framework for IOT and smart agriculture using automation.

Figure 9. Results screen shot

Figure 10. Soil moisture values

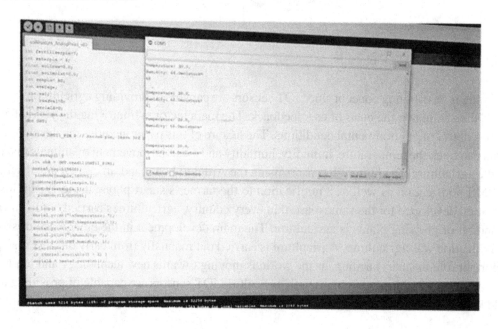

Figure 11. Graph representing different humidity

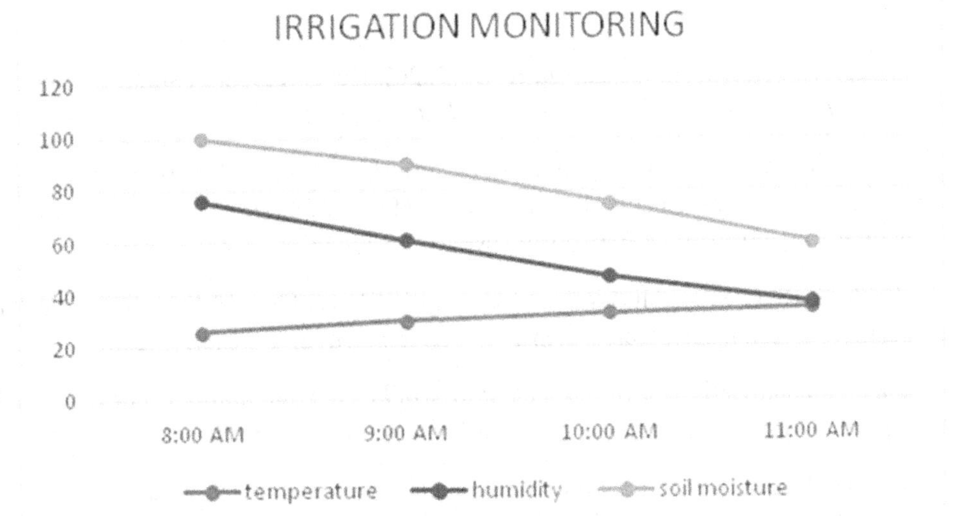

IRRIGATION MONITORING

FUTURE SCOPE

In future we are going to add wifi module instead of bluetooth so that we can be able to communicate over far distance .we can also place colour sensor that help to detect the ripe fruit or raw fruit based on its colour. We can also affix a camera to monitor the disease of the plant virtually and take necessary measures to protect plants. This IoT-based system helps farmers and gardeners to collect the data from plants with little to no effort and helps to keep plants in a healthy state and to produce a better yield. This kit helps to maintain different types of crops in good health. This project has the potential to be used in different areas and it can be updated, it is easy to use, build, and cost-effective.

REFERENCES

Anupkumar, T., Sharma, N., Srinath, A., & Dileep, G. S. (2019). Topology Optimization for Diffusive Heat Transfer in a Domain with Internal Heat Generation using Optimality Criteria. *International Journal of Vehicle Structures & Systems*, *11*(4), 447–451. doi:10.4273/ijvss.11.4.22

Bhavana, D., Kishore Kumar, K., & Bipin Chandra Medasani, S. K. B. (2021). Hand Sign Recognition using CNN. *International Journal of Performability Engineering*, *17*(3), 314–321. doi:10.23940/ijpe.21.03.p7.314321

Bhavana, D., Kumar, K. K., Kaushik, N., Lokesh, G., Harish, P., Mounisha, E., & Tej, D. R. (2020). Computer vision based classroom attendance management system-with speech output using LBPH algorithm. *International Journal of Speech Technology*, *23*(4), 779–787. doi:10.100710772-020-09739-2

Bhavana, D., Rajesh, V., & Kishore, P. V. V. (2016). A new pixel level image fusion method based on genetic algorithm. *Indian Journal of Science and Technology*, *9*(45), 1–8. doi:10.17485/ijst/2016/v9i45/76691

Chaitanya, P., Kotte, D., Srinath, A., & Kalyan, K. B. (2020). Development of smart pesticide spraying robot. *International Journal of Recent Technology and Engineering*.

Gutiérrez, J., Villa-Medina, J. F., Nieto-Garibay, A., & Porta-Gándara, M. Á. (2013). Automated irrigation system using a wireless sensor network and GPRS module. *IEEE Transactions on Instrumentation and Measurement*, *63*(1), 166–176. doi:10.1109/TIM.2013.2276487

Howell, T. A., Evett, S. R., O'Shaughnessy, S. A., Colaizzi, P. D., & Gowda, P. H. (2012). Advanced irrigation engineering: Precision and precise. *Journal of Agricultural Science and Technology A*, *2*(1A), 1.

Jyothi, Debanjan, & Anusha. (2020). *ASIC Implementation of Fixed-Point Iterative, Parallel, and Pipeline CORDIC Algorithm. In Soft Computing for Problem Solving* (pp. 341–351). Springer.

Jyothi, G. N., & Sriadibhatla, S. (2019). Asic implementation of low power, area efficient adaptive fir filter using pipelined da. In *Microelectronics, electromagnetics and telecommunications* (pp. 385–394). Springer. doi:10.1007/978-981-13-1906-8_40

Jyothi, G. N., & Sridevi, S. (2018). Low power, low area adaptive finite impulse response filter based on memory less distributed arithmetic. *Journal of Computational and Theoretical Nanoscience*, *15*(6-7), 2003–2008. doi:10.1166/jctn.2018.7397

Kekan, A. HKumar, B. R. (2019). Crack depth and crack location identification using artificial neural network. *Int. J. Mech. Product. Eng. Res. Develop*, *9*(2), 699–708. doi:10.24247/ijmperdapr201970

Kumar, G. N. S., & Srinath, A. (2020). Induction Motor for Pedestrian Transportation in Benz Circle Vijayawada. *Journal of Information Science and Engineering*, *36*(2).

Murali, G., Nagavamsi, V., Srinath, A., & Prakash, M. A. (2020). *Battery Thermal Management System Using Phase Change Material on Trapezoidal Battery Pack with Liquid Cooling System*. Academic Press.

Nandurkar, S. R., Thool, V. R., & Thool, R. C. (2014, February). Design and development of precision agriculture system using wireless sensor network. In *2014 First International Conference on Automation, Control, Energy and Systems (ACES)* (pp. 1-6). IEEE. 10.1109/ACES.2014.6808017

Prasad, P. I., Basha, S. G., & Janjanam, N. (2020). *Design, Fabrication and Evaluation of Thermal Performance of Parabolic Trough Collector with Elliptical Absorber Using Zn/H2O Nanofluid*. Academic Press.

Ram Prasad, A. V. S., Ramji, K., Kolli, M., & Vamsi Krishna, G. (2019). Multi-Response Optimization of Machining Process Parameters for Wire Electrical Discharge Machining of Lead-Induced Ti-6Al-4V Alloy Using AHP–TOPSIS Method. *Journal of Advanced Manufacturing Systems*, *18*(02), 213–236. doi:10.1142/S0219686719500112

Rao, P. K. VKumar, B. RSaiteja, ASrikar, NSreenivasulu, V. (2018). Experimental Investigation of Thermal stability of carbon nanotubes reinforced Aluminium matrix using TGA-DSC Analysis. *International Journal of Mechanical and Production Engineering Research and Development*, 8(3), 161–168. doi:10.24247/ijmperdjun201818

Sepaskhah, A. R., & Ahmadi, S. H. (2012). A review on partial root-zone drying irrigation. *International Journal of Plant Production*, 4(4), 241–258.

Venkatarao, K., & Anup Kumar, T. (2019). An experimental parametric analysis on performance characteristics in wire electric discharge machining of Inconel 718. *Proceedings of the Institution of Mechanical Engineers. Part C, Journal of Mechanical Engineering Science*, 233(14), 4836–4849. doi:10.1177/0954406219840677

Vishnu, A. V., Kumar, P. J., & Ramana, M. V. (2018). Comparison among Dry, Flooded and MQL Conditions in Machining of EN 353 Steel Alloys-An Experimental Investigation. *Materials Today: Proceedings*, 5(11), 24954–24962. doi:10.1016/j.matpr.2018.10.296

Chapter 5
Mobile Geo–Fencing Triggers for Alerting Entries Into COVID–19 Containment Zones Using IoT

M. V. Ramana Rao
Osmania University, India

Thondepu Adilakshmi
Vasavi College of Engineering, India

M. Gokul Venkatesh
Sidhartha Medical College, India

Jothikumar R
iD https://orcid.org/0000-0003-0806-7368
Department of Computer Science and Engineering, Shadan College of Engineering and Technology, India

ABSTRACT

In a thickly populated nation like India, it is hard to forecast community transmission of COVID-19. Hence, a number of containment zones had been recognized all over the country separated into red, orange, and green zones, individually. People are restricted to move into these containment zones. This chapter focuses on informing the public about the containment zone when they are in travel and also sends an alert to the police when a person enters the containment zone without permission using the containment zone alert system. This chapter suggests a containment zone alert system by means of geo-fencing technology to identify the movement of public, deliver info about the danger to the public in travel and also send an alert to the police when there is an entry or exit detected in the containment zone by the use of location-based services (LBS). By creating a fence virtually called geo-fence at the containment zones established based on the government info, this system monitors public movements like entry and exit to fence.

DOI: 10.4018/978-1-7998-6870-5.ch005

I. INTRODUCTION

THE COVID-19, an abbreviation for "Coronavirus Disease-2019", is a breathing disease brought about by the serious intense lung disorder coronavirus-2 (SARS-CoV-2), an infectious infection having a place with a group of single-abandoned, positive-sense RNA infections known as coronavirus. Much like the flu infection, SARS-CoV-2 assaults the respiratory framework, causing infirmities, for example, hack, fever, fatigue, and shortness of breath. While the specific wellspring of the infection is obscure, researchers have planned the genome grouping of the SARS-CoV-2, which normally determines its quality sources from bats and rodents Cascella, M., (2021). The COVID-19 was first answered to influence human life in Wuhan City, in the Hubei territory of China in December 2019. Since at that point, the COVID-19 has fanned out quickly all through the rest of the world, denoting its quality in 213 nations furthermore, free domains. As per the WHO, the current worldwide tally1 of affirmed coronavirus cases remains at 2,285,210 while the loss of life has arrived at 155,124. The quick ascent in the quantity of COVID-19 occurrences around the world has provoked the requirement for guaranteed countermeasures to check the terrible impacts of the COVID-19 flare-up. Steffens, I. (2020) The later center east breathing condition coronavirus (MERS-CoV) contamination of 2012. The SARS-CoV flare-up started in the Guandong area of China and later spread to more than 37 nations around the world, causing over 8000 diseases and around 774 passings Singhal, T. (2020). COVID-19 shows with scientific highlights going from the asymptomatic express (no indications) to intense respiratory trouble disorder (ARDS) and various organ brokenness disorder (MODS). As per the aftereffects of an ongoing report led by the WHO in a joint effort with China, of the 55,924 labs affirmed COVID-19 cases that remained analyzed, a larger part displayed clinical attributes such as fever, dry hack, weariness, and sputum creation. With this just a bunch of patients displayed indications, for example, sore throat, cerebral pain, myalgia, and windedness, while indications, for example, sickness, nasal blockage, hemoptysis, looseness of the bowels, and conjunctival blockage were found to be exceptionally uncommon Chamola, V., et al (2020).

In a thickly peopled nation like India, it is extremely hard to forestall the public transmission in any event, during lockdown without social mindfulness and prudent steps occupied by the individuals. As of late, a few regulation zones had been distinguished all through the nation and partitioned into red, orange, and green zones, separately. The red zones show the disease hotspots, orange zones mean certain contamination, and green zones demonstrate a territory with no contamination Mallik, R., et al (2020). Containment zones were significant for distinguishing pockets that need basic intercessions for centered administration of COVID-19. In urban zones, regardless of whether there remained a solitary or under five COVID-19 positive cases, the road ought to be divided as a regulation zone, the request said. On the off chance that the road is long, a sensible bouncemight be carried under the control zone to limit development. In-country zones, "the residence" of the villa was to be differentiated as a regulation zone. In urban regions, if the bunch (at least five COVID-19 positive cases) fell inside the district, the district wherein the living arrangements of the positive cases were found ought to be delineated as the regulation zones. "If the home occurs to be the outskirt of the region, at that point the abutting regions may likewise be remembered for the regulation zone," the rule said. In-country regions, the entire town or the gathering of towns anywhere the bunching of cases was accounted for ought to be delineated as regulation zone/zones. The rundown would be reconsidered on a week after week premise or prior. In light of the field criticism and extra investigation at State-level, These States may assign extra red or orange zones as proper.

However these containment zones are monitored by Police, immobile there stays an opportunity that individuals may accidentally pace into these zones. In this circumstance where individuals can travel in the city, these regulation zones represent a danger of contagion to these city tenants. Accordingly, advising individuals about the area regarding the containment zones can assist them with bypassing and stay away from these zones and in this manner lessen the chances of community transmission. Also when individuals wanted to step into these zones or when one from the zones step outside these zones it is mandatory to inform the police. Hence this Chapter proposes a Containment Zone Alert Structureusing geofencing technology to provide information or alert notification regarding the containment zones if the user enters these containment zones. Also, this system provides the alert notification to cops when an individual enters this zone or an individual from the containment zone moves outside the range. Geofencing customer is utilized to make geofences round the containment zones and the alert is sent through the notification manager who utilizes to give warnings. Section II is explained about Related work. In Section III the proposed methodology based on Geofencing technology is explained. In Section IVthe architecture of the Containment Zone Alert System is reported. In Section VSystem Processing flow is delivered. In Section VI results and Section, VII the conclusion has been delivered.

II RELATED WORK

The expression "geofencing" is utilized from aboutyear 2000. This showed up in investigation writing by Munson and Gupta (2002). Geofencing is the center advancements for area-based administrations including promoting, following, what's more, chance administration. Szczytowski, P. (2014) proposed a methodology because of joining geofencing with long-range interpersonal communication frameworks (SNS) to compose unstructured data gathered from SNS. Yelne, S., & Kapade, V. (2015) structured assistance is a application which is installedand running on an android working framework dependent on geofencing. Discovery exactness with force utilization are significant for geofencing applications. Nakagawa, T., et al. (2013) proposed a strategy for location recognition and their initiation recurrence is dictated through speed near the objective spot. Alsaqer, M., et al. (2015) researched exactness and Energy-utilization of Esri's geo-trigger assistance now little, outside, geo-fenced territories. Prabhu, J., et al., (2020)IoT role in prevention of COVID-19 and health care workforces behavioural intention in India-an empirical examination.Josephine, M. S., et al (2020) proposed the concept of monitoring and sensing COVID-19 symptoms as a precaution using electronic wearable devices.Priya, K. Banu, et al.,(2020) has discussed about Pediatric and geriatric immunity network mobile computational model for COVID-19

III THE PROPOSED METHODOLOGY

a)Geofencing

Geofencing is area-based assistance, organizations use to connect with their crowd by sending applicable messages to smartphone users who enter a pre-characterized area or geographic territory. There are smart companies that send item offers or explicit advancements to purchasers' smartphones when they trigger a hunt in a specific geographic area, enter a shopping center, neighborhood, or store. Geofencing is a scheme that creates a virtual fence in a particular zone. The requestcollects a geofence in a hazardous region and provides chance data to the user. In request to characterize a fence and facilitate (latitude and longitude) of a particular spot are prerequisite. A geofence zone is characterized by the organizing and range as shown in figure 1. A geofence is set with the region Developers, A. (2016).The framework doesn't providethe occasion until the user moves into the area more from the limit in addition to a framework characterized region separation. This pad esteem forestalls the framework to produce various occasions while the client is heading out near the limit. The region separation is controlled by the equipment and the area advancements that are at present accessible Suyama, A., & Inoue, U. (2016).Geo-fencing empowered far off checking of geographic territories encompassed by a virtual fence (geofence), and programmed location when it followed portable articles that entered or left these regions Abbas, A. H., et al (2019).Geofencing, what's more, foundation following in this way empowers an expansive scope of new applications, particularly in the zone of data importance. Notwithstanding, their presentation is likewise related to new obstructions, for instance, the battery utilization at cell phones and security worries of the user Küpper, A., (2011).

Figure 1. Geofence technology

b) Containment Zone Alert System

The Tamil Nadu government has informed a rundown of 711 geological areas across 36 regions as COVID-19 "containment zones" to control the spread of the pandemic in the State. "On the off chance that there are more than five [COVID-19] positive cases in a territory then it will be measured as groups," a rule given here by the administration assumed. Containment zones were significant for distinguishing pockets that need basic intercessions for centered administration of COVID-19. In our proposed Containment Zone Alert System, the virtual perimeter is created for the containment zone as geographical area, a geofence is dynamically generatedacross the containment zone by selecting a set of boundaries calculating the radius around the point of location of the containment zone. These sets of boundaries calculated with a radius around a point location will be predefined. To define a fence across the containment zone, the coordinate around the zone is required like (latitude and longitude). A round territory is measured by the coordinate and radius. A containment zone is measured as a circular area and it is set to geofence.

Figure 2. Containment Zone Alert System

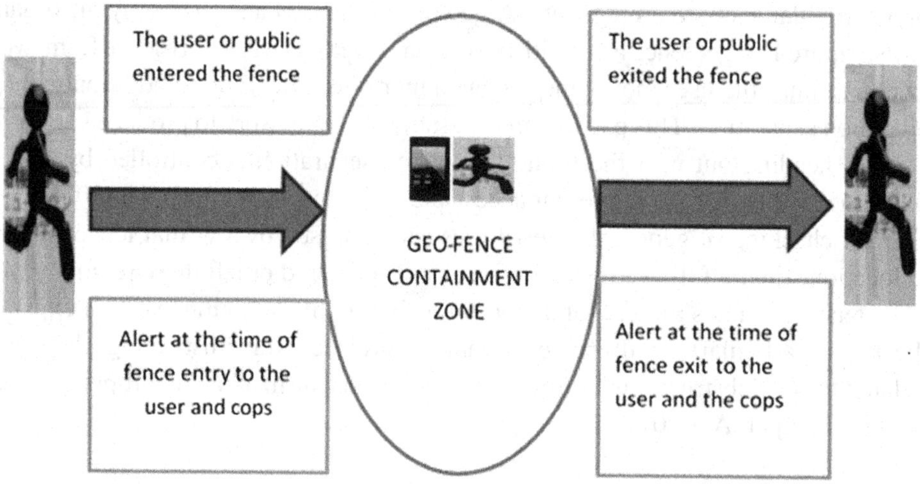

The geofencing created around the containment zone will involve location-aware devices of a location-based service(LBS) which monitors the user entering or exiting the containment zone as shown in figure 2. When the user is entering the fence the alert system will trigger an alert to the device of the user and data is sent to the mobile and telephone about the location that it is the containment zone. So that user can restrict himself getting into the containment zone. This is mostly used for persons who travel for business or work. Also when the public is moving into the containment zone without permission, the alert notification about the mobile phone entered into the zone will be sent to the police who are the geo-fence operator. Using geofencing for containment zone the alert notification will be sent to the cop's mobile phone or email account when a person moves out from the containment zone or exited from the containment zone without permission.

IV SYSTEM ARCHITECTURE

The framework utilizing geofencing is conceivable to convey the data about the containment zone to the consumer, who just entered the fence and also about the person's mobile information to the cops who entered or exited the fence. In this assessment, we execute geofencing with the mainposition structure of iOS. The system gives an identification of the passages and ways out of the consumer with the perception of a particular geographical district. The topographical district is a territory characterized with hover including predetermined sweep nearby a recognized position on the earth. Each time when the user or public moves out of the limit given to the zone, the framework produces an occasion for our request. This allows the warning of the containment zone info to both users and the cops. That is, by utilizing the perceptions of the topographical containment zone. It is possible to perceive the entry and exit movements of user and other people.

The System Design is shown in Figure 3, Which is composed of the client-side, server-side, and the data sources. On the client-side, all client is an application program installed on ios. IOS is an operating system of smartphones used by geofence users and the cops involved in the containment zones. It

Figure 3. System Design

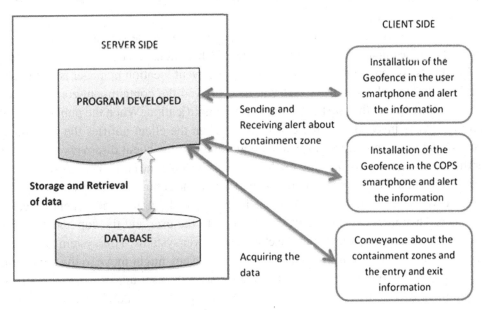

associates with the Internet, which acquires the data from the database of server side. It characterizes a geofence dependent on data from the database server and alerts the user about the containment zone, also notifies the cops about the person entering into and exiting from the containment zone. The client is executed by utilizing the program developed, and tried by the iOS test system and genuine smartphone. The server procures containment zone data from data sources by running the web application on Linux. It dissects the data and saves the conclusion in a database. The database is utilized to characterize a containment zone by the client. The info about the containment zone is provided by the government based on COVID-19 infections. This information is updated regularly by the government.

V SYSTEM PROCESSING FLOW

As an illustration, Consider that the containment zones have been increased due to the spread of novel COVID-19 in different areas. As a result of this, a warning is issued to the areas which are declared as containment zones that entry and exit are restricted. Users mostly the business persons who are in travel might enter this containment zones unknowingly. Hence the Containment Zone Alert System will be used by the user, This system will alert the user when the user is about to enter the containment zone. When a person enters the containment zone wanted an alert notification will be sent to the cop. Also When a person exits from the containment zone an alert warning will be directed to the cop. For this, the servers program acquires the information about the containment zone through the internet or by the updations given by the government about the zone. After obtaining the information about containment zones, they are stored in the database. The data are updated in the server periodically whenever needed and the client access the server periodically to monitor the info about the updations. When a client is requesting the information from the server, the server program retrieves the information from the database based on the client request and server gives the result about the containment zone which defines a fence by calculating the set of boundaries.

The depiction of the containment zone that is fence is made on the server-side and the client collects the fence. Then the client starts observing the user, who enters and exits the containment zone by invoking the methods. When the user arrives the fence the event location manager is appealed with the entry method, so that the client notifies that you have arrived the containment zone. When a person enters the fence, aclient alerts the cop by sending the alert notification. When the public exited the fence the location manager is invoked with the exit method. Then the client notifies the cop about the exit information. When there is no information about entry and exit the client discovers no notice from the server after this it calls the stop monitoring method to stop checking the containment zone about the user entry and exit. Figure 4 shows the system processing flow of the containment alert system, which starts with a flow of server which acquire and update the information about the containment zone, define and register the fence. Then alert by sending the notification to the user and the cop about the movement information of the containment zone using the client and server systems. This system is mostly used for persons who travel for business purposes in different areas. They might move to the containment zone without knowing information about it, Using this alert system the user may restrict themselves moving into the containment zone. This system is also helpful to the police to identify the persons moving into the containment zone without permission.

VI RESULT

By expending our proposed system, the information delivery about the containment zone can be confirmed using geofencing. A user will be carrying a real smartphone and a fence is defined. The fence was installed in the containment zone based on the size of the containment zone. The fence was determined for the containment zone were the COVID-19 positive cases appear.The range of the fence is measured from the center of the containment zone . The measurement and the alert system will be within this range. As shown in Figure 5 the user or the public may enter the fence, the person from the containment zone may exit the fence. In this case, the WiFi should be on in the smartphone of both the user and cops. In case the WiFi is on, the user will get the information about the containment zone when user is outside the fence and at the position of 25m-35m away the fence.

When a person is exited from the fence the cops will receive the information about the phone when the exit position is at a distance of higher than 100 m calculated from the fence. Also when a person enters the fence the cops will be alerted at the position of the person,calculated the radiusof the person aroundthe fence is 50m. As shown in Figure 6 the notification about the fence entry and exit is sent through the mobile when WiFi is ON. All these are done using geofencing technology.

VII CONCLUSION

By disbursing our proposed Containment Zone Alert System, the user and the cops will get the information and the notification alert about the person's movement in the containment zone. Due to the increase of COVID-19 in different areas containment zones are sealed and no one is allowed to enter or exit from those zones, In this pandemic situation, it is mandatory to restrict people moving into the containment

Figure 4. System Processing Flow

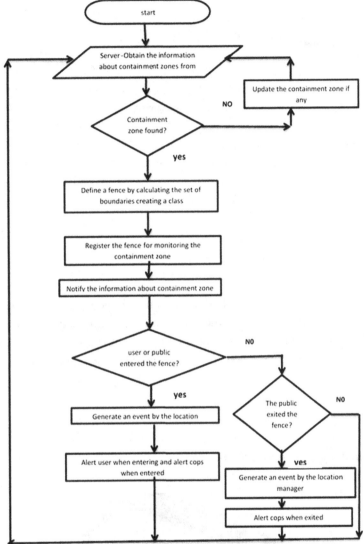

zones to control the spread of the disease. Hence we proposed a Containment Zone Alert System by means of geofencing that can be estimated based on the system in the containment zone. This Chapterrecommended that this system reports that this is containment zone when the consumer arrives the fence, WiFi should be ON when user is about to enter the zone. When the user is 20-30m away from the fence which is containment zone, users mobile will get a notification alert about the containment zone. So that user may restrict himself entering into the containment zone. When a person exits outside the containment zone information is delivered to the cop around when mobile is 100m outside the fence. Also when a person enters the containment zone his mobile information will be sent to the cops when he is around the radius using the Geofencing technology. Hence each containment zone can be created as a fence or virtual geographical area to restrict the movements of the public to break the chain of the spread of this pandemic disease. For a large scale of containment zones, the fence can be created with several

Figure 5. Notice of the movement in the containment zone

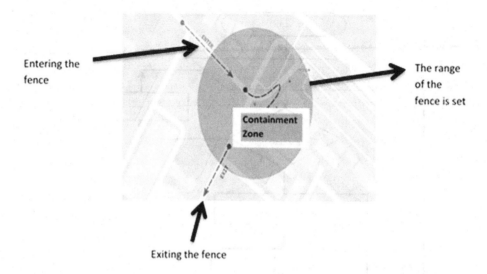

kilometers of length. In future work, the method to evaluate a large number of containment zones with increased length size is mandatory to restrict the people moving into the containment zones.Our future system will be based on the containment zone which will be able to describe and collect several zones as fences, at the same time to support cops to restrict the movement of the public in these containment zones.

Figure 6. Notification about the fence in the mobile

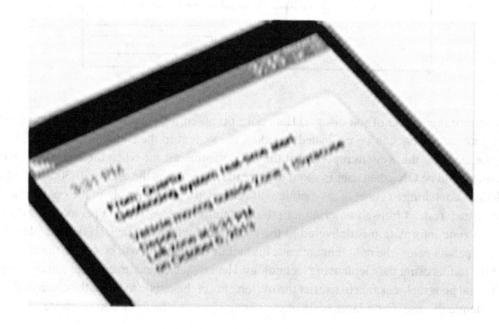

REFERENCES

Abbas, A. H., Habelalmateen, M. I., Jurdi, S., Audah, L., & Alduais, N. A. M. (2019, November). GPS based location monitoring system with geo-fencing capabilities. In. AIP Conference Proceedings: Vol. 2173. *No. 1* (p. 020014). AIP Publishing LLC. doi:10.1063/1.5133929

Alsaqer, M., Hilton, B., Horan, T., & Aboulola, O. (2015). Performance assessment of geo-triggering in small geo-fences: Accuracy, reliability, and battery drain in different tracking profiles and trigger directions. *Procedia Engineering*, *107*, 337–348. doi:10.1016/j.proeng.2015.06.090

Cascella, M., Rajnik, M., Aleem, A., Dulebohn, S., & Di Napoli, R. (2021). *Features, evaluation, and treatment of coronavirus (COVID-19)*. StatPearls.

Chamola, V., Hassija, V., Gupta, V., & Guizani, M. (2020). A comprehensive review of the COVID-19 pandemic and the role of IoT, drones, AI, blockchain, and 5G in managing its impact. *IEEE Access: Practical Innovations, Open Solutions*, *8*, 90225–90265. doi:10.1109/ACCESS.2020.2992341

Developers, A. (2016). *Creating and monitoring geofences*. Academic Press.

Josephine, M. S., Lakshmanan, L., Nair, R. R., Visu, P., Ganesan, R., & Jothikumar, R. (2020). Monitoring and sensing COVID-19 symptoms as a precaution using electronic wearable devices. *International Journal of Pervasive Computing and Communications*.

Küpper, A., Bareth, U., & Freese, B. (2011, October). Geofencing and background tracking–the next features in LBSs. *Proceedings of the 41th Annual Conference of the Gesellschaft für Informatik eV*.

Mallik, R., Hazarika, A. P., Dastidar, S. G., Sing, D., & Bandyopadhyay, R. (2020). Development of an android application for viewing covid-19 containment zones and monitoring violators who are trespassing into it using firebase and geofencing. *Transactions of the Indian National Academy of Engineering*, *5*(2), 163–179. doi:10.100741403-020-00137-3

Munson, J. P., & Gupta, V. K. (2002, September). Location-based notification as a general-purpose service. In *Proceedings of the 2nd international workshop on Mobile commerce* (pp. 40-44). 10.1145/570705.570713

Nakagawa, T., Yamada, W., Doi, C., Inamura, H., Ohta, K., Suzuki, M., & Morikawa, H. (2013, September). Variable interval positioning method for smartphone-based power-saving geofencing. In *2013 IEEE 24th Annual International Symposium on Personal, Indoor, and Mobile Radio Communications (PIMRC)* (pp. 3482-3486). IEEE. 10.1109/PIMRC.2013.6666751

Prabhu, J., Kumar, P. J., Manivannan, S. S., Rajendran, S., Kumar, K. R., Susi, S., & Jothikumar, R. (2020). IoT role in prevention of COVID-19 and health care workforces behavioural intention in India-an empirical examination. *International Journal of Pervasive Computing and Communications*.

Priya, K. B., Rajendran, P., Kumar, S., Prabhu, J., Rajendran, S., Kumar, P. J., ... Jothikumar, R. (2020). Pediatric and geriatric immunity network mobile computational model for COVID-19. *International Journal of Pervasive Computing and Communications*.

Singhal, T. (2020). A review of coronavirus disease-2019 (COVID-19). *Indian Journal of Pediatrics*, *87*(4), 281–286. doi:10.100712098-020-03263-6 PMID:32166607

Steffens, I. (2020). A hundred days into the coronavirus disease (COVID-19) pandemic. *Eurosurveillance*, 25(14), 2000550. doi:10.2807/1560-7917.ES.2020.25.14.2000550 PMID:32290905

Suyama, A., & Inoue, U. (2016, June). Using geofencing for a disaster information system. In *2016 IEEE/ACIS 15th International Conference on Computer and Information Science (ICIS)* (pp. 1-5). IEEE. 10.1109/ICIS.2016.7550849

Szczytowski, P. (2014). Geo-fencing based disaster management service. In *Agent technology for intelligent mobile services and smart societies* (pp. 11–21). Springer.

Yelne, S., & Kapade, V. (2015). Human protection with the disaster management using an android application. *International Journal of Scientific Research in Science, Engineering and Technology*, 1(5), 15–19.

Chapter 6
Fine Tuning Smart Manufacturing Enterprise Systems:
A Perspective of Internet of Things–Based Service–Oriented Architecture

Senthil Murugan Nagarajan
ⓘ https://orcid.org/0000-0001-9284-7724
VIT-AP University, India

Muthukumaran V.
ⓘ https://orcid.org/0000-0002-3393-5596
REVA University, India

Vinoth Kumar V.
MVJ College of Engineering, India

Beschi I. S.
St. Joseph's College, India

S. Magesh
ⓘ https://orcid.org/0000-0003-2876-7337
Maruthi Technocrat Services, India

ABSTRACT

The workflow between business and manufacturing system level is changing leading to delay in exploring the context of innovative ideas and solutions. Smart manufacturing systems progress rapid growth in integrating the operational capabilities of networking functionality and communication services with cloud-based enterprise architectures through runtime environment. Fine tuning aims to process intelligent

DOI: 10.4018/978-1-7998-6870-5.ch006

management, flexible monitoring, dynamic network services using internet of things (IoT)-based service oriented architecture (SOA) solutions in numerous enterprise systems. SOA is an architectural pattern for building software business systems based on loosely coupled enterprise infrastructure services and components. The IoT-based SOA enterprise systems incorporate data elicitation, integrating agile methodologies, orchestrate underlying black-box services by promoting growth in manufacturer enterprises workflow. This chapter proposes the integration of standard workflow model between business system level and manufacturing production level with an IoT-enabled SOA framework.

1 INTRODUCTION

The sustainable competitiveness of a manufacturing industry mainly depends on its ability such as delivery, quality, flexibility, and cost. These capabilities are maximized by attempting Smart Manufacturing System (SMS) using the most advance technologies and it recommends widespread and rapid flow

Figure 1. Overview of Smart Manufacturing System

usage of digital information between the manufacturing systems (Tao & Qi, 2017) (Davis et al., 2012). Extraordinary production agility, efficiency, and quality are driven by SMS across various companies and factories to improve competitiveness for a long-term. Particularly, communication and information technology is used by SMS along with the intelligent-software applications such as optimizing the use materials, energy, and labour for high quality and customized products for delivering on-time. Furthermore, the SMS respond to supply chains and market demands quickly (Elhoseny et al., 2016) (Reis & Gonçalves, 2018). Figure 1 shows the overview of smart manufacturing system for the recent technology.

Some of the key capabilities such as quality, sustainability, productivity, and agility are included in smart manufacturing systems. The agile manufacturing succeeded by enabling some technologies like supply chain integration, distributed intelligence, and model based engineering. The measure of agility is done using some metrics such as time to make changeovers, cycle time, change order per year, new product rate, and percentage of on-time delivery (Mohammadi & Mukhtar, 2018). The quality is measured based on some metrics like returns or rejects per year, material authorizations per year, and yield for the product customization and innovation. The production output ratio is defined to be productivity process in which it typically measured using size and number of products taken or improved. The impacts related to sustainability increased for smart manufacturing systems due to agility and productivity is increasing and a better understanding and control is much needed (Xu et al., 2018) (Theorin et al., 2017).

The next generation of internet of things (IoT) in industry is the biggest potential to have lot of benefits which breakdown the farming of enterprise. This technology allowed to integrate enterprise systems, supply chain management (SCM) systems, customer relationship management (CRP) systems, product lifecycle management (PLM) systems, and production systems (Maksuti et al., 2019). Figure 2 shows the overview of IoT technology integrated with enterprise systems.

Figure 2. Overview of Smart Manufacturing Enterprise System

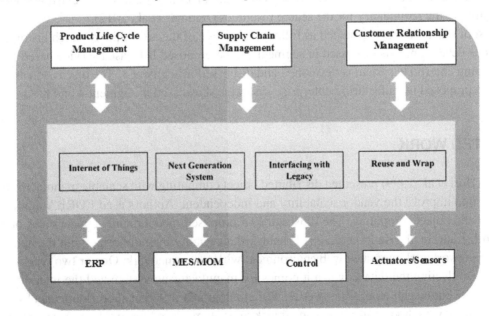

Nowadays, the systems based on smart technology are independently managed with each other in which the enterprises holistic view is prohibited. This holistic approach believed to obtaining high efficiency enormously by 26% when it comes for enterprise systems. When using the smart enterprise control, it is not meant to replace with new complete system by removing current automation systems. However, it denotes the connection between enterprise and current automation system (Leyh et al., 2016) (Ordanini & Pasini, 2008). So for this reason, entire manufacturing enterprise system is optimized and enables business control with a high degree. Enterprise will be more efficient when there is a tight integration and also makes more profitable due to volatile market condition as it gives responsiveness and greater flexibility (Borangiu et al., 2015).

In this modern society, Service Oriented Architecture (SOA) concept have become an enormous development with significant traction. The origination of SOA is deliberately defined from Information and Communication Technologies (ICT) that had major impact and this technique can be transposed. The communications based on SOA are asynchronous in nature. Various additional technologies, standards, products, best practices, and capabilities are emerging based on SOA (Lewis & Smith, 2008). The early usage of SOA is fastened by some architects which can lead to misapply, abuse, overuse, and misname the concepts that indulge in project risk. The impacts based on the emerging technologies need to understood by the researchers and lift up the balanced considerations when evaluating the SOA architecture. Based on the technical capabilities and risk posture of organization, it will critical when analysing these considerations (Valipour et al., 2009).

New requirements have brought by IoT which substantially calls various approaches for the improvement of traditional SOA. When it comes to discovery, the present challenge is about dealing with enormous things which produce large amount of data on interest where particularly real-world measurements are collected from sensor devices (Bhadoria et al., 2018). From old SOA, a specific request is fulfilled by selecting it finally even if there are several services registered. The discovery of IoT leads to potential selection of huge number of things the will provide combination of data for the existing query which helps in limiting the redundancy for the scarce resources (Jammes et al., 2012).

The rest of this article is segmented as follows: The state of the art for the research history on SOA with various development is discussed in segment 2. The proposed IoT based service based fine-tuned manufacturing enterprise system is developed and discussed in segment 3. The discussion and conclusion for the proposed manufacturing enterprise system is showcased in segment 4 and 5.

2. RELATED WORK

Hori et al. (Hori et al., 1999) presented distributed object computing with scalable manufacturing enterprise system to improve the vendor scalability and independent. Authors used CORBA model for data integration and object computing. The illustration of proposed model mainly concentrates on production control. Kadri et al. (Kadri et al., 2013) analysed air quality by monitoring with the help of wireless sensor network and integrated web environment with sensor network. One or two types of sensors are used for collecting the data as when it comes for manufacturing system and the integration which is needed to it. Fumangali et al. (Zhong et al., 2013) integrated different shop floors data by using the means of ontology method. A use-case named picking system is used for ontology for the modelling of manufacturing systems.

The production order and RFID data is used by Zhong et al. (Rolón & Martínez, 2012) to derive schedule of appropriate operation. The modification of this schedule is done based on the RFID data to find whether the worker is operating the machine or not which could make changes in original schedule. Agent based modelling and efficient scheduling simulation is suggested by Rolon et al. (Chen et al., 2008). Authors used resource-agent, order-agent and communication for this modelling. Each and every agent monitor and plan accordingly for executing the cycle. Intelligent manufacturing enterprise system is proposed by Chen et al. (Wang et al., 2014) by integrating data warehouse, data mining, and OLAP (On-Line Analytical Processing) with MES for the information which are useful from the collected data. This proposed model by authors only describes the delivery of information extraction for the enterprise information system. Also, authors have applied the context aware computing for the development of MES.

Authors in (Zhang et al., 2004) proposed a system for managing the IoT environments about the consequent changes and complexity. Authors have proposed a model by incorporating accompanying developments for accomplishing the objective and it mainly includes strong, expandable, dependable, object-oriented, and adaptable for encouraging the reuse and interface of the framework segments. Authors in (Zhu et al., 2019) discussed the sharing and collection information of industrial IoT for enabling more dynamic application in the field of manufacturing to various elements and advance planning and scheduling. Authors also discussed the paradigms of SOA which is a suitable architecture for smart manufacturing enterprise system. The authors shown the reason behind using the SOA is that its low costs for the maintenance stage of the architecture.

Perry and Lycett 2003 (Park et al., 2019) discussed the design of SOA for running the generated data from heterogeneous and distributed elements which requires the review of internal organization for providing integrated solutions. Ramollari, Dranidis, and Simons 2007 (Ramollari et al., 2007) analyzed that massive distributed applications are enabled by SOA to be flexible and adaptive through connections with 'loosely-coupled' for achieving extensive integration, management, processing, and development. Hossain and Muhammad 2016 (Hossain & Muhammad, 2016) analyzed the centralized coordination element and applied in two architectures and used cost and time that are reduced and provide more effective services for the deployment of SOA and its maintenance stage. Komoda 2006 (Komoda, 2006) defined SOA as a design framework for constructing computational systems by combining some services. The unified definition is difficulty to find for these services where it can be defined as abstract elements for providing capabilities based on the needs of users.

Yaqoob et al. 2017 (Yaqoob et al., 2017) analyzed the process of DFS that it will be much slow for developing the energy efficient machines which shorten the considered characteristics of each thread present in smart manufacturing enterprise. In addition, dyeing a product or repeated-dyeing again analyzing the discovering defects which can occur between 15 to 20% time which can cause problem of decrease in product quality and increase in energy consumption. Mcgovern et al. (McGovern et al., 2004) introduced service oriented architecture which formally separate the services in an architectural style. This introduced system has the functionality which can provide service to consumers and accomplished the mechanism known to be service contract.

According to (De Deugd et al., 2006) reliance investigation, inclusion examination, interface investigation, heterogeneity investigation, intricacy examination, consistence investigation, cost and advantage investigation have been depicted as ways how EA can uphold IT/business arrangement, business progression arranging, security the executives, innovation hazard the board, venture portfolio arranging, business measure overhaul, quality and consistence the executives, post consolidation coordination, the presentation of business of-the-rack programming, sourcing choices, IT administration and IT tasks the executives just as IT union. According to (Gholami et al., 2010) Genuine associations, be that as it may, are described by an enormous number and a wide assortment of curios. To diminish intricacy and structure the turn of events/advancement measure, "engineering" layers have been presented. As per ANSI/IEEE Sexually transmitted disease 1471-2000, engineering is characterized as the "crucial association of a framework, encapsulated in its parts, their connections to one another and the climate, and the standards overseeing its plan and advancement".

According to (Papazoglou & Van Den Heuvel, 2006) Coordination alludes to the meaning of a certain grouping of administration calls between at least two inaccessible administrations. Administration Orientation too clings to the guideline of cooperation. Entirety cross-hierarchical business cycles can be arrangement based on SOAs. Specialized spryness is accomplished because of the free coupling of various administrations. SOAs are expressly planned for change and empower its clients to fabricate or adjust applications in a small amount of the time required when utilizing customary ideas of programming plan. Last, decentralized possession also, control is key to the SOA theory. The epitome of uses with the assistance of uniform interfaces permits proprietors of the various administrations to change inner activities as indicated by their souls' longing while at the same time maintaining a strategic distance from any interference.

According to (Muthukumaran & Manimozhi, 2021; Deverajan et al., ; Muthukumaran & Ezhilmaran, 2020; Nagarajan et al., 2021) Enterprise Architecture (EA) is a term utilized for encouraging the incorporation of procedure, business, data frameworks and innovation towards a shared objective and dominating authoritative multifaceted nature through the turn of events what's more, use of building depictions. Enterprise Engineering (EA) has created to bring the data framework plan and business prerequisites together. EA examinations an association right from its nonexclusive key segments to its nitty gritty IT foundation. Subsequently, EA is more than engineering since it includes administration as well as a guide for adjusting IT ventures to business needs. According to (Muthukumaran, 2021) The idea of SOA has instigated EA methodological changes. The mix of SOA and EA presenting the idea of Service-arranged Enterprise Architecture (SOEA) which has features their synergic relationship. This new approach permits EA and SOA to finish each other for better help of coordinated business needs.

3 PROPOSED WORK

This section presents the fine tuning smart business and manufacturing level concepts which depict the workflow standards by sharing the resources between the enterprise levels to improve the growth using IoT-based SOA standards.

3.1 Traditional Smart Manufacturing System Using IoT

Enabling IoT technology using the new industry 4.0 standard revolutionized organizing the production processes via internet platforms. Smart manufacturing systems are connected with cloud-based architectures through the internet as a mode of communication. The flow of inputs from machineries and shipping details are send to stakeholders via cloud infrastructure. This enables the information's need to synchronize in all the stages of business manufacturing system which makes the production to go in order by collecting information in real-time.

Further managing the internet and manufacturing systems may also lead to internet faults or device damages when not proper protection is installed with equipment. This has to be taken care by the manufacturing business system analyst. The Figure 3 describes the traditional and conventional smart manufacturing system.

Figure 3. From traditional to conventional smart manufacturing systems

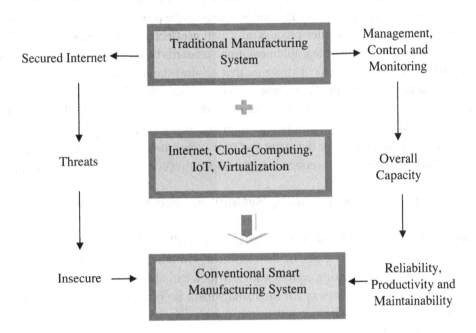

Networks, procedures and activities that are inadequately secured pose huge risks to business systems. They are vulnerable to a number of attacks and challenged by intrusion, denial of controls of systems, loss of data sensitive changes. They can remotely control devices or manufacturing processes once attackers gain access to a sensitive programme. These problems need to be efficiently taken care by IoT technologies providing secured services with internet.

Figure 4. Graphical components for constructing BPMS workflow

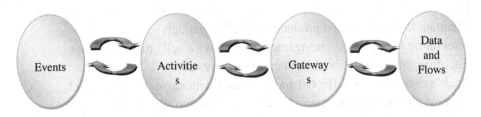

3.2 Modelling Smart Business Workflow System

The model of smart business system enables the workflows such as resource management, adaptability; optimization, scalability, and efficiency are evolving in modern day to day life with noticeable changes. The suitable industry 4.0, workflow standards and procedures have become a common and important subject. The business system level modelling describes various tools and structure to evaluate the role of business functions. The Business Process Management System (BPMS) is the standardized modeling structure that formalizes the suitable task, connection, relationship entity that tends to manage process-related entity as services. All the services are integrated as single process, if needed the sub-services are constructed which will measure the inter-process information effectively. The important consideration by BPM is planning structure and suitable implementation standards between the consumer and production manager. The BPMS creates a graphical structure that ensures the design is in hierarchical order to be followed with the following graphical components, which is shown in Fig. 4.

The events will ensure the start, end and intermediate details. The activities will focus on user scripting, messages between two parties, sending and receiving services, creating sub-process, segregation through loops. The gateways monitor the event and type of communication between different ends. Finally, data and flows will provide the sequence type, association connectivity between different entities.

3.2.1 Manufacturing Workflows during System and Production Level

It is relatively easy to understand modelling languages, like BPMS, but they are not entirely appropriate for representing very complex cases of development. On the other hand, the bases of a previous comparison are useful in representing complex logic which can also represent distributed systems and output flows. A framework is necessary for the management according to the following processes and services specifications.

1. Business System Level:
    a. Providing graphical representation of workflows for monitoring the process
    b. Providing workflows for easy understanding
    c. Providing common scripting language for communication
    d. Providing modelling design and workflows in dynamic real-time environment
2. Manufacturing Production Level
    a. Collaborating and representing distributed production workflows
    b. Structuring and managing complex tasks
    c. Provide timing limitations and enabling the flows to be in hierarchical order
3. Smart Manufacturing Enterprise Level
    a. Use standard predicted modeling techniques
    b. Customize the framework to be supported in both static and dynamic behavioural scenarios
    c. Maintain the business and production levels separately
    d. Integrating the dynamic management system with agile methodologies
    e. Enabling automatic hierarchical deployment system
    f. Enable changes to be reflected and corrected with current environment
    g. Support allocation of resources in real time with loosely coupled features
    h. Enhance the interoperability and integrity for heterogeneous based systems

    i.   Check the usage of SOA-driven workflow is adapted in all levels of management

3.3 Integrating Workflow Services

Industrial data is a very significant resource that if properly managed, can be more important for global manufacturing business processes and the source of tremendous wealth. Due to its large, complex, and unstructured nature, handling this information requires high processing and storage capacities. With the support of three phases, such as physical, middleware and application, the lifecycle of industrial data can be described as shown in Fig. 5.

Figure 5. Fine tuning smart manufacturing systems

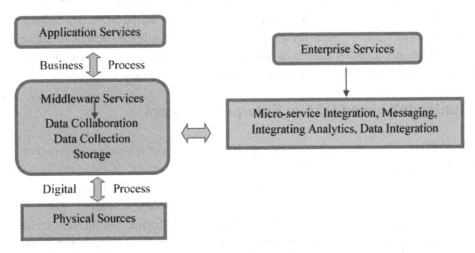

In real-world environments, numerous physical components of smart factories generate raw industrial data with different data types, formats, and varied dimensions. All valid data sources such as sensors are included in the physical devices component, Data, databases, and third-party applications created by the Web. This element is known as the discovery of data as well. This information becomes a variable after the digitalization and aggregation processes are introduced. In digital world, middleware and application components give various Services to administer it. Interoperability across different factories is handled by the middleware component.

Smart production has characteristics such as deep integration, enormous volume of data, and high correlation compared to conventional manufacturing processes. Consequently, the majority of manufacturers still face numerous challenges in obtaining industrial data for the collection, transmission, processing and storage of real-time and scalable data. By enabling service exchange and value co-creation, service platforms increase resource density. This is done through the suitable resources in a given context of value creation) and the bundling of resources in a location (or transporting resources to a location). Service platforms allow players inside and across service platforms to scan, pick, match and combine resources. Technologies for middleware are a constituent element that links all sorts of things, objects, and other structures.

The orchestration and choreography are two workflow based concept define the automation of enterprise system. It processes the production order from the production point of view, and implements the manufacturing accordingly. In view of that the choreographer of the Workflows based on predefined templates are performed according to precise methods. The manufacturing designs use these templates, and the workflow depending on the development steps created by choreographers which are the true achievements in development of the services. It performs the associated activities and tasks accordingly. Dependent on the performance equipment and processes will be organised to complement the relevant workflow phase in each levels. The main functional enterprise service of workflow processes supports architecture, control, connection, discovery, management level, orchestration functions.

The output of the workflow choreographer will specifies the specification of the product. The workflow choreographer creates the performance on the design basis that includes precisely the relevant steps of the workflow. The levels for manufacturing can be changed dynamically during development, according to the resources available. From a logical point of view, IoT-based SOA is a resource integration enabler and provides new possibilities for aggregating and integrating resources to create new service technologies and enhance existing offerings. The IoT-based SOA allows customer-centric information to be obtained, which in turn contributes to Innovative offers of service.

3.4 Fine Tuning IoT-based SOA in Enterprise System

Service networks increase resource density by enabling exchange of resources and co-creation of value. This is done through the quest for suitable resources in a given context of value creation) and the bundling of resources in a location (or transporting resources to a location). Service platforms allow players inside and across service platforms to scan, pick, match and combine resources. Three logics, namely linking, exchanging and integrating, are the basis for service platforms. Middleware technologies are a part of all sorts of objects and things and other systems that are related. Figure 6 shows the fine-tuned architecture for smart manufacturing enterprise system.

Therefore, from a strategic point of view, many value propositions are focused on the provision of service channels that promote access to resources and enable cooperation with other actors and thus provide opportunities for the integration of resources. Contributions often concentrate on mere IoT elements that do not give adequate importance to the coordination and integration of individuals, systems and things. For the rest, we thrive on a more holistic view of service networks that accept more abstract definitions of service e.g. Service structures with regard to service environments and co-creation of value. Four component subsystems or layers, namely sensing, network, service and interface layer, are based on the architecture. Implementing corresponding structures and logics, provide the environment for interaction and resource integration activities.

4. DISCUSSION

A smart manufacturing system is provided by enterprise services and is applicable across many industries. The paper highlighted the respective capabilities of linking smart objects across digital infrastructures, allowing new offerings and developments in digital services. A competitive market powered by current advances in smart technology with linking agile methodologies is updated by fine tuning services. Combined with scalable service structures and structural perspectives, namely service systems, models

Figure 6. Fine tuning service oriented enterprise model

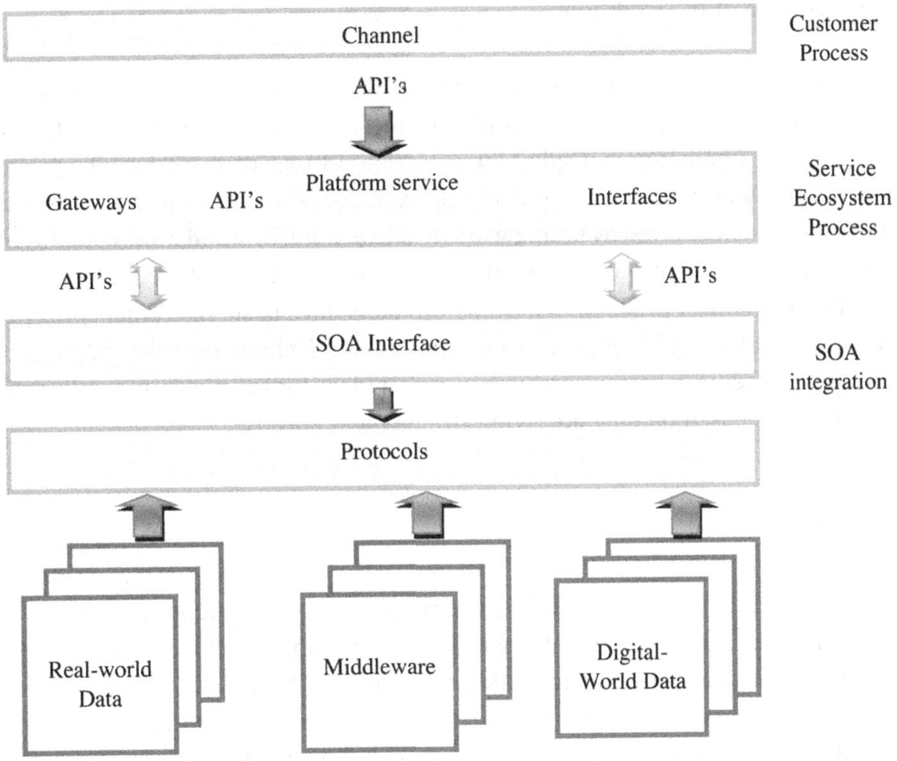

are of strategic interest as they help resource arrangements to achieve complex service offerings and developments in service.

Different approaches to solution integration were considered during the development of the workflow engine and as always, finding the optimal solution depends on the application field. Given the manufacturing requirements of Industry 4.0, we aimed to provide a viable solution both for participants in industry and manufacturing at system and production levels. Workflow enterprise management has taken the closely connected approach from workflow specification to workflow execution, i.e. to create acceptable workflow implementations by using workflow specifications. This technique is an effective process implementation model, as it removes the dividing line between the specification and implementation of the workflow using fine tuning process.

In most of the workflow process integrating agile methodologies will support activity based approach for fine tuning the communication workflow between system and production level design. The specification and design activity of workflow consists of following elements:

- Workflows: A complete or partial ordering of a series of tasks
- Tasks: A partial or complete order of events, details of human activities or other assignments
- Manipulated objects: A collection of storage related materials
- Roles: A placeholder for a human skill or operation of an information system required to perform a specific task

- Agents: They may be individuals or information systems for conducting tasks and communicating during the execution of the workflow.

Tools for workflow testing simulate a workflow by enabling sample data to be entered and triggering events such as completion of the mission, expiration of the deadline, and exceptions. To uncover logic errors and get estimates, simulation is needed for completion times of workflow. To predict potential bottlenecks in a workflow through the analysis of the workflow specification, workflow analysis tools are required. The research is achieved by taking workflow execution or simulation statistics into account. Analysis software, for instance, may collect workflow output statistics and recommend changes to the workflow specification to boost productivity. Some products have basic instruments for testing and/or analysis, but they are usually insufficient. A major future research challenge is linking different systems to a system of systems. There are various architectures and ideas for IoT platforms and applications that provide different features based on the underlying strategy and purpose.

5 CONCLUSION

Latest innovations related to product and service integration, e.g. the interconnection between business and services (Industry 4.0) and the advent of smart utility technology (service systems) creates new opportunities through the introduction of advanced services, such as the integration of energy management system tools as a service component of their product offerings. Service-led fine tuning services, underlying architectural designs with black box services and enabling agile technologies were highlighted in the paper. Integrating fine tuning services helps businesses to maximise their goods' lifetime value with proper updation. Service ecosystems provide attractive strategic options, especially for manufacturers and suppliers selling capital goods, to switch from selling products to offering solutions. This needs new capabilities to be created, such as integrating and configuring resources from different service systems (service exchange). Companies will make the leap to the strategic use of IoT technologies as a central element of value development and a source of competitive advantage by using IoT-based SOA service strategies. The trend is clearly evident towards a more service-oriented industry. Companies have to change their product / service portfolio accordingly in order to protect their competitiveness. Further research should make strategic formulations concrete and explore the interdependencies between innovation in IoT-based SOA infrastructures.

REFERENCES

Bhadoria, R. S., Chaudhari, N. S., & Vidanagama, V. T. N. (2018). Analyzing the role of interfaces in enterprise service bus: A middleware epitome for service-oriented systems. *Computer Standards & Interfaces*, *55*, 146–155. doi:10.1016/j.csi.2017.08.001

Borangiu, T., Morariu, C., Morariu, O., Drăgoicea, M., Răileanu, S., Voinescu, I., ... Purcărea, A. A. (2015, February). A Service Oriented Architecture for total manufacturing enterprise integration. In *International Conference on Exploring Services Science* (pp. 95-108). Springer. 10.1007/978-3-319-14980-6_8

Chen, Z., Liu, S., & Wang, X. (2008, September). Application of context-aware computing in manufacturing execution system. In *2008 IEEE International Conference on Automation and Logistics* (pp. 1969-1973). IEEE. 10.1109/ICAL.2008.4636484

Davis, J., Edgar, T., Porter, J., Bernaden, J., & Sarli, M. (2012). Smart manufacturing, manufacturing intelligence and demand-dynamic performance. *Computers & Chemical Engineering, 47*, 145–156. doi:10.1016/j.compchemeng.2012.06.037

De Deugd, S., Carroll, R., Kelly, K., Millett, B., & Ricker, J. (2006). SODA: Service oriented device architecture. *IEEE Pervasive Computing, 5*(3), 94–96. doi:10.1109/MPRV.2006.59

Deverajan, G. G., Muthukumaran, V., Hsu, C. H., Karuppiah, M., Chung, Y. C., & Chen, Y. H. Public key encryption with equality test for Industrial Internet of Things system in cloud computing. *Transactions on Emerging Telecommunications Technologies*, e4202.

Elhoseny, H., Elhoseny, M., Abdelrazek, S., Bakry, H., & Riad, A. (2016). Utilizing service oriented architecture (SOA) in smart cities. *International Journal of Advancements in Computing Technology, 8*(3), 77–84.

Gholami, M. F., Habibi, J., Shams, F., & Khoshnevis, S. (2010, March). Criteria-Based evaluation framework for service-oriented methodologies. In *2010 12th International Conference on Computer Modelling and Simulation* (pp. 122-130). IEEE. 10.1109/UKSIM.2010.30

Hori, M., Kawamura, T., & Okano, A. (1999, October). OpenMES: scalable manufacturing execution framework based on distributed object computing. In *IEEE SMC'99 Conference Proceedings. 1999 IEEE International Conference on Systems, Man, and Cybernetics (Cat. No. 99CH37028)* (Vol. 6, pp. 398-403). IEEE. 10.1109/ICSMC.1999.816585

Hossain, M. S., & Muhammad, G. (2016). Cloud-assisted industrial internet of things (iiot)–enabled framework for health monitoring. *Computer Networks, 101*, 192–202. doi:10.1016/j.comnet.2016.01.009

Jammes, F., Bony, B., Nappey, P., Colombo, A. W., Delsing, J., Eliasson, J., . . . Till, M. (2012, October). Technologies for SOA-based distributed large scale process monitoring and control systems. In *IECON 2012-38th Annual Conference on IEEE Industrial Electronics Society* (pp. 5799-5804). IEEE. 10.1109/IECON.2012.6389589

Kadri, A., Yaacoub, E., Mushtaha, M., & Abu-Dayya, A. (2013, February). Wireless sensor network for real-time air pollution monitoring. In *2013 1st international conference on communications, signal processing, and their applications (ICCSPA)* (pp. 1-5). IEEE. 10.1109/ICCSPA.2013.6487323

Komoda, N. (2006, August). Service oriented architecture (SOA) in industrial systems. In *2006 4th IEEE international conference on industrial informatics* (pp. 1-5). IEEE.

Lewis, G. A., & Smith, D. B. (2008, September). *Service-oriented architecture and its implications for software maintenance and evolution. In 2008 Frontiers of Software Maintenance*. IEEE.

Leyh, C., Schäffer, T., Bley, K., & Forstenhäusler, S. (2016). Assessing the IT and software landscapes of Industry 4.0-Enterprises: the maturity model SIMMI 4.0. In *Information technology for management: New ideas and real solutions* (pp. 103–119). Springer.

Maksuti, S., Tauber, M., & Delsing, J. (2019, October). Generic autonomic management as a service in a soa-based framework for industry 4.0. In *IECON 2019-45th Annual Conference of the IEEE Industrial Electronics Society* (Vol. 1, pp. 5480-5485). IEEE.

McGovern, J., Ambler, S. W., Stevens, M. E., Linn, J., Jo, E. K., & Sharan, V. (2004). *A practical guide to enterprise architecture*. Prentice Hall Professional.

Mohammadi, M., & Mukhtar, M. (2018, September). Service-oriented architecture and process modeling. In *2018 International Conference on Information Technologies (InfoTech)* (pp. 1-4). IEEE.

Muthukumaran, V. (2021). Efficient Digital Signature Scheme for Internet of Things. *Turkish Journal of Computer and Mathematics Education, 12*(5), 751–755.

Muthukumaran, V., & Ezhilmaran, D. (2020). A Cloud-Assisted Proxy Re-Encryption Scheme for Efficient Data Sharing Across IoT Systems. *International Journal of Information Technology and Web Engineering, 15*(4), 18–36. doi:10.4018/IJITWE.2020100102

Muthukumaran, V., & Manimozhi, I. (2021). Public Key Encryption With Equality Test for Industrial Internet of Things Based on Near-Ring. *International Journal of e-Collaboration, 17*(3), 25–45. doi:10.4018/IJeC.2021070102

Nagarajan, S. M., Muthukumaran, V., Murugesan, R., Joseph, R. B., & Munirathanam, M. (2021). Feature selection model for healthcare analysis and classification using classifier ensemble technique. *International Journal of System Assurance Engineering and Management*, 1-12.

Ordanini, A., & Pasini, P. (2008). Service co-production and value co-creation: The case for a service-oriented architecture (SOA). *European Management Journal, 26*(5), 289–297. doi:10.1016/j.emj.2008.04.005

Papazoglou, M. P., & Van Den Heuvel, W. J. (2006). Service-oriented design and development methodology. *International Journal of Web Engineering and Technology, 2*(4), 412–442. doi:10.1504/IJWET.2006.010423

Park, K. T., Im, S. J., Kang, Y. S., Noh, S. D., Kang, Y. T., & Yang, S. G. (2019). Service-oriented platform for smart operation of dyeing and finishing industry. *International Journal of Computer Integrated Manufacturing, 32*(3), 307–326. doi:10.1080/0951192X.2019.1572225

Ramollari, E., Dranidis, D., & Simons, A. J. (2007, June). A survey of service oriented development methodologies. In *The 2nd European Young Researchers Workshop on Service Oriented Computing* (Vol. 75). Academic Press.

Reis, J. Z., & Gonçalves, R. F. (2018, August). The role of internet of services (ios) on industry 4.0 through the service oriented architecture (soa). In *IFIP International Conference on Advances in Production Management Systems* (pp. 20-26). Springer. 10.1007/978-3-319-99707-0_3

Rolón, M., & Martínez, E. (2012). Agent-based modeling and simulation of an autonomic manufacturing execution system. *Computers in Industry, 63*(1), 53–78. doi:10.1016/j.compind.2011.10.005

Tao, F., & Qi, Q. (2017). New IT driven service-oriented smart manufacturing: Framework and characteristics. *IEEE Transactions on Systems, Man, and Cybernetics. Systems, 49*(1), 81–91. doi:10.1109/TSMC.2017.2723764

Theorin, A., Bengtsson, K., Provost, J., Lieder, M., Johnsson, C., Lundholm, T., & Lennartson, B. (2017). An event-driven manufacturing information system architecture for Industry 4.0. *International Journal of Production Research*, *55*(5), 1297–1311. doi:10.1080/00207543.2016.1201604

Valipour, M. H. AmirZafari, B., Maleki, K. N., & Daneshpour, N. (2009, August). A brief survey of software architecture concepts and service oriented architecture. In *2009 2nd IEEE International Conference on Computer Science and Information Technology* (pp. 34-38). IEEE.

Wang, C., Bi, Z., & Da Xu, L. (2014). IoT and cloud computing in automation of assembly modeling systems. *IEEE Transactions on Industrial Informatics*, *10*(2), 1426–1434. doi:10.1109/TII.2014.2300346

Xu, L. D., Xu, E. L., & Li, L. (2018). Industry 4.0: State of the art and future trends. *International Journal of Production Research*, *56*(8), 2941–2962. doi:10.1080/00207543.2018.1444806

Yaqoob, I., Ahmed, E., Hashem, I. A. T., Ahmed, A. I. A., Gani, A., Imran, M., & Guizani, M. (2017). Internet of things architecture: Recent advances, taxonomy, requirements, and open challenges. *IEEE Wireless Communications*, *24*(3), 10–16. doi:10.1109/MWC.2017.1600421

Zhang, L., Wang, X., & Dou, W. (2004, December). A K-connected energy-saving topology control algorithm for wireless sensor networks. In *International Workshop on Distributed Computing* (pp. 520-525). Springer. 10.1007/978-3-540-30536-1_57

Zhong, R. Y., Dai, Q. Y., Qu, T., Hu, G. J., & Huang, G. Q. (2013). RFID-enabled real-time manufacturing execution system for mass-customization production. *Robotics and Computer-integrated Manufacturing*, *29*(2), 283–292. doi:10.1016/j.rcim.2012.08.001

Zhu, T., Dhelim, S., Zhou, Z., Yang, S., & Ning, H. (2019). An architecture for aggregating information from distributed data nodes for industrial internet of things. In *Cyber-Enabled Intelligence* (pp. 17–35). Taylor & Francis. doi:10.1201/9780429196621-2

Chapter 7
A Quantum Technology–Based LiFi Security Using Quantum Key Distribution

Vinoth Kumar
MVJ College of Engineering, India

V. R. Niveditha
ⓘ https://orcid.org/0000-0002-6315-1784
Dr. M. G. R. Educational and Research Institute, Chennai, India

V. Muthukumaran
REVA University, India

S.Satheesh Kumar
REVA University, India

Samyukta D. Kumta
REVA University, India

Murugesan R.
REVA University, India

ABSTRACT

Light fidelity (Li-Fi) is a technology that is used to design a wireless network for communication using light. Current technology based on wireless fidelity (Wi-Fi) has some drawbacks that include speed and bandwidth limit, security issues, and attacks by malicious users, which yield Wi-Fi as less reliable compared to LiFi. The conventional key generation techniques are vulnerable to the current technological improvement in terms of computing power, so the solution is to introduce physics laws based

DOI: 10.4018/978-1-7998-6870-5.ch007

on quantum technology and particle nature of light. Here the authors give a methodology to make the BB84 algorithm, a quantum cryptographic algorithm to generate the secret keys which will be shared by polarizing photons and more secure by eliminating one of its limitations that deals with dependency on the classical channel. The result obtained is sequence of 0 and 1, which is the secret key. The authors make use of the generated shared secret key to encrypt data using a one-time pad technique and transmit the encrypted data using LiFi and removing the disadvantage of the existing one-time pad technique.

I. INTRODUCTION

The Vernam one time pad algorithm of classical cryptography is considered as the safest or algorithm with the highest security because it has the generation of key size based on the input and XORing them with the input gives the ciphertext that can be decrypted by the receiver if he has the key. There are two problems arising here one is that the key should be physically shared between sender and receiver by some other medium like physical meeting hence delivery of secret key is a crucial problem and other issue is that the key should be of the length of input which might be too large(Gisin et al., 2002). The Quantum key cryptography is the solution to this problem. It provides the method to share the secret key between the two parties. It relies on quantum physics rules and does not provide security using computing power (Mailloux, Grimaila, Hodson et al, 2015). It provides its best security due to the laws of physics the Heisenberg-uncertainty principle and the no-cloning theorem (Mailloux et al., 2016). With the Heisenberg-uncertainty principle, the attacker cannot know the states of qubits without changing it whereas with the no-cloning principle it cannot copy the states of the qubits hence when anyone trying to attack the sender Alice and receiver Bob would get to know about the attacker due to the changes in the quantum states (El Rifai, 2016).

Nowadays Wi-Fi is used widely without knowing whether it is secured or not. Whether any security algorithm is been provided to it. In mission-critical situations Wi-Fi technology is prone to attacks hence the light as a medium of communication is been used in the data transfer because it is considered as secure means technically named as LiFi. The security of LiFi under various situations is defined hence we send the message using light without any encryption or if encrypt we use classical algorithms other than quantum. Here we introduce a method of combining quantum security with LiFi giving technology with the best security (Jayasuruthi et al., 2018). In this paper, we are explaining the concept of how the quantum cryptography can be used using light as a medium of data transfer and how to secure the LiFi system using the best security algorithms of quantum cryptography such as BB84. The framework used to demonstrate the security provided by quantum cryptography is a modelling framework using the LiFi concept. Hence network security is the key output obtained from the concept introduced here.

The paper is organized as follows: Section II of the paper describes the QKD systems and BB84 protocol. Section III describes the LiFi system and its advantages over WiFi. In Section IV we present the concept of the security of LIFI using quantum. Section V gives the results obtained after the simulation Section VI gives the conclusions,and future work.

II. QKD SYSTEMS

A. Basics of Quantum Key Distributions

The Quantum key distribution is the method of key transfer to the two parties in communication. It is the key exchange mechanism using the photon for the encryption of very sensitive data between the two parties. The key generated by the QKD is not only for quantum cryptographic algorithms but it can also be used by the classical algorithms such advance encryption standard (AES) and Data encryption standard (DES) to increase their security level (Mailloux, Grimaila, Hodson et al, 2015). QKD mainly follows the theory of quantum physics by using the photon system for communication. The physics are represented using the bra (I) and key (>) representation (Harun et al., 2018).

The mapping of the classical to qu bits can be written as

$$0 \rightarrow |0> 1 \rightarrow |1> \tag{1}$$

The superposition states are notated as:

$$|W> \equiv \alpha|0>+ \beta|1>= [\alpha \ \beta] \tag{2}$$

Where W is the state of superposition, $|0>$ and $|1>$ are the qu bits and α and β are the complex numbers. The $[\alpha \ \beta]$ is the vector where $(0>$ equals to $[1 \ 0]$ and $(1>$ equals to the $[0 \ 1])$.

The probability of a and B are satisfied by

$$|\alpha|^2 + |\beta|^2 = 1 \tag{3}$$

Where $[a]^2$ is probability of obtaining $|W>$ in $|0>$ and $[b]^2$ is of $|W>$ in $|1>$.

The QKD system has a quantum and classical channel. The quantum is used to transmit qubits and must have a transparent optical path. The classical channel (public) can be any IP channel and must be

Figure 1. QKD Mechanism

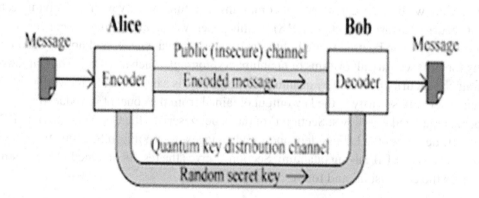

a dedicated channel based on the timing requirements. Here in this paper we propose the method where we can use the light as the quantum channel for both the channels. Hence classical channel dependency is been removed shown in Figure 1.

In the classical cryptography method the timestamps are used to prevent the third party replay attacks but they require the clock synchronizations which is not a practical approach. The classical further cannot detect the existence of attacks such as eavesdropping which on the contrary the quantum channel eliminates (Maithili et al., 2018).

B. BB84 Algorithm

Although there are many QKD algorithms such as B92, E91 but this paper focus on the BB84 algorithm because it is widely known and is much secure than the other algorithms. The genetics or history of bb84 begins from back 1960s when Wiesner developed the idea of quantum conjugate coding. He had described the two applications named quantum money and quantum multiplexing for the fraud-proof banking notes and transmitting messages such that on reading messages delete the previous one respectively(Wiesner, 1983).

Then Bennett and Brassard in 1984 gave the first QKD BB84 protocol. (Bennett & Brassard, 1984)(). The bb84 protocol describes the means for the two communicating parties to generate the shared secret key. It is based on the polarization-based mechanisms using the qu bits using one of the four polarization states: horizontal, vertical, diagonal, or anti diagonal. The Dirac notation for all the qu bits is in the Table 1 given below (Mailloux, Morris, Grimaila et al, 2015).

Table 1. Polorization States

Bits	Polarization	Basis	Dirac Notation
0	Rectilinear	Horizontal	$|H\rangle$
1	Rectilinear	Vertical	$|V\rangle$
0	Diagonal	Diagonal	$\frac{1}{\sqrt{2}}(|H\rangle+|V\rangle)$
1	Diagonal	Anti-Diagonal	$\frac{1}{\sqrt{2}}(|H\rangle-|V\rangle)$

The phase encoding of the BB84 systems is described using the figure (Mailloux, Grimaila, Hodson et al, 2015). The phase difference is θa between the two paths chosen between the four positions namely {0, π} and {$\pi/2$, $3\pi/2$}. The interferometer, a photon positioned over the phase difference θa. Bob transmits the arriving photon through the infer meter identical to Alice with a phase difference of θb, randomly chosen between {0, $\pi/2$}. These configurations are equivalent to the young's double-slit experiment with a single photon in the time domain. The Bobs interferometer is shown in the figure. Either the detector 1 or 2 clicks according to the interference. The pattern is dependent on the phase value.

Figure 2. Phase shift encoding

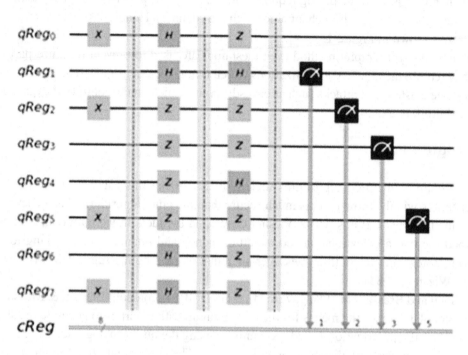

The figure 2 shows the detection probability is maximum or minimum at the DET for combinations of the f (θa, θb), such as at a peak for (0, 0) and (π/2, π/2) and at a bottom for (π, 0) and (3π/2, π/2). Which

Figure 3. Photon detector probability

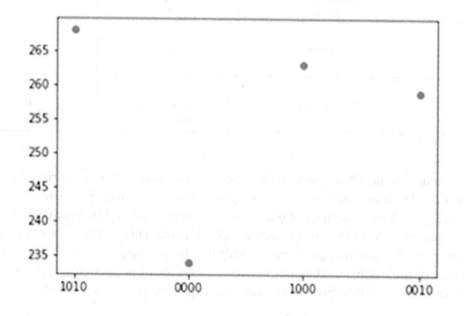

mean she photon detected for r (0, 0) and ($\pi/2$, $\pi/2$) and by DET 2 for (π, 0) and ($3\pi/2$, $\pi/2$) provided that the extinction ratio of the interference is perfect?

Using the setup of this working of BB84 describes as:

1. The photon count transmitted from Alice to bob
2. Bob tells Alice which he detected
3. Alice tells bobs she chose θa from $\{0, \pi\}$ or from $\{\pi/2, 3\pi/2\}$.
4. The deterministic detection tells Alice regards $\theta a = 0$ or from $\pi/2$ as the bit "0" and $\theta b = \pi$ or from $3\pi/2$ as the bit "1," and Bob regards the D1 click as a bit "0" and the D 2 click as a bit "1."

Thus, in the above steps the bit information is not disclosed hence created bit can be the secret key shown in Figure 3 (Inoue, 2006).

Assuming the ideal cases the Bob measures the same as used by Alice we then obtain a high degree of accuracy. However, if Bob gets incorrect guesses, then random results obtained will be lost previously. This leads to the collapse of the polarization state associated (Loepp & Wootters, 2006)().

This protocol ensures that the eavesdropper necessarily introduces detectable flaws when attacking because she does not know the states prior. Thus, by examining the QBER (Quantum bit error rate) Alice and bob determine the eavesdropper is trying to invade their conversation. Thus, by randomly selecting the states we secure the secret key transmission and communication channel. (Gisin et al., 2002).

III. Li-fi TECHNOLOGY

It is a technology that is based on light as the name indicates and not on the radio waves mechanisms use by other wireless technologies. It is basically on illumination. This technology uses the LED for transmitting the data wirelessly. It is the fastest and cheapest method and has speed around 250 times faster than Wi-Fi.

A. Working

The main focus here is the use of the high intensity LED for data transfer. Data can be audio, video, or text. The intensity is dependent on the ON/OFF capability of the LED. If the LED is at ON position, then we transmit signal one and in OFF position we transmit zero so these 0s and 1s are basically the binary bits which is the basis of the data. The modulation is too fast a human eye cannot catch into observe.

The photo detector receives the signal from the source and coverts back to original data this is known as visible light communication. At the sender side the photo resistor is used which converts the incoming data signal into the light form and sends it to the receiver which receives the signal converts to the original data type and hence communication happened between the sender and receiver.

The efficiency provided by the LiFi includes the two features: Availability and security. The light is available in every part of the world hence it is possible to implement this communication in any part of the proof even in the airplanes where Wi-Fi cannot be used. The security is given because it cannot go into solid walls like radio waves hence provide the abundance of network privacy(Monisha & Sudheendra, 2017).

Figure 4. LiFi technology

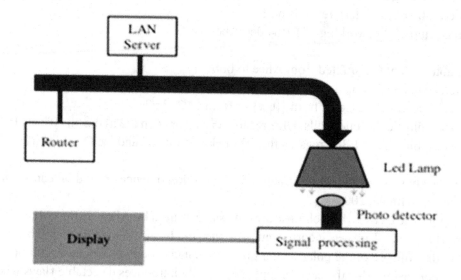

The figure 4 represents the working of LiFi having the LAN server providing the input message which converts to light signal using the LED Lamp and the receiver side has the photo detector to get the signals which get converted to bits that is displayed on the screen.

B. Advantages of LiFi over WiFi

It is possible to have more than 10GBPS allow a high definition motion picture be too downloaded in less than 30secs. Most LEDs are used hence cost well organized and are fast and easy. LiFi has a bandwidth of 10000 more than Wi-Fi. Table 2 represents the comparison of LiFi advantages over Wi-Fi.

Table 2. Advantages of lifi vs. wifi

Feature	LIFI	WIFI
Operation	Light	Radio Waves
Interference	No Issues	Have issues due to less access points
Technology	IrDA	WLAN 802.11a/b/g/n/ac/d
Data Speed	1 Gbps	WLAN – 150 Mbps
Distance	10 Meters	32 Meters
System Components	LED, Photo Resistor	Routers, Laptops and Desktops
Frequency	10K times radio spectrum	2.4GHz, 4.9GHz and 5GHz

C. Issues with Radio Waves and LiFi Advantages over these Issues

1. Capacity: The increasing no of the mobile user or computer users has made the spectrum congested and capacity is less and the less bandwidth becomes more expensive. LiFi on the other hand is based on the light-medium hence bandwidth is unlimited hence no congestion issue occurs.
2. Efficiency: Million of work points consume high energy for radio waves transmission. Almost 1.4 million cellular radio station bases are there. Here efficiency is only 5% hence LiFi is more advantageous over it because it has more availability and security because light has more network privacy.
3. Security: Radio waves can pass through walls hence are less safe, more prone to attacks whereas light cannot pass through walls they stay inside the boundaries of the wall and hence are safe from outside attacks.

The LiFi is insecure inside a room, confined area because of which its security is still under risk. Though it is more secure than Wi-Fi the security of LiFi is still prone to eavesdropping by inside environment attackers. These are mainly high risk in any military operation or such which is using LiFi for communication.(Monisha & Sudheendra, 2017)

IV. IMPLEMENTATION

The implementation of the LiFi quantum is mainly integrating the LiFi technology with the QKD algorithm than is BB84 by modifying the QKD system.

A. Quantum key Distribution and BB84

The implementation of the LiFi security is by changing the classical channel to the quantum channel itself for the key distribution. Basic QKD systems use the classical or public channel here we define the change by intruding the quantum channel for both as the data sharing and the key sharing between the two parties. Thus, the working of QKD goes like Bob and Alice both with communicate with each other through the quantum channel itself and the key sharing also takes place from the quantum channel itself. The four qubits' states are there in the BB84 algorithm hence we first with the quantum channel generate the shared secret using the qubits states and on a match with 50% probability, we get the shared secret. Both the parties have that and then the sender sends the message encrypted with the shared secret in the same quantum channel. The receiver can decrypt the message using the shared secret and hence the communication using the quantum channel in both the cases. The communication between Alice and bob takes place.

1. Alice decided the qubits state between $\{0, \pi\}$ or from $\{\pi/2, 3\pi/2\}$ and send bits through a quantum channel
2. Bob receives the bits, and he also chooses qubits states between $\{0, \pi\}$ or from $\{\pi/2, 3\pi/2\}$.
3. Bob sends his states to Alice and based on match the key has been decided which is the secret key.

B. LiFi Implementation

The communicating parties are using the modified QKD system for communication. The quantum channel used in the communication is mainly the light channel which is the use of LIFI technology in the QKD system. The classical channel is replaced but the light of communication and similarly the quantum channel for secret key sharing the light channel. So, the two parties can communicate using light. The security to the LiFi is provided using the BB84 algorithm. Hence the LIFI is secured using the QKD system and BB84 has feature enhancement because of the light channel for communication which has more features than radio waves.

Figure 5. QKD reference architecture

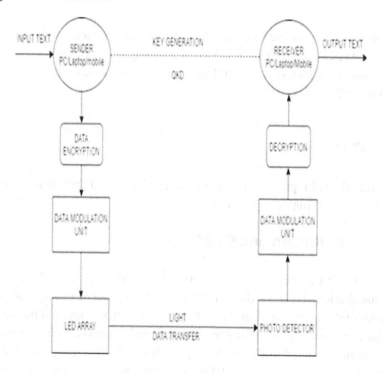

C. QKD Architectural Model

The combination of the LiFi and QKD algorithm of BB84 can be implemented using the new QKD architecture which can be modeled as shown in the figure 5. This is the simulation model but has the features used to provide security to the wireless communication of the light and the feature enhancement of the BB84 algorithm using the light as the quantum channel. Figure 5 represents the simulation model.

1. The sender and receiver as the Alice and bob for communication.
2. The QKD system is used for the key generation and sharing which is basically the light channel used as the quantum channel of the BB84 algorithm.
3. The sender encrypts the message using the key generated by using the encryption mechanism.

4. The Data modulation unit converts the data into 0s and 1s to send via light.
5. The LED array sends the data through light and the photodetector receives the signal.
6. The decryption occurs at the receiver end using the same shared secret key and he revives the original text.

D. Attacks and Security

1. Bean splitting attack: - An eavesdropper tries to steal information between the communication happing between Alice and bob. However, the attack will fail because the photons do not reach bob and thus no key is generated. If eavesdropper attacks during the message sharing time also Bob and Alice would know about the attack because bob would not receive the bits at all. There is a high probability than bob and eve detect the same photon at the same time due to the passions distribution bit the leaked information, as a result, it can be extinguished by privacy amplification.

2. Intercept- Resend Attack: - Another way to eavesdrop is to intercept the message signal and measure its state, resend a fake message to bob based on its measurement. But this also will not work because the eavesdropper eve dose does not know the states $\{0, \pi\}$ or from $\{\pi/2, 3\pi/2\}$ which to choose in one go. $\pi/2$ and $3\pi/2$ are in ambiguous states when Eve tries to distinguish 0 and π, and 0 and π are in ambiguous states when Eve tries to distinguish $\pi/2$ and $3\pi/2$. Thus eve at any cost cannot know the states and hence cannot send the same copy of Alice signal to bob. Hence now if Bob creates a key with the improper signal, then due to bit mismatches in some tests both will get to know about the eavesdropping. The probability of Eve being ambiguous is 0.5. Thus, bit mismatch becomes 0.25.

   The bit-mismatch induced probability is $\alpha/4$, where α is a fraction of the intercepted signal. When $\alpha/4 << e$, where e is the error rate caused by system imperfections eavesdropping is masked by the system error, and α of the information is leaked to Eve. An extreme case is $\alpha/4 = e$, where Eve fully utilizes the error rate assumed by Alice and Bob.

3. Photon- Number –splitting attack: The PNS is using the strongly attenuated light. Eve probes the no. of photons using the QND. When there are two photons in a signal, she extracts one and sends all other to bob. To avoid the transmission loss, we have to keep the probability less than the system transmittance rate. Thus, detection rate decreases for Bob in proportion to the square of transmittance reaching the cut off value of detector noise (Inoue, 2006).

V. EXPERIMENTAL DATA AND RESULTS

The below graphs are the experimental data based on the parameterization on the single photon used in the transmission.

A) Channel Losses

This graph shows that the key generation rates are reduced by the increasing channel losses. This is in the optic channel where the short, medium, and long-range applications are being used. The coefficients of attenuation 0.25, 0.35, and 2 DB/km respectively shown in Figure 6.

Figure 6. Channel losses

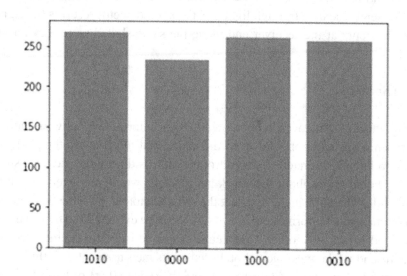

B) Final Bits During Eve's Drop

From this graph we can see that if the channel has been attacked Bob can get or may not get the signal due to the light channel which cannot give perfect bits length send by Alice. So, there is a little bit of error in the transmission.

In the Intercept of resending attack, we get the same length bits hence cannot find the attack with this parameter. In this attack, Eve will generate a new string and send it to bob So now the probabilities of an attacker to get detected are 50%(increased). The probability that the string will be like Alice one is 50%. Hence, they detect the error rate lower than the maximum error rate (e < max).

In the third case of beam splitting attack by the eavesdropper the attack is much lower than the other two cases. This is because of the random number attack on the Alice bits by Eve. Hence more errors can be detected by Alice and Bob in the 'error correction' process. The error rate (e) is still lower than the maximum error rate (e < max) but still Alice and Bob can detect 50% error. Thus, both the attacks are being detected by the two communicating parties hence the security of the LiFi is been achieved in both the cases shown in Figure 7.

VI. CONCLUSION

This paper focuses on a methodology to successfully generate a shared secret key and implement a secured communication between two communicating parties with the help of LiFi (Light Fidelity) technology. The steps used in order to generate the key are based on BB84 quantum encryption. The algorithm has many limitations. Our proposed paper eliminates one of its limitations which is based on the dependency on the classical channels. The main advantage is that it provides more security than the classical key distribution algorithm, is faster as compared to asymmetric key generation algorithms, and more reliable than WiFi (Wireless Fidelity).

Figure 7. Eve's Drop

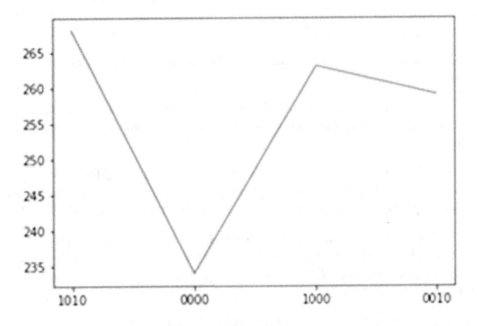

The proposed method can further be improved by eliminating other limitations which mainly include reducing cost and implementing it for larger distances. Since, LiFi requires security, and the quantum cryptographic algorithms provide better security hence working these two together can make the network communication highly robust and secure.

REFERENCES

Bennett, C. H., Bessette, F., Brassard, G., Salvail, L., & Smolin, J. (1992). Experimental quantum cryptography. *Journal of Cryptology*, 5(1), 3–28. doi:10.1007/BF00191318

Bennett, C. H., & Brassard, G. (1984). Quantum cryptography: Public key distribution and coin tossing. *Proc. IEEE Int. Conf. Comput., Syst. Signal Process.*, 475–480.

El Rifai, M. (2016). *Quantum secure communication using polarization hopping multi-stage protocols* (Ph.D. dissertation). School Elect. Comput. Eng., Univ. Oklahoma, Norman, OK.

Gisin, N., Ribordy, G., Tittel, W., & Zbinden, H. (2002, January). Quantum cryptography. *Reviews of Modern Physics*, 74(1), 145–195. doi:10.1103/RevModPhys.74.145

Harun, N. Z., Zukarnain, Z. A., Hanapi, Z. M., & Ahmad, I. (2018). Evaluation of Parameters Effect in Multiphoton Quantum Key Distribution Over Fiber Optic. *IEEE Access: Practical Innovations, Open Solutions*, 6, 47699–47706. doi:10.1109/ACCESS.2018.2866554

Inoue, K. (2006, July-August). Quantum key distribution technologies. *IEEE Journal of Selected Topics in Quantum Electronics*, 12(4), 888–896. doi:10.1109/JSTQE.2006.876606

Jayasuruthi, L., Shalini, A., & Vinoth Kumar, V. (2018). Application of rough set theory in data mining market analysis using rough sets data explorer. *Journal of Computational and Theoretical Nanoscience, 15*(6-7), 2126–2130. doi:10.1166/jctn.2018.7420

Loepp, S., & Wootters, W. K. (2006). *Protecting Information*. Cambridge Univ. Press. doi:10.1017/CBO9780511813719

Mailloux, L. O., Grimaila, M. R., Hodson, D. D., Baumgartner, G., & McLaughlin, C. (2015, January). Performance evaluations of quantum key distribution system architectures. *IEEE Security and Privacy, 13*(1), 30–40. doi:10.1109/MSP.2015.11

Mailloux, L. O., Hodson, D. D., Grimaila, M. R., Engle, R. D., McLaughlin, C. V., & Baumgartner, G. B. (2016). Using modeling and simulation to study photon number splitting attacks. *IEEE Access: Practical Innovations, Open Solutions, 4*, 2188–2197. doi:10.1109/ACCESS.2016.2555759

Mailloux, L. O., Morris, J. D., Grimaila, M. R., Hodson, D. D., Jacques, D. R., Colombi, J. M., Mclaughlin, C. V., & Holes, J. A. (2015). A Modeling Framework for Studying Quantum Key Distribution System Implementation Nonidealities. *IEEE Access: Practical Innovations, Open Solutions, 3*, 110–130. doi:10.1109/ACCESS.2015.2399101

Maithili, K., Vinothkumar, V., & Latha, P. (2018). Analyzing the security mechanisms to prevent unauthorized access in cloud and network security. *Journal of Computational and Theoretical Nanoscience, 15*(6), 2059–2063. doi:10.1166/jctn.2018.7407

Monisha, M., & Sudheendra, G. (2017). Lifi-Light Fidelity Technology. *2017 International Conference on Current Trends in Computer, Electrical, Electronics and Communication (CTCEEC)*, 818-821. 10.1109/CTCEEC.2017.8455097

Nielsen, M. A., & Chuang, I. L. (2010). Quantum Computation and Quantum Information. Cambridge Univ. Press.

Vinoth Kumar, V., Karthikeyan, T., Praveen Sundar, P. V., Magesh, G., & Balajee, J. M. (2020). A Quantum Approach in LiFi Security using Quantum Key Distribution. *International Journal of Advanced Science and Technology, 29*(6s), 2345–2354.

Wiesner, S. (1983). Conjugate coding. *ACM SIGACT News, 15*(1), 78–88. doi:10.1145/1008908.1008920

Chapter 8
Role of Artificial Intelligence of Things (AIoT) to Combat Pandemic COVID–19

Arti Jain
https://orcid.org/0000-0002-3764-8834
Jaypee Institute of Information Technology, Noida, India

Rashmi Kushwah
https://orcid.org/0000-0003-1232-0426
Jaypee Institute of Information Technology, Noida, India

Abhishek Swaroop
Bhagwan Parshuram Institute of Technology, India

Arun Yadav
National Institute of Technology, Hamirpur, India

ABSTRACT

COVID-19 is caused by virus called SARS-CoV-2, which was declared by the WHO as global pandemic. Since the outbreak, there has been a rush to explore Artificial Intelligence (AI) and Internet of Things (IoT) for diagnosing, predicting, and treating infections. At present, individual technologies, AI and IoT, play important roles yet do not impact individually against the pandemic because of constraints like lack of historical data and the existence of biased, noisy, and outlier data. To overcome, balance among data privacy, public health, and human-AI-IoT interaction is must. Artificial Intelligence of Things (AIoT) appears to be a more efficient technological solution that can play a significant role to control COVID-19. IoT devices produce huge data which are gathered and mined for actionable effects in AI. AI converts data into useful results which are utilized by IoT devices. AIoT entails AI through machine learning and decision making to IoT and renovates IoT to add data exchange and analytics to AI. In this chapter, AIoT will serve as a potential analytical tool to fight against the pandemic.

DOI: 10.4018/978-1-7998-6870-5.ch008

1. INTRODUCTION AND CHALLENGES

COVID-19, as is initially known 2019-nCoV is a disease which is caused by a virus called SARS-CoV-2 (Shereen et al., 2020). At first, it is detected in Wuhan, China during December 2019. This disease is declared by the World Health Organization (WHO) (Lupia et al., 2020) as a global pandemic, dated March 11, 2020. The COVID-19 virus is transmitted via infected respiratory droplets, and whenever there is a close contact with an infected individual. COVID-19 is a highly contagious virus that remains alive on passive objects for hours and it has no official treatment or vaccine available so far. Exponential spread of the pandemic and lakhs of death at the global level has forced many countries all over the world to force lockdown (Lockdown, 2020) and other containment measures. The cost of the pandemic for human lives and economic havoc is beyond imagination.

To prevent the harsh effects of the pandemic, several containment measures are undertaken by the worldwide governments such as lockdowns, quarantines (Quarantine, 2020), and social distancing (Social Distancing, 2020).

- Lockdown: Lockdown is a protocol that is usually initiated by a person in position of authority and is followed by the people during an emergency situation which means that people must stay where they are, to protect them inside their home.
- Quarantine: Quarantine is a separation and restriction in movement of people who are not ill but are believed to be exposed to COVID-19 infection which is intended to prevent the spread of pandemic. Suspected people are quarantined in the community-based facilities or in their homes.
- Social distancing: Social distancing is a non-pharmaceutical measure to thwart the outbreak of pandemic while maintaining physical distance and reducing meetings among people, in order to avoid close contact with each other. It includes a safe distance of 6 feet or 2 meters between one and other person to avoid gathering of people in large groups.

To meet the effectiveness of the measures mentioned above, several prerequisites are to be considered such as good health hygiene, regular hand cleaning, closing of schools, colleges, malls, parks and all non-essential businesses; shut down of public transports to reduce the pandemic spread. Apart from these obligations, different challenges of COVID-19 are to be handled with utmost care.

The various challenges of COVID-19 are divided into seven different categories (Figure 1). The challenges are, namely- identifying origin of pandemic (patient zero), preventing cross-infection between medical staff and patients, containing spread of the virus, quarantining potentially infected patients, treating critically ill patients, tracing origin of outbreak (local hotspots), and ensuring enough medical resources.

All these challenges along with an accelerated spread of COVID-19 has resulted into severe problems in the health-care systems such as lack of adequate healthcare professionals, testing devices, ventilators, safety devices etc. (Song et al., 2020). These problems have reached to a level that governments are unable to scale up the solution in proportionate to the pandemic outbreak. Therefore, a feasible solution is the demand of time which is to be easily scalable, automated, tuned, cost efficient and lesser dependent upon the man power.

Artificial Intelligence (AI) is a potential data analytic tool to fight against the COVID-19. Also, Internet of Things (IoT) is a prospective tool for data collection and ability to transfer data over a network. So far, AI serves diverse applications in Machine Learning (ML) (Thrall et al., 2018), natural language

Figure 1. Various Challenges of COVID-19

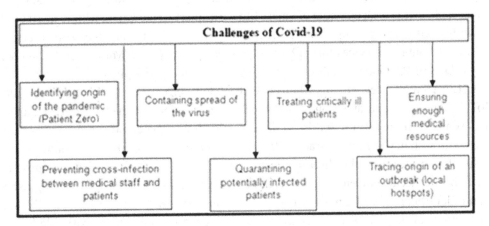

processing (Matuszek, 2018), computer vision (Lemley et al., 2017), and pattern recognition (Zadeh et al., 2018). And, IoT serves miscellaneous applications (Kushwah et al., 2020) (Kushwah et al., 2019) in wearables, connected health such as telehealth, or telemedicine (Ansari et al., 2020). At present, these two technologies work at separate, individual levels; and so, both are not significantly impactful against the pandemic. It is because both the AI and IoT are hampered due to lack of data (historical data), existence of biased, noisy and outlier data. In order to overcome such constrictions, balance among privacy of data, public health-care, and human-AI-IoT interactions are must. It is believed that binding of AI and IoT together as- Artificial Intelligence of Things (AIoT) is going to be a handy, analytical tool to fight with the pandemic- COVID-19.

The chapter is structured as follows. Section 2 presents the related work in COVID-19. Section 3 takes care of the technology used by AIoT. Section 4 presents the role of AIoT in combating COVID-19. Finally, Section 5 concludes the chapter.

2. RELATED WORK

Artificial Intelligence of Things (AIoT) network proliferates, several industries (e.g., supply chains, sales and marketing), service enterprises (e.g., products and services delivery), and consumer products (e.g., support models- Grofers and Big Baskets) are motivated to be transformational through integration of AI with IoT technologies. The human generated and machine oriented unstructured, valuable data are growing at a rapid pace to drive substantial opportunities for both the AI and IoT supported systems towards real-time data analytic solutions and enhanced decision-making processes. The exploration of AIoT is only limited by our imagination.

According to Schabenberger (Schabenberger, 2018) the AIoT harnesses the possibilities of enabling data towards collective learning to automate decision-making tasks in the connected world. With this, AIoT is capable of solving several real-life problems such as human asset management (Sharpe et al., 2019). The key applications of AIoT are in the field of monitoring (Lin, Chuang, Yen, Huang, Chen et al, 2019), location tracking (Lai et al., 2019), decision making (Yu et al., 2019), supply chain (Koncar et al., 2020) and remote access for the diagnosis (Lin, Chuang, Yen, Huang, Huang et al, 2019) purpose. In another application, AIoT use cases are retail product oriented, as in consumer appliances. For example, Sharp (Vanus et al., 2015) is an example of smart-life with smart-home where people learn to live while interacting with smart housing equipments. AIoT is likely to expand in the healthcare data services which involve IoT and AI data as services through data analytics giants (AIoT Market, 2020), namely- SAS, SAP, Alteryx, Dell EMC, Google, Splunk Technology, Teradata, and VMware.

Lin et al. (Lin, Chuang, Yen, Huang, Chen et al, 2019) have proposed ECG monitoring system based on the AIoT design. It incorporates front-end device (solar and wireless charging circuits, analog front-end of ECG sensing circuit). With 4G, ECG data is uploaded to the cloud server, and is trained with decision tree. The system gives an accuracy of 98.22%. With this, long-term ECG monitoring is feasible for diagnosis of the cardiovascular-disease. Zhang et al. (Zhang et al., 2020) have proposed a smart-sock based TriboElectric NanoGenerator (TENG) for the AIoT based healthcare. It distinguishes human activities and identification with an accuracy of 89%. On extracting suitable features, the smart-sock analyzes symptoms of the Parkinson's disease with 82% accuracy and is also profitable for healthcare monitoring. The smart-sock has diversified functions which are beneficial for the smart healthcare e.g., energy harvesting, pattern recognition during walking, gait sensing etc.

Muthu et al. (Muthu et al., 2020) have hybridized the AI and IoT through wearable sensors and data analysis in the healthcare. They have designed the GARIC which collects the patient information and trains using AI based Boltzman belief network. The GWAS is responsible to predict the patient's disease and undergoes advisory from his consultant doctor, also receives alerts through emails, whatapp, sms etc. This in turn increases the efficacy and accuracy of doctor during treatment of his patient. In addition, patient pain is reduced, recovery rate is uplifted, healthier planning and safer execution process is performed with the penetration of virtual reality (Dai et al., 2020).

Mdhaffar et al. (Mdhaffar et al., 2017) have proposed IoT based LoRaWAN architecture for health monitoring which collects data from medical sensors and send it to the AI model for analysis. The method includes blood sugar, glucose and temperature to measure and send to the AI model. Al-Tudjman and Deepak (Al-Tudjman & Deepak, 2020) have concluded that IoT based applications have crafted innovation for evaluating blood gas analyzers, thermometers, smart beds, glucose meters, ultrasounds, and many more devices to make patient comfortable during the COVID-19 pandemic. They have mentioned applications of IoT for the clinical operations- drug delivery, patient evaluation, lab test and overall medical management during the pandemic. Singh et al. (Singh et al., 2020) have elaborated that IoT based technology is a pathway to manage hospital in a streamlined and coherent mode during COVID-19 pandemic. The IoT technology is responsible for monitoring drug for patient, provides proper distribution process for drugs and equipment to the hospital. Shaikh and Nikilla (Shaikh & Nikilla, 2012) have investigated the IoT technology for the patient savior while monitoring his body temperature, heartbeat etc. that too from remote location, once the patient is discharged from the hospital. Archana and Indira (Archana & Indira, 2013) have proposed the soldier monitoring and health indication systems to monitor heart rate, location of solider and temperature. The proposal uses polar heart rate transmitter and RMC01 receiver as a heartbeat sensor. Islam et al. (Islam et al., 2015) have surveyed the IoT based technologies to receive patient's data and have suggested various health monitoring networks for collecting, storing, processing and executing the patient's information for better decision to cure them. Also, they commented that the proposed system has capability to provide better results in terms reducing the monitoring and decision time.

Zhao et al. (Zhao et al., 2019) have provided systematic review on health monitoring using Deep Learning (DL) technology. They have discussed DL architecture on Auto-Encoder, Convolutional Neural

Network (CNN) and Recurrent Neural Network (RNN). Nweke et al. (Nweke et al., 2018) have undergone the deep learning technology along with the IoT sensors to monitor health related features automatically within the mobile devices. They proposed three deep learning models i.e generative model, disseminative models and hybrid models. Their study combines generative and discriminative models to design hybrid models and extracts refined features for better monitoring of the patients. Da Costa et al. (Da Costa et al., 2018) have mentioned the concept of IoT with concrete aim to identify physical condition of patients who are in critical life condition, so as to keenly monitor them and provide valuable consultation within optimal time constraint. The main objective of the system is the collection of data, process the data and present it in a systematic way with the deep learning technology. Alam et al. (Alam et al., 2017) have used the concept of data fusion that is stored by the IoT sensors. The objective of fusion is to clean the data while removing ambiguity and repetitive features and applies deep learning methods to find the results. This process increases the quality of features that are received by the IoT sensors and provides better health monitoring of the patients.

After brainstorming the literature review over applications of AI and IoT in the healthcare sector, several significant research findings are identified. Firstly, most of the researchers have applied AI based machine learning or deep learning technologies along with the IoT enabled features for the collection of patients records, storage, processing and analyzes of patients' condition. Secondly, AIoT reduces the response time of doctors, increases the cure rate of patients, increases availability of doctors even within remote locations digitally. On the basis of above discussions, the following Research Questions (RQs) are designed and discussed throughout this paper.

RQ1: Whether AIoT can work for COVID-19 patients also?
RQ2: Which category/classification of machine learning is workable in this scenario?
RQ3: How much life does AIoT can save and increases the efficiency of doctors to cure patients?

The above stated RQs are answered in the Section 3 and Section 4.

3. TECHNOLOGY USED BY AIOT

Artifical Intelligence of Things based diagnosis system uses the Fifth Generation (5G) mobile networks technology because it is significantly superior as compared with the earlier generations (4G, 3G, 2G…) of the mobile networks (Ni et al., 2018). The 5G networks have a lot of advantages. For example, 5G peak downlink data rate and peak uplink data rate are 20 Gbps and 10 Gbps respectively. Additionally, 5G improves overall efficiency of network and reduces the network delay. In a simple network, 5G provides node-to-node latency of less than 5ms. The AIoT system, enabled with the 5G networks, provides an attractive operating model for the COVID-19 pandemic and efficiently offers various new services for the diagnosis, prediction and treatment of the patients (Peeri et al., 2020).

AIoT amalgamates (Bai et al., 2020) basic functions of IoT infrastructure as well as of AI technology (Figure 2). Originally, IoT supports Radio-Frequency Identification (RFID) technology which uses varied communication technologies such as sensors, remote control, machines and people to achieve intelligent management of information. In addition, vivid platforms, devices, chipsets, servers, and programs etc. contribute IoT infrastructure to improve the capabilities of human-machine interaction, data management, data analytics, and decision making to IoT (Pham et al., 2019). The most important feature of IoT is ubiquitous connectivity which is based on the agreed communication protocol. The IoT has three important processes which are comprehensive perception, reliable transmission and intelligent processing of data. Within the AI technology, ML provides IoT an ability to learn from the data into useful information, transforms IoT enabled assets into the learning-machines which facilitates improved decision-making.

Figure 2. Technology used by AIoT

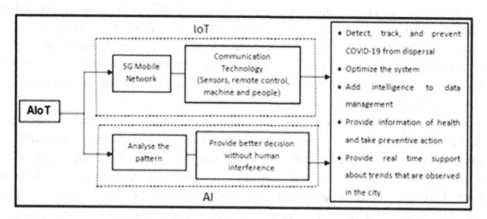

In other words, Artificial Intelligence adds value to the IoT through machine learning and deep learning via enhanced decision-making; and Internet of things adds value to the AI through connectivity, signals and exchange of data. Within the AIoT, AI improves IoT while transforming IoT originated data into some actionable insights, and vice versa. There is an ultimate aspiration of Application Programming Interfaces (API) to support interoperability between IoT infrastructure and AI technology for better leveraging of the AIoT capabilities to add on new dimension towards COVID-19. The aim of AIoT in the COVID-19 is to create a decision-oriented data analytical model which is supported by the information technology e.g., reliable communication, artificial intelligence, and machine learning. The health care monitoring system integrates IoT sensors with the AI technology. One benefit of AIoT based health prototype is that the information of patient can be accessed from anywhere and at any point of time. The patient can be reviewed, monitored and consulted with the doctor while on move. The patient kith can be alarmed about symptoms of Corona virus, dropping in the oxygen level, severity of heart attack, level of the lung infection, availability of ICU bed in a hospital, stock of medicine and injection etc. The AIoT is tech savvy with respect to sensing, notification, and logging functions, and can easily be extended to support instantaneous healthcare through remote sensing based mobile monitoring system.

In a nutshell, AIoT is transformational for both the AI and IoT technologies. The AIoT is instrumental for solving AI in IoT services as well as for IoT in AI applications.

4. ROLE OF AIOT IN COMBATING COVID-19

AIoT not only collects the data of COVID-19 using various IoT devices, sensors, actuators but also analyzes the patterns and makes best decisions without any human interference. The result gives unbelievable insight to various professionals and doctors to help people consider their health risks related to COVID-19 (McCall, 2020). IoT, a network of interconnected systems that provide ubiquitous connectivity is used to detect, track, and prevent COVID-19 from dispersal. AIoT also uses data analytics which is used to optimize the system, adds intelligence to data management and provides higher performance. With the AIoT technique, people can provide information of their health and support city administration to take preventive actions (Lin, Chuang, Yen, Huang, Chen et al, 2019). This data which is given by the citizens about their symptoms, travel and contact history is stored in the IoT devices, which then used by AI and data analytics techniques to provide a real-time report about the trends observed in the city.

The following points highlight important roles of AIoT in combating COVID-19 (Figure 3) with suggestive exemplars.

- Tracking of infected person: Multinational technology companies can build innovation for AIoT based solution to screen larger population in an effective manner. Their system can detect changes in the body temperature while people are on move. AIoT can aid in examining hundreds of people per minute that too without disrupting the flow of people in crowded areas and easily recognize the sick ones.
- Ensuring compliance to quarantine: Once an infected person goes into quarantine, AIoT can be helpful to undertake the patient compliance. A person who is to be continued in quarantine, or who has completed his/her quarantine can be monitored by the public health personnel.
- Thermal imaging supports contagion monitoring: Thermal imaging using AIoT can be in the form of a small but affordable package such as visual camera which can be worthwhile to monitor and alarm about the patient temperature. AIoT based solution can guard against unplanned breakdown, service disruption, or equipment failure; supports generation of energy and its distribution, along with the facility of well-equipped storage.
- Remote monitoring and diagnosis: For real-time, AIoT based human-computer interface can be preferred for remote monitoring and diagnosis which can be developed and deployed while combining technological aspects. AIoT integrates heterogeneous, multi-source information and comprise of technologies- sensor based, monitor online, wireless transmission and fault diagnosis etc. which can provide feedback to the potential user too.

- Remote imaging diagnosis platforms enhance collaboration: To collaborate, interact and consult with physicians, scientists and specialists at different care settings, AIoT based remote imaging platform can be provided. The platform can endow clinical and diagnostic services which are easily accessible, cost effective and highly secure for medical image segmentation, quality visualization in 3D etc. Furthermore, this AIoT platform can prove to be a rich resource for clinical research and investigations.
- Smart medical robots to help care for quarantined patient: AIoT based smart medical robots can be set-up to assist healthcare professionals with various tasks for quarantined patients. Some of these tasks can be temperature scanning, heart-rate monitoring, premise cleaning and sanitization etc. The AIoT based robots can perform remarkably well to answer complex queries of patients' visitors, and can also provide suitable safety measures according to the health severity.
- Epidemic prevention patrol drone: For epidemic prevention, AIoT based patrol drone can undergo pilot training to reduce risks that are borne by the direct contact among people. AIoT based drone

Figure 3. Important Roles of AIoT in Combating COVID-19

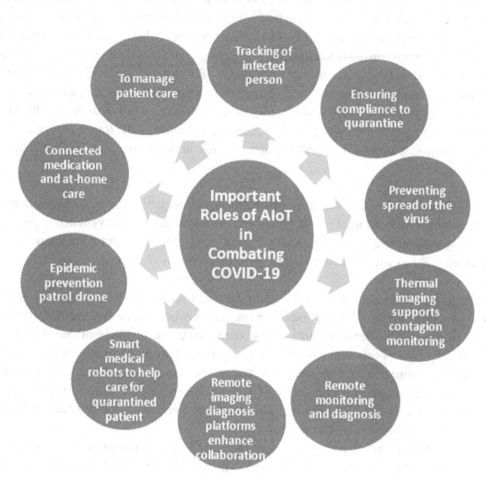

can fly to high-altitude for alert about epidemic prevention program, spray disinfectant water over the target area, regular inspection and surveillance of epidemic control such as check people for wearing masks and obliging social distancing etc.

- Connected medication and at-home care: AIoT based stay-at-home medical devices and connected medication ensures uninterrupted care for out-patients as well as reduces strain on medical personnel. AIoT for patients include exercise reminder, alert for medication, monitor glucose level, and immediate response to heart attack etc. For medical staff, real-time track of patient medication course and health progress that too with reduced recurring hospital visits by the patient.
- To manage patient care: To manage patient healthcare, AIoT based decision making can mimic human intelligence. AIoT based systems may be installed in make shift isolation camps and make-shift hospitals in order to provide better health care with minimize human interaction.

The COVID-19 virus can be prevented from further spreading, if governments collectively build a huge global network of virus-detection intelligent AIoT sensors using mobile phones and surveillance cameras to identify the disease; monitor and diagnose their citizens with smart AIoT devices that have already been contracted by the COVID-19; alert and alarm citizens through AIoT from community spread of virus; instruct and impose guidelines and preventive measures through AIoT from Corona virus spread; circulate in a channelized fashion the requirements of aid in oxygen supply, hospital beds, experienced doctors, medications and injections; and arrangement of cremation for unclaimed dead bodies due to Covid through AIoT.

5. CONCLUSION AND FUTURE WORK

In this chapter, it is revealed that Artificial Intelligence of Things (AIoT) has the potential to optimize AI systems, IoT network operations, exploration of data through real-time analytics and enhanced decision-making processes. AIoT technology has open up enormous possibilities for the healthcare system, and influences of 5G on healthcare are manifold and far-reaching. There is an ultimate aspiration of API to support interoperability between IoT infrastructure and AI technology for better leveraging of the AIoT capabilities to add on new dimension towards COVID-19. The AI and IoT technologies are combined to reorganize the healthcare, and pandemic effected zones in the field of virtual reality, telemedicine, automatic patient monitoring and diagnosing devices for suspected and infected patients. We hope to improve the safety conditions of medical professionals, healthier and hygiene environment to patients, and treating the COVID-19 infections with utmost care using AIoT as a latest tool. The AIoT is destined to bring a more appropriate and protected environment to the people all around the world against the pandemic.

In the present work, important roles of AIoT to tackle COVID pandemic using suggestive exemplars are highlighted. In future, some of these key roles can be modeled and implemented using AIoT. These solutions can be adapted and deployed for other virus-related pandemics too in future.

REFERENCES

AIoT Market. (2020). https://mindcommerce.com/artificial-intelligence-of-things/

Al-Tudjman, F., & Deepak, B. D. (2020). Privacy-aware energy-efficient framework using the internet of medical things for COVID-19. *IEEE Internet Things Magazine*, *3*(3), 64–68. doi:10.1109/IOTM.0001.2000123

Alam, F., Mehmood, R., Katib, I., Albogami, N. N., & Albeshri, A. (2017). Data fusion and IoT for smart ubiquitous environments: A survey. *IEEE Access: Practical Innovations, Open Solutions*, *5*, 9533–9554. doi:10.1109/ACCESS.2017.2697839

Ansari, S., Aslam, T., Poncela, J., Otero, P., & Ansari, A. (2020). Internet of things-Based healthcare applications. In *IoT Architectures, Models, and Platforms for Smart City Applications* (pp. 1–28). IGI Global. doi:10.4018/978-1-7998-1253-1.ch001

Archana, R., & Indira, S. (2013). Soldier monitoring and health indication system. *International Journal of Science and Research*.

Bai, L., Yang, D., Wang, X., Tong, L., Zhu, X., Bai, C., & Powell, C. A. (2020). *Chinese experts' consensus on the Internet of Things-aided diagnosis and treatment of coronavirus disease 2019. In Clinical eHealth* (Vol. 3). Elsevier.

Da Costa, C. A., Pasluosta, C. F., Eskofier, B., da Silva, D. B., & da Rosa, R. R. (2018). Internet of Health Things: Toward intelligent vital signs monitoring in hospital wards. *Artificial Intelligence in Medicine*, *89*, 61–69. doi:10.1016/j.artmed.2018.05.005 PMID:29871778

Dai, H. N., Imran, M., & Haider, N. (2020). Blockchain-enabled internet of medical things to combat COVID-19. *IEEE Internet Things Magazine*, *3*(3), 52–57. doi:10.1109/IOTM.0001.2000087

Islam, S. R., Kwak, D., Kabir, M. H., Hossain, M., & Kwak, K. S. (2015). The internet of things for health care: A comprehensive survey. *IEEE Access: Practical Innovations, Open Solutions*, *3*, 678–708. doi:10.1109/ACCESS.2015.2437951

Koncar, J., Grubor, A., Maric, R., Vucenovic, S., & Vukmirovic, G. (2020). Setbacks to IoT implementation in the function of FMCG supply chain sustainability during COVID-19 Pandemic. *Sustainability*, *12*(18), 7391. doi:10.3390u12187391

Kushwah, R., Batra, P. K., & Jain, A. (2020). Internet of things- Architectural elements, challenges and future directions. In *6th International Conference on Signal Processing and Communication*, (pp. 1-5). IEEE. 10.1109/ICSC48311.2020.9182773

Kushwah, R., Kulshreshtha, A., Singh, K., & Sharma, S. (2019). ECDSA for data origin authentication and vehicle security in VANET. In *Proceedings of the Twelfth International Conference on Contemporary Computing (IC3)*, (pp. 1-5). IEEE. 10.1109/IC3.2019.8844912

Lai, Y. H., Chen, S. Y., Lai, C. F., Chang, Y. C., & Su, Y. S. (2019). Study on enhancing AIoT computational thinking skills by plot image-based VR. *Interactive Learning Environments*, 1–14. doi:10.1080/10494820.2019.1580750

Lemley, J., Bazrafkan, S., & Corcoran, P. (2017). Deep learning for consumer devices and services: Pushing the limits for machine learning, artificial intelligence, and computer vision. *IEEE Consumer Electronics Magazine*, *6*(2), 48–56. doi:10.1109/MCE.2016.2640698

Lin, Y. J., Chuang, C. W., Yen, C. Y., Huang, S. H., Chen, J. Y., & Lee, S. Y. (2019). An AIoT wearable ECG patch with decision tree for arrhythmia analysis. In *Proceedings of the 2019 IEEE Biomedical Circuits and Systems Conference (BioCAS)*, (pp. 1-4). IEEE. 10.1109/BIOCAS.2019.8919141

Lin, Y. J., Chuang, C. W., Yen, C. Y., Huang, S. H., Huang, P. W., Chen, J. Y., & Lee, S. Y. (2019). Artificial intelligence of things wearable system for cardiac disease detection. In *Proceedings of the 2019 IEEE International Conference on Artificial Intelligence Circuits and Systems (AICAS)*, (pp. 67-70). IEEE. 10.1109/AICAS.2019.8771630

Lockdown. (2020). https://www.theguardian.com/world/2020/apr/04/lockdown-could-last-weeks-more-across-europe-officials-warn

Lupia, T., Scabini, S., Pinna, S. M., Di Perri, G., De Rosa, F. G., & Corcione, S. (2020). 2019 novel coronavirus (2019-nCoV) outbreak: A new challenge. *Journal of Global Antimicrobial Resistance, 21*, 22–27. doi:10.1016/j.jgar.2020.02.021 PMID:32156648

Matuszek, C. (2018). Grounded language learning: Where robotics and NLP meet. *Proceedings of the International Joint Conference on Artificial Intelligence*, 5687-5691. 10.24963/ijcai.2018/810

McCall, B. (2020). COVID-19 and artificial intelligence: Protecting health-care workers and curbing the spread. *The Lancet. Digital Health, 2*(4), e166–e167. doi:10.1016/S2589-7500(20)30054-6 PMID:32289116

Mdhaffar, A., Chaari, T., Larbi, K., Jmaiel, M., & Freisleben, B. (2017). IoT-based health monitoring via LoRaWAN. In *IEEE EUROCON 2017-17th International Conference on Smart Technologies*, (pp. 519-524). IEEE. 10.1109/EUROCON.2017.8011165

Muthu, B., Sivaparthipan, C. B., Manogaran, G., Sundarasekar, R., Kadry, S., Shanthini, A., & Dasel, A. (2020). IOT based wearable sensor for diseases prediction and symptom analysis in healthcare sector. *Peer-to-Peer Networking and Applications, 13*(6), 2123–2134. doi:10.100712083-019-00823-2

Ni, J., Lin, X., & Shen, X. S. (2018). Efficient and secure service-oriented authentication supporting network slicing for 5G-enabled IoT. *IEEE Journal on Selected Areas in Communications, 36*(3), 644–657. doi:10.1109/JSAC.2018.2815418

Nweke, H. F., Teh, Y. W., Al-Garadi, M. A., & Alo, U. R. (2018). Deep learning algorithms for human activity recognition using mobile and wearable sensor networks: State of the art and research challenges. *Expert Systems with Applications, 105*, 233–261. doi:10.1016/j.eswa.2018.03.056

Peeri, N. C., Shrestha, N., Rahman, M. S., Zaki, R., Tan, Z., Bibi, S., Baghbanzadeh, M., Aghamohammadi, N., Zhang, W., & Haque, U. (2020). The SARS, MERS and novel coronavirus (COVID-19) epidemics, the newest and biggest global health threats: What lessons have we learned? *International Journal of Epidemiology, 49*(3), 1–10. doi:10.1093/ije/dyaa033 PMID:32086938

Pham, H. T., Nguyen, M. A., & Sun, C. C. (2019). AIoT solution survey and comparison in machine learning on low-cost microcontroller. In *Proceedings of the 2019 International Symposium on Intelligent Signal Processing and Communication Systems*, pp. 1-2. IEEE. 10.1109/ISPACS48206.2019.8986357

Quarantine. (2020). https://ncdc.gov.in/WriteReadData/l892s/90542653311584546120.pdf

SchabenbergerO. (2018). https://iotpractitioner.com/what-is-the-aiot-exclusive-interview-with-oliver-schabenberger-at-iot-slam-live-2018/

Shaikh, R. A., & Nikilla, S. (2012). Real time health monitoring system of remote patient using ARM7. *International Journal of Instrumentation, Controle & Automação, 1*(3-4), 102–105.

Sharpe, R., van Lopik, K., Neal, A., Goodall, P., Conway, P. P., & West, A. A. (2019). An industrial evaluation of an Industry 4.0 reference architecture demonstrating the need for the inclusion of security and human components. *Computers in Industry, 108*, 37–44. doi:10.1016/j.compind.2019.02.007

Shereen, M. A., Khan, S., Kazmi, A., Bashir, N., & Siddique, R. (2020). COVID-19 infection: Origin, transmission, and characteristics of human coronaviruses. *Journal of Advanced Research, 24*, 91–98. doi:10.1016/j.jare.2020.03.005 PMID:32257431

Singh, R. P., Javaid, M., Haleem, A., & Suman, R. (2020). Internet of things (IoT) applications to fight against COVID-19 pandemic. *Diabetes & Metabolic Syndrome, 14*(4), 521–524. doi:10.1016/j.dsx.2020.04.041 PMID:32388333

Social Distancing. (2020). https://www.health.gov.au/sites/default/files/documents/2020/03/coronavirus-covid-19-information-on-social-distancing.pdf

Song, Y., Jiang, J., Yang, D., & Bai, C. (2020). *Prospect and application of Internet of things technology for prevention of SARIs. In Clinical eHealth* (Vol. 3). Elsevier.

Thrall, J. H., Li, X., Li, Q., Cruz, C., Do, S., Dreyer, K., & Brink, J. (2018). Artificial intelligence and machine learning in radiology: Opportunities, challenges, pitfalls, and criteria for success. *Journal of the American College of Radiology, 15*(3), 504–508. doi:10.1016/j.jacr.2017.12.026 PMID:29402533

Vanus, J., Smolon, M., Martinek, R., Koziorek, J., Zidek, J., & Bilik, P. (2015). Testing of the voice communication in smart home care. Human-Centric Computing and Information Sciences, 5(15), 1-22. doi:10.118613673-015-0035-0

Yu, K. M., Chen, Y. C., Liu, C. H., Hsu, H. P., Lei, M. Y., & Tsai, N. (2019). IFPSS: intelligence fire point sensing systems in AIoT environments. In *Proceedings of the 2019 International Conference on Image and Video Processing, and Artificial Intelligence*. International Society for Optics and Photonics. 10.1117/12.2550315

Zadeh, L. A., Tadayon, S., & Tadayon, B. (2018). *U. S. Patent Application No. 15/919,170*. US Patent Office.

Zhang, Z., He, T., Zhu, M., Shi, Q., & Lee, C. (2020). Smart triboelectric socks for enabling artificial intelligence of things (AIoT) based smart home and healthcare. In *Proceedings of the 33rd International Conference on Micro Electro Mechanical Systems (MEMS)*, (pp. 80-83). Vancouver, Canada. IEEE. 10.1109/MEMS46641.2020.9056149

Zhao, R., Yan, R., Chen, Z., Mao, K., Wang, P., & Gao, R. X. (2019). Deep learning and its applications to machine health monitoring. *Mechanical Systems and Signal Processing, 115*, 213–237. doi:10.1016/j.ymssp.2018.05.050

Chapter 9
Recent Trends in Wearable Device Technology for Health State Monitoring

E. D. Kanmani Ruby

Veltech Rangarajan Dr. Sakunthala R&D Institute of Science and Technology, India

P. Janani

Vel Tech Rangarajan Dr. Sagunthala R&D Institute of Science and Technology, India

V. Mahalakshmi

Vel Tech Rangarajan Dr. Sagunthala R&D Institute of Science and Technology, India

B. Sathyasri

Vel Tech Rangarajan Dr. Sagunthala R&D Institute of Science and Technology, India

S. Vishnu Kumar

Vel Tech Rangarajan Dr. Sagunthala R&D Institute of Science and Technology, India

ABSTRACT

The latest innovative technology products in the market are paving the way for a new growth in the medical field over medical wearable devices. Globally, the medical market is said to be segmented on the basis of global medical wearable report by its type, application level, regional level, and country level. In this medical advisory, these devices are classified as diagnostic, therapeutic, and respiratory. The regions covered include Europe, Asia-Pacific, and the rest of the world. These wearable devices are technically embedded with electronic devices which the users are able to adhere to their body parts. The main function of these wearable devices is said to be collecting users' personal health data (e.g., such devices include measurement on fitness of body, heartbeat measurement, ECG measurement, blood pressure monitoring, etc.).

DOI: 10.4018/978-1-7998-6870-5.ch009

I. INTRODUCTION

Wearable devices should be portable, should be smart hence it is so called mobile electronic device. Initially these wearable devices were used in Military applications. Nowadays with increased number of population rate around the world these devices are used in medical domain and even used for entertainment. (Shaffer, 2002)In rural areas where people are not provided with good hospital facility, these wearable devices are used to get time to time data in the case of health care providers. It would be main advantage for aged people, because it could predict the vital signs of the patient early and paves way for early stage detection of diseases or prevention of diseases necessarily e.g. checking oxygen saturation level, glucose level, blood pressure etc. This is a main advent for disease management. This situation comes inside when no hospital environment can be provided for the patient for long time. Hence these wearable devices play a vital role in post-operative conditions.(Mukhopadhyay, 2015)Wearable devices in the current trend are emerging with necessary tools in the way of Internet of Things. With the advancements in sensors, wireless connectivity, active materials and batteries these wearable devices are emerging in the way with the latest key trends in many industries. (Patel et al., 2012)After the advent of smart phone and Digital money, revolution of these wearable devices has made the next frontier in digital transformation. In this latest COVID19 situation a new wearable De novo has made its projection as a new Innovation for COVID19 scenario.(Park & Jayaraman, 2003) It is a Patient monitoring solution to detect the early phase of the pandemic developed by Masimo SafetyNet-COVID-19.It helps to coordinate the continuous monitoring of people through a variety of sensors and user applications. These provide inherited and secure monitoring.(Bandodkar & Wang, 2014)

The main aim of the medical field nowadays is to enable patient-enabled decisions that improve the health of the patient with reduction in cost and manual monitoring. To reach this goal the medicine is said to be relying on data which are pin pointed on large set of data related to clinical subgroups and paves way in discovering disease mechanism.(Stoppa & Chiolerio, 2014) On the aspect of these medicines the analysis data are said to be only good as the quality of data, therefore high quality measurements are said to be generating high - information for content data.(Robinson et al., 1966) Combination of these analysis and measurement of medical data will be considered as one of the part of field digital medicine. (Snyder et al., 1964)Wearable technologies play a vital role in the advancements of this precision medicine by enabling information of high content information that is very clinically relevant. (Caplan & Jones, 1975)Wearable device technology promotes the digital measurement of parameters in real time. This technology has emerged as a major component in the current world scenario and as a main part of fitness markets with accelerated monitoring devices and heart-monitors based on photo-plethysmography which is widely available in the market. (Obrist et al., 1978) These wearable devices represent wide range of personalized apps related to these health monitoring devices.(Buchman et al., 2002) The main challenge for these devices is that only a very few devices have been approved for healthcare monitoring use. Whatsoever maybe, these abiding opportunities to adapt to wearable technologies for use in healthcare. Ability to continuously monitor healthcare related parameters of individuality's health state results in deriving high volume of data which ad-verses in both opportunities and challenges of these health parameter data analysis.(Parati et al., 1988) Hence for medical parameters analysis technologies like Artificial intelligence, data mining and machine learning provides new tools for the analytical measurements. (Seccareccia et al., 2001)

In the medical settings four main primary vital signs such as Respiration rate, Heart rate, Blood pressure and Temperature is often done by manual measurement or any physical examination being done.

(Camm, 1996)Whereas the continuous measurement is often with in-patient setting along with limited portability. Hence the main key aspect of wearable healthcare and this precision medicine will device high quality, clinical relevant data to the users. (Asada et al., 2003)Here the factors which are been reviewed are (1) Respiration rate (2) Blood Pressure (3) Glucose level (4) Heart Rate (5) Body Temperature (6) Sweating and (7) Sensation. Although four main parameters are being measured in addition to sweating and sensation activity is also being measured. (Pansiot, 2007)In depth, a brief review has been done and assessed for various methods of measurement modularity's which are in use currently. To be frank enough to say out of these technologies some can be considered for measurable wearable devices while some are not compatible with continuous convenient and are said to be non-invasive. (Chance, 1992)

Figure 1. Wearable Device Parametric measurement

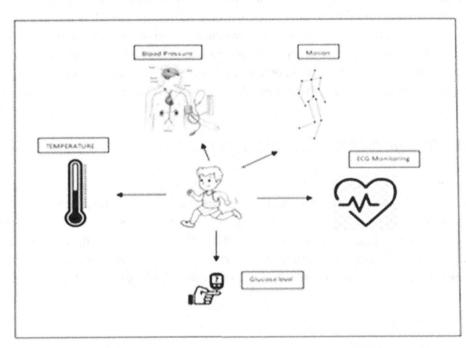

II.IOT BASED HEALTHCARE

IoT based healthcare devices provides excellence in providing the parameters like monitoring patients remotely, improvised healthcare access, cost reduction, time saving and more over detection of early stage deterioration.(Poon et al., 2006) These are mainly focusing on reduction in risk mainly. The future of healthcare industry with the wearable devices is aimed at wellness management and prevention of diseases at the early stage promoting for a value based healthcare services. With this the medical industry service provider are shifting their interest from consumable/device providers to care management center/ organizations. (Allen, 2007)Companies operating with the IoT space include many connected devices put together. (Konig, 2000)

IoT technology has not only enhanced the independence in medical industry but also diversified the ability of humans to interact with the external environment . It helps with the forth coming algorithms and techniques used in medical field which further contributes to global communication . As the IoT connects N number of devices, wireless sensors, home appliances and most of the electronic devices over the internet . The main advantage is duly because of showing high accuracy in data, low power consumption, low cost and ability to predict the future events. In the healthcare industry the sensors used for medical; diagnosis are mostly embedded or wearable on the human body to collect the relevant data from the user. Adding to this the environment information can also be recorded like temperature, humidity, date and time. These data gives an precise inference in the health monitoring data. Data storing and delivering from an IoT devices plays a major role as large amount of data are acquired/recorded from various different sources like sensors,mobiles, tablets, email, software and other related applications. These type of devices are said to be made available for all doctors, caretakers, physicians and medical authorized parties. Sharing of all these data along with the healthcare providers through cloud allows quick diagnosis of the user or the patients. The cooperation between the users, patients and communication module is maintained for effective and secure transmission . These iot subsystem acts as a dashboard for many medical caregivers where user has control over the data, data visualization etc. Many countries now have adopted the maximum use of IoT in healthcare systems with latest technology and policies.

III. CHALLENGES IN WEARABLE TECHNOLOGIES

Wearable devices should be reliable with human movement. Where it should be ergonomic and must be easy to use. Provided with no hindrance and should be unobtrusive to human activity. All these applications rely in the battery life for which now-a-days all wearable devices come up with rechargeable batteries, which are washable and reusable. (Anderson & Parrish, 1981)But even though these devices are providing data with mobile ecosystem data security is very much least seen in these devices. Therefore, detailed reviews of these wearable technologies are as follows. (Faber et al., 2004)

Heart Rate

A. Heart Rate Measurement

The heart is the main part of the human body. Considered to be a biomedical organ, It pumps blood around the vascular stem supplying nutrients to other organs. It is said to be dependent on the level of physical activity done by the user. (Roggan et al., 1999)Hence this is said to be an physiological indicator of health state monitoring. Generally, the heart rate is 40 - 60 bpm for athletes and 60 - 100bpm for children and adults. While working out or doing exercise may reach up to 200 bpm.

Generally heart rate detection is basically said to be prevailing under mechanical, electrical or any of the optical signals that which is associated with human heart beat. (Kim & Yoo, 2006)By placing an artery near near the heart surface the heart rate can be measured. Of all this the electrodes promotes the electrical activity of the heart muscle that which is detected when attached to the chest. i.e Electrocardiogram or the ECG.(Wang et al., 2007) Rise in changes of the impedance of the issue can be caused due to the blood flow and this can be measured with a variation of 4 electrodes resistant method. (Birse, 2004)where the electrodes are being placed at the neck or waist of the user.Heart beat signal is normally detected by a stethoscope with the help of an ultrasound sensor. (Waller, 1887)The vessels inside the human body give rise to optical transmission or reflection that which is recorded by the PPG (photo plethysmography)(Rivera-Ruiz, 2008)

B. Heart Rate Measured using Wearable Devices

The most promptly used wearable device for heart rate are photo plethysmography and electrocardiogram. (Hurst, 1998) Along with the latest home healthcare services and tale-medicine services detection of heart rate and several other methods are under development. (Burke, 1998)

Figure 2. Various methods of measuring the Heart rate and cardiac functions.

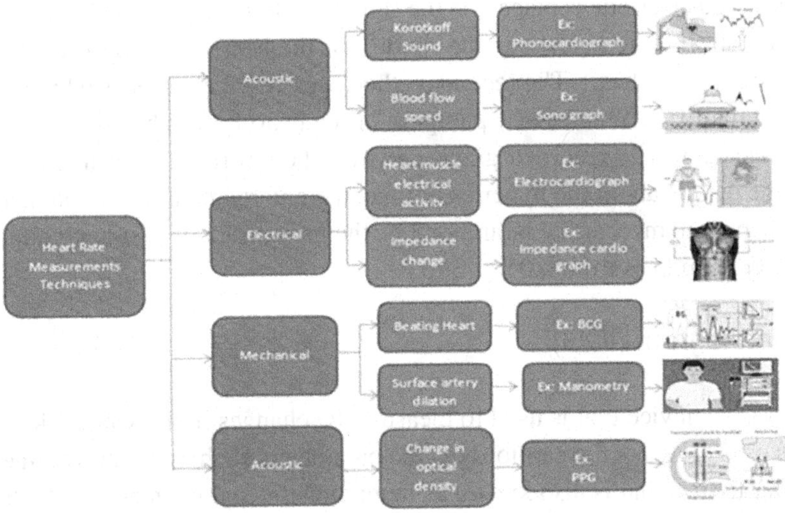

C. Photo Plethysmography (PPG)

The changes in the tissue or the organs inside the human body is identified using the PPG. (Eldridge, 2014)For eg: The pulmonary function changes in the blood volume can be used to assess the cardiac function . the PPG along with the optical light source acts as a detector giving RED or GREEN.(Alvarez et al., 2013) With respect to the wearable sensors, PPG is also said to be measuring the changes in blood volume which looks after the contraction of vessels in demis and hypo-demis. (Chi et al., 2010) Therefore, this PPG provides an insight to the cardiovascular functions of blood oxygen level, stroke volume, heart rate and resistance in vascular tissues.(Lee & Chung, 2009)

PPG is said to be measured using an infrared source. The diagnostic is around 700- 1000nm. This is said to be identified using the melanin that is excreted from the skin and other proteins in it. (Parak & Korhonen, 2014)Blood oxygen level saturation (SpO_2) usually gives you the two light sources which are used with emission wavelengths at 660nm and 940nm. More number of PPG sensors are said to be embedded into a wrist -worn device that is said by wirelessly transmitting the data. (Dias, 2009) SNR (Signal to noise ratio) is a part and parcel of the blood tissues. (Summers et al., 2003)These PPG sensors can also be placed near the ear lobe where the increase in vasculature can increase the SNR ratio even though it requires eye glasses or Headphones to bc mounted for the sensor and electronics. (Lee, 2013) The disadvantage considered here is near the ear lobe the temperature is very much decreased which results in vasoconstriction which inversely decreases the SNR. (Ruiz, 2013)

D. Electrocardiogram (ECG)

As the heart function like a 2 stage electrical pump, it is possible to measure the electrical activity by placing the electrodes over the skin surface of the heart. After the introduction of monitors in 1970s the measured ECG signals were recorded using the most trending device for cardiovascular measurements.

Basically there are two different methods to measure the ECG signal, in which the first method uses three electrodes. Each one electrode is placed on the right arm, left arm and the left chest. The recorded waveform is used to measure the heart rate and rhythm of the heart rate only it does not provide the complete information about the heart. The second method uses 10-12 electrodes which is a conventional method of measuring the ECG. When every pumping of the heart beat is heard in the muscle the electrodes record these changes. The recorded graph provides the information of heart rhythm and heart rate. It also shows the details about the hypertension i.e, the enlargement of human heart due to heavy blood pressure and about the myocardial infarction i.e., the probability of past heart attack. (Chen,) The amplitude of the ECG monitor is from 0.05v - 3mv.

E. Impedance Cardiogram (ICG)

The ICG is a noninvasive device that is used to measure the changes in the whole electro conductivity of the thorax that is used to assess the cardio-dynamic parameters such as stroke volume, heart rate and resistance in vascular tissues that is also known as hemodynamic parameters. In which low amplitude of current with high frequency is passed through the electrode's that is placed over the chest and the potential difference were measured using the electrodes placed inside. Thus the ICG records the difference of impedance that is caused due to the variation of blood volume and the velocity of the aorta due to each and every heartbeat.

Now-a-days the modern ICG device uses four electrode system to measure the vital signs.(Shu et al., 2015) This is mainly used to measure the frequency range from 20-100khz along with an sinusoidal of 1-5mA. These ICG devices can be attached to the user's chest, waist or neck and even it can be infused along with smart clothing. (Gong, 2014)

F. Other Methods for Heart Rate Detection

The above discussed methods like ECG, PPG, ICG are sincerely devised for heart rate parameter measurements. For eg: the change in conductance is used to detect the arterial rate. (Cotes, 2009)

RESPIRATION

A. Respiratory Rate and Pulmonary Measurement

Another major part of human body is the respiratory systems which includes nose and nasal cavity, mouth, larynx, trachea, bronchi and bronchitis etc.(Goodman, 1978) Here the respiratory rate is one of the four vital signs and is usually measure by the number of times the chest is expanded / contracted per minute (BPM) has been discussed. (Murphy, 1981)This respiration rate varies from 30-60 bpm for newborn and infants of 0-12 months old. This respiration rate can be continuously monitored and can provide important data regarding health state monitoring. Respiration rate more than 20BPM caused asthma for adults like fever, dehydration, stress, pain, anger, lung cancer etc. (Selley et al., 1989)

The PFT test (pulmonary function test) include various measurements like volume, capacity, flow rate, efficiency etc. (Otis et al., 1950)This includes a spirometry test which measures the inhalation/exhalation volume during breathing phenomenon which depends upon the amount of intake of oxygen in one minute. (Al-Khalidi et al., 2011)The diffusion capacity of the lungs exchange oxygen with the vascular system. Using a pulse oximeter involves measurement of SpO2. (Clark & von Euler, 1972)

B. Wearable Devices for Measurement of Respiration Rate

Enhanced wearable devices for respiratory measurements are given which is mainly dependent on expansion and contraction of abdomen and chest during breathing . (Netzer et al., 2001)

Figure 3. Measurement of pulmonary function and Respiration rate

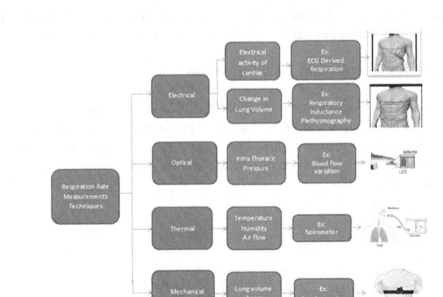

1.Spirometry

During the breathing phenomenon the flow rate and volume rate can be obtained from wearable devices, as they usually include a face-mask. The commonly known device is said to be ergo spirometer mask which is used to detect the amount of air breathed in and breathed out during exercises. Spiro-meters are said to be measuring the airflow using ultrasonic transducers. These inhaled and exhaled air is usually measured using a flow meter as there can be changes in the humidity and temperature of the surroundings. Small scale model of these sensors and methods which are used for reliable measurements near the mouth or nose may cause convenience of a face-mask which leads to continuous monitor of the respiration rate. (Sackner et al., 1989)

These sensors usually consist of two brands one containing a loop wire which and is said to be placed around the chest and the other is said to be placed around the abdomen. (Seymour, 2010)Where

2.Respiratory Inductance Plethysmography (RIP)

The expansion and contraction of the chest and abdomen is detected by Plethysmography. Sensors are said to be consisting of two bands of which each is said to be containing a wire loop which is placed around the chest and one is around the abdomen. (Bianchi et al., 2013)The change in the magnetic field which is induced in the wire loop results in expansion and contraction with the breathing phenomenon. The cross section of the loop is said to be in a frequency shift of the signal which is proportional to the cross-sectional area. Gas volume is calculated by the sum of the signal from the two bands which is calibrated against a known gas volume. (Mathew et al., 2012)High accuracy is got only when elastomeric plethysmography and impedance plethysmography is taken.

3.Respiration Rate from Photo plethysmography (PPG)

The PPG Signal measures the blood volume in tissue this intern is useful for heart rate measurement. (Flisberg et al., 2002)The breathing generates small changes in heart along with the blood pressure.

C. Other Methods for Respiration Rate Detection

A wide range of motion based respiration devices which are still under development with low cost and flexibility. (Cabot, 2008)The changes with the contraction of the chest and abdomen is measured from the respiratory rate that can be estimated further. The bi-axial accelerometer this can determine the respiration rate under certain static conditions like the user standing, lying, sitting or doing any body movement .(Stemp & Ramsay, 2006)

BLOOD PRESSURE

A. Measurement of Blood Pressure

Blood ejected is the flow of arteries from the heart into the aorta it will results in Pulsate. In normal heart function, the maximum and minimum of the blood pressure is between various from a maximum **systolic** and minimum **diastolic** value. (Fohr, 1998)The detection of the blood pressure is relying on the transduction of acoustic, mechanical, and optical signals. Occluding an artery, is a Non-invasive measurement of blood pressure it involves bronchial artery in the arm, using an inflatable cuff. While deflating the cuff, the pressure will be onset of turbulent pulsate flow. The unimpeded continuous flow is an estimate of the diastolic pressure. systolic pressure, is the artery to measure the pressure corresponding.(Hanly et al., 1989)

There are three methods of detect pulsate flow:

1. Palpitation.
2. Listening a characteristics of tapping sounds
3. (Korotkoff sounds) stethoscope (auscultatory method).
4. Measuring the oscillations in the pressure cuff (Oscillometric method).

The auscultate method is non-invasive, and it have large variability in measuring between a clinician. Oscillometric method is the Automated blood pressure measurement. The total peripheral resistance and Blood pressure is caused by the cardiac output, and is dependent on stress, drugs, disease, daily activity, sleep and exercise. The cardiovascular risk is a physiological indicator of Blood pressure. The systolic pressures of 120 – 139 mm Hg. The diastolic pressures of 80 - 89 mm Hg.

B.Wearable Devices for Measurement of Blood Pressure

Hypertension diseases are certainly measured from blood pressure measurements and are done more frequently. This helps in the cardiovascular diseases and these wearable devices which are used to diagnose the blood pressure are mostly wanted for monitoring the Oscillometric pressure or the pulse rate. (Dolfin, 1983)

1.Oscillometric Blood Pressure

The Korotkoff sounds using a stethoscope is a blood pressure measurement which detects the blood pressure. Wearable device for these are most usually adhered to the human wrist, shoulders or finger in a Oscillometric method. Without removing clothing, these devices can be used. The heart measurement must be the same height. (Varady et al., 2002)High accuracy devices are widely used in technologically robust. The maximum of the waveform are said to be calculated from the amplitude but not accurately detected. A long-term monitoring wearable Oscillometric devices can require numerous components

Figure 4. Measurement's of Blood Pressure

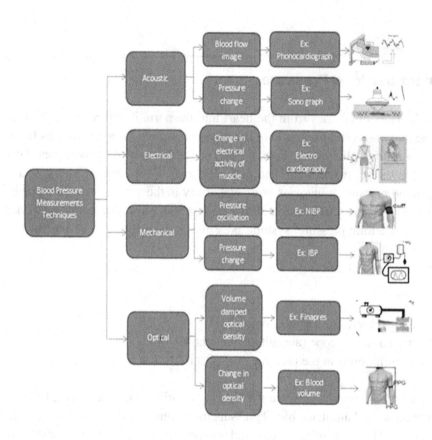

and hence it is not convenient. The discontinuous of the blood pressure is measured more than two to four times than the normal one. (Warren et al., 1997)

2.Pulse Transit Time (PTT)

It is defines as the pulse pressure which is put along the aeterial tree depending on various factors combined along with the blood pressure The ECG chest harness of a PTT is often implemented using a finger-type optical PPG sensor for seeing heart rate measurement.(Watanabe et al., 2005) The time difference between the R wave and pulse wave the R wave is from the ECG and the pulse wave from the PPG sensor. the number of factors in PTT is dependent on, and it must be calibrated for each individual by obtaining a calibration curve over a range of pressures. (Wilhelm et al., 2003)

3. Other Methods for Blood Pressure Detection

With the help of a Camera which works at high speed small changes in the skin colour can be detected which occurs due to blood pulsation that allows the contact-less image based PTT and PPG generation of signals. The blood flow is detected by the distal locations on the skin. (Yumino & Bradley, 2008)

TEMPERATURE

A. Measurement of Temperature

Human body's normal temperature is around 98.63 ^{0}F (37.0 ^{0}C) is said to be a seminal study that is derived from more than 1 million auxiliary measurements of 25,000 healthy adults reported recently. But this is discussed to be of measurement errors due to equilibration and thermometers used where not accurate. The body temperature of adults who are healthy are considered to be range in 97.8^{0}F. (36.5 ^{0}C) to 99^{0}F (37.2^{0}C). But here factors may vary depending on age, gender, nutritious body, hydration, smoking status, ovulation, skin or surface temperature is typically several degrees lower.(Toy et al., 2005)

B. Wearable Sensors for Temperature Measurement

Mostly the wearable devices for human body exposure are thermistor based. These are said to be highly sensitive at ambient temperatures. For which these devices are made in such a way that they easily adhere to the human body surface like wrist, chest, feet, ears, forehead etc. Even smart capsules can be used that are used to transmit data through temperature while present in the gastrointestinal system. (Wasserman, 1994)When heat is reflected over the surface heat exchange takes place hence it is said to be on-air temperature, exposure to sunlight, barometric pressure and altitude. For athletes these parameters are said to be monitored like fatigue, various emotional state, ovulation cycle, sleep state etc. Out of these most of the devices include in measuring the skin temperature.(Crapo, 1994)

PHYSICAL ACTIVITY

A. Measurement of Physical Activity

Human body movements are said to be produced by the musculoskeletal system are defined as body movements called as the sixth vital sign, whereas other vital signs like any kind of physical activity provide information about healthcare and its state. For example, particularly like diabetic information, cardiovascular diseases, obesity etc. Declining of these measures might affect in multiple illness and other symptoms related to functional impairment. (Reinhard et al., 1979)

Inside the hospital premises, the physical activity is said to be very important where it promotes mobility which improves the readmission rates and overall outcomes. Where daily energy (joules or calories) which provides activity measurements most appropriately. Wearable devices like worn in the wrist are actively used to monitor. Sensor which are present in the foot called foot pressure sensors can also be used for accessing the ambulatory movement. (Hollmann & Prinz, 1997)The global positioning system(GPS) as well as Indoor positioning systems are used for monitor physical activity. Electromyography is the activity which is used to measure the muscle contraction by incorporating electrodes inside the smart clothing that enables translation of this wearable device concept.

B. Wearable Accelerometer-Based Devices

Wearable device measures parameters which are based on accelerometer measuring one or more axis, where the gyroscope is used for angular measurements.(Li et al., 2017)Accelerometer s which are miniaturized are small and robust so that the sensors can be easily worn on the wrist, ankle, chest etc. Even though these are being used to measure physical activity there is no substandard in giving data directly. (Jang, 2014)Whereas in the fitness band this accelerometer data is usually taken as a step count, energy volume or any workout exercises. These are helpful in clinical relevant settings also.

C. Other Methods

The measurement of the energy expenditure in assessment of all the metabolic functions are relevant to a wide range of diseases. Whereas the energy expenditure can be measured directly by calorimeter or by the oxygen consumption and CO_2 production. Inside the face-mask portable calorimeter are embedded such that they are easy to take measurements outside the laboratory. The best solution for these type of problem is that to use heat gauges on the skin for detecting the body flux. This method has risk factors that would affect the body by heat exchange with the environment that can occur through convention of air or water whereas in conduction of materials when contact with skin can cause radiation and evaporation. (Mogera et al., 2014)

Figure 5. Methods of measurement of physical activity

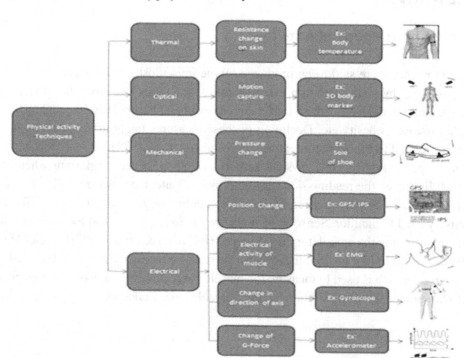

OTHER PARAMETERS

A. Sweat

Sweat is most commonly used for testing CF for laboratory purposes and further analysis. This testing can also be used for drugs, alcohol, checking the pH levels, hormonal change etc.

In recent trends of the wearable devices the sensors embedded with them are used to measure bio-marker in sweat which is considered to be a major challenge in the development process. Therefore, wearable titration sensors have been reported that requires subsequent analysis on a specific instrument.

When the user is said to be sweating it is primarily composed of water 99% along with salts, lactate, urea etc. The sweat in which the salt composition is comprised of sodium which is 0.9gL, potassium of 0.2gL, calcium of 0.015gL and magnesium of 0.0013gL. All these are interconnected on many other factors which includes heat, exercise activities which produces high sweat rate. When calculating the pH level of sweat it is moderately acidic to neutral levels which comes between 4.5 to 7.0. In a adult when the maximum range is said to be calculated it is of 0.8- 4 litres per hour. (Kang et al., 2006)

B.Emotional State Monitoring

When considering the health state monitoring the awareness under emotion, mood affect are important health conditioning factors. These emotions are said to be brief, consistent, physiological, behavioral and other neural responses to internal pr any other external events as they manifest the depression, anxiety, stress, fatigue, sleep disorders, drowsiness and other acclivity changes which affects the sleep pattern.

Emotional state can be identified from various physiological parameters like wearable sensors embedded with Heart rate (HR), respiration rate(RR), heart rate variability(HRV), Blood pressure(BP),e lectromyogram(EEG), electroencephalogram (EEG), electro-oculogram(EOG),plethysmograph(PPG), galvanic skin response(GSR) and skin temperature. With latest trending smartphones which has the range for these emotion detection parameters like alarm, voice message, text message, GPS location sharing, social networking etc. With recent updates in technology in smartphones they can detect and respond to the end user para-maters via mobile application services. If the sleep pattern is disturbed it leads to high rate of emotion .

CONCLUSION

Blooming wearable sensors technology into the healthcare industry is very slow when compared to the development of many wearable wear-outs which fits to our lifestyle and fitness market. These advances in the wearable sensor technology provides tremendous up-liftment in healthcare industry deployment particularly in precision medicine. In this chapter, the reviewed methods of wearable device technologies and parameter measurement modalities are most important for body measurements like heart rate, respiration rate, blood pressure, temperature, physical activity, emotional state, sweat etc. Also the relevant physiology had been discussed and reviewed along with the technical issues associated with measurement of these parameters. This review provides a clean and comprehensive part of the parametric design considerations for measuring the major parameters using wearable technology.

REFERENCES

Al-Khalidi, F. Q., Saatchi, R., Burke, D., Elphick, H., & Tan, S. (2011). Respiration rate monitoring methods: A review. *Pediatric Pulmonology, 46*(6), 523–529. doi:10.1002/ppul.21416 PMID:21560260

Allen, J. (2007). Photoplethysmography and its application in clinical physiological measurement. *Physiological Measurement, 28*(3), R1–R39. doi:10.1088/0967-3334/28/3/R01 PMID:17322588

Alvarez, R. A., Penín, A. J. M., & Sobrino, X. A. V. (2013). A comparison of three QRS detection algorithms over a public database. *Procedia Technology, 9,* 1159–1165. doi:10.1016/j.protcy.2013.12.129

Anderson, R. R., & Parrish, J. A. (1981). The optics of human skin. *The Journal of Investigative Dermatology, 77*(1), 13–19. doi:10.1111/1523-1747.ep12479191 PMID:7252245

Asada, H. H., Shaltis, P., Reisner, A., Sokwoo Rhee, & Hutchinson, R. C. (2003). Mobile monitoring with wearable photoplethysmographic biosensors. *IEEE Engineering in Medicine and Biology Magazine, 22*(3), 28–40. doi:10.1109/MEMB.2003.1213624 PMID:12845817

Bandodkar, A. J., & Wang, J. (2014). Non-invasive wearable electrochemical sensors: A review. *Trends in Biotechnology, 32*(7), 363–371. doi:10.1016/j.tibtech.2014.04.005 PMID:24853270

Bianchi, W., Dugas, A. F., Hsieh, Y.-H., Saheed, M., Hill, P., Lindauer, C., Terzis, A., & Rothman, R. E. (2013). Revitalizing a vital sign: Improving detection of tachypnea at primary triage. *Annals of Emergency Medicine, 61*(1), 37–43. doi:10.1016/j.annemergmed.2012.05.030 PMID:22738682

Birse, R. M. (2004). rev. Patricia E. Knowlden. In Oxford Dictionary of National Biography. Academic Press.

Buchman, T. G., Stein, P. K., & Goldstein, B. (2002). Heart rate variability in critical illness and critical care. *Current Opinion in Critical Care, 8*(4), 311–315. doi:10.1097/00075198-200208000-00007 PMID:12386491

Burke, E. (1998). *Precision heart rate training*. Human Kinetics.

Cabot, R. C. (2008). Case 12-2008: A newborn infant with intermittent apnea and seizures. *The New England Journal of Medicine, 358*(16), 1713–1723. doi:10.1056/NEJMcpc0801164 PMID:18420504

Camm, A. J. (1996, March 1). Heart rate variability: Standards of measurement, physiological interpretation, and clinical use. *Circulation, 93*(5), 1043–1065. doi:10.1161/01.CIR.93.5.1043 PMID:8598068

Caplan, R. D., & Jones, K. W. (1975). Effects of work load, role ambiguity, and type A personality on anxiety, depression, and heart rate. *The Journal of Applied Psychology, 60*(6), 713–719. doi:10.1037/0021-9010.60.6.713 PMID:1194173

Chance, B. (1992). *User-wearable Hemoglobinometer for measuring the metabolic condition of a subject*. Google Patents.

Chen, W. (2015). Wearable solutions using bioimpedance for cardiac monitoring. In Recent Advances in Ambient Assisted Living- Bridging Assistive Technologies, E-Health and Personalized Health Care. Academic Press.

Chi, Y. M., Jung, T.-P., & Cauwenberghs, G. (2010). Dry-contact and noncontact biopotential electrodes: Methodological review. *IEEE Reviews in Biomedical Engineering, 3*, 106–119. doi:10.1109/RBME.2010.2084078 PMID:22275204

Clark, F., & von Euler, C. (1972). On the regulation of depth and rate of breathing. *The Journal of Physiology, 222*(2), 267–295. doi:10.1113/jphysiol.1972.sp009797 PMID:5033464

Cotes, J. E. (2009). Lung function: Physiology, measurement and application in medicine. John Wiley & Sons.

Crapo, R. O. (1994). Pulmonary-function testing. *The New England Journal of Medicine, 331*(1), 25–30. doi:10.1056/NEJM199407073310107 PMID:8202099

Dias. (2009). Measuring physical activity with sensors: a qualitative study. *MIE*, 475-479.

Dolfin, T. (1983). Effects of a face mask and pneumotachograph on breathing in sleeping infants 1–3. *The American Review of Respiratory Disease, 128*, 977–979. PMID:6650989

Eldridge, J. (2014). Recording a standard 12-lead electrocardiogram. An approved methodology by the Society of Cardiological Science and Technology (SCST). *Clinical Guidelines by Consensus.*

Faber, D. J., Aalders, M. C. G., Mik, E. G., Hooper, B. A., van Gemert, M. J. C., & van Leeuwen, T. G. (2004). Oxygen saturation-dependent absorption and scattering of blood. *Physical Review Letters, 93*(2), 028102. doi:10.1103/PhysRevLett.93.028102 PMID:15323954

Flisberg, P., Jakobsson, J., & Lundberg, J. (2002). Apnea and bradypnea in patients receiving epidural bupivacaine-morphine for postoperative pain relief as assessed by a new monitoring method. *Journal of Clinical Anesthesia, 14*(2), 129–134. doi:10.1016/S0952-8180(01)00369-5 PMID:11943527

Fohr, S. A. (1998). The double effect of pain medication: Separating myth from reality. *Journal of Palliative Medicine, 1*(4), 315–328. doi:10.1089/jpm.1998.1.315 PMID:15859849

Gong, S. (2014). *A wearable and highly sensitive pressure sensor with ultrathin gold nanowires* (Vol. 5). Nat. Commun.

Goodman. (1978). Relationship of smoking history and pulmonary function tests to tracheal mucous velocity in nonsmokers, young smokers, ex-smokers, and patients with chronic bronchitis 1–3. Am. Rev. Respir. Dis., 117, 205-214.

Hanly, P. J., Millar, T. W., Steljes, D. G., Boert, R., Frais, M. A., & Kryger, M. H. (1989). Respiration and abnormal sleep in patients with congestive heart failure. *Chest, 96*(3), 480–488. doi:10.1378/chest.96.3.480 PMID:2766808

Hollmann, W., & Prinz, J. P. (1997). Ergospirometry and its history. *Sports Medicine (Auckland, N.Z.), 23*(2), 93–105. doi:10.2165/00007256-199723020-00003 PMID:9068094

Hurst, J. W. (1998). Naming of the waves in the ECG, with a brief account of their genesis. *Circulation, 98*(18), 1937–1942. doi:10.1161/01.CIR.98.18.1937 PMID:9799216

Jang. (2014). *Wearable respiration measurement apparatus.* US 8,636,671 B2.

Kang, Y., Ruan, H., Wang, Y., Arregui, F. J., Matias, I. R., & Claus, R. O. (2006). Nanostructured optical fibre sensors for breathing airflow monitoring. *Measurement Science & Technology*, *17*(5), 1207–1210. doi:10.1088/0957-0233/17/5/S44

Kim, B. S., & Yoo, S. K. (2006). Motion artifact reduction in photoplethysmography using independent component analysis. *IEEE Transactions on Biomedical Engineering*, *53*(3), 566–568. doi:10.1109/TBME.2005.869784 PMID:16532785

Konig, K. (2000, November). Multiphoton microscopy in life sciences. *Journal of Microscopy*, *200*(2), 83–104. doi:10.1046/j.1365-2818.2000.00738.x PMID:11106949

Lee, S. (2013). A low-power and compact-sized wearable bioimpedance monitor with wireless connectivity. *Journal of Physics: Conference Series*.

Lee, Y.-D., & Chung, W.-Y. (2009). Wireless sensor network based wearable smart shirt for ubiquitous health and activity monitoring. *Sensors and Actuators. B, Chemical*, *140*(2), 390–395. doi:10.1016/j.snb.2009.04.040

Li, S., Lin, B.-S., Tsai, C.-H., Yang, C.-T., & Lin, B.-S. (2017). Design of wearable breathing sound monitoring system for real-time wheeze detection. *Sensors (Basel)*, *17*(12), 171. doi:10.339017010171 PMID:28106747

Mathew, J., Semenova, Y., & Farrell, G. (2012). A miniature optical breathing sensor. *Biomedical Optics Express*, *3*(12), 3325–3331. doi:10.1364/BOE.3.003325 PMID:23243581

Mogera, U., Sagade, A. A., George, S. J., & Kulkarni, G. U. (2014). Ultrafast response humidity sensor using supramolecular nanofibre and its application in monitoring breath humidity and flow. *Scientific Reports*, *4*(1), 4103. doi:10.1038rep04103 PMID:24531132

Mukhopadhyay, S. C. (2015). Wearable sensors for human activity monitoring: A review. *IEEE Sensors Journal*, *15*(3), 1321–1330. doi:10.1109/JSEN.2014.2370945

Murphy, R. L. (1981). Auscultation of the lung: Past lessons, future possibilities. *Thorax*, *36*(2), 99–107. doi:10.1136/thx.36.2.99 PMID:7268687

Netzer, N., Eliasson, A. H., Netzer, C., & Kristo, D. A. (2001). Overnight pulse oximetry for sleep-disordered breathing in adults: A review. *Chest*, *120*(2), 625–633. doi:10.1378/chest.120.2.625 PMID:11502669

Obrist, P. A., Gaebelein, C. J., Teller, E. S., Langer, A. W., Grignolo, A., Light, K. C., & McCubbin, J. A. (1978). The relationship among heart rate, carotid dP/dt, and blood pressure in humans as a function of the type of stress. *Psychophysiology*, *15*(2), 102–115. doi:10.1111/j.1469-8986.1978.tb01344.x PMID:652904

Otis, A. B., Fenn, W. O., & Rahn, H. (1950). Mechanics of breathing in man. *Journal of Applied Physiology*, *2*(11), 592–607. doi:10.1152/jappl.1950.2.11.592 PMID:15436363

Pansiot, J. (2007). Ambient and wearable sensor fusion for activity recognition in healthcare monitoring systems. *IFMBE Proc.BSN*, 208-212. 10.1007/978-3-540-70994-7_36

Parak & Korhonen. (2014). Evaluation of wearable consumer heart rate monitors based on photopletys-mography. *IEEE EMBC*, 3670-3673.

Parati, G., Di Rienzo, M., Bertinieri, G., Pomidossi, G., Casadei, R., Groppelli, A., Pedotti, A., Zanchetti, A., & Mancia, G. (1988). Evaluation of the baroreceptor-heart rate reflex by 24-hour intra-arterial blood pressure monitoring in humans. *Hypertension*, *12*(2), 214–222. doi:10.1161/01.HYP.12.2.214 PMID:3410530

Park, S., & Jayaraman, S. (2003). Enhancing the quality of life throug wearable technology. *IEEE Engineering in Medicine and Biology Magazine*, *22*(3), 41–48. doi:10.1109/MEMB.2003.1213625 PMID:12845818

Patel, S., Park, H., Bonato, P., Chan, L., & Rodgers, M. (2012). A review of wearable sensors and systems wit application in rehabilitation. *Journal of Neuroengineering and Rehabilitation*, *9*(1), 21. doi:10.1186/1743-0003-9-21 PMID:22520559

Poon, C. C., Yuan-Ting Zhang, & Shu-Di Bao. (2006). A novel biometrics method to secure wireless body area sensor networks for telemedicine and m-health. *IEEE Communications Magazine*, *44*(4), 73–81. doi:10.1109/MCOM.2006.1632652

Reinhard, U., Müller, P. H., & Schmülling, R.-M. (1979). Determination of anaerobic threshold by the ventilation equivalent in normal individuals. *Respiration*, *38*(1), 36–42. doi:10.1159/000194056 PMID:493728

Rivera-Ruiz, M. (2008). Einthoven's string galvanometer: The first electrocardiograph. *Texas Heart Institute Journal*, *35*, 174. PMID:18612490

Robinson, B. F., Epstein, S., Beiser, G. D., & Braunwald, E. (1966). Control of heart rate by the autonomic nervous system studies in man on the interrelation between baroreceptor mechanisms and exercise. *Circulation Research*, *19*(2), 400–411. doi:10.1161/01.RES.19.2.400 PMID:5914852

Roggan, A., Schädel, D., Netz, U., Ritz, J.-P., Germer, C.-T., & Müller, G. (1999). The effect of preparation technique on the optical parameters of biological tissue. *Applied Physics. B, Lasers and Optics*, *69*(5-6), 445–453. doi:10.1007003400050833

Ruiz, J. C. M. (2013). Textrode-enabled transthoracic electrical bioimpedance measurements-towards wearable applications of impedance cardiography. *J. Elect. Bioimp.*, *4*(1), 45–50. doi:10.5617/jeb.542

Sackner, M. A., Watson, H., Belsito, A. S., Feinerman, D., Suarez, M., Gonzalez, G., Bizousky, F., & Krieger, B. (1989). Calibration of respiratory inductive plethysmograph during natural breathing. *Journal of Applied Physiology (Bethesda, Md.)*, *66*(1), 410–420. doi:10.1152/jappl.1989.66.1.410 PMID:2917945

Seccareccia, F., Pannozzo, F., Dima, F., Minoprio, A., Menditto, A., Lo Noce, C., & Giampaoli, S. (2001). Heart rate as a predictor of mortality: The MATISS project. *American Journal of Public Health*, *91*(8), 1258–1263. doi:10.2105/AJPH.91.8.1258 PMID:11499115

Selley, W., Flack, F. C., Ellis, R. E., & Brooks, W. A. (1989). Respiratory patterns associated with swallowing: Part 1. The normal adult pattern and changes with age. *Age and Ageing*, *18*(3), 168–172. doi:10.1093/ageing/18.3.168 PMID:2782213

Seymour, R. S. (2010). Scaling of heat production by thermogenic flowers: Limits to floral size and maximum rate of respiration. *Plant, Cell & Environment*, *33*, 1474–1485. doi:10.1111/j.1365-3040.2010.02190.x PMID:20545882

Shaffer, D. W. (2002). What is digital medicine? *Studies in Health Technology and Informatics*, 195–204. PMID:12026129

Sherwood, A., McFetridge, J., & Hutcheson, J. S. (1998). Ambulatory impedance cardiography: A feasibility study. *Journal of Applied Physiology (Bethesda, Md.)*, *85*(6), 2365–2369. doi:10.1152/jappl.1998.85.6.2365 PMID:9843565

Shu, Y., Li, C., Wang, Z., Mi, W., Li, Y., & Ren, T.-L. (2015). A pressure sensing system for heart rate monitoring with polymer-based pressure sensors and an anti-Interference post processing circuit. *Sensors (Basel)*, *15*(2), 3224–3235. doi:10.3390150203224 PMID:25648708

Snyder, F., Hobson, J. A., Morrison, D. F., & Goldfrank, F. (1964). Changes in respiration, heart rate, and systolic blood pressure in human sleep. *Journal of Applied Physiology*, *19*(3), 417–422. doi:10.1152/jappl.1964.19.3.417 PMID:14174589

Stemp, L. I., & Ramsay, M. A. (2006). Oxygen may mask hypoventilation: Patient breathing must be ensured. *APSF Newsletter*, *20*, 80.

Stoppa, M., & Chiolerio, A. (2014). Wearable electronics and smart textiles: A critical review. *Sensors (Basel)*, *14*(7), 11957–11992. doi:10.3390140711957 PMID:25004153

Summers, R. L., Shoemaker, W. C., Peacock, W. F., Ander, D. S., & Coleman, T. G. (2003). Bench to bedside: Electrophysiologic and clinical principles of noninvasive hemodynamic monitoring using impedance cardiography. *Academic Emergency Medicine*, *10*(6), 669–680. doi:10.1111/j.1553-2712.2003.tb00054.x PMID:12782531

Toy, P., Popovsky, M. A., Abraham, E., Ambruso, D. R., Holness, L. G., Kopko, P. M., McFarland, J. G., Nathens, A. B., Silliman, C. C., & Stroncek, D. (2005). Transfusion-related acute lung injury: Definition andreview. *Critical Care Medicine*, *33*(4), 721–726. doi:10.1097/01.CCM.0000159849.94750.51 PMID:15818095

Varady, P., Micsik, T., Benedek, S., & Benyo, Z. (2002). A novel method for the detection of apnea and hypopnea events in respiration signals. *IEEE Transactions on Biomedical Engineering*, *49*(9), 936–942. doi:10.1109/TBME.2002.802009 PMID:12214883

Waller, A. D. (1887). A demonstration on man of electromotive changes accompanying the heart's beat. *The Journal of Physiology*, *8*(5), 229–234. doi:10.1113/jphysiol.1887.sp000257 PMID:16991463

Wang, L., Lo, B. P. L., & Yang, G.-Z. (2007). Multichannel reflective PPG earpiece sensor with passive motion cancellation. *IEEE Transactions on Circuits and Systems*, *1*(4), 235–241. doi:10.1109/TBCAS.2007.910900 PMID:23852004

Warren, R., Horan, S. M., & Robertson, P. K. (1997). Chest wall motion in preterm infants using respiratory inductive plethysmography. *The European Respiratory Journal*, *10*(10), 2295–2300. doi:10.1183/09031936.97.10102295 PMID:9387956

Wasserman, K. (1994). Coupling of external to cellular respiration during exercise: The wisdom of the body revisited. *American Journal of Physiology. Endocrinology and Metabolism, 266*(4), E519–E539. doi.10.1152/ajpendo.1994.266.4.E519 PMID:8178973

Watanabe, K., Watanabe, T., Watanabe, H., Ando, H., Ishikawa, T., & Kobayashi, K. (2005). Noninvasive measurement of heartbeat, respiration, snoring and body movements of a subject in bed via a pneumatic method. *IEEE Transactions on Biomedical Engineering, 52*(12), 2100–2107. doi:10.1109/TBME.2005.857637 PMID:16366233

Wilhelm, F. H., Roth, W. T., & Sackner, M. A. (2003). The LifeShirt an advanced system for ambulatory measurement of respiratory and cardiac function. *Behavior Modification, 27*(5), 671–691. doi:10.1177/0145445503256321 PMID:14531161

Yumino, D., & Bradley, T. D. (2008). Central sleep apnea and Cheyne- Stokes respiration. *Proceedings of the American Thoracic Society, 5*(2), 226–236. doi:10.1513/pats.200708-129MG PMID:18250216

Chapter 10
IoT–Based Delineation and Evolution of Kid's Safety Portable Devices

Hemasundari H.
Dr. M. G. R. Educational and Research Institute, India

Swapna B.
https://orcid.org/0000-0002-7186-2842
Dr. M. G. R. Educational and Research Institute, India

ABSTRACT

Currently there are numerous portables in the retail which assist tracking the day-by-day actions of kids and furthermore help discover the kid utilizing Wi-Fi and Bluetooth directions available on the gadget. Bluetooth gives off an impression of being an untrustworthy mode of correspondence connecting the parent and kid. Along these lines, the pivotal motive of this chapter is to get an authorized corresponding medium that links the kid's portable and the parents. The genitor can send a book with explicit tags, for example, "location," "temperature," "UV," "SOS," "BUZZ," and so forth. The portable device will response with a book encompassing the continuous precise region of the kid, which after monitoring hand down headings to the youngster's region on Google Maps software will similarly provide the atmospheric temperature and UV ray emission file with a goal that genitors can pursue if temperature or UV emission isn't reasonable to the kid.

1. INTRODUCTION

As the parents are purchasing in the shopping malls with their kids, the indoor play region is increasing as a convenient place for them to visit. Parents frequently leave kids to these regions and then leave to purchase in other shops in the mall. Inevitably, they are not with their children always. However, shopping complexes have a multiple type of visitors. If parents are not with their children and if kids face hazard or struggling in the unknown surrounding, their children will not be able to call for help from

DOI: 10.4018/978-1-7998-6870-5.ch010

them within a limited amount of time. To permit children to call for help when they face problems or complex risks in unfamiliar surroundings, digital interactive devices are merged with the IOT to create an assistance mechanism for kids while they are in the play regions. This can permit children to call for guidance from parents or employee in the play regions and slowly improve children's risk management skill for detecting problems in unknown environments. This may also avert problems from worsening and will let the parents to make children play independently in play regions peacefully even if their parents are not next to them.

The children from 2 to 7 ages have weaker logical thinking skills so they are not fully grown with their environmental, formal, object – relating and behavioral cognitions. Indoor play regions are comprised of various space for activity, big game props, and a difficult visitor population. Thus, kids gaming in play regions generally experience mishaps, worry, trips and falls, and more risks. These issues will give output in children's sense of hazardous situations, and crying. The children should improve their disaster management skills while they face hazards, which improves their security while playing. The Kids should get a good experience and satisfaction while playing.

Without regarding children, there is a need to spot out other persons in play regions which are involved with the support requesting behavior, device requirements, factors, and solving methods of children when they meet danger in play regions. The related persons in play regions should also be identified. This study planned the following guide to the service items of sensory gadget:

Observation on children's aid requirements and service flow: Initially, 4 researchers involved in 2 weeks monitoring kids in play regions. A sum of 42 children were observed who are in the ages between 2 and 7. Mean observation time for each was 45 - 60 minutes. When children are cried and asked for assistance in the play regions their interplays with parents and hands employed are recorded. Assistance searching for behavior of the children and experience gathered when interplaying with parents, employee around the offsite regions, and employee of child care at the site are also monitored. Then, the age of children, encountered problems, people, solutions, and time were all also noted.

The records which were collected are separated to examine the base reason children look for assistance and the people included. The parents, workers and childcare staffs who aptly handled children's appeal for assistance throughout the observation procedure were interrogated. These permitted researchers to grasp the answer used and the queries that kids came across, therefore getting guidance and relevant service demands from parents or workers in the play regions. The intention is to resolve same problems that can serve as guidance for the service flow delineate. Delineate of service flow design: This study then examined problems came across by children in playing surroundings which is not familiar to them such as accidents and various hazards. Scenario research is used to combine children's assistance requirements and parent's/play area employee role in servicing. A blueprint for service flow has been planned up to appeal structure analysis and the appropriate collaborative device design for children.

Outline of assistance sensory devices:

The technical outline for the indispensable sensory device and the service of the gadget template were developed. The interactive assistance device prototype for the Children was introduced, which includes device applications, engineering specifications, methodology, and technique of interaction.

II. LITERATURE SURVEY

The Internet of Things (IoT) insinuates a game plan of interrelated, web related articles that can assemble and move data over a distant association without human mediation (Rahmath et al., 2021, pp. 768-772). The augmentation of Internet-related automation into a lot of new application locales, IoT is furthermore expected to create a great deal of data from grouped regions, with the ensuing requirement for an energetic mixture of the data, and development in the need to rundown, store, and system such data even more sufficiently. IoT is the best phase of the current Smart City and Smart Energy Management Systems (Swapna et al., 2020, pp. 1-8).

IoT refers to the set of devices and systems that stay interconnected with real-world sensors and actuators to the Internet. IoT includes many different systems like smart cars, wearable devices (Moustafa et al., 2015, pp. 10-11) and even human implanted devices, home automation systems (Nasrin et al., 2014, pp. 212-216) and lighting controls; smartphones which are increasingly being used to measure the world around them. Similarly, wireless sensor networks (Silva, 2014 pp. 67-68) that measure weather, flood defenses, tides and more. There are two key aspects to the IoT: the devices themselves and the server-side architecture that supports them (Jun Zheng et al., 2011, pp. 30-31).

IoT plays a major work in this paper for children safety. In the proposed system Arduino Uno microcontroller is used to transmit and receive data from the elements like camera or temperature sensors. Microprocessor will have only a CPU whereas microcontroller will have CPU, memory, input/output and all integrated into one chip. In this we are using Arduino microcontroller (Braam et al., 2015, pp. 87-91). The wearable IoT device tasked with acquiring various data from the all the different modules connected. It comprises of Arduino Uno based on the ATmega328P microcontroller. It receives the data from its various physically connected modules, anatomizes this data and refines the data in a more user understandable format to the different available user interfaces (Tushar Bhoye et al., 2018, pp. 1-10).

A microcontroller is a small computer on a single metal oxide semiconductor. It will be embedded into the system to control the function by interpreting the data which it receives from the I/O peripherals. Bluetooth is a wireless technology which allow the exchange of data between various devices, it uses wavelength to transmit information (Bhagwat, 2001, pp. 96-103). When the information or data is received the scanning device which is present in the Bluetooth will send a response with the information. Since it is untrustworthy, we are moving into advanced work in this paper. Bluetooth and Wi-Fi are an unreliable medium for far communications; therefore, we used GPS enabled communication for a reliable medium between people (Badamasi, 2014, pp. 1-4).

Arduino Uno is a very valuable addition in electronics which consist of USB interface, 14 digital I/O pins, 6 analog pins and Atmega328 microcontroller. It will emit electromagnetic radiation and looks for a change in the signal that returns. It allows the designers to control and sense the external electronic devices in the real world. Proximity sensor detects the presence of nearby objects without any physical contact. In the proposed system the temperature sensors and UV sensors are connected to the Arduino board. The Arduino board is programmed using IDE software (Swapna et al., 2020, pp. 2208-2213). Different types of sensor like load cell sensor, Smell sensor (Swapna et al., 2019, pp. 1-10). A sensor is an input device that provide an output signal with respect to a specific physical quantity (Swapna et al., 2020, pp. 1900-1912).

Many smart watches are available in the market for kids care, but most of the parents are not aware and not purchasing those items for their children security. So, awareness is to be given to all parents those who admitted their children in the school. The child's position information is periodically sent through

GSM to the parent's smart phone. This can help the parents and the school authorities to monitor the children when they leave the school or they go missing. But lot of details like heart beat rate, pulses and temperature are collected here which does not provide the valuable details to the parents (Poonkuzhlai et al., 2021, pp. 10839 – 10847). LCD works by blocking light. At the same time electrical current will cause the light crystal molecules to align and allow varying levels of light to pass through to second substrate and create the colors and images that we see (Swapna, 2018, pp. 24-36).

III. IMPLEMENTATION AND RESULT ANALYSIS

The experimental trials were used for detuning the different parts of the planned portable gadget.

Camera Module

For monitoring of kid's surrounding, to have a better portrait of the area, this wearable can as well consist a camera element fused it. Since the main aim of this wearable project is the GSM element which is a finer substitute than Bluetooth, WI-FI or ZigBee due to the small scope and connectivity problems of these technologies. The GSM module is useful for this project since the cellular scope is huge and all the transmission between gadget and user is through messages, therefore cyberspace connection is not essential. But anyway, the Arduino GSM shield uses GPRS which allows the board to use the internet if needed.

Even though the camera element can hold up video streaming, since it has a restriction of using only message, only four wire connections can be implemented. The red and black wires are connected directly to +5V and GND to the Arduino uno board. The RX pin used for transmitting data through Arduino uno and Arduino gsm board and the TX pin used for receiving incoming data through from the elements. The 10 K resistor divider, data pins of camera are 3.3v logic, and it would be preferable if we can divide the 5V to 2.5V. Typically the yield from the computerized 0 pin is 5V high; the manner in which we related the resistance of the input camera in such a way that (white wire) it will never go beyond 3.3V.

To converse with the camera, the Arduino uno will utilize two computerized pins and a product sequential port. As the camera or the Arduino Uno do not have sufficient onboard memory to save shots and store it for a limited time, therefore an outer storage source microSD board will be used to store the pictures in a temporary manner. The camera tasks on a standard baud rate of 38400 bauds. The camera can assemble data in a similar way as the GPS part. It can be on reserve moderating force sitting tight for the specific watchword "Preview" to be transmitted from the client's phone to GSM shield order the camera to click a depiction of the encompassing and spare the document briefly on the outside microSD card. After which Arduino Uno will get to spared picture from microSD stockpiling and move that to the GSM element that transmit it to client by means of SMS/MMS content.

Android App

The idea in making the Android app has been obtained from having a mechanized robot to answer to text message reply from the user. It may give the client with pre-defined reply options by clicking the button. The client need not remember the particular words to send. A preprogrammed robot with a collection of predefined words like "LOCATION," "SNAPSHOT," "SOS," is used by the presenter. More specific keywords like, "HUMIDITY," "ALTITUDE," etc. are used at the same time for the future facet of the gadget

GPS Location Sensor

After testing the wearable gadget on numerous occasions with rehashed SMS writings. The GPS area sensor had the option to react back with exact scope and longitude directions of the portable gadget to the client's mobile, which will make the client would tap on got Google maps URL which will in turn open google maps application to show the particular area. In every one of the situations the GPS module tried, it could react back to the client's mobile inside a moment.

Though for the mobile (blue dab) is showing the wearable to be available in the city, which is hardly off from the precise area. This negligible miss coordinate in the pin-point area of the wearable would turn be able to out for being deadly in a genuine situation, where the parent might be misled to an inappropriate area of kid. In this way, the Parallax PMB-648 GPS module demonstrates to be fruitful in giving the exact area high exactness and decent reaction time. The only drawback may be the GSM module might not understand various valuable words transmitted in a single message.

Figure 1. Block Diagram for Kids Wearable Device

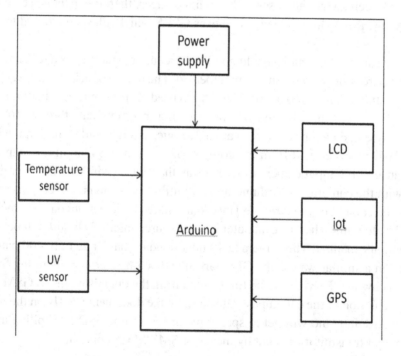

On a par to the GPS location sensor, the Temperature, and UV sensors were trialed multiple times under various temperatures and greater intensities of sunlight. Both the sensors performed exceptionally well to the test pardoned. The response time to 442 get a answer back to the keywords "TEMPERATURE" and "UV" was less a minute. The temperature sensor was also implemented to undergo greater temperatures and on comparison with a reading in a thermostat in the room which may vary with the reading in sensor by +O.2°C to -O.2°C. The UV sensor was also measured under various intensities of sunlight.

The UV sensor reacted faster to the sunlight intensity change. The acknowledge time to get an answer back to the keywords "UV" was less a minute.

Figure 2. Mobile app for LOCA NON sensor

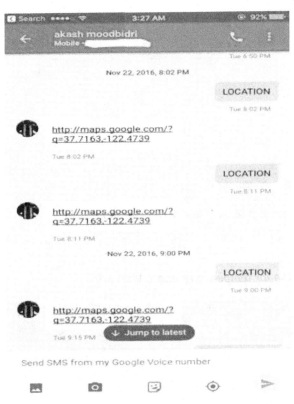

Alike to the GPS location sensor, the Temperature, and UV sensors were dragged separately at various temperatures and in soaring intensities of sunlight. Both the sensors exemplarily performed well. The response time to get back the words such as "TEMPERATURE" and "UV" was less than a minute.

The temperature sensor was exposed to higher temperatures and collated with a thermostat kept inside the room which varies temperature reading from +0.2°C to -0.2°C. The UV sensor was also measured under different sunlight strength. The UV sensor quickly reacted the variations in the sunlight strength. The response time to get an answer back to the words "UV" was less a minute.

SOS LIGHT AND DISTRESS ALARM BUZZER:

The light and buzzer changes from the above sensors in the message trigger mechanism. By sending a message with either "SOS" or "BUZZ," this would activates the light and buzzer to pardon an resulting

Figure 3. Display of Google maps with latitude and longitude coordinates.

Figure 4. Cellphone SMS app for temperature and UV sensor

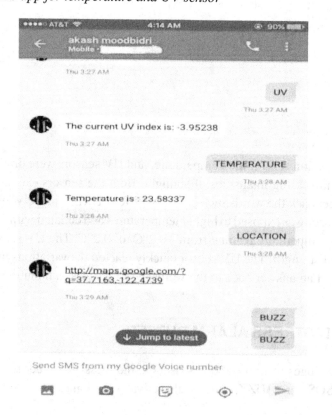

function instead of sending readings back to the client's mobile in the same way of the other sensors. After getting the right words, the SOS light and Alarm Buzzer would basically play out the specific assignment of blazing the SOS light and give a sound as hazard alert which might take longer than their sensor partners. After consummation of their particular capacities, the reaction is transmitted back to the client' PDA expressing: "SOS Signal Sent" and "Playing Buzzer."

Figure 5. Mobile Messaging app for SOS signal

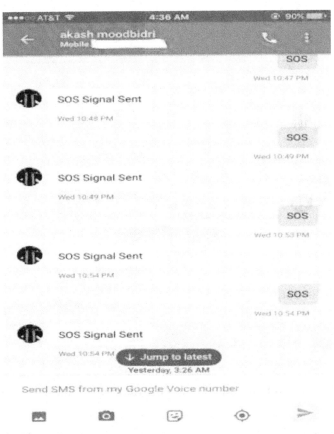

IV. CONCLUSION

The child safety portable gadget is equipped for going about as a brilliant IoT gadget. It furnishes parents with the ongoing region, including temperature, UV radiation record and SOS light alongside anguish care bell for their children surrounding and the volume to find their children or ready on lookers in trying to protect or give comfort to the child. The smart kid wellbeing portable may be developed significantly more later on by using deeply minimized Arduino elements, for example, the LilyPad Arduino which can be seamed to textures. Additionally, a more force productive equipment should be made which will fit to hold the battery for a more prolonged time.

Figure 6. Cellphone SMS app for Buzzer

REFERENCES

Badamasi, Y. A. (2014). The working principle of an Arduino. Electronics. *Computer and Computation, 1*, 1–4.

Bhagwat, P. (2001). Bluetooth: Technology for short-range wireless apps. *IEEE Internet Computing, 5*, 96–103.

Bhoye, T., More, S., Gatkawar, R., & Patil, R. (2018). Child Safety Wearable Device. *IJARIIE, 4*, 2395–4396.

Braam, Huang, Chen, Montgomery, Vo, & Beausoleil. (2015). Wristband Vital: A wearable multi-sensor microsystem for real-time assistance via low-power Bluetooth link. *Internet of Things (WFIoT), 2015 IEEE 2nd World Forwn on, 1*, 87-91.

Moustafa, H., Kenn, H., Sayrafian, K., Scanlon, W., & Zhang, Y. (2015). Mobile wearable communications. *IEEE Wireless Communications, 22*, 1O–11.

Nasrin, S., & Radcliffe, P. (2014). Novel protocol enables DIY home automation. *Telecommunication Networks and Applications*, *1*, 212–216.

Poonkuzhlai, P., Aarthi, R., Yaazhini, V. M., Yuvashri, S., & Vidhyalakshmi, G. (2021). Child Monitoring and Safety System Using Wsn and Iot Technology. *Annals of R.S.C.B.*, *25*, 10839–10847.

Rahmath,, S., Nisha, Shyamala, C., Sheela, D., Abirami, M., Harshini, M., & Keerthana, R. (2021). Women and Children Safety System with IOT. *Turkish Journal of Computer and Mathematics Education*, *12*, 768–772.

Silva, F. A. (2014). Industrial Wireless Sensor Networks: Applications, Protocols, and Standards. *IEEE Industrial Electronics Magazine*, *8*, 67–68.

Swapna, Kamalahsan, & Sowmiya, Konda, & SaiZignasa. (2019). Design of smart garbage landfill monitoring system using Internet of Things. *IOP Conference Series. Materials Science and Engineering*, *561*, 012084–012091.

Swapna, & Shubhashree, George, & Hemasundari. (2020). IoT Based Intelligence Parking Management System using LOT Display. *International Journal of Advanced Science and Technology*, *29*, 2208–2213.

Swapna,, B., Gayathri,, S., Kamalahasan,, M., & Hemasundari,, H., SiraasGanth, M., & Ranjith, S. (2021). E-healthcare monitoring using internet of things. *IOP Conference Series. Materials Science and Engineering*, *872*, 1–8.

Swapna, B., & Manivannan, S. (2018). Analysis: Smart Agriculture and Landslides Monitoring System Using Internet of Things (IOT). *International Journal of Pure and Applied Mathematics*, *118*, 24–30.

Swapna, B., Manivannan, S., & Nandhinidevi, R. (2020). Prediction of soil reaction (pH) and soil nutrients using multivariate statistics techniques for agricultural crop and soil management. *International Journal of Advanced Science and Technology*, *29*, 1900–1912.

Zheng, J., Simplot-Ryl, D., Bisdikian, C., & Mouftah, H.T. (2011). The internet of things. *Communications Magazine IEEE*, *49*, 30-31.

Chapter 11
Blockchain–Enabled Internet of Things for Unsupervised Structural Health Monitoring in Potential Building Structures

Wael Mohammad Alenazy

CFY Deanship, King Saud University, Riyadh, Saudi Arabia

ABSTRACT

The integration of internet of things, artificial intelligence, and blockchain enabled the monitoring of structural health with unattended and automated means. Remote monitoring mandates intelligent automated decision-making capability, which is still absent in present solutions. The proposed solution in this chapter contemplates the architecture of smart sensors, customized for individual structures, to regulate the monitoring of structural health through stress, strain, and bolted joints looseness. Long range sensors are deployed for transmitting the messages a longer distance than existing techniques. From the simulated results, different sensors record the monitoring information and transmit to the blockchain platform in terms of pressure points, temperature, pre-tension force, and the architecture deems the criticality of transactions. Blockchain platform will also be responsible for storage and accessibility of information from a decentralized medium, automation, and security.

1 INTRODUCTION

Structural Health Monitoring is a platform for monitoring the health of a building structure, which may not be consistently monitored over a manual or physical methodology. The structures will undergo normal wear and tear, consequence of varying operations and thus effects of faults occurring during regular operations. All structures can be monitored effectively for pre-emptive fault diagnosis and recovery operations to avert life and property losses. Solutions intended for monitoring capital based machineries, assets monitoring and real time fault diagnosis should be cost effective, reliable and trust worthy. Assets of an industry have to be invested with huge capitals and classified as potential areas which need

DOI: 10.4018/978-1-7998-6870-5.ch011

to be facilitated with utmost protective and pre-emptive measures. Structures, on the other hand, also contribute to development of society in terms of bridges, railways, airports, energy generation systems from renewable sources and elements upgrading the socio-economic factors. These structures have to be monitored at all times, nevertheless, with an automated method of monitoring and fault diagnosis Yli-Ojanperä et al (2019), Sharma, P. K., & Park, J. H. (2018), Performance, H., & Scale Working Group. (2018), Serpanos, D., & Wolf, M. (2018). All these structures are potential elements, indulging in huge costs for installation and maintenance at a regular interval. Assembling these structures are performed by bolted joints for fastening to compose the entire architectural structure and this is the predominant approach carried out in many structures. Nonetheless, this approach of bolted joints cannot assure a one-time deployment and are prone to faults due to loosening its pre-tension over time, normal wear and tear, and operational effects.

Similarly, Industry 4.0 is the upgradation of industries to a whole new level of manufacturing systems with cyber inflicted monitoring and operative measures. Recent trends of communication, information, Internet of Things, artificial intelligence and edge computing have notably transformed the approaches of a manufacturing industry. The scope of incorporating such technologies into industry has exponentially grown ever since the use of Industrial Internet of Things (IIoT) and Blockchain technologies. By 2030, humongous investments and research is expected to be carried out in the industry 4.0 structuring, monitoring and management workflows. Cyber physical systems are one stop solutions for automating business decisions, regulating business flows, establishing a communication interface between manufacturers, logistics and hence customers. The cyber physical systems can rightfully be called as Process Automation Systems, as they integrate different devices and interfaces for enabling Industry 4.0 infrastructures. Devices are made to update their operations and status to decision making modules for deciding on the activities at run-time. IIoT devices are usually bound by factors like processing power, energy limitations, autonomy and transmission ranges. All such factors are considered in the design and implementation of Industry 4.0 architectures.

The said approach of defining the architecture of industries proclaim various benefits ranging from transparency to simplified and unified processing. However, there are various paradigms which have failed in delivering the expected outcomes. Machine to Machine communication was followed based on a Publish-Subscribe mechanism, unable to meet the expectations through an intermediate node, resulting in unwanted communications overhead. Real time systems demand immediate and round the clock transmission of messages from different machineries. The same should reach the destined nodes/ devices within the stipulated deadline. Publish-Subscribe approach will possess difficulties in communicating the information within the prescribed deadlines. A blockchain based process automation system can effectively address the problem identified in conventional and recent systems built for upgrading industrial IoT frameworks. A decentralized architecture is proposed for Machine to Machine communications, without any intermediate bodies to regulate in the form of brokers, transparent communication is ensured without any inherent major or minor changes. The intention of the proposed system is to organize the updates from sensory devices to update the status of machines, for appropriate decision making, safe operations, trace-back facilitated faults and real time communications.

There are numerous protocols to streamline the flow of information between the machineries, networks are organized to address variable traffic with performance peaks at all times. Primary requirements of a PAS architecture are time, trust-worthiness, flexibility and scalability. The protocols generally implemented in other environments such as smart homes and business applications cannot be directly utilized, owing the constraints of a smart industry. Process automation deals with a bigger role, adaptive and real time scenarios necessitate some different modus operandi. The process of communication commences with information registered by sensors, equipment, which are further processed by Programmable Logic Controllers and monitored by interfaces. Necessary information which demand utmost attention from the top level management is forwarded to the Plant Management and Corporate Management in the top of the hierarchy. Supervision Control and Data Acquisition defines the architecture of interfaces for managing the collected information and organizing them accordingly to determine if they need top level interventions. Enterprise Resource Planning software is a dedicated platform for handling the operations from the top end, where the intricate details about processing are masked for simplified accessibility. The sensors and actuators are usually distributed over a large area, over machinery, different departments of an industry to capture real time and time-critical information with great synchronicity. Interfaces are defined for easier understanding and better control over the devices from a remote server. All this information will collectively work for the goal and profit associated instruction defined by the ERP at top level of hierarchy.

Industrial Internet of Things (IIoT) devices are deployed in a harsh environment with an intent of detecting and reporting vulnerabilities. When machineries in an industry are equipped with IIoT devices, the communication is processed in real-time, synchronous and observed for fault occurrences. Blockchain on the other hand, aims to construct an architecture for decentralizing the operations happening within an industry, making consistent and irreplaceable transactions. Consensus algorithms are responsible for organizing the nodes, devices and the data collected from them at prescribed intervals for validation, reliability and sustainability. Predominant algorithms deployed in blockchain platforms are Proof of Work, Proof of Stake, Proof of Authority. Apart from the management of real time operations, the proposed system will also facilitate the process of detecting faults in machineries. When devices are subjected to real time operations, environmental conditions, fatigue, vibrations, stress and corrosion, bolts are subjected to failures. Inspections are mandatory to avoid life and property losses within the working environment. An unnoticed fault or a loose bolt can be serious enough to cost dearly, and has to be identified at source before the mandated inspections. In order to restrict the cost of inspections, blockchain application can be extended to monitor, log the level of fastening in critical machineries. The proposed system composes a sensory device to monitor the pre-tension load of bolts fastened in critical machinery, protocols to regulate the contracts and other operations in an industry and upgrade the industry into 4.0.

The following sections are organized into the following sections. Section II describes similar architectures and conventional state of art techniques for the same purpose. Section III defines the architecture of proposed system and sensory devices for monitoring faults and diagnosis. Section IV combines the sensory node, IIoT application with Blockchain interfaces and Section V illustrates the performance of the said system. Section VI concludes the article with necessary references.

2 RELATED WORK

Various research in design and implementation of structural health monitoring applications have been carried out. Similarly, process automation and production systems are integrated through dedicated networks in protocols. When two devices or equipment intends to communicate with each other, the automated connection establishment and protocols involved in transfer of messages can smoothen the function and operations. This is where blockchain comes into the play. For smarter and automated industries, higher level of integration is required for capturing, validating, processing and transmitting information collected from different entities of an industry. A digitised environment in industry 4.0 has to be reliable list the different operations performed in one industry as a whole. Raw data collected from different segments of an industry need further processing before they are forwarded into a decision making module. Having said this, this section analyses different techniques for Optimisation and maintenance of individual operations collected in form of raw data Yli-Ojanperä et al (2019), Sharma, P. K., & Park, J. H. (2018), Performance, H., & Scale Working Group. (2018). Networking of heterogeneous devices and multi-agent systems require suitable protocol for including IoT devices within industrial applications. Protocols such as Open Platform Communications Unified Architecture (OPC-UA) implemented as a vertical integration technique and the results were investigated in in real time. In case of a centralised network, cute server cost and computations other equipment for interconnection and qualified professionals for handling the architecture restrict the implementation of PAS systems in industries.

Chances of server failures in case of a centralised server, worsen the situation in intermittent handling and recovery. Applying blockchain in the design of a decentralized network will facilitate better communication and decision making. Smart contracts defined for achieving betterments in control and Management of supply chain, product and process automation, data collection in aggregation, security and privacy, etc. Vertical integration of PAS hierarchy can improve the incorporated operations of blockchain architectures and cyber-physical industries Serpanos, D., & Wolf, M. (2018). The proposed system facilitates the monitoring process of field devices with the assistance of PLCs, register information exchange happening in in blockchain and the system, maintain a clean log of devices, their performance and hence fault diagnosis. Process control in this proposed methodology can be challenging as smart contracts are implemented. In existing state of art technologies, their execution time cannot be defined as a standard unit due to varying network latency and frequent transactions occurring in a blockchain platform. The next challenge subjecting the transactions to a safe and fail proof technique to meet proper deadlines in real time applications.

In case of structural monitoring and blockchain technologies, various techniques have been proposed for identifying the strength of bolted joints, their looseness levels and alarming mechanisms to alert respective teams. Such systems are categorized according to their regions of deployments namely on premise and remote deployment. On premise deployments are associated with manual inspections by a professional and reporting the status, remote monitoring involves design and implementation of automated methods for detections. The current monitoring strategies are predominant in using on premise inspections. Bolts are identified for looseness and tightness based on a series of images. Image processing techniques are later applied to detect the level of looseness. In an approach, feature extraction technique was used with Hough Transform, for determining the rotational angle of every bolt in the assembly. When there is a change in predefined angles, it is noted that the bolt is loose *Accenture*, (2015). A classifier in a linear support vector machine was introduced for sensing the looseness, exhibiting a better detection rate LeDoux, F. S., et al (1983). A common difficulty was the detection of looseness in extremely less variations in rotational angles and presence of bolt cracks were eliminated from the detection strategies.

Later techniques incorporated the analysis through wave forms, being acoustic waves and ultrasonic waves where actuators played a major role. Piezoelectric Lead ZirconateTitanate transducers Sidorov, M., et al (2019) were deployed in sending and receiving wave forms within a stipulated time interval. These sensors were later substituted by concave and convex annular disks Park, J., et al (2015) to remove the problem of saturation, found prominently in PZT sensors. In some other techniques, a washer based on PZT sensors were deployed for detecting bolt looseness and tightness Da Xu, L., et al (2014). Transducers connected to the head of a bolt will be transmitted with an ultrasonic wave and the time taken for receiving the reverse signal is measured. Pulse transmission time was used as an indication for looseness Li, S., et al (2019), Müller, J. M., et al (2018). All designs demanded huge electrical connectivity and external sources to process the collected information. Complex circuits are also associated with an increased expenditure, failing its preference in such architectures. The next approach was suggesting a manual health check-up like strategy, for understanding different sounds with probabilities. Every unique sound will be related to an abnormality based on frequencies. A smartphone was deployed Zhao, X., et al (2019) for recording sounds received from bolts, which is processed to categorize the frequency level and detect the looseness levels. An accelerometer was used in place of the smartphone with almost a same functionality Parvasi, S. M., et al (2016). Machine learning and entropy measurements later contributed to a wider scale of implementation of using accelerometers. The true level of tightness or looseness cannot be accurately derived from the recordings and hence monitoring became a challenge in implementing such techniques Yin, H., et al (2016).

Fiber Bragg Grating was the next option to detect looseness of bolts with fiber optics and resonant microstructures accordingly. Bragg wavelength was measured in a transmission of broadband light, illustrating the period of the structure. Compression of the structure will alter the wavelength, in turn, indicating the level of looseness. The benefits of registering Bragg wavelength was numerous as they are immune to other interferences, possess high sensitivity and thus being more reliable than discussed techniques Huo, L., et al (2017). Despite the benefits, fiber optic infrastructures are costlier than the sensors, and cannot be deployed exclusively for structural health monitoring applications. The FBG sensors were integrated in other combinations for deriving new strategies, along with PZT sensors and fasteners but still experienced the mentioned shortcomings. Having listed out the on premise inspecting strategies, the remote sensing techniques are limited owing to associated constraints. Custom designed Tensense 2.0 and washers were implemented in a few techniques along with the difficulties of covering smaller assemblies and distance of transmission. Kumar, Sandeep, et al. has applied the concept of Blockchain in Agriculture and have a Study on Blockchain Technology, Benefits, and Challenges.

3 PROPOSED METHODOLOGY

The proposed system follows the design with mechanical and circuit components merged within the Blockchain platform. Numerous sensors are deployed on a building structure for monitoring the health in certain frequencies. All such sensors will be reliant on maximum stress, strain and temperature of the units, being influenced by natural or artificial occurrences. Important buildings of governments, bridges, nuclear power plants, dams, hospitals and other structures emphasize the need of structural health monitoring applications. In that case, an industry is constituted with number of assemblies connected with bolted joints, which are the primary points of getting affected. BoltSafe sensors, deployment and their control over on premise and remote application constitutes the mechanical design of proposed method. The system is evaluated with series of tests to promise a reliable and integral system for measuring pre-tension force over every bolted joint in the assembly. The circuits are built with measuring modes, transmission mode, tracking and alerting schemes for various values measured in regular intervals. The implementation of BoltSafe sensors with the Process Automation System enables the cyber physical system to expedite structural health monitoring and intelligent decision making systems. The mechanical components of the proposed system intend to perform and ensure the following entities.

1. Track and monitor the pre-tension load of all bolts in an assembly
2. Periodical monitoring of load values and transmission
3. Ensuring safety of the assembly and equipment at all times
4. Scalability
5. Blockchain integration for secure and better accessibility across different stakeholders of the industry

3.1 Physical Design

BoltSafe sensor is a custom made sensor in the shape of a washer, available in common sizes for measuring the tension load of bolts deployed in industrial machineries. Loads of every bolt can be measured during installation, the load can be monitored in regular intervals and thus the health of structural health monitoring can be ensured at all times. The sensor is proved to be able to withstand harsh environment conditions and deliver simplified means of data collection. BoltSafe sensors are pre-built with an Application Specific Integrated Circuit, dedicated for sensing, recording and transmission of critical information. All sensors are provided with a unique serial number, indicating the ways of identifying different assemblies and immediate attention to respective joints. A digitally available monitoring system records the temperature and residual pre-tension load of all bolts at any given time. Once the calibration of sensors is performed during the installation, the system eliminates further need for recalibration throughout its deployment. Periodic monitoring circuits of BoltSafe sensors do not need any additional circuitry for monitoring purposes and can be deployed with connectionless modes of inspections. The following figure 1 illustrates the component of BoltSafe sensor to be deployed in our proposal.

Figure 1. BoltSafe Sensor

The BoltSafe sensor is monitored using a Network Adapter CM 1000 incorporated into a Power Data Interfacing Unit, capable of accommodating up to 30 network adapters. Each adapter will be capable of handling 8 Boltsafe sensors each, giving a network of 240 real time detection capability. Potential machinery can be deployed with a PDI unit in order to deliver a better remote sensing ability. Scalability of the approach is ensured with simple plug and play interfaces. Entries in the PDI unit can be transmitted through a wireless medium to reach Blockchain platform for storage and retrieval. The values captured in defined intervals are transmitted to the blockchain platform for validation and alarming necessary officials for manual inspection or remedial actions.

3.2 Blockchain Integration

Process Automation Systems built with blockchain architectures and Industrial IoT components can be regulated with smart contracts, despite the challenges. Smart contracts expedite the process of intelligent decision making from correlated and aggregated information collected during the actual functioning of an industry. Management and control of flow ofinformation can be organised and thus made easier for accession. Entries from PDI-NT units are processed as transactions in a blockchain environment. It has to be noted that a blockchain platform is similar to a cloud computing platform. Yet the abundant storage space, simplicity in handling the resources and autonomy through validations can be achieved with a comparatively lesser cost. Usually, the blockchain platform will be the ultimate destination of all IP based message transmissions, deeming it the best option for creating interfaces for different levels of supervision. A blockchain client will be responsible for correlating information obtained from sensors, process control and hence manipulates a decentralized working atmosphere. Physical devices (BoltSafe sensors) are made to communicate with interfaces through smart contracts. Records collected can be mediated as assets to enhance autonomy through predefined rules and decision making strategies.

Figure 2. BoltSafe PDI-NT interface with Network Adapters and Sensors

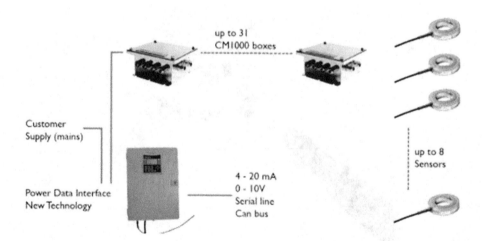

The proposed methodology comprises of blockchain platform, the physical sensors and user interfaces along with following components. A blockchain platform can be considered as a network of devices and interfaces which communicate by adhering the regulations of a consensus algorithm. Nodes are present within the network to coordinate and maintain the order in mediations, validation and other processes involved with handling data. RESTAPIs, being a subordinate of nodes, acts as the validator of messages with JSON and HTTP standards. A ledger is a virtual notebook for registering all operations with a timestamp in its dedicated database. Blockchain platforms ensure that all communication between nodes and other layers of the hierarchy follow certain standards and no compromise is made on security. Transactions, before being stored on databases, are validated with a recognised device ID and states. Similar to individual sensor IDs, all operations are labelled with a predefined operational code to validate the particular operations. The following Figure 3 illustrates the processes defined for transmission of messages within a blockchain platform.

The architecture is evaluated based on the pre-tension of bolted joints in the assembly, during installation and in regular intervals. Linearity of the collected information have to be validated with the common threshold values defined during the set up process. The system, as understood in a harsh environment, is subjected to environmental conditions which may induce certain effects and necessary countermeasures. The proposed architecture is implemented over a universal testing machine for evaluating the compression ratio of the BoltSafe sensors. Washers are calibrated to its optimal level prior to actual simulation of an industry operation namely gradual loading and de-loading. Reading recorded by the BoltSafe sensor is measured after units of pressure are released by the testing machine.

Figure 3. Entities and Devices communicating with a IIoT platform

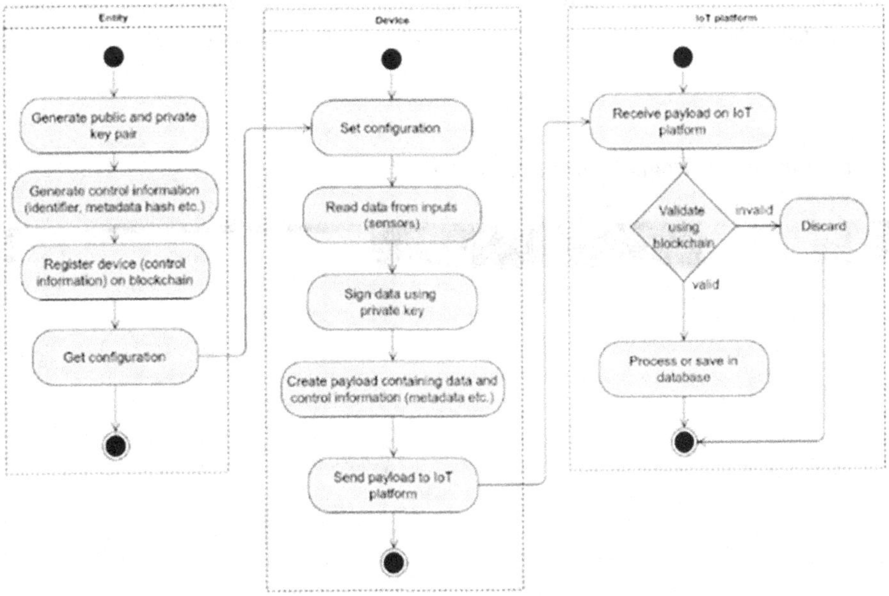

Considering the environmental conditions, temperature rapidly increase or decrease, weather conditions may contribute to rusting accompanying the symptoms of normal wear and tear. Temperature variations can affect the system, bolt assembly, circuits and of course the performance of entire system. It is observed that different strain sensors act differently to different temperature and evidently nonlinear. The manufacturer of the BoltSafe sensors has considered a nominal sensor drift to be considered as a compensation. BoltSafe sensors are backed up with a temperature sensor, to compensate the drift, incorporate an offset value to attain the final response curve with minimal deviations. Steel is also said to undergo elongations at specific temperatures. Pre-tension force cannot be measured as accurately possible, with minimal adjustments and consideration of thermal expansion, a near-optimal value can be derived.

3.2.1 Proof of Concept

The entire system emphasized on the Structural Health Monitoring application with sensory devices and Blockchain platform, serving with a unique property of security and immutability. The consensus algorithm defines the protocols for message transmission between sensors and nodes, nodes and interfaces and how the same can be extended into similar industries with inter-related operations. Pain points and points of failures are addressed pre-emptively, limiting the modifications over the table with stringent rules. A publically available Blockchain framework will ensure transparency and immediate attention to critical elements of a building structure. Ethereum test network is the background of the proposed simulation, with a publically available smart contract in the backend. The code is written in Solidity and compilation was performed using Truffle. The entire code is compiled and fed into Ethereum network. A front end application is built on HTML and javascript for providing a simplified interfacing experience to the managers and end users. Smart contracts will be customized to carry the required information from different entities as described in Figure 3. Sensor ID, its location, residual force and temperature are transmitted as the critical information through the smart contract. All such information will be accompanied by a timestamp and hash value to indicate the values of transaction. The following Figure 4 illustrates the interface on the web browser for device/ sensor registration and how the information is processed.

Figure 4. Interface for Sensor Registration and Approval

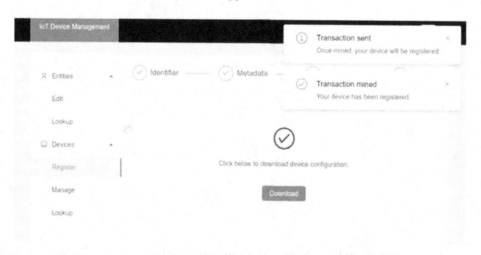

The information processed through consensus algorithm will be in the following format as depicted in Figure 5, securing algorithm, BoltSafe Sensor registered ID, temperature and pre-tension residual force available in the respective bolt.

Figure 5. Sensor information and transmission data format

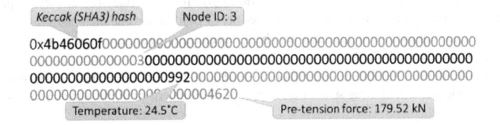

The input of blockchain application is protected with necessary algorithms to prevent any modifications. Transaction hash is an important element to achieve this security feat, indicating a successful transmission of data from sensory devices to interfaces and secure information. The modifications will not be permitted at all or every modification will be carrying the information of registered users with time and location. The cost incurred for transactions and storage are measured with respect to fuel, represented by gas. The transaction cost for every message transfer will approximately cost 0.00008 ETH. If the system can be migrated to an IOTA environment, the cost of storage and maintenance will considerably lesser than the incurred costs.

Figure 6. Reports from BoltSafe Sensors

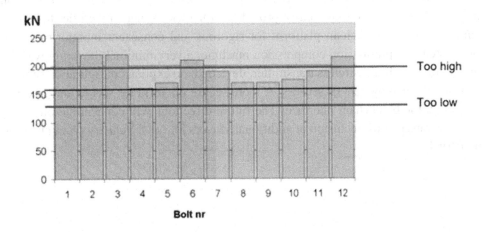

4 RESULTS AND DISCUSSIONS

The system is evaluated against the comparisons of previous techniques and from the results evidently states that this methodology is remarkably better in reporting the health status of critical industries. Blockchain platform with necessary incorporations have upgraded the functionalities to the next level of architecture namely Industry 4.0. Having the blockchain and process automation systems incorporated together, the proposed system achieved better security, reliability and scalability as a cyber-physical system. The following figure 6 illustrates the functionality of BoltSensors and their alarming capability beyond the defined threshold. The threshold will be defined according to the industry and in this case, the threshold is denoted as 200kN.

The following Table 1 lists out the expectations and requirements of a structural health monitoring system, the performance of existing state of art techniques in comparison of the proposed method.

Table 1. Proposed Method vs Existing methods

Conditions	Remote Access	Non-Invasive	Continuous	Blockchain Platform	Pre-defined calibration
Proposed	Satisfied	Satisfied	Satisfied	Satisfied	Satisfied
Tensense M30	Satisfied	Satisfied	Satisfied	NA	Satisfied
FBG	NA	Satisfied	Satisfied	NA	Satisfied
PZT	NA	Satisfied	NA	NA	Satisfied
Percussion	NA	Satisfied	NA	NA	NA
Mekid	Satisfied	NA	NA	NA	NA

CONCLUSION

The presented article listed out the possibilities of building a structural health monitoring application for an industry enabled with blockchain and IIoT devices. Remote monitoring was facilitated through IIoT sensors and blockchain enabled reliability and security. The presented design enabled the enrichment and extension of the system throughout the smart industries and was affirmative to any size of bolted joint assemblies. The BoltSafe sensor provided with a wide range of facilities to be incorporated and programmed into a blockchain platform for rigorous and consistent monitoring of pre-tension loads. When tested in a pre-tension compression machine, a very minimal deviation was exhibited in the pressure. With an in-built wireless transmitter, device id, no explicit configuration of protocols was necessary during the deployment of sensors. A more secure platform for secure and 24x& access was facilitated through the blockchain interface. The future directions will concentrate on conserving the cost incurred for transactions, inclusion of additional sensors for other industry types and thus ensuring a smart industry 4.0.

REFERENCES

Da Xu, L., He, W., & Li, S. (2014). Internet of things in industries: A survey. *IEEE Transactions on Industrial Informatics*, *10*(4), 2233–2243. doi:10.1109/TII.2014.2300753

Huo, L., Chen, D., Kong, Q., Li, H., & Song, G. (2017). Smart washer—A piezoceramic-based transducer to monitor looseness of bolted connection. *Smart Materials and Structures*, *26*(2), 025033. doi:10.1088/1361-665X/26/2/025033

Hyperledger Performance & the Scale Working Group. (2018). *Hyperledger blockchain performance metrics white paper*. https://www.hyperledger.org/resources/publications/blockchain-performance-metrics

Kumar, S., Maheshwari, V., Prabhu, J., Prasanna, M., & Jothikumar, R. (2021). Applying Blockchain in Agriculture: A Study on Blockchain Technology, Benefits, and Challenges. In *Deep Learning and Edge Computing Solutions for High Performance Computing* (pp. 167–181). Springer.

LeDoux, F. S., Levine, A. S., & Kamp, R. N. (1983). New York State Department of Transportation bridge inspection and rehabilitation design program. *Transportation Research Record: Journal of the Transportation Research Board*, (899), 35–38.

Li, S., Zhao, S., Yang, P., Andriotis, P., Xu, L., & Sun, Q. (2019). Distributed consensus algorithm for events detection in cyber-physical systems. *IEEE Internet of Things Journal*, *6*(2), 2299–2308. doi:10.1109/JIOT.2019.2906157

Müller, J. M., Kiel, D., & Voigt, K. I. (2018). What drives the implementation of Industry 4.0? The role of opportunities and challenges in the context of sustainability. *Sustainability*, *10*(1), 247. doi:10.3390u10010247

Park, J., Kim, T., & Kim, J. (2015, August). Image-based bolt-loosening detection technique of bolt joint in steel bridges. In *6th international conference on advances in experimental structural engineering*. University of Illinois, Urbana-Champaign.

Parvasi, S. M., Ho, S. C. M., Kong, Q., Mousavi, R., & Song, G. (2016). Real time bolt preload monitoring using piezoceramic transducers and time reversal technique—A numerical study with experimental verification. *Smart Materials and Structures*, *25*(8), 085015. doi:10.1088/0964-1726/25/8/085015

Serpanos, D., & Wolf, M. (2018). Industrial internet of things. In *Internet-of-Things (IoT) Systems* (pp. 37–54). Springer. doi:10.1007/978-3-319-69715-4_5

Sharma, P. K., & Park, J. H. (2018). Blockchain based hybrid network architecture for the smart city. *Future Generation Computer Systems*, *86*, 650–655. doi:10.1016/j.future.2018.04.060

Sidorov, M., Nhut, P. V., Matsumoto, Y., & Ohmura, R. (2019). Lora-based precision wireless structural health monitoring system for bolted joints in a smart city environment. *IEEE Access: Practical Innovations, Open Solutions*, *7*, 179235–179251. doi:10.1109/ACCESS.2019.2958835

Winning with the industrial Internet of things. (2015). *Accenture*. Available: https://www.accenture.com/t20160909T042713Z—w—/us-en/_acnmedia/Accenture/Conversion-Assets/DotCom/Documents/Global/PDF/Dualpub_11/Accenture-Industrial-Internet-of-Things-Positioning-Paper-Report-2015.pdfla=en

Yin, H., Wang, T., Yang, D., Liu, S., Shao, J., & Li, Y. (2016). A smart washer for bolt looseness monitoring based on piezoelectric active sensing method. *Applied Sciences (Basel, Switzerland), 6*(11), 320. doi:10.3390/app6110320

Yli-Ojanperä, M., Sierla, S., Papakonstantinou, N., & Vyatkin, V. (2019). Adapting an agile manufacturing concept to the reference architecture model industry 4.0: A survey and case study. *Journal of Industrial Information Integration, 15*, 147–160. doi:10.1016/j.jii.2018.12.002

Zhao, X., Zhang, Y., & Wang, N. (2019). Bolt loosening angle detection technology using deep learning. *Structural Control and Health Monitoring, 26*(1), e2292. doi:10.1002tc.2292

Chapter 12
An Advanced Wireless Sensor Networks Design for Energy–Efficient Applications Using Condition–Based Access Protocol

Vijendra Babu D.

Aarupadai Veedu Institute of Technology, Vinayaka Mission's Research Foundation, India

K. Nagi Reddy

Lords Institute of Engineering and Technology, India

K. Butchi Raju

GRIET, India

A. Ratna Raju

Mahatma Gandhi Institute of Technology, India

ABSTRACT

A modern wireless sensor and its development majorly depend on distributed condition maintenance protocol. The medium access and its computing have been handled by multi hope sensor mechanism. In this investigation, WSN networks maintenance is balanced through condition-based access (CBA) protocol. The CBA is most useful for real-time 4G and 5G communication to handle internet assistance devices. The following CBA mechanism is energy efficient to increase the battery lifetime. Due to sleep mode and backup mode mechanism, this protocol maintains its energy efficiency as well as network throughput. Finally, 76% of the energy consumption and 42.8% of the speed of operation have been attained using CBI WSN protocol.

DOI: 10.4018/978-1-7998-6870-5.ch012

1. INTRODUCTION

With the rapid advancement in technology, remote sensing and measurement are becoming increasingly necessary. Data can be stored, analyzed, structured, and interpreted as never before thanks to the internet's ability to process vast volumes of data and sophisticated optical signal processing techniques Adler, R.,(2005), Baweja, G. (2002). This opens up the possibility of extensive environmental and animal habitat surveillance, as well as sophisticated industrial machinery, aerospace vehicle platforms, consumer electronics, and the home climate. From manual meter reading to unified data collection devices, sensing has progressed to a new age of distributed Wireless Sensor Networks (WSN). WSN will now provide an intelligent interface for data collection and analysis without the need for human interaction WSN hold great promise in a variety of device domains, thanks to recent advancements in MEMS Sensor technology. The wireless sensor network's topology and connectivity protocol was chosen based on energy consumption considerations Carley, T., (2003), Chatterjee, (2001).

2. MOTIVATION FOR WSN IN CBM

Two key components of a successful CBM system are dispersed data acquisition and real-time data analysis. As a result, distributed data acquisition can suffice for both machine management and tracking system learning. Both a component to operate the machinery and a component to probe or classify the mechanism are required in control theory. WS play a key role in enabling this functionality. The cost of wiring, which ranges from \$10 to \$1,000 a foot in wired systems, also limits the number of sensors that can be used. Wireless sensors can now navigate previously inaccessible sites, rotating equipment, dangerous or restricted areas, and mobile properties. If a sensor has to be shifted, these are simple to switch.

Com Duarte-Melo. E. J. (2003) panies also use manual methods to calibrate, weigh, and maintain their equipment. This time-consuming approach not only raises operating costs but also exposes the device to human error. Reduced manning levels, especially on US Navy shipboard systems, necessitate the installation of automated maintenance management systems. The CBM of a limited machinery space is the subject of our application. As a result, a single-hop WSN method piques our curiosity Heinzelman, W., (2000).

2.1 Limitation of Existed Method

Existed models are facing energy efficiency and delay parameters, the following discussing techniques are unable to providing clear operations. Therefore energy efficiency, speed of operations and delay need to be improve so proposed model is most effective compared to earlier methods.

3. DESIGN REQUIREMENTS

It is critical to comprehend the criteria that are applicable to sensor applications in order to design an effective WSN architecture (Heinzelman et al. 2000). The following are the specifications for implementing WSN for CBM and other similar applications.

Sensing that never stops. Variations and malfunctions of critical production systems and machinery must be constantly tracked. A minor change in output may have a negative impact on overall product quality or the health of manufacturing equipment Kannan, R., (2003), Soumya, N., (2019).

Transmission of data on a regular basis For the diagnosis of imminent errors and faults, CBM systems depend on historical records. There are complex processes that are always learning as they operate. Periodic data transfer helps update historical data, and improves the device's general usefulness for both diagnosing and prognosing system errors, as well as computing equipment's remaining usable lifespan.

Addressing Emergencies and Setting Alarms Unexpected malfunctions or deviations outside of specified tolerance bands can occur. As a result, a procedure for defining tolerance bands for each sensing module is needed. They must be able to respond to new circumstances and incorporate new information into their knowledge base. CBM systems' intrinsic adaptability necessitates a related feature from the WSN architecture Saikumar K, (2020).

Scalability is an important factor. Any sensing nodes can malfunction or have their batteries drained over the course of service. In order to Processes and facilities are being tracked more carefully and reliably, more sensing nodes might be needed? The WSN should be modular such that increases in the number of nodes do not disrupt the system's operation Saikumar, K., (2020).

4. SENSOR NET HARDWARE FOR CROSSBOW

We used the MOTE-KIT5040 Professional Kit from Crossbow Machinery, for our implementation, which is a Mote Development integrated 8-node package. It includes MICA2 and MICA2DOT Motes from Crossbow's new generation. TinyOS is compliant with all Motes Shorabi, K. (1999), Singh, S. (1998).

5. SYSTEM ARCHITECTURE

We now look at the actual architecture, network topology, and protocol design to solve the specifications that have been specifically defined for the specified application domain. Figure 1 depicts our device, with the letter 'SN' denoting a sensor node. Finally, the equipment's residual usable life (RUL) must be estimated and the findings shown to maintenance staff. Any sensor with wireless networking capability can be used as a sensor node in the network. Compatibility between various modules produced by different suppliers is a big issue with too many sensor and network product manufacturers on the market (Lewis 2004).

create smart sensors and connect them to networks. The basic architecture of IEEE 1451 is depicted in Figure 2.

6. ENERGY CONSIDERATIONS

The secret to efficient energy-aware device architecture for wireless sensor networks is knowledge of the application environment as well as the physical layer capabilities of network hardware. When choosing a WSN topology and coordination protocol, energy considerations should be taken into account Ye, W., (2002), Zhao, F., (2002).

Figure 1. Overview of the system.

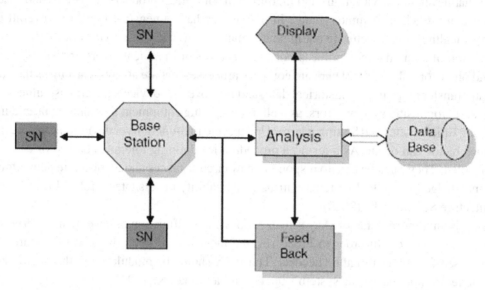

Figure 2. Smart sensor networks IEEE 1451 specification.

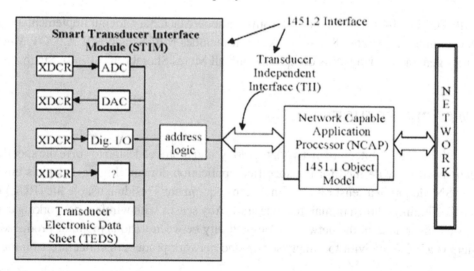

6.1 Radio Battery Consumption Model

For our radio model, we considered three modes of operation: send, receive, and sleep. A standard transceiver is depicted in Figure 3 as a symbolic block diagram. In the figure, I_{tx} represents the total current drawn by the radio when the transmitter unit is dynamic (in transmit mode), I_{rx} This sign represents the current drawn by the radio.

The batteries that are located on the sensor nodes power the radios on the nodes. The power of a battery is estimated in ampere-hours, which is the amount of energy it can carry. As a result, we model the

Figure 3. Symbolic radio model.

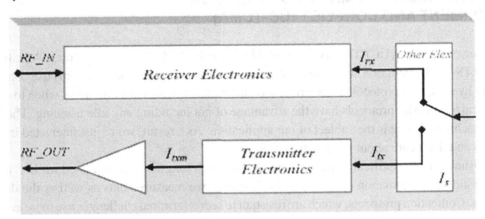

radio's energy usage in ampere-hours in order to create a clear relationship between the radio's energy consumption and the battery's energy consumption. The Battery Consumption Equation, seen below, is an expansion of Shih et al. (2001)'s work: it calculates how much battery energy the radio uses in an hour. Using this per-hour measurement, you can conveniently calculate total battery usage.

6.2 Topology—Battery Consumption Perspective

Our network's multiple sensor nodes send data to a single sink for processing, decision-making, and storage. As a result, the many-to-one model emerges as a natural form of communication. However, since we're looking for applications in narrow machinery spaces, networking isn't a problem. As a result, this section focuses on energy issues. Multi-hop aims to reduce the amount of energy dissipated at each node by reducing the distance over which each node would transmit. However, by doing so, the network's overall energy consumption may increase.

The receiver noise figure is assumed to be 7.5 dB. Antennas with a gain of 1 dB are employed. The fade margin is set to 20 dB. (99 percent Rayleigh probability). Packets are 456 bits long and contain 38 bytes (excluding the preamble). The system's target is to read 90% of packets on the first attempt. 868 MHz is the operating frequency.

For our system, we developed a single-hop topology, keeping in mind the energy limitations, latency requirements, required simplicity at nodes, and to avoid all control overheads. Furthermore, since we're only interested in a limited machine domain, multi-hop isn't needed to maintain communication.

6.3 MAC Protocol from the Point of View of Battery Use

MAC is used in a broadcast network to manage channel control such that data can be sent from a source to a target. The battery consumption formula is used in this section to calculate the impact of the MAC protocol's swapping of radio modes (sleep/receive) on the node's battery consumption.

This demonstrates that switching modes can be performed as infrequently as practicable, and that brief texts should be prevented. This incorporates scheduling and contention mechanisms to minimize switching frequency while accessing a common communication medium.

7. A NEW HYBRID TDMA MAC PROTOCOL FOR DEVELOPMENT AND CONFLICT (UC-TDMA)

We add user-configured (UC) TDMA, a novel TDMA hybrid MAC schedules framework built on a reworked RTS-CTS contention structure, in this article. Short-range wireless systems use considerably less energy by using this protocol. Since nodes are told ahead of time, via a timetable, when to anticipate incoming traffic, TDMA protocols have the advantage of not including any idle listening. The CBM of a limited machinery space is the subject of our application. As a result, we're just interested in a single-hop system and don't care about propagation times.

The consumer must specifically specify the order in which data would be obtained while using CBM. This would aid in the formation of relationships between two measurements as well as the drawing of conclusions. Contention protocols, which are resistant to secret terminal challenges use more energy than TDMA. Even if there is no contention or scheduling, there is always the possibility of a collision. The proper operation of a TDMA scheme necessitates close time synchronization among different nodes. To maintain a high level of precision in synchronization, wireless nodes must exchange messages frequently.

Figure 4. each node has a Finite State Machine operating.

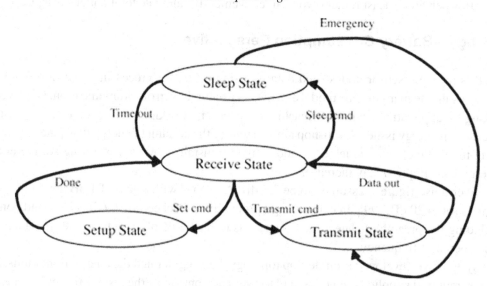

Distribution and an updated RTS-CTS mechanism. We also look for minimum protocol processing across nodes and low node conflict over-heads. The most of these properties are due to the single-hop topology and unified regulator provided to the access point.

7.1 Node-Based Processing

One of the most critical facets of our network organization and protocol is to reduce the amount of computation needed at the nodes in order to allow node-to-BS communication. The base station will move between states in emergency warning conditions or at times set by the user. Nodes join the receive state as they turn on because it needs less initialization energy than the transmit state.

7.2 Processing at Base Station

Many of the protocol's energy-intensive activities are done by provided united power and a resourceful base center, the base station. The flowchart of the UC-TDMA protocol operating at the base station is shown in Figure 6. The base station's functions are described in detail in the following four subsections.

Figure 5. Frame for UC-TDMA.

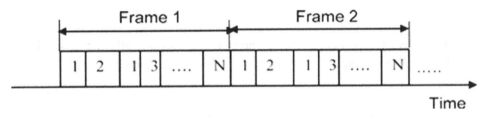

7.2.1 Scheduling of TDMA Slot Allocation for memory-constrained sensor networks, TDMA-based interfaces aren't often the best choice, despite the fact that they naturally avoid collisions and save bandwidth. It is also simpler to keep track of all nodes' time slots at the base station, and work with them when required. This solution eliminates the need for nodes to manage a table or schedule time slots, saving both memory and complexity at the nodes.

The user will specify the order in which nodes can enter the channel, as well as the time span for their respective slots, using the GUI described in Section 9. Nodes can enter the channel more than once in a given time period, depending on the program, and the duration of each slot can vary. The TDMA frame for our network is seen in Figure 5. Note that this differs from the standard TDMA approach in that we are sacrificing justice at each node in order to meet our application's needs.

The nodes' wake-up times are calculated by the base station. When nodes enter or leave the network, these times change, as well as on a regular basis to account for clock drift between nodes. To update the times, we present a stream-lined matrix formulation.

This makes it possible to deal with node clock drifts, which happen when each node wakes up a little earlier than in its turn to submit data. As a consequence, we set the nodes' sleep cycles to sleep as long as possible thus switching as least as possible.

7.2.3 TDMA Protocol that is configured by the user. In this part, we'll go through a new contention scheme that the base station should use (BS). Our updated RTS-CTS mechanism is affected by our plug-powered base station. After determining that no other node is interacting with it, BS produces a simulated RTS on behalf of that node at the time it acquires the channel according to its timetable. The startup of transmitter electronics wastes a lot of energy because these power over-heads are small packets. In addition, the amount of processing needed at the node to receive the channel is virtually negligible. The node actually sleeps and wakes up on a set schedule.

7.2.4 Setup and operation of a network. The movement of the whole network service is depicted in Figure 6. To begin monitoring the system with a sensor network, we must first physically mount the sensors in the right positions on the mechanism. Each sensor has a 16-bit node form that corresponds to the physical quantity it measures (such as vibration, temperature, friction, strain, and so on). The application software is executed on a hand-held computing system (such as a PDA or laptop) that is connected to the base station via a serial port. As the network is turned on, all nodes are in receive mode.

The first few blocks of the flow map in Figure 6 explain the network configuration. The user describes the functionality of different nodes in the network using a user interface. The user will set the values of different parameters for each node individually. If a node is discovered to be absent, the base station notifies the user and restores all of the node's parameter arrays. Equation (2) or (3) is then used to measure the sleep time for each node (3).

$$y(n) = u(n) + w(n) . \text{---} \tag{1}$$

Equation 3 and 4 clearly explains that entire channels and its users allowed range. This numerous value used to allow the limited number of users from channel 1.....channel m.

$$x = \sum_{n=0}^{N} |y(n)|^2 . \text{---} \tag{2}$$

Here x = entropy calculation model, which is expressed as mathematically in equation 2 & N = number of channels available. It is used to determine the channel is available or not. This availability is calculated based on entropy compared to the threshold value. Here H_0 .are the primary users in channel zero, H_1 .is primary users in channel one, respectively.

$$y(n) = w(n) : H_0 . \text{---} \tag{3}$$

$$y(n) = u(n) + w(n) : H_1 . \text{---} \tag{4}$$

The equation 4 gives the number of primary user's presence; the above mathematical computations clearly show the availability of the spectrum and primary users count in channels. For secondary users and its transmission performance can be determined using P_D.

$$p_D = p_r (\chi \geq \lambda \mid H_1) . \text{--} \tag{5}$$

$$p_D = \frac{1}{2} erfc \left(\frac{\lambda - u}{\sqrt{2w}} \right) . \text{--} \tag{6}$$

Here P_D = Probability detection of primary users, when equation 5 and 6 satisfies the condition, then we estimate the secondary user's transmission else doesn't allow the secondary users.

Figure 6. UC-TDMA MAC protocol flow table.

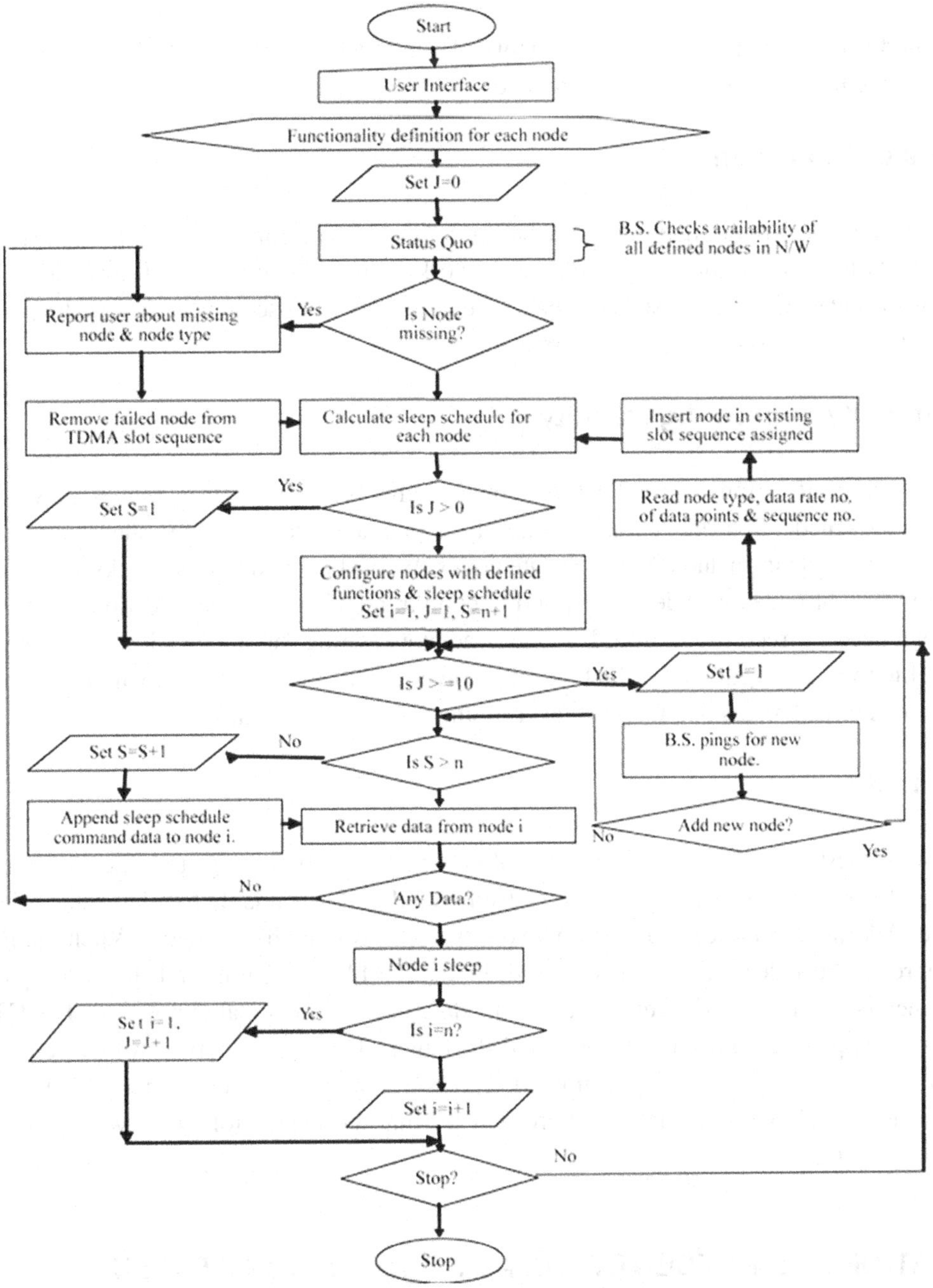

We can now collect data from of the sensor devices after setting up the device and reconfiguring all of the nodes. Using the user-configured TDMA frame, the base station decides when the node has access to the channel and for how long. The BS then establishes a connection with the selected node using a changed RTS-CTS protocol to prevent any impacts.

8. SYSTEM FEATURES

This segment discusses the additional functionality of our system design that render the wireless sensor network effective for condition-based maintenance.

8.1 Modes of Operation

In general, two network operating modes have been considered. The continuous mode is the first choice. This is particularly valuable for newly installed networks. where the question of how much data from multiple sensor nodes can be accessed is often encountered. As a consequence, this mode gives you the most complete picture of the computer you're using.

8.2 Adaptability and Configurability

Application data specifications for a collection of measurements can alter during regular network activity in non-continuous mode. To achieve this adaptability, Using new node variables, the access point determines a new sleeping pattern and for appropriate nodes. We've believed that networks aren't compelled to adjust the order of nodes in order to respond to evolving device data specifications. After deciding that there is a need for reconfiguration, BS ceases acknowledging the nodes with sleep commands in order to enable the reconfiguration, which often involves a shift in the series in which nodes transmit. As a result, the network takes one loop to start running with the new setup.

8.3 Scalability

The collapse of existing nodes and the addition of new nodes are two things that must be addressed in order to meet the scalability criteria. A node is deemed failed if it is unable to relay data at the agreed-upon period. Adding new nodes to a CBM network is not a normal phenomenon. When the BS finds a new node, it reads the node type, sequence number in the TDMA time frame, and other node parameters. These parameters are then packed into arrays at the data packet's designated position. The CTS signal for each node is appended with a newly computed sleep plan. The channel will now be filled with nodes from a new slot series. When the file is extracted, the node is put to sleep for the period of time stated in the current calendar, allowing the network to resume normal service the following week. Take a glance at Figure 6 as an illustration.

9. OPERATION AND PRESENTATION AUTHENTICATION OF WSN

9.1 Application of WSN

In a heating and air-conditioning plant, ARRI used the UC-TDMA protocol. Our WSN can be mounted in this small computer room in about 30 minutes using our special interface GUI. The operation contained four Crossbow sensor nodes, one base station, one phone, and Lab VIEW model 7.1. On Figure 7's equipment, all of the sensor nodes were in ideal physical arrangements.

Figure 7. Farm for heating and cooling.

9.2 Graphical User Interfaces

As seen in Figures 8 and 9, we built two different GUIs: a network setup wizard and an submission GUI. Many of the node parameters are saved in a separate configuration file by the network setup wizard, which is then used by the device GUI to customize the network. It's possible to make and store several configuration files for later usage. The following parameters are set for each node. 1.

Figure 8. The network setup wizard as used on a computer screen.

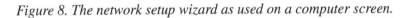

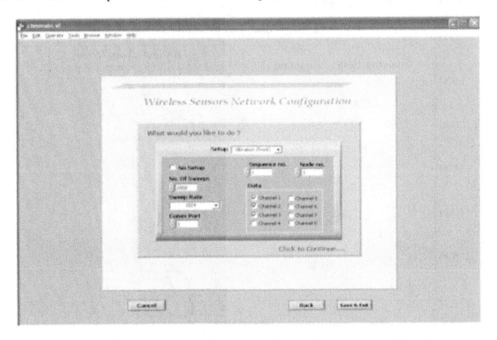

Figure 9. A screenshot of the application's user interface

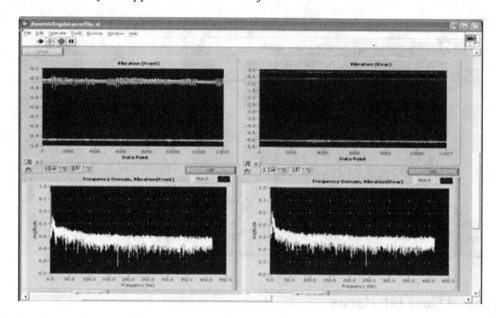

The number of available networks was its number of lines that communicate while a node sends a packet. Each node may have up to eight channels allocated to it. 5. Node Number: The form of actual quantity that each node is calculating is defined by this 16-bit number. The application programmer generates the UC-TDMA frame using the Sweep rate, Number of Sweeps, and Sequence Number arrays.

Figure 10. battery energy usage of a sensor node.

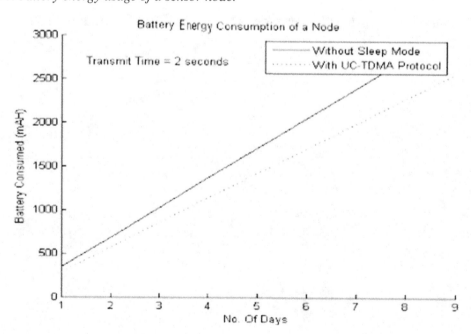

The sweep rate and amount of sweeps decide the duration of the TDMA slot for that specific node. The sequence number, on the other hand, determines the slot's position in the picture.

Each sensor's data is stored in a data file chosen by the user over the course of the network's service. This information can be used by some other application software for more in-depth data processing using methods such as fuzzy logic, neural networks, and other techniques. For data collection, the network was run in continuous mode using the UC-TDMA protocol.

Figure 11. The nodes in the network on a node's battery

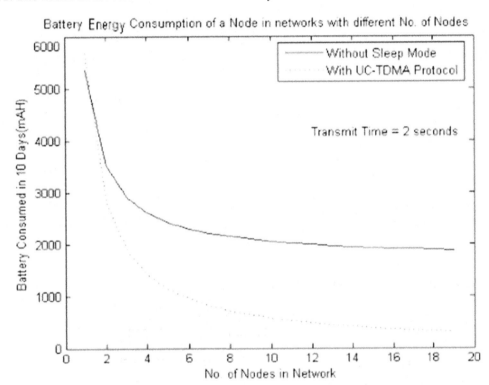

Equations (2) and (3) (3), we inferred the battery usage to greater amounts of nodes created on the real streams dignified in our WSN with 4 nodes (3). Figure 11 depicts the predicted result of cumulative the no of nodes in a continuous-mode network. The amount of days each node turns to transmit mode per hour reduces as the number of nodes in the network grows. As a consequence, over the course of ten days, the node's battery energy usage decreases. It can be shown that as the number of nodes in the network rises, the UC-TDMA protocol becomes more effective. It graphs the battery energy used by a node over the course of ten days vs the node's data updating rate.

Figure 13 displays the effects of a five-day battery life test. This demonstrates that UC-TDMA outperforms the continuous transmission scheme, and that its battery life increases as the number of nodes in the device grows.

Figure 12. Battery energy use is affected by the data updated.

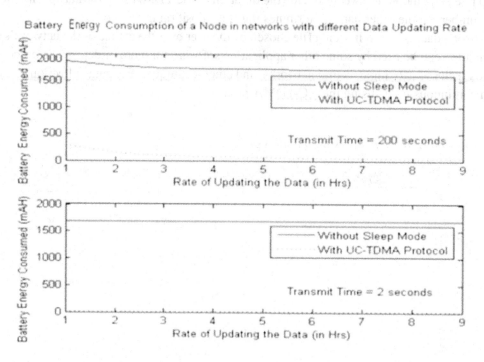

Figure 13. Series life test on the WSN that has been applied

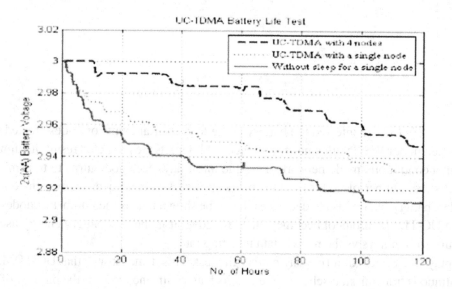

CONCLUSION

In this research work CBA protocol is designed for Wireless Network applications. The proposed model providing accurate network energy efficiency and lifetime. Another test was performed on the deployed WSN to check the stability of UC-TDMA and to determine the realistic operation lifetime of the node battery power subsystem. Three sets of experiments were carried out, each lasting five days, Finally 76% energy consumption and 42.8% speed of operation has been attained using CBI WSN protocol

REFERENCES

Adler, R., Buonadonna, P., Chhabra, J., Flanigan, M., & Krishnamurthy, L. Kushalnagar, N., Nachman, L., & Yarvis, M. (2005). Design and deployment of industrial sensor networks: Experiences from the north sea and a semiconductor plant. *Proceedings of ACM SenSys.*

Baweja, G., & Ouyang, B. (2002). Data acquisition approach for real-time equipment monitoring and control. *IEEE 2002 Advanced Semiconductor Manufacturing Conference (ASMC '02)*, 223–227. 10.1109/ASMC.2002.1001608

Carley, T., Ba, M., Barua, R., & Stewart, D. (2003). Contention-free periodic message scheduler medium access control in wireless sensor/actuator networks. RTSS, 24th IEEE, 298–304.

Chatterjee, M., & Das, S. (2001). Performance evaluation of a request-TDMA/CDMA protocol for wireless networks. *J. Interconn. Netw.*, 2(1), 49–67. doi:10.1142/S0219265901000257

Duarte-Melo, E. J., & Liu, M. (2003). Data-gathering wireless sensor networks: Organization and capacity. Comput. Netw. (COMNET). *Special Issue on Wireless Sensor Networks*, 43(4), 519–537.

Heinzelman, W., Chandrakasan, A., & Balakrishnan, H. (2000). An application-specific protocol architecture for wireless microsensor networks. *IEEE Transactions on Wireless Communications*, 1(4), 660–670. doi:10.1109/TWC.2002.804190

Kannan, R., Kalidindi, R., Iyengar, S. S., & Kumar, V. (2003). Energy and rate based MAC protocol for wireless sensor networks. *SIGMOD Record*, 32(4), 60–65. doi:10.1145/959060.959071

Saikumar, K., & Rajesh, V. (2020). Diagnosis of coronary blockage of artery using MRI/CTA images through adaptive random forest optimization. *Journal of Critical Reviews*, 7(14), 591–600.

Saikumar, K., Rajesh, V., Hasane Ahammad, S. K., Sai Krishna, M., Sai Pranitha, G., & Ajay Kumar Reddy, R. (2020). Cab for Heart Diagnosis with RFO Artificial Intelligence Algorithm. *International Journal of Research in Pharmaceutical Sciences, 11*(1).

Shorabi, K., & Pottie, G. J. (1999). Performance of a novel self-Organization protocol for wireless ad hoc sensor networks. *Proceedings of IEEE VTC*, 1222–1226.

Singh, S., & Raghavendra, C. S. (1998). PAMAS—Power aware multi-access protocol with signal- ing for ad hoc networks. *Proceedings of ACM SIGCOMM'98*, 5–26.

Soumya, N., Kumar, K. S., Rao, K. R., Rooban, S., Kumar, P. S., & Kumar, G. N. S. (n.d.). 4-Bit Multiplier Design using CMOS Gates in Electric VLSI. *International Journal of Recent Technology and Engineering*.

Ye, W., Heidemann, J., & Estrin, D. (2002). An energy-efficient MAC protocol for wireless sensor networks. Proceedings of IEEE Infocom, 1567–1576.

Zhao, F., Shin, J., & Reich, J. (2002). Information-driven dynamic sensor collaboration for target tracking. *IEEE Signal Processing Magazine*, *19*(2), 61–72. doi:10.1109/79.985685

Chapter 13
Effective Image Fusion of PET–MRI Brain Images Using Wavelet Transforms

Magesh S.
https://orcid.org/0000-0003-2876-7337
Maruthi Technocrat E-Services, India

Niveditha V. R.
Dr. M. G. R. Educational and Research Institute, India

Radha RamMohan S
Dr. M. G. R. Educational and Research Institute, India

Amandeep Singh K.
Dr. M. G. R. Educational and Research Institute, India

Bessy Deborah P.
https://orcid.org/0000-0002-6315-1784
Dr. M. G. R. Educational and Research Institute, India

ABSTRACT

Image processing concepts are used in the biomedical domain. Brain tumors are among the dreadful diseases. The primary brain tumors start with errors which are the mutations that take place in the part of DNA. This mutation makes the cells breed in a huge manner and makes the healthier cells die. The mass of the unhealthier cells is called tumor, or the unwanted cell growth in the tissues of the brain are called brain tumors. In this chapter, images from the positron emission tomography (PET) scan and MRI scan are fused as a one image, and from that image, neural network concepts are applied to detect the tumor. The main intention of this proposed approach is to segment and identify brain tumors in an automatic manner using image fusion with neural network concepts. Segmentation of brain images is needed to segment properly from other brain tissues. Perfect detection of size and position of the brain tumor plays an essential role in the identification of the tumor.

DOI: 10.4018/978-1-7998-6870-5.ch013

1. INTRODUCTION

Brain tumors are the diseases which cannot be detected easily, so many people who are affected by the brain tumor and are becoming prey to death. So this dreadful disease should be predicted in the earlier stages. In image fusion is a process where all the main data from the images are collected and combined as a one image. The single image is more accurate and informative, in medical domain image fusion concept is carried out in diagnosing the diseases. PET images are used for detecting very complex diseases. It makes use of the tracer which is called as radioactive substance. It allows the doctors to detect the disease. The tracer are swallowed or injected in a view of the arm on the particular area. The image from the tomography is considered for the image processing technique. Another image which is taken for fusing is the images from the MRI (Magnetic Resonance Imaging) reports. It is a type of scanning that makes use of the powerful magnetic fields and radioactive waves to create the images of physiological process and anatomy process. Using the concept of Image fusion, images taken out from MRI scan and PET tomography are considered for the process. In existing methods wavelet transform techniques are used to identify the brain tumor. The major disadvantages of existing methods are the amount of noise is high and provides low accuracy result. This proposed works image fusion is carried out for the detection of the brain tumors with neural networks concept. Current system can be implemented by using MATLAB software.

2 LITERATURE REVIEW

Brundashree .R et al (2015) combined PET images and MRI images from the brain are utilizing wavelet transforms for tumor detection. After applying wavelet decomposition technique and fusion in gray level acquire better result. This is done by modifying basic data in Gray Matter (GM) region and fixing the data in the zone of white Matter (WM). Authors utilize three different datasets: normal axial data file, normal coronal dataset and the dataset from the brain disease Alzheimer's for checking and evaluation. The exhibition of the combination of spectral discrepancy (SD) strategy and the value of average gradient (AG) strategy provides a proof to be enhanced result with numerically and visually. The image fusion helps to collect valued data from various pictures into form a single image. The mixture of PET and MRI brain disease images for the current concept creates the better normal tendency by which the spatial aim of picture is elevated. The spectral discrepancy for the dataset 1 and 3 produces same result as the current strategy, dataset 2 is changes to bring downwards the qualities. Henceforth the spatial aim of the picture are improved when using with the proposed technique .

Breast tissues material parameters and direction of the gravitational forces are estimated by using biomechanical model. Remaining deformation details are finding out by using non-rigid intensity based image registration.

Introduced A-NSIFT algorithm enhances the interest key point extraction procedure. Compare to conventional NSIFT, A –NSIFT is 200 time faster in interest point extraction. Further key point matching a probability density estimation Gaussian mixture model is used for alignment. From the experimental result it is concluded that the given module presents the faster 3D alignment even though there is poor initialization.

But the most challenging thing is to integrate or fuse two different medical images. The referred paper propose an efficient algorithm to integrated both MR and ultrasound images. Robust PaTch-based correlations Ratio (RaPTOR) used to register the 3D volumetric data of both ultrasound and MR data. The referred system proposed a decent approach towards the image registration module which speeds up the performance.

A similarity based image registration module in which correlation among the pixel is computed by Sum-of-Squared-Difference (SSD) is presented. Along with SSD the aid of the low rank matrix theory is applied. Application of the theory will compensate the intensity distortion which leads to the rank regularised SSD (RRSSD). The proposed architectures effectively perform intensity distortion correction and image registration at a time. Experimental result concluded that designed algorithm effectively meets the clinically acceptable image registration output.

When complex and spatially dependent intensity relation between images to be registered it is difficult to apply MI based registration. A new similarity measure on self-similarity using a patch based paradigm is proposed. Unlike other registration techniques, self-similarity registration make use of statistical dependency, the innovative measure is able to take the internal structural relationship into account. The system gives better alignment accuracy.

Before applying the registration, histogram specification is used for transformation. Using SURF features points are extracted and preliminary registered image is obtained. Optical flow algorithm is applied to registered image to get the accurate result. The system is evaluated using multimodal brain image.

Shakir H et.al has proposed an image registration using DWT followed by Gaussian pyramids. The dwt is applied to target and reference image, LL, LH, HL and HH sub-bands are used for the registration and also Gaussian pyramids also used for registration. The quality of the registered image is calculated using the maximum MI and CC values.

Zhang Li et.al (2007) proposed an autocorrelation based image registration techniques, in which local structure of given input images are analysed in terms of local similarity structure. Autocorrelation of LOcal STructure (ALOST) algorithm is planned to calculate the local structure in terms of standard deviation. The referred algorithm is less sensitive to the intensity distortions. Due to the silent features of designed algorithm, the referred algorithm meets highest correlation.

Finding similarity between fixed image and moved image is tough in image registration. Varying in spatially intensity was always important and most challenging in image registration.

Umer Javed et al (2014), presents fusion strategy of MRI image with PET images utilizing local characteristics and fuzzy concept. The aim of this current method is to increase integrate valuable data available in MRI pictures and PET pictures. Image characteristics are obtained and integrated with logic of fuzzy concept to calculate value for every pixel. Result shows that the current method creates improved outcomes than the other state-of-art concept .

Over most recent couple of decades, 3 dimensional reconstructions of clinical pictures become advanced method in healthcare domain. Remodeling of 2 dimensional images of some informational collections into 3 dimensional amount, by means of enrollment of 2D areas had turn into the majority fascinating subject. In recent years, MRI images are utilized for various medical domain data analysis. Vidhya K et al (2019), proposed a framework uses MRI pictures are used in a similar view, various method or procured by various picture modalities to build the data .

Clinical images with fusion concept represent the way toward registering and integrating numerous pictures from solo or different imaging features to increase the imaging worth and Multi-modular clinical image fusion calculations. And gadgets have indicated prominent accomplishments in improving clinical precision of choices dependent on medical pictures. A.P. James et al (2014), describes the medial images fusion look into dependent on (1) the broadly utilized image combination techniques, (2) imaging features, and (3) parts of images that under investigation. This survey despites the detail that exists a few open finished innovative and logical difficulties. The combination of medical images has end up being valuable for clinical dependability of utilizing clinical imaging for clinical diagnostics and examination .

AT used for registration, the proposed algorithm AT is tested over MR brain image and brain images with lesion and also test with the brain image with the brain images with various noise levels. The registered image helps in reduces the time for decision making in diagnosis and treatment.

We used Magnetic Resonance Image (MRI) and Positron Emission Tomography as input images, so fused them based on combination of two dimensional Hilbert transform and Intensity Hue Saturation method. Evaluation metrics that we apply are Discrepancy as an assessing spectral features and Average Gradient as an evaluating spatial features and also Overall Performance (O.P) to verify properly of the proposed method

Medical images fusion joins the corresponding pictures from various quality and improves the features of fusion pictures that issues high anatomical and useful data to specialists. Siva Kumar N et al., proposed a hybrid approach to merge the pictures got from Computer Tomography technique and MRI images utilizing Wavelet with Curve let Transform concepts. Coding concept of sub-band and scale finding techniques are utilized for image features filtering. Scaling tasks on the images are used to decomposition. A Ridgelet Transform procedure was utilized with radon Transform to do picture from 1D - 2D in reconstruction phase of combined image data. The output image was valuated using Entropy metric (H), Root Mean Square Error metric (RMSE) and Peak Signal to Noise Ratio metric (PSNR) values. Performance of the current technique provides enhanced result than previous methods .

3 PROPOSED ARCHITECTURE

The images from MRI and PET are considered for the image fusion process and neural networks to makes the detection of brain tumor is an easiest way. The key intention of proposed framework is to develop a novel image fusion based system for identifying of brain tumors. In this current research work, image fusion technique contains two important tasks such as (I) enhancement of an image and (II) image mixture both depends on neural network. To improve the image quality Lagrange's interpolation approach is used. MRI image and PET images are combined based on enhancement of an image and fusion concept that has been developed and modeled using MATLAB software. The following figure 1 represents the framework of proposed tumor detection system.

Let various approximations $I_{MRI,k}$ value of the MRI picture I_{MRI} value (with M into N dimension) be received by consecutive convolutions of a filter value f, that is[2],

$$I_{MRI,k+1} = I_{MRI,k} * f \qquad (1)$$

Here $I_{MRI,0}$ equals to I_{MRI} value & f represented as a filter data of bicubic B-spline. k^{th} wavelet level $W_{MRI,k}$ of the image I_{MRI} represented as,

$$W_{MRI,k} = I_{MRI,k+1} - I_{MRI,k} \qquad (2)$$

Figure 1. Proposed Framework for Tumor Detection System

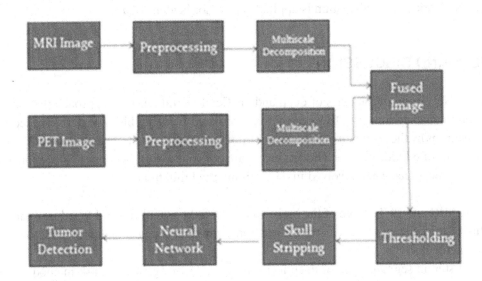

The data of an image I $_{MRI}$ is divided into low and high frequency is represented as $I_{MRI,L}$ and $I_{MRI,H}$.

$$I_{MRI} = I_{MRI,L} + I_{MRI,H} = I_{MRI,L} + \sum_{k=0}^{K} W_{MRI,k} \tag{3}$$

In the equation (3) k represents the total value of the division level

The data of an PET image I $_{PET}$ is divided into low and high frequency is represented as $I_{PET,L}$ and $I_{PET,H}$.

$$I_{PET(\beta)} = I_{PET,L(\beta)} + \sum_{k=0}^{K} W_{PET,k(\beta)} \tag{4}$$

Here $\beta \in \{R, G, B\}$, as represents PET images are considered as in pseudocolor [6].

The MRI image and PET image is taken and given as input. During preprocessing phase where the mage is converted from RGB to GREY scale. After that the processes of decomposition takes place and image is fused. Thresholding is one of the simple methods for image segmentation. This depends on a threshold value to convert an image with gray level into a binary representation of an image.

The result from the Thresholding process normally it is an binary image. Mathematically, the operation of the thresholding can be expressed as,

$$g(x,y) = \begin{cases} 1 \: if \: f(x,y) > T \\ 0 \: if \: f(x,y) \leq T \end{cases}$$

Here T represents thresholding value, if g(i, j)=1 for image data of the concern objects, and g(i,j)=0 for background image components.

On the fused image apply threshold value to find the region of interest where tumor is present. Then Deep Belief Network (DBN) algorithm is applied and tumor is identified.

4 RESULTS AND DISCUSSION

DBN, or DeepNet is one of the class of deep and artificial neural network approach that has been effectively used to analyze images. DBN concept is stimulated by using biological tasks in connectivity model with neurons by the arrangement of visual cortex. It contains various layers like input layer, output layer and number of e hidden layers. Hidden layers are convolutional, pooling or fully connected type. The following process has been involved in determining the brain tumor.

1. First the MRI and PET image of the brain which should be analyzed and fused is obtained
2. For fusion the concept of wavelet transformation technique. Using deep learning metod the images are segmented
3. The first step in segmentation of deep learning is the binary gradient mask method in which the binary values of the fused image is calculated

Figure 2.

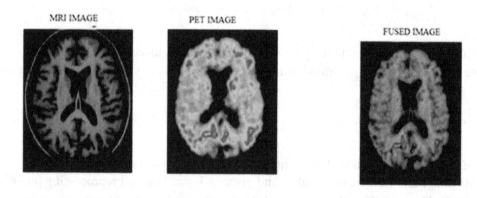

Figure 3. Images after applying applying image processing concepts

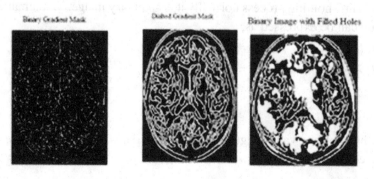

4.   Then Dilated gradient mask method where the layers of the brain fused image is dilated for finding the brain tumor location

Figure 4. Segmented Images

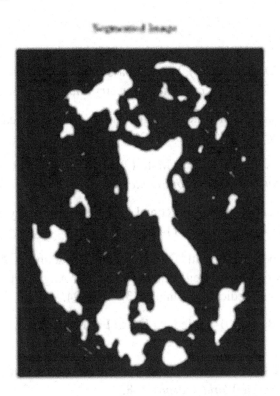

5.   Implementing the deep net classification algorithm the tumor location is identified clearly

The following figure 2 shows MRI image, PET image and combined image.

The figure 4 shows the resultant segmented image after using neural network concepts. The white marks represents the affected area of the given brain image.

5 CONCLUSION

Brain tumor is one of the major fatal type diseases. Detection of this disease is difficult in earlier stage. Image fusion is the process of combing two or more images as a single one; the images are gathered from the sensors of various wavelengths. Normally MRI images are used to recognize the disease in affected areas. PET images are used to identify complex diseases. In this research work the MRI images and PET images are combined as a single one. From fusion image affected area can be identified easily by using various image processing techniques. Here wavelet transform and neural network concepts are applied on the fused image to identify the affected area on the brain images. The major benefit in this proposed work is identifying affected area accurately than other existing methods. This method is a radiopharmaceutical commonly used for PET scans. The PET-FDG images of normal, grade II, and grade IV astrocytoma, respectively. It can be seen that different imaging modalities provide complementary information for the same region.

REFERENCES

Amolins, K., Zhang, Y., & Dare, P. (2007). Wavelet based image fusion techniques—An introduction, review and comparison. *ISPRS Journal of Photogrammetry and Remote Sensing, 62*(4), 249–263.

Bhatnagar, G. (2013). Human visual system inspired multi-modal medical image fusion framework. *Expert Systems with Applications, 40*(5), 1708–1720.

Brundashree, Kakhandaki, Kulkarni, Rod, & Matt. (2015). The PET-MRI Brain Image Fusion Using Wavelet Transforms. *International Journal of Engineering Trends and Technology, 23*(6), 304-307.

Daneshvar, S., & Ghassemian, H. (2011). MRI and PET image fusion by combining IHS and retina-inspired models. *Information Fusion, 11*(2), 114–123.

Han, L., Hipwell, J. H., Eiben, B., Barratt, D., Modat, M., Ourselin, S., & Hawkes, D. J. (2014). A Nonlinear Biomechanical Model based Registration Method for Aligning Prone and Supine MR Breast Images. IEEE Transactions on Medical Imaging, 33(3), 682-694.

James, A. P., & Dasarathy, B. V. (2014). Medical Image Fusion: A survey of the state of the art. *Information Fusion.*

Javed, Riaz, Ghafoor, Ali, & Cheema. (2014). MRI and PET Image Fusion Using Fuzzy Logic and Image Local Features. *The Scientific World Journal*, 1–8.

Saravanan, D., Nirmala Sumitra Rajini, S., & Dharmarajan, K. (2020). Efficient image Data Extraction using Image Clustering Technique. *Test Engineering and Management, 82*, 14574–14579.

Sivakumar, N., & Helenprabha, K. (2017). Hybrid medical image fusion using wavelet and curvelet transform with multi-resolution processing. *Biomedical Research, 28*(6).

Vidhya, Patil, & Patil. (2019). An Efficient MRI Brain Image Registration and Wavelet Based Fusion. *International Journal of Recent Technology and Engineering, 8*(4), 10209-10218.

Yang, L., Guo, B. L., & Ni, W. (2008). Multimodality medical image fusion based on multiscale geometric analysis of contourlet transform. Neurocomputing, 72, 203–211.

Yu, D., Yang, F., Yang, C., Leng, C., Cao, J., Wang, Y., & Tian, J. (2016). Fast Rotation-Free Feature-Based Image Registration using Improved N-SIFT and GMM-Based Parallel Optimization. *IEEE Transactions on Biomedical Engineering, 63*(8), 1653–1664.

Chapter 14
IoT–Enabled Non–Contact–Based Infrared Thermometer for Temperature Recording of a Person

Swapna B.

ⓘ https://orcid.org/0000-0002-7186-2842

Dr M. G. R. Educational and Research Institute, India

ABSTRACT

Laboratories are essential facilities provided in professional institutes for scientific and technological work. The lab must ensure their accuracy to regulatory requirements and maintain their data records so that the laboratory environment can be monitored properly. The laboratory environment temperature (LET) is monitored to ensure proper regulation and maintenance of indoor conditions and also to correlate the collected samples with these conditions. The LET data collection must be stored to influence the quality of the results and to ensure the stability of the laboratory environment. Hence, an IoT solution is presented to supervise real-time temperature and is known as iRT. This method helps to monitor ambient object supervision in real time. It is composed of a hardware prototype to collect the temperature data and to use web application to provide the history of temperature evolution. The result gained from the study is promising, and it provides a significant contribution to IoT-based temperature monitoring systems.

I. INTRODUCTION

Covid – 19 is the pandemic disease which affects different people in different ways. Most infected people will develop mild to moderate illness and recover without hospitalization. High body temperature leads to fever, cold, covid, etc., Our proposed work is used to monitor the body temperature automatically with the help of Internet of Things, Processor and Temperature Sensors. The normal human body temperature is 98.6°F (37°C). The temperature 100.4°F (38°C) leads to fever which is caused by an infection or ill-

DOI: 10.4018/978-1-7998-6870-5.ch014

ness. We need to take foods, herbals, fruits and vegetables which are used to increase the immunity of human lives. Challenges are measuring the capable of producing of the crops in the land from year to year, understanding the change in the silt structure, optimize water use and improve plant health. With Internet of Things, we have been benefiting from the automation. We are doing the same with agriculture industry. It is also used to the delivery & supply chain, cattle management, water conservation, etc.

With IoT's intervention in agriculture and recent technology which will enable even small land holders to look at automation as an affordable option and the producers to provide quality product and helped them to gain additional profits.

IoT has developed from the combination of wireless technologies and the Internet. Internet of Things (IoT) is a domain in which items, creatures, or individuals are furnished with a kind of identifiers and the capacity to exchange information over a system without demanding human-to-human or human-tocomputer interaction. The proposed IoT allows objects to be sensed and/or controlled remotely across existing network Infra-structure, creating opportunities for more direct integration of the physical world into computer-based systems and resulting in improved efficiency, accuracy, and economic benefit. IoT board included with SIM900 GPRS modem to actuate web association is additionally furnished with a processor to process all info UART information to GPRS based online information.

IoT Based Contactless Body Temperature Monitoring using Raspberry Pi with Camera and Email Alert. To solve this issue, temperature devices are often used to measure body temperature. These devices have non-contact IR temperature sensors which can measure the body temperature without any physical contact.

II. LITERATURE REVIEW

Recent studies proved that people spend 90% of their day in indoors which makes indoor environment a major concern for all. One of the primary environmental health risks is Indoor air quality (IAQ). For a healthy living environment, it is crucial to monitor IAQ in a real time basis. Especially for school going children who are more susceptible to harmful environmental pollutants than adults (De Gennaro G et al., 2014, pp. 467-482). These pollutants can result in the development of respiratory problems. Which is why it becomes important to control and maintain the IAQ and the exposure to these environmental pollutants. A survey was even conducted in Portugal to check and monitor the IAQ in classrooms and the results showed high levels of PM_{25}, PM_{10} and bacteria which even exceeded the WHO IAQ guidelines (Madureira J et al., 2015, pp. 145-156).

A large share of global energy consumption lies in thermal comfort in buildings such as air conditioning. It is also known that as climate is changing, temperature will rise gradually so changes in air conditioning will vary from place to place and climate to climate. Therefore, it's important for the energy distribution systems to take these future climate analyses into consideration to meet the thermal requirements (Yang L et al., 2014, pp. 164-173).

A very popular model used for assessment of thermal conduct in buildings is Predicted Mean Vote (PMA). It requires data on climate, clothes and heat production. For example, effects of body motion and air movements must be accounted for in order to get accurate results. So, this paper deals with the parameters and requirements for PMV (Havenith G et al., 2002, pp. 581-591).

A survey detailing about thermal comfort and indoor air quality in schools was conducted in Portugal by measuring some of the physical environmental parameters. The survey was also directed for the occupants regarding the temperatures during different climates. And they also ended up analyzing the CO_2 concentration present (Pereira LD et al., 2015, pp. 256-268)

This paper focuses on the topic of Internet of Things (IoT) and the various devices invented based on the IoT. It deals with the future visions for IoT and how it could help in making what is known as a 'smart' living environment (Gusto D, 2010, pp. 22-34).

It can be said that living habitat must be monitored in order to keep track of the IAQ and control the harmful environmental pollutants. In the recent years, these systems are made based on the Internet of Things (IoT) and Wireless Sensor Networks (WSN). The IoT is an ideal emerging technology to provide new evolving data and the required computational resources for creating revolutionary apps. One field of application of IoT is in the field of monitoring (Gubbi J et al., 2013, pp. 1645-1660).

The Future of Internet (FI) is very huge when talking about the Smart Cities. The IoT and Internet of Services (IoS) play main roles in the development of this FI idea and these particular components are the building factors towards the progress of our digital economy and smart cities (Hernández-Munoz JM et al., 2011, pp. 447-462).

Buildings are very crucial when considering smart strategies for public health as people stay indoors more. And they also play an important role for Smart Cities which develops different strategies to face public health challenges. Hence monitoring of IAQ becomes very essential and this paper shows the review of the systems made using IoT for IAQ in the past 5 years (Marques G et al., 2018, pp. 1169-1177).

A specific syndrome caused due to the impact of IAQ is 'Sick Building Syndrome' which involves problems related due to the skin, eyes, nervous system, upper and lower parts of the respiratory tract, eyes and the nervous system etc. Based on all these observations and results we can definitely consider Indoor Air Quality (IAQ) as a vital parameter for the safety of people in both their living and work places (Pitarma R et al., 2017, pp. 1-10).

On the topic of Indoor Environmental Conditions, Laboratories are also an important workplace to be considered. Also used as classrooms, laboratories tend to contain a large number of polluting sources. Therefore, it becomes important to monitor the environmental conditions in laboratories for experimental work and to maintain a healthy and productive workplace (Marques G et al., 2019, pp. 168-177).

This paper presents a solution called iAQs for monitoring environmental parameters in indoor based on the IoT. It is a prototype that collects data and even has social network compatibility as it accesses the Facebook application (Marques G et al., 2019, pp. 424-432).

This paper presents yet another solution to monitor and improve the indoor quality of air. It presents a hardware prototype with a WiFi system which offers many more advantages over the existing monitoring system especially due to its low cost. It incorporates temperature and humidity sensors, communication units and various other physical environmental parameters (Marques G et al., 2017, pp. 785-794).

'Particulate Matter' (PM) is a complex mixture of of organic and inorganic solid and liquid substances suspended in the air. It is a very harmful pollutant which could end up causing developments of respiratory, cardiac diseases, and even cancer of lungs. Hence a system is presented based on IoT called iDust for PM monitoring (Marques G et al., 2018, pp. 821-834). This paper aims to develop a smart object which is able to monitor the Indoor Environmental Quality (IEQ) by interacting with the air exchange unit and lighting system known as 'open-source smart lamp' (Salamone F et al., 2017, pp. 1021-1033).

The development of IoT has made it possible to collect massive data and also improve the energy consumption of buildings. This paper shows a survey to compare the existing models and the future research ideas and also deals with techniques for the development of smart cities (Akkaya K et al., 2015, pp.58-73).

With the help of IoT many technological advancements have been made possible and it has also helped in providing several improvements in living services and systems. These Indoor Air Quality monitoring systems based on IoT incorporate sensors, microcontrollers, processing units and even a web application for data storage purposes. It is even safe to say that most of these systems provide very reliable and accurate results (Marques G et al., 2014, pp. 170-182).

As we've established it is important to monitor the IAQ because it has a huge impact on the public health conditions and this means monitoring and controlling the physical parameter concentration values such as CO_2 which is an important index of IAQ. The monitoring of CO2 levels is crucial to detect IAQ issues as level of CO_2 over 100ppm indicates an indoor air quality problem. So, this paper presents a solution based on IoT called iAirCO2 for CO_2 monitoring (Marques G et al., 2019, pp. 1-12).

Infrared thermometer is a new type of non-contact thermometer which has many advantages over the traditional mercury thermometers. These thermometers give fast responses, incorporate safety and also have other characteristics like over-temperature alarm and LCD display which make it safer to use than mercury thermometers (Zhang J, 2017, pp. 10-22). Mean, median, standard deviation are measured by using regression line method for body temperature measurement (Tuladhar et al., 2021, pp. 31-36). Forehead temperatures are measured by using non-contact infrared thermometer which is useful for humans (Adrian Shajkofci, 2021, pp. 31-36). The proportional integral derivative (PID) algorithm and the pulse width modulation (PWM) technology used to measure the body temperature (Yaonan. Tong et al., 2021, pp. 1-14). The contactless body temperature monitoring (CBTM) of the inpatient department (IPD) is applied using a 2.4 GHz microwave band via the Internet of Things (IoT) network (Wasana Boonsong et al., 2021, pp. 1-7).

III. ENVIRONMENTAL IMPACT ASSESSMENT AND RISK ANALYSIS

A real-time data on the number of those infected with the spread of virus transmitted in the environment is lacking. This data is a needle to plan and choosing is focused. There is an inability to remotely monitor and equip the service key machinery, is a part of infrastructure. The economic impact of leisure, hospitality and retail businesses have to close their doors to the public and other industries to ensure their staff can work remotely. A sophisticated medical system to monitor patients remotely to balance with the availability of beds are needed very much. The spread of the virus through physical contact and in public places has to be avoided by using IRT proposed prototype.

IV. METHODOLOGY

Patient health monitoring is done by the doctors to monitor patient's health. The Patient's body temperature is pivotal to be monitored. Unfortunately, the present system requires doctors to see patients face to face, to check patient's body temperature. "The Non-Intrusive Human Body Temperature Acquisition and Measurement System" is a unique computer-based health system to monitor the patient's temperature.

Figure 1 consists of a processor which senses the temperature via the DF Robot IR thermometer sensor which incorporates an MLX90614 sensor module which has high accuracy of 0.5 degree Celsius and wide range of frequency 0 to 50 degree celsius. By detecting infrared radiation energy and wavelength distribution it measures the surface temperature of the patient. The Infrared radiation temperature encom-

Figure 1. Proposed Block Diagram for IRT Measurement

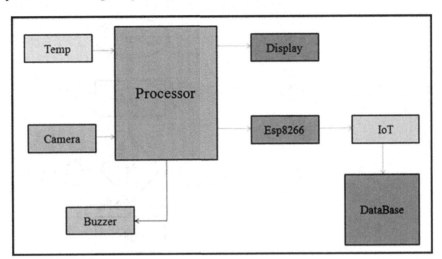

passes of an optical system, photoelectric detector, amplifier, signal processing and output module. The optical system ensures the collection of infrared radiation. The infrared radiation energy is converted into electrical signals when it converges on the photoelectric detector. After converging it will be processed by the amplifier and hence signal processing circuit converts the signal into the value of temperature.

A. Software Unit

In software section the front-end design is made with the Microsoft visual studio software form the tool of ASP.Net which is able to monitor through the webpage which utilizing the cloud technology and collecting the data's and stored in the database additionally vice versa update via through android app. Further the data obtained through the webpage.

Figure 2 is a flowchart showing the working of the IoT solution. When we use this system, it first senses the temperature using the sensor modules to get the accurate readings and then acts according to the degree of temperature sensed by it. If the temperature of the lab environment is maximum i.e., quite high, it rings a buzzer to indicate the high readings. And if the temperature sensed by the system is min i.e., low it displays the accurate temperature readings by using IoT and accesses its web applications. This is the working of the IoT solution, iRT.

V. RESULTS AND DISCUSSION

This prototype consists of an infrared thermometer sensor module and an MLX90614 sensor which is used to measure the surface temperature by detecting infrared radiation energy and wavelength distribution. The MLX90614 has a low noise amplifier integrated into the signal processing chip to support auto calibration.

The processor also consists of ESP8266 microcontroller which is a Wi-Fi chip with integrated antenna switches, RF balun, power amplifier, low noise amplifier, filters and power management modules. It sup-

Figure 2. Flow chart for IoT based Body Temperature monitor

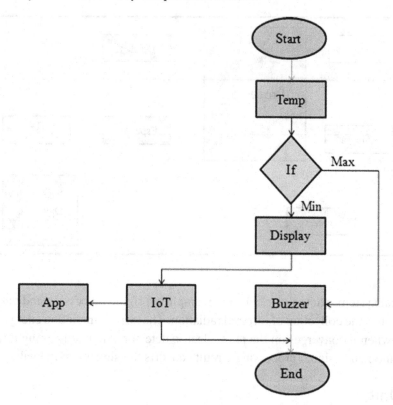

ports built-in Wi-Fi potentialities and is used for both processing and communication. It uses the .NET Web services and it stores the monitored data in a SQL Server Database. Data stored in the database are used to predict the health-related issues to reduce the risk factors in future. Internet of Things place an important to store the data with the help of cloud server. Infrared Thermometer sensor is used to sense the human body temperature and stores the data through processor which is interfaced with the sensors and internet of things device.

A buzzer is located inside a system based IoT to alert and notify the temperature readings. Body temperature of the body should be measured and monitored regularly with accuracy, compatibility and punctilious as shown in Figure 3. It is critical to measure temperature accurately as it has a major and huge impact on diagnosis and treatment. By using this prototype, measure the human body temperature and send this information to email, sms and stored in the cloud using Internet of Things. Temperature shown in the LCD screen immediately scan the human body which will useful to avoid the spreading of pandemic disease. When the temperature is higher than the normal, alarm sound comes in the prototype. Hence, we should take care of our health without fear. This prototype helps to monitor the human body temperature to prevent the pandemic disease. It helps to monitor the temperature in advance to avoid the spreading of pandemic infectious disease from one human to other human.

Figure 3. IoT Based Temperature monitoring prototype

VI. CONCLUSION

IRT device can be fixed in lobbies, hallways, and other access points to help businesses, institutions, hospitals, malls to create a safer environment for their workforce, customers, members, and patrons, patients etc. Healthcare institutions, institutions, malls, shops, and many businesses ventures find it challenging to perform temperature monitoring and screening at high speed. IoT can use the processors with thermal camera to seamlessly detect people who have fever, and showing abnormalities due to rise in body temperature. This designed app will generate an alert if a feverish human is ascertained. This reduces weariness due to scanning, to human work force and saves time. This product gives relief from pandemic covid-19. User friendly device will be very useful to the society to prevent the spreading of pandemic virus to human. Temperature monitoring device helps to reduce the severe problem created by infectious virus for human.

REFERENCES

Akkaya, K., Guvenc, I., Aygun, R., Pala, N., & Kadri, A. (2015). IoT-based occupancy monitoring techniques for energy-efficient smart buildings. *IEEE Wireless Communications and Networking Conference Workshops, 1*, 58–73.

Boonsong, W., Senajit, N., & Prasongchan, P. (2021). Contactless Body Temperature Monitoring of In-Patient Department (IPD) Using 2.4 GHz Microwave Frequency via the Internet of Things (IoT) Network. *Wireless Personal Communications, 1*, 1–7.

De Gennaro, G., Dambruoso, P. R., Loiotile, A. D., Di Gilio, A., Giungato, P., Tutino, M., Marzocca, A., Mazzone, A., Palmisani, J., & Porcelli, F. (2014). Indoor air quality in schools. *Environmental Chemistry Letters, 12*(4), 467–482. doi:10.100710311-014-0470-6

Giusto, D. (2010). The internet of things 20th Tyrrhenian workshop on digital communications. Springer.

Gubbi, J., Buyya, R., Marusic, S., & Palaniswami, M. (2013). Internet of Things (IoT): A vision, architectural elements, and future directions. *Future Generation Computer Systems, 29*, 1645–1660.

Havenith, G., Holmér, I., & Parsons, K. (2002). Personal factors in thermal comfort assessment: Clothing properties and metabolic heat production. *Energy and Building, 34*(6), 581–591. doi:10.1016/S0378-7788(02)00008-7

Hernández-Muñoz, J. M., Vercher, J. B., Muñoz, L., Galache, J. A., Presser, M., & Hernández Gómez, L. A. (2011). Smart Cities at the Forefront of the Future Internet. Lecture Notes in Computer Science, 6656.

Madureira, J., Paciência, I., Rufo, J., Ramos, E., Barros, H., Teixeira, J. P., & de Oliveira Fernandes, E. (2015). Indoor air quality in schools and its relationship with children's respiratory symptoms. *Atmospheric Environment, 118*, 145–156. doi:10.1016/j.atmosenv.2015.07.028

Marques, G., Ferreira, C. R., & Pitarma, R. (2019). Indoor Air Quality Assessment Using a CO2 Monitoring System Based on Internet of Things. *Journal of Medical Systems, 43*(3), 1–12. doi:10.100710916-019-1184-x PMID:30729368

Marques, G., & Pitarma, R. (2017). Monitoring Health Factors in Indoor Living Environments Using Internet of Things. *Recent Advances in Information Systems and Technologies, 570*, 785–794. doi:10.1007/978-3-319-56538-5_79

Marques, G., & Pitarma, R. (2018). IAQ Evaluation Using an IoT CO2 Monitoring System for Enhanced Living Environments. *Trends and Advances in Information Systems and Technologies, 746*, 1169–1177. doi:10.1007/978-3-319-77712-2_112

Marques, G., & Pitarma, R. (2019). Smartwatch-Based Application for Enhanced Healthy Lifestyle in Indoor Environments. *Computational Intelligence in Information Systems, 888*, 168–177. doi:10.1007/978-3-030-03302-6_15

Marques, G., & Pitarma, R. (2019). Using IoT and Social Networks for Enhanced Healthy Practices in Buildings. *Information Systems and Technologies to Support Learning, 111*, 424–432. doi:10.1007/978-3-030-03577-8_47

Marques, G., & Pitarma, R. (2019). A Cost-Effective Air Quality Supervision Solution for Enhanced Living Environments through the Internet of Things. *Electronics (Basel)*, *8*(2), 170–182. doi:10.3390/electronics8020170

Marques, G., Roque Ferreira, C., & Pitarma, R. (2018). A System Based on the Internet of Things for Real-Time Particle Monitoring in Buildings. *International Journal of Environmental Research and Public Health*, *15*(4), 821–834. doi:10.3390/ijerph15040821 PMID:29690534

Pereira, L. D., Cardoso, E., & da Silva, M. G. (2015). Indoor air quality audit and evaluation on thermal comfort in a school in Portugal. *Indoor and Built Environment*, *24*(2), 256–268. doi:10.1177/1420326X13508966

Pitarma, R., Marques, G., & Ferreira, B. R. (2017). Monitoring Indoor Air Quality for Enhanced Occupational Health. *Journal of Medical Systems*, *41*(2), 1–10. doi:10.100710916-016-0667-2 PMID:28000117

Salamone, F., Belussi, L., Danza, L., Galanos, T., Ghellere, M., & Meroni, I. (2017). Design and Development of a Nearable Wireless System to Control Indoor Air Quality and Indoor Lighting Quality. *Sensors (Basel)*, *17*, 1021–1033.

Shajkofci, A. (2021). Correction of human forehead temperature variations measured by non-contact infrared thermometer. *IEEE Sensors Journal*, *11*, 1–6. doi:10.1109/JSEN.2021.3058958

Tong, Zhao, Yang, Cao, Chen, & Chen. (2021). A hybrid method for overcoming thermal shock of non-contact infrared thermometers. *Physics. Med-Ph*, *1*, 1-14.

Tuladhar, L. R., Shrestha, S., Ghimire, N., Acharya, N., & Tamrakar, E. T. (2021). Clinical Efficiency of Non-Contact Infrared Thermometer over Axillary Digital Thermometer and Mercury in Glass Thermometer with Paracetamol. *Nepal Medical College Journal*, *23*(1), 31–36. doi:10.3126/nmcj.v23i1.36224

Yang, L., Yan, H., & Lam, J. C. (2014). Thermal comfort and building energy consumption implications – A review. *Applied Energy*, *115*, 164–173. doi:10.1016/j.apenergy.2013.10.062

Zhang, J. (2017). Development of a Non-Contact Infrared Thermometer. *Proceedings of the International Conference Advanced Engineering and Technology Research*, *1*, 10-22.

Chapter 15
Predictive Modeling for Classification of Breast Cancer Dataset Using Feature Selection Techniques

Leena Nesamani S.

Dr. M. G. R. Educational and Research Institute, India

S. Nirmala Sigirtha Rajini

Dr. M. G. R. Educational and Research Institute, India

ABSTRACT

Predictive modeling or predict analysis is the process of trying to predict the outcome from data using machine learning models. The quality of the output predominantly depends on the quality of the data that is provided to the model. The process of selecting the best choice of input to a machine learning model depends on a variety of criteria and is referred to as feature engineering. The work is conducted to classify the breast cancer patients into either the recurrence or non-recurrence category. A categorical breast cancer dataset is used in this work from which the best set of features is selected to make accurate predictions. Two feature selection techniques, namely the chi-squared technique and the mutual information technique, have been used. The selected features were then used by the logistic regression model to make the final prediction. It was identified that the mutual information technique proved to be more efficient and produced higher accuracy in the predictions.

INTRODUCTION

Machine learning is the task of solving a problem based on the available data patterns instead of being programmed directly. Machine learning models are deployed in various processes such as classification, regression, clustering, anomaly detection, ranking, and recommendations and forecasting.

DOI: 10.4018/978-1-7998-6870-5.ch015

Predicting the class to which an instance of data falls into is known as classification. Classification algorithms operate on labeled data where each label determines the class or the category to which the data belongs to. Classification may be either binary classification or multi class classification. The former predicts the unlabeled data into either one of the two available classes and the later makes predictions among N class or category of labels. Regression on the other hand is the process of trying to predict a label which is a continuous value from related set of features. The regression algorithm works on a set of labeled features and uses a function to predict the value of an unlabelled data.

Clustering is the task of grouping instances of data into different groups based on their similarity. The individual groups are called clusters and the members of the cluster share similar characteristics which are specific to a particular cluster. Anomalies are rare or infrequent events or observations that are misleading and dissimilar from the rest of the observations. Anomaly detections help in identifying fraudulent transactions, finding abnormal clusters, identify patterns that exhibit network intrusion, outlier identification, etc. In Ranking, the labeled data are grouped into instances and assigned scores which are used by the ranker to assign ranks for the unseen instances. Recommendation refers to the task of recommending products or services to the user based on their historical data. Making future predictions based on the past time-series data is known as Forecasting.

Machine learning model are mostly based on predictive modeling. In predictive modeling the model is trained on historical data in-order to make predictions on the new unseen data. The performance of the machine learning model depends on the efficiency of the algorithm that is chosen to handle the problem. Machine learning models perform well when they are provided with the right data. Feature Selection procedures help a lot in this aspect. They help not only to reduce the computational cost but also improve the performance of the model as well. Feature selection or variable selection is the process of selecting a subset of variables or features from the total dataset to build machine learning models. It is the key to construct faster, simpler and reliable Machine Learning models. Simpler models are easier to interpret and have shorter training time. It is easier to understand the model that uses ten variables rather than a model that uses hundred variables. Reducing the number of variables also reduces the computational cost and speeds up model building. Feature Selection also enhances generalization and hence improves model overfitting. Often many of the variables are the noise with no or very little predictive value. The Machine learning models learn from this noise reducing generalization and causing overfitting. By eliminating this noise we can substantially improve generalization and reduce overfitting. Reducing the number of variables also reduces the data errors that may be incurred during data collection. Variable redundancy could be removed by selecting only the necessary features and removing the highly correlated feature without losing important information.

FEATURE SELECTION METHODS

Feature selection procedure involves a combination of search procedures which selects different subset of features and an evaluation measure that scores each subset of features (Azhar, 2019). This is computationally expensive and different subset of features may produce optimal performance for different machine learning models. This means that there is no one set of optimal features but different sets of optimal features based on the machine learning algorithm that is intended to be used.

Feature selection methods may be classified into supervised or unsupervised. This classification is based on whether the target variable is considered or not in the process of feature selection. Unsupervised methods does not consider the target variable in removing redundant variable using the correlation method. Supervised methods are used to remove the variables that irrelevant to the target variable. Methods like filter methods and wrapper methods are known to be supervised in nature.

Figure 1. Feature Selection

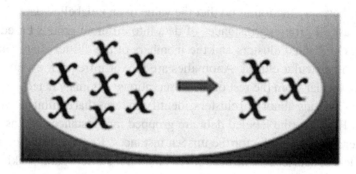

Filter methods relies on the characteristics of the features themselves in the selection of variables. They do not involve any machine learning algorithm but simply rely only on the feature characteristics. Filter methods are model agnostic and are computationally inexpensive. But they tend to produce lower performance when compared to the other feature selection methods. On the other hand they are very well suited for quick screening and fast removal of irrelevant features from a data set.

Wrapper methods use a predictive Machine Learning algorithm to select an optimal feature subset. Wrapper methods build a machine learning algorithm for each of the subset and select the subset of features that produce the highest performance. This makes them computationally very expensive but tends to produce the best performing subset of features for the given Machine Learning algorithm. It also implies that the subset of selected features may not produce the optimum result for a different machine learning algorithm.

Embedded method is an unsupervised method that performs feature selection as part of the model creation process. The models contain built-in feature selection procedures that select and include only those variables that produce maximum accuracy. The embedded method considers the interaction between the features and the model.

Dimensionality reduction differ from the feature selection method by creating a new projection of data with a completely new set of variables in contrast to the feature selection methods discussed above which remove the variables from the dataset.

Choosing the Best Method

Selecting the best method for the feature selection procedure is a million dollar question. There are no such rules as to select one. On the contrary we may move about in the selection process by looking at the variable type that we are intended to handle for the problem. The variables include both the input and output variables. This scenario is depicted in Fig 2, where the methods to be chosen for feature selection is based on the variable types involved.

When both the input and the output variables are numerical in nature the predictive modeling problem is a regression. The Pearson correlation coefficient technique is employed for a linear correlation. When the correlation appears to be nonlinear, the Spearman's rank coefficient technique is deployed. When the input variable is numerical and the output variable is categorical in nature the predictive modeling problem is a classification. ANOVA correlation coefficient technique is employed when the correlation is linear and the Kendall's rank coefficient is used in case of nonlinear correlation.

Figure 2. Choosing the correct method for feature selection

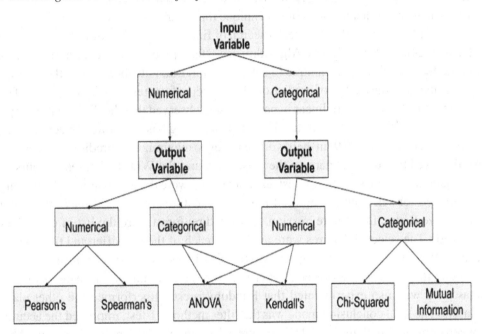

In some rare cases of regression where the input is a categorical variable and the output is numerical, the ANOVA method is used in the case of nonlinear coefficient and the Kendall's technique is for a linear correlation. When the predictive modeling problem employs categorical data for the input as well as the output the Chi-Square test and the Mutual Information techniques comes handy.

LITERATURE REVIEW

Asim et al.,(Khan & Arora, 2019) have extracted the texture features from the thermal images for the classification of the breast cancer. The texture features were extracted using Gabor filters at various orientation and scale levels. To extract the necessary features from the texture features the Gaussian modulated sinusoid that uniformly covered the spatial domain was used. The Gabor filters are a counterpart to performing wavelet transforms on an image at given spatial frequency domain. As a result twenty features were selected for the classification for breast cancer into cancerous or non cancerous.

In the work carried out by Gayathri et al.,(Gayathri Devi & Sabrigiriraj Feature Selection, 2018) the authors have thoroughly studied the feature selection techniques and have concluded that online feature selections techniques will be the best method suited for high dimensional data that follow a sequential training strategy. Online classification Applications involve high dimensional data where batch learning feature selections methods cannot be employed directly. Since the available online feature selection techniques suffer due to the scalability issues in terms of the high dimensional data, this has inspires the author to explore various feature selection methods like the Scalable and Accurate Online Approach (SAOLA), the Online Streaming Feature Selection (OSFS) technique and the Scalable and Accurate Online Approach (SAOLA) for feature selection in huge datasets and a few of Hadoop based classifiers. The results obtained are far more better than the existing online feature selection techniques but have not catered to the scalability measures of high dimensional data which throws forth new research ideas relating to identifying new online feature selection technique that addresses the scalability issue on big data.

Satyabrata et al.,(Aich et al., 2018) in their research of finding the quality of red and white wine, employed two techniques namely the Genetic Algorithm based feature selection technique and the Simulated Annealing based feature selection technique to select the important attributes from the original dataset that contribute to increase the quality of prediction. The selected features were input to a set of classifiers that were probabilistic, linear and nonlinear in nature. It was observed that the SVM classifier performed well with an accuracy of 95% to 98% on the simulated annealing based features selection technique.

Dhanya et al.,(Dhanya et al., 2019) in their work of trying to maximize the prediction accuracy of breast cancer, have deployed two datasets namely the Wisconsin and the WDBC datasets. Recursive feature elimination, sequential feature selection, f-test and correlation were the four feature selection algorithms have been employed for selecting the optimal number of features that would maximize the accuracy of prediction. Naïve Bayes, Logistic regression, and Random forest were the list of classifiers used in this research work. The selected features were applied to each of the classifier and their performances being measured. Both the datasets were applied to the classifiers with and without feature selection. It was observed that the classifiers performed better with the reduced number of features rather than the original datasets. It was also experimented that Random forest outperformed the other classifiers in both the datasets. The final conclusion made was the filter method - f-test, improved the accuracy of the classifiers when compared with the rest of the feature selection methods and was best suited for smaller datasets (Wisconsin). And the wrapper method – sequential forward selection, made the classifiers perform well than the other feature selection methods and was best suited for larger data sets (WDBC).

In the breast cancer prediction work carried out by Quang et al.,(Quang et al., 2019) as an initial step, data was prepared and made ready for classification through various preprocessing steps such as checking for missing values, class imbalance, normalization, correlation and the train/test split. In order to ensure better generalization of the solutions, feature selection techniques such as scaling and principal component analysis have been employed. This ensures that only the essential features are fed into the classifiers. Various classifiers were employed to evaluate the model and it was observed that four models namely, Logistics Regression, Ensemble Voting Classifier, SVM Tuning, Ada Boost gave accuracy over 98%. It was concluded that ensemble models gave the best performance in terms of recall, precision, F1 score, ROC-AUC and computational time.

In the Breast cancer prediction work carried out by Leena et al.,(Nesamani & Rajini, 2020; Nesamani et al., 2021; Umadevi et al., 2020) it was identified that a reduced feature set out performed the other existing works. It was also observed that ensemble models provided better results than the individual machine learning classifiers. Bhavani et al., (Bavani et al., 2019) suggests that the decision tree classifiers were better in the heart disease prediction system carried out by their team.

The modal proposed by Li et al.,(Li et al., 2020) employs a three layered feature selection technique for the accurate diagnosis of breast cancer from mammogram images. The first layer of feature selection tries to reduce the noise in the data. The second layer is aimed at identifying the pathological correlations in the different modalities which contains heterogeneous features. In the final layer an adaptive feature selection technique concatenated all the features to perform an effective breast cancer prediction system.

The extraction and selection of important features from medical images are very important as they are of higher dimensions. Ji et al.,(Ji, 2019) used a feature selection technique maximum relevance and minimum redundancy algorithm which used weighted value improvement in the classification of rheumatoid arthritis. Feature selection techniques recognize the most relevant features that will correctly classify the given dataset.

Sudha et al., (Sudha et al., 2019) employed a minimum distance method called the lion optimization algorithm to select the most relevant feature set that are mandatory for the accurate classification of breast cancer. This technique proved to be more accurate than the harmony search and the cuckoo search. Feature selection and dimensionality reduction are important for achieving good results in any machine learning problem. Kumar et al.,(Kumar & Inbarani, 2016) proposed a neighborhood rough set classifier (NRSC) method to classify multiclass motor images. They used a particle swarm optimization algorithm for selecting the most relevant features that are necessary for the classification process. Kumar et al.,(Kumar et al., 2019) proposed a model that used the fusion of features from the PET-CT images. A supervised CNN was used which leant the combined features from the multiple modalities and made a successful classification in the lung cancer dataset.

PROPOSED METHODS

Data Set: The dataset used for this research work is the Breast Cancer dataset which is a multivariate categorical dataset containing a total of 286 instances with 9 attributes, among which some of them are linear in nature and the rest are nominal in nature. This dataset is given by the Oncology Institute of Oncology and is available at the UCI Machine Learning Repository. Patients who have recovered from breast cancer can be classified into two categories namely the recurrent and non recurrent based on whether they will be affected again or not.

All the variables in the dataset are categorical in nature, and a few among them are ordinal and the rest are not.

METHODS

A. Pearson's Chi-Squared Method

In the breast cancer dataset both the input variables and the target variable are categorical data and the problem here is a classification predictive modeling where it tries to classify the data into either the recurrent or a non recurrent class. The test for independence between the input variable and the target variable is obtained using the Pearson's Chi squared statistical method. A test statistics χ^2 is calculated between the observed and the theoretical values. The chi-squared test is used to test whether the distribution of the categorical observed variables and the expected variables differ from each there. The value of the test statistics is obtained from the following formula:

$$x^2 = \sum_{i=1}^{n} \frac{\left(O_i - E_i\right)^2}{E_i} \tag{1}$$

Where,

χ^2 is the cumulative test statistics
O is the observed frequencies
E is the expected / theoretical frequencies

The Value of the statistics is interpreted as below: If the value χ^2 is greater than a critical value then the null hypothesis is rejected and the result is significant which also interprets the variable to be dependent. Otherwise the value is insignificant and do not reject the null hypothesis in which case the variable is independent. The value can also be interpreted in terms of the p-value and a significant value (alpha) where variable is said to be dependent if the p-value is less than or equal to alpha and independent otherwise.

B. Mutual Information Method

Mutual information is the name given to Information Gain when it is deployed in the procedure of variable selection. In probability theory it is calculated as the statistical dependence between any two random variables. If (X,Y) are the random variables then mutual information is given by

$$I\left(X;Y\right) = D_{KL}\left(P_{(X,Y)} \| P_X \otimes P_y\right). \tag{2}$$

Where,

P(X,Y) is their joint distribution
P_X and P_Y are the marginal distributions
D_{KL} is the reference probability distribution or divergence

X and Y are independent when the information gain is zero and a non-negative value indicates they are dependent.

C. Logistic Regression

Logistic regression is a machine learning technique that claims it origin from the field of statistics. It is a statistical model which uses a logistic function to estimate the values of a logistic model. The logistic function here is a sigmoid function which takes a real value (t) as input and outputs the value between 0 and 1.

Figure 3. A logistic function σ(t)

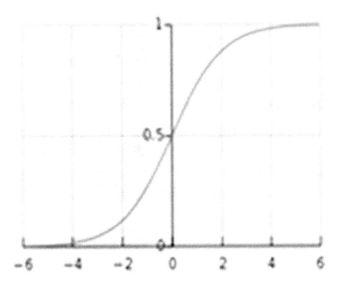

$$\sigma\left(t\right) = \frac{e^t}{e^t + 1} = \frac{1}{1 + e^{-t}} \tag{3}$$

Regression analysis refers to the process of calculating the relation between a dependent variable and one or more independent variables by calculating the probabilities with a logistic function (Lei, 2018). For a model having two predictor variables (x_1 and x_2) and one response variable Y, the relationship between them is given by the formula:

$$\ell = \log_b \frac{p}{1 - p} = \beta_0 + \beta_1 x_1 + \beta_2 x_2 \tag{4}$$

Where,

p, is the probability of the event
β_i is the model parameters

Once the values of the β_i are fixed then probability, Y=1 or Y=0 can be calculated.

RESULT AND DISCUSSION

The breast cancer data set that is considered for this work consists of categorical data. The dataset is split into train and test sets to fit and evaluate the model. 67% of the data is used for training and 33% is used for testing. The categorical variables are encoded to integer using ordinal encoding. The target

Figure 4. Bar chart of features Vs the Chi squared value.

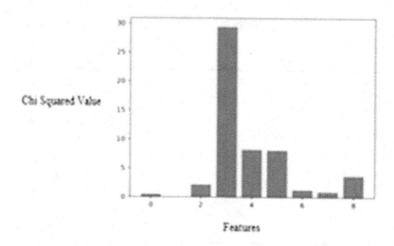

variable is label encoded to enable the binary classification. As a part of feature engineering the most relevant features are selected first using the chi-square statistical method. It was identified that the third feature was the most relevant one. The top four features were selected for this work.

The mutual information method was applied next to select the most relevant features. It was observed that features 2,3,5 and 6 were the most relevant features.

Figure 5. Bar chart of features Vs the Mutual Information value.

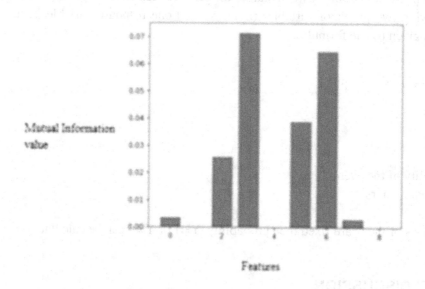

A Logistic regression models was created to classify the patients into the recurrence and non recurrence class by including all the features in the dataset. Later two other logistic regression models were built using the features selected from the chi squared method and the mutual information method.

The results of the three models were compared. It was observed that the model which used all the features gave an accuracy of 75%, the model created with the features selected from the chi squared method yielded an accuracy of 74% and final model built the features extracted using the mutual information method produced an accuracy of 76%.

A Logistic regression models was created to classify the patients into the recurrence and non recurrence class by including all the features in the dataset. Later two other logistic regression models were built using the features selected from the chi squared method and the mutual information method. The results of the three models were compared. It was observed that the model which used all the features gave an accuracy of 75%, the model created with the features selected from the chi squared method yielded an accuracy of 74% and final model built the features extracted using the mutual information method produced an accuracy of 76%.

Table I. Comparison of feature selection methods

Classifier	Feature selection method	No. of features used	Accuracy %
Logistic Regression	No method	All	75
	Chi squared	2,4,5,8	74
	Mutual Information	2,3,5,6	76

CONCLUSION

Feature engineering play a major role in predictive modeling. It is one of the major aspects that affect the accuracy of model. Best feature selection methods tend to produce good models. The choice of selecting the best method depends on multiple factors which include the characteristics of the features and the model being used. In this work, the focus was on using a categorical dataset and trying to employ the best feature selection method to obtain higher accuracy of prediction. The categorical variables were encoded into integer values using ordinal encoding. Chi-square method and mutual information method were applied individually on the categorical dataset and the results were evaluated by employing a logistic regression classifier on the subset of features. The dataset was also examined on the logistic regression classifier with all the features. It was observed that the best feature selection method for a model can only be identified by examining different subset of features from the dataset for the machine learning algorithm. The mutual information technique proved to give a better result for a categorical dataset on a regression problem. The work could be further extended to study the behavior of different feature selection techniques on numerical datasets for classification problems by examining different subsets of features to fit different models. The scope of this research could be extended to different types of datasets to identify and filter out the necessary features that would improve the performance and accuracy of any machine learning problem.

REFERENCES

Aich, Al-Absi, Hui, & Sain. (2018). Prediction of Quality for Different Type of Wine based on Different Feature Sets Using Supervised Machine Learning Techniques. *ICACT Transactions on Advanced Communications Technology, 7*(3).

Azhar, M. A. (2019). *Comparative Review of Feature Selection and Classification modeling, COMP-118-241-Ver-2*. IEEE.

Bavani, Rajini, Josephine, & Prasannakumari. (2019). Heart Disease Prediction System based on Decision Tree Classifier. Jour of Adv Research in Dynamical & Control Systems, 11(10).

Dhanya, Paul, Akula, Sivakumar, Jyothisha, & Nair. (2019). Comparative Study for Breast Cancer Prediction using Machine Learning and Feature Selection. *Proceedings of the International Conference on Intelligent Computing and Control Systems*.

Gayathri Devi, S., & Sabrigiriraj Feature Selection, M. (2018). Online Feature Selection Techniques for Big Data Classification: - A Review. *Proceeding of 2018 IEEE International Conference on Current Trends toward Converging Technologies*. 10.1109/ICCTCT.2018.8550928

Ji, H. (2019). Research on Feature Selection and Classification Algorithm of Medical Optical Tomography Images. *2019 IEEE/CIC International Conference on Communications in China (ICCC)*, 561-566. 10.1109/ICCChina.2019.8855804

Khan & Arora. (2019). Classification in Thermograms for Breast Cancer Detection using Texture Features with Feature Selection Method and Ensemble Classifier. *2nd International Conference on Issues and Challenges in Intelligent Computing Techniques (ICICT)*.

Kumar, A., Fulham, M., Feng, D., & Kim, J. (2019). Co-learning feature fusion maps from PET-CT images of lung cancer. *IEEE Transactions on Medical Imaging, 39*(1), 204–217. doi:10.1109/TMI.2019.2923601 PMID:31217099

Kumar, S. U., & Inbarani, H. H. (2016). PSO-based feature selection and neighborhood rough set-based classification for BCI multiclass motor imagery task. *Neural Computing & Applications, 28*(11), 3239–3258. doi:10.100700521-016-2236-5

Lei, L. (2018). Research on Logistic Regression Algorithm of Breast Diagnose Data by Machine Learning. *International Conference on Robots & Intelligent System*. 10.1109/ICRIS.2018.00049

Li, G., Yuan, T., Li, C., Zhuo, J., Jiang, Z., Wu, J., Ji, D., & Zhang, H. (2020). Effective Breast Cancer Recognition Based on Fine-Grained Feature Selection. *IEEE Access: Practical Innovations, Open Solutions, 8*, 227538–227555. doi:10.1109/ACCESS.2020.3046309

Nesamani & Rajini. (2020). Evaluation of Ensemble Machines in Breast Cancer Prediction. Advances in Parallel Computing, 37, 391-395.

Nesamani, Rajini, Josphine, & Salome. (2021). Deep Learning-Based Mammogram Classification for Breast Cancer Diagnosis Using Multi-level Support Vector Machine. In Advances in Automation, Signal Processing, Instrumentation, and Control. Springer.

Quang, Nguyen, Do, Wang, Heng, Chen, Ang, Philip, Singh, Pham, Nguyen, & Chua. (2019). Breast Cancer Prediction using Feature Selection and Ensemble Voting. *International Conference on System Science and Engineering (ICSSE)*.

Sudha, M. N., Selvarajan, S., & Suganthi, M. (2019). Feature selection using improved lion optimisation algorithm for breast cancer classification. *IJBIC*, *14*(4), 237–246. doi:10.1504/IJBIC.2019.103963

Umadevi, S., Chandrakala, T., Bhavani, P., Leena Nesamani, S., & Ezhilmathi, T.M. (2020). Solutions to Critical Issues on Big Data using Machine Learning Techniques. *International Journal of Advanced Science and Technology*, *29*(7S), 3054-3059.

Chapter 16
Credit Card Fraud Detection Using K-Means and Fuzzy C-Means

Arti Jain

 https://orcid.org/0000-0002-3764-8834

Jaypee Institute of Information Technology, Noida, India

Archana Purwar

Jaypee Institute of Information Technology, Noida, India

Divakar Yadav

 https://orcid.org/0000-0001-6051-479X

National Institute of Technology, Hamirpur, India

ABSTRACT

Machine learning (ML) proven to be an emerging technology from small-scale to large-scale industries. One of the important industries is banking, where ML is being adapted all over the world by employing online banking. The online banking is using ML techniques in detecting fraudulent transactions like credit card fraud detection, etc. Hence, in this chapter, a Credit card Fraud Detection (CFD) system is devised using Luhn's algorithm and k-means clustering. Moreover, CFD system is also developed using Fuzzy C-Means (FCM) clustering instead of k-means clustering. Performance of CFD using both clustering techniques is compared using precision, recall and f-measure. The FCM gives better results in comparison to k-means clustering. Further, other evaluation metrics such as fraud catching rate, false alarm rate, balanced classification rate, and Mathews correlation coefficient are also calculated to show how well the CFD system works in the presence of skewed data.

DOI: 10.4018/978-1-7998-6870-5.ch016

1. INTRODUCTION

Credit card is a plastic card, which is handy, and is provided to customer through which he is offered credit to make payment for products and services, usually when there is no cash in hand. The customer is facilitated to pay back his credited money to the card-issuing bank afterwards. The usage of credit card was originated in USA during 1920s'. At present, the credit card holders are spread across the globe, especially in countries like Unite Kingdom, Spain, Greece, Belgium, Italy, Ireland, and Germany etc. The cards are associated with well-known credit card merchants such as American Express, Barclaycard, and Citibank who offer varied services to customers, depending upon portfolios- card fee, credit limit, late payment fee, exchange rate fee, rate of interest and so on. The U.S. Census Bureau and Nilson Report[1] show that there are a total of 1.06 billion credit cards in 2017, and projection of 1.2 billion by the 2022. The economical internet and smart devices execute the credit card transaction easy, handy and contented. Because of easiness with the payment mode, the fast usage of credit card is more or less unavoidable especially in the e-commerce transactions. The wide spectrum of credit card as a transaction medium for the purchase or shopping purposes is exponentially been increased. However, it is quite unfortunate that the fraudulent usage of this digital card is becoming a vulnerable source of crime.

1.1. Credit Card Fraud, Fraudster and CFD System

Credit card fraud (Delamaire et al., 2009) refers to a theft in which fraudulent- wrong, suspect or illegitimate source of funds are involved in a transaction. The fraud occurs through any of the credit productions such as home loan, personal loan, retail, shopping etc. Such crime is committed by the fraudster within a short span of time while gaining immense money that too without many hazards.

Card Fraudsters can be classified as credit card information buyer: fraudster with little or no professional programming skills, black hat hacker: fraudster with professional programming skills, and physical credit card stealer: fraudster who steals credit card and write out information on it. The fraudster misuses not only the consumer's personal information but also his financial information. According to the global market survey[2], the fraud losses that are incurred by the banks and the merchants hit $21.84 billion during 2015 and, by the 2022 it can surpass with the global growth rate of 45%. Identification of fraudulent credit card transaction is quite challenging, since the crime is detected too late after the incidence.

The fraud can be perpetrated in several ways, for example (Akhilomen, 2013; Sakharova, 2012) - card theft, counterfeit card, spyware, key-logger, site cloning, imposter, skimming, sniffing, eavesdropping, phishing and dumpster diving etc. Overall, the credit card fraud has two classifications- (i) online credit card fraud, and (ii) offline credit card fraud. In the online fraud, or cyber credit card fraud or no card present fraud), crime is committed without the presence of the card. The card information is operational to execute a fraudulent transaction. While in the offline fraud, crime is committed in the presence of card. The credit card is either stolen or counterfeit in most of the cases to perform a fraudulent transaction. A survey of more than 160 companies has highlighted that online credit card frauds are 12 times higher than their offline counterparts (Behera and Panigrahi, 2015). Thus, online credit card frauds are currently rampant and results into monetary loss of card holders as well as reputational risks of the related banks. If the fraudulent instance is considered as legitimate then it causes a big loss to both the consumers and merchant banks. The potential credit card frauds (Correia et al., 2015) are identified as follows:

- Consecutive Withdrawals: Consecutive withdrawals of varying amounts for a credit card are made.

- CVV Attack: Several attempts are made towards incorrect credit card- Card Verification Value (CVV). For example, at least three attempts with incorrect CVV are made within few minutes.
- Flash Attack: Wide numbers of transactions for credit card are made within a short span of time period.
- Small Followed by Big Amounts: Small transaction is succeeded by a bigger credit card transaction within a short temporal window of say 5-10 minutes.
- Multiple ATM Withdrawals: Higher numbers of withdrawals from the ATM are made.
- Card Usage near Expiration Date: There is a sudden credit card usage which is near its expiration date.
- Transactions in Distant Places: Consecutive attempts to use a credit card at different distant locations are made.

It is crucial to prevent and detect the credit card frauds. The card issuers seek innovative methods, and keep watchful eyes to monitor their credit reports. The credit card fraud can be prevented via anti-fraud techniques, for example- card activation, card and address verification, customer education, real-time point-of-sale authorization etc. Apart from this, financial institutions[3] employ diverse fraud prevention models and fraud management techniques. However, in actuality, fraudsters are highly adaptive, very inventive, and extremely intellectual. They devise newer, sophisticated strategies to intrude the preventive models and commit crimes. The card holders are aware about the fraud only when fraudulent transaction is executed. Then there is no means for tracking the fraudulent transaction which is disguised as genuine transaction. Once the fraud prevention mechanisms are failed, detection of the credit card fraud becomes essential. It becomes important to develop a computerized detection model which can help to detect whether the transaction is fraud or not. Therefore, it necessitates the proposed Credit card Fraud Detection (CFD) system which maintains reliability of the card payment system through customer behavioral analysis. The Research objectives (RO1 and RO2) of the Credit card Fraud Detection system are mentioned below: -

RO1: To reduce losses that are incurred to card merchants, issuing banks and card holders due to credit card payment frauds.

RO2: To investigate the CFD system while using machine learning models e.g., K-Means and Fuzzy C-Means clustering.

Card fraud detection via manual mode is literally impossible; however, an automation of the credit card fraud detection via contemporary technology tackles the problem and fulfills the objectives. In this research, the CFD system is implemented which deduces real time frauds, minimizes false alarms and is more secure to distinguish between transactions types- fraudulent or legitimate. The CFD system delivers stable and positive experiences to valuable credit card customers.

2. RELATED WORK

Credit card fraud detection (Salazar et al., 2012) is a classification problem which is traditionally being taken care using data mining techniques. Data mining (Sharma and Panigrahi, 2012) deploy mathematical, Machine Learning (ML) techniques for identification and extraction of interesting patterns;

subsequently discovers information from big data since the patterns are hidden, reliable and actionable. The ML based supervised model classifies transactions as fraudulent or legitimate e.g., decision tree (Save et al., 2017), Hidden Markov Model (HMM) (Bhusari and Patil, 2011), Support Vector Machine (SVM) (Dheepa and Dhanapal, 2012), K-Nearest Neighbor (Malini and Pushpa, 2017) and Random Forest (Van Vlasselaer et al, 2015) etc. The ML based unsupervised model considers outliers as potential fraudulent transactions e.g., Self-Organizing Map (Zhang et al., 2009). There are other optimization algorithms e.g., Genetic Algorithm (GA) (Patel and Singh, 2013), Neural Networks (Akhilomen, 2013) and Multi-Layer Perceptron (MLP) (Mishra and Dash, 2014) etc. This section discusses some of these credit fraud detection algorithms.

Maes et al., 2002 have outlined Artificial Neural Networks (ANN) and Bayesian Belief Networks (BBN). The BBN has faster training period with better results, whereas ANN has faster detection process. Kim and Kim, 2002 have observed a clearly skewed distribution of the legitimate and fraudulent data. They have generated the weighted fraud score from fraud density of the transactional data which in turn reduces the number of false detections.

Chiu and Tsai, 2004 have elaborated the fraud pattern mining algorithm to mine the fraud association rules using features information in fraud transactions which is relatively accurate but inherently slow. Chen et al., 2005 have employed SVM and ANN via personalized approach, without any transactional data. But their system is neither fully automated nor independent of the users' expertise level.

Meador and Moore, 2008 have patented a transaction system which includes a central station to receive the credit card transactional data. Whenever the station receives a transactional request, it checks for the location consistency w.r.to the normal usage of the Account Identification Document (AID). If the location is consistent with the AID, transaction is approved. Otherwise, central station contacts the card holder via mobile communication device and transaction is approved once the authentication of the person is successfully completed.

Yu and Wang, 2009 have discussed distance-based outlier method for the credit card fraud which has predicted fraudulent transactions while computing distance and setting the outlier threshold. Zhang et al., 2009 have suggested historical behavior-based credit card fraud system where past patterns of customers are used for the fraud detection using unsupervised SOM.

Bhusari and Patil, 2011 have built HMM model to maintain transactions' log to check whether every transaction is fraudulent or legitimate but has generated high false alarms and false positives. Duman and Ozcelik, 2011 have stated the card problem using high performance, and powerful GA which reduces the number of wrongly classified transactions but is quite expensive. Patidar and Sharma, 2011 have tried to handle the credit card fraud task with the neural networks technique where network is trained from the already occurred events and results are enhanced as the time progresses while constituting relationship among I/O, or pattern searching within dataset.

Dheepa and Dhanapal, 2012 have applied SVM to find the credit card frauds while looking into low false alarms and high fraud catching rate to determine discrepancies within the behavior transactional patterns, and to predict suspicion. They have used 576 legal samples and 15 fraud samples for the SVM with kernel functions. Ramaki et al., 2012 have proposed an ontology graph approach where distance between an input transaction and the previously recorded patterns are computed. If the transactional distance is higher than threshold then input transaction behaves as fraudulent. The ontology-based approach has lower computational overload, lesser storage for transactional data, quicker real-time fraud detection, and has achieved 89.4% precision of detecting credit card fraud. Salazar et al., 2012 have employed a robust non-linear signal processing method via discriminant classifiers and non-Gaussian mixture models which comprise of vivid stages such as feature extraction, train and classify, decision fusion, and result presentation respectively.

Akhilomen, 2013 has used anomaly detection strategy or deviation detection to detect patterns that do not validate to the well-known normal patterns' behavior. Whenever suspicious fraud classification arises, further investigation is done while contacting the card holder related to the transaction. Duman and Elikucuk, 2013 have implemented the fraud detection for Turkish bank while applying Migrating Birds Optimization (MBO) algorithm which has performed better than other classical data mining algorithms. Hormozi et al., 2013 have discussed Artificial Immune System (AIS) by parallelization of the Negative Selection Algorithm (NSA) over cloud computing through the MapReduce framework. The NSA with mapper reduces false negative rate to 15% while increases true positive or detection rate to 84% respectively.

Pawar et al., 2014 have worked with Principal Component Analysis (PCA) as outlier detection where training set comprises of standard German credit card fraud dataset from the UCI repository with 20 attributes- numerical or categorical types, and 1,000 instances. The test set comprises of 2-dimensional synthetic data which is generated with the 100 data instances, where normal instances are 80 and outlier instances are 20 respectively. The method is workable for applications where memory and computation limitations exist.

De Montjoye et al., 2015 have studied credit card metadata of more than one million people for three months, and have observed that 4 spatial-temporal points uniquely re-identify 90% of them. Also, women folk are better re-identifiable than their men counterparts for the credit card records. Behera and Panigrahi, 2015 have illustrated a three-phase fraud detection system which combines fuzzy clustering along with neural learning. The first phase performs user authentication and card verification. The second phase executes Fuzzy C-Means (FCM) clustering to find customer's card usage patterns based on his past transactions' history. The third phase then applies neural networks to determine whether the suspicious transaction is a fraud or an occasional deviation by the customer himself. The combined usage of clustering along with learning detects frauds more effectively while minimizing the misclassification of transactions which in turn minimizes the generation of false alarms. Correia et al., 2015 have presented an extension to the open-source PROTON tool which accompanies development of latest features, methods, operands, and event processing patterns etc. for handling uncertainty aspect of the fraud detection.

Zareapoor and Shamsolmoali, 2015 have introduced bagging ensemble classifier with stable performance and lesser execution time over other classifiers. The credit card dataset is chosen from the University of California, San Diego (UCSD) as a part of competition organized by the FICO. The dataset comprises of hundred thousand credit card instances where every instance has twenty fields. The ratio of legitimate vs. fraudulent transactions is about 100:3, indicates that the dataset is highly imbalanced. The whole dataset is distributed among 3 categories- training (70%), validation (15%) and test (15%) phases respectively. Two balanced metrics- BCR and MCC are class imbalance handling criteria between fraudulent and legitimate instances. Ensemble classifier then classifies suspicious transactions and achieves- correctly classified (94.91%) and incorrectly classified (5.09%) transactions respectively.

Hegazy et al., 2016 have performed LINGO cluster data mining algorithm instead of Apriori algorithm. LINGO generates more meaningful frequent patterns to facilitate customers' behavior and separates legal or fraud transactions while eliminating the problem of imbalanced dataset. Kamaruddin and Ravi, 2016 have employed a One-Class Classification (OCC) strategy through hybrid Particle Swarm Optimization (PSO) and Auto-Associative Neural Network (AANN). The OCC strategy is quite workable as the legitimate transactions are in abundance while fraudulent (class of interest) are in scarcity.

Mishra and Ghorpade, 2018 have used logistic regression, SVM, decision trees, random forest and stacked classifier over the European dataset to achieve high recall of about 91%. Carcillo et al., 2018 have discussed several challenges, namely- concept drift, class imbalance and verification latency. Concept drift refers to the fact that customers' behaviors evolve over time and so are the changes in the fraudsters' strategies. Class imbalance refers to the fact that legitimate transactions far outnumber the fraudulent transactions. Verification latency refers to the fact that only few transactions are possibly been checked by the professional investigators.

Varmedja et al., 2019 have analyzed MLP, Random Forest, Logistic Regression and Naïve Bayes for the credit card fraud detection. Xie et al., 2019 have portrayed that frequency-based features are not enough while rule-based is more effective to distinguish between the legitimate and fraudulent transactions.

Arya and Sastry, 2020 have proposed the DEAL- a predictive framework using Deep Ensemble learning for real-time streaming. The DEAL is highly adaptive and robust towards data imbalance and latent transaction patterns. Gianini et al., 2020 have managed a set of rules using game theory and Zhu et al., 2020 have proposed WELM algorithm to achieve high fraud detection performance.

3. CHALLENGES IN CFD SYSTEM

The CFD system detects fraudulent behavior while relying on automated analysis of the customers' credit card transactions. There are some challenges (Zareapoor and Shamsolmoali, 2015) with the CFD system that are discussed herewith.

- *Real-data not shared:* The real-data for the credit card fraud detection is not shared to maintain privacy of users.
- *Size of dataset:* Credit card transactions are performed in millions per day which is quite tedious to process efficiently. Analyzing such pile of massive transactional data requires very high computing power for the competent techniques.
- *Skewed dataset:* Credit card dataset is highly skewed and is extremely imbalanced, where maximum instances are legitimate, only few transactions are fraudulent. In real scenarios, around 98% of the credit card transactions are legitimate while rest 2% transactions are fraudulent. The system thus undergoes very low false positive rate.
- *Classifier training:* Whenever all the legal transactions participate during training phase, the card fraud detection system requires excessive training time. Alternatively, within the training phase choose only few transactional data but certain behaviors are not characterized there, since related data is not selected in the classifier.
- *Missing or unreliable factors:* Several factors such as income and age of the card holders are either missing or unreliable in the credit card fraud detection system.
- *Fraudster behavior:* Fraudsters are highly adaptive, very inventive, and extremely intellectual as they devise newer strategies to intrude the fraud detection system. So, frauds are not predictable even by the human experts because of their increased complexity and sophistication.
- *Evaluation metrics:* Evaluation metrics such as accuracy and error rates are not workable for the credit card fraud detection because of the skewed dataset. However, other metrics such as Fraud Catching Rate (FCR), False Alarm Rate (FAR), Balanced Classification Rate (BCR) and Mathews Correlation Coefficient (MCC) are taken care.

4. PROPOSED CFD SYSTEM

This section details the proposed CFD system, which comprises of different phases such as CFD Luhn's Algorithm, CFD Feature Extraction, CFD Data Normalization, and CFD ML Algorithms.

4.1. CFD Luhn's Algorithm

In general, the Luhn's (Luhn, 1960) algorithm is also known as modulus 10 algorithm, which is named after its creator Hans Peter Luhn, an IBM scientist. The algorithm is simply a checksum formula, which is in the wide public domains and distinguishes valid numbers from mistyped or incorrect numbers. The algorithm validates a variety of identification numbers, for example- National Provider Identifier (USA), Social Insurance Number (Canada), ID Number (Israel), Social Security Number (Greece), Credit Card Number (Across Globe), and Survey Codes- McDonald's, Taco Bell etc. The algorithm is not designed against malicious attacks, rather to protect against accidental errors.

In this research, the CFD Luhn's algorithm validates the credit card, so that the valid cards are distinguished from the mistyped or incorrect card numbers, helpful during test dataset. The credit card is valid if it passes the CFD Luhn's test.

4.2. CFD Feature Extraction

The CFD system endeavors to defend transactions from illegal usage, and maximizes correct predictions by analyzing spending behavior of the credit card holders. For this, appropriate features are selected to represent customers' behavior such as amount of transaction, date of transaction, time of transaction and frequency of transaction. These features are individual elements (e_k) with a transaction (a_i) in a behavior set(A), as is represented in equation (1).

$$a_i = \{e_k\},\ k = 1\ to\ n,\ a_i \in A\ (1)$$

Some of these features are described as- e_1: amount of transaction, e_2: date of transaction, e_3: time period or duration of transaction, and e_4: frequency of transaction etc.

The stated features describe behavior patterns of the credit card holder, for instance, as in equation (2), $a_t \in A$ such that:

$$a_t = \{10000,\ 10/10/2019,\ 20{:}00,\ 1\} \tag{2}$$

The transaction (a_t) represents that the amount of transaction (money withdrawn) is *$10,000*, date of transaction is *October 10, 2019*, time of transaction is *8 PM*, and *1* represents first transaction on the given date respectively.

During pre-processing, wherever feature values are missing in the dataset, they are filled with the default system generated values to avoid missing gaps. Since, feature values vary in magnitudes as well as types, so these values are normalized to reduce the influence of diversified features. To do so, all the feature values are mapped into the interval 0 and 1.

4.3. CFD Data Normalization

For the CFD data normalization, Fuzzy Logic is used to measure the customer transaction behavior patterns through the membership functions. This section discusses membership functions for features- amount of transaction, date of transaction, time of transaction and frequency of transaction.

4.3.1. Membership Function- Amount of Transaction

Membership function of legal transaction amount for customer behavioral pattern is given in equation (3). Where, $f_1(e_1)$ is a membership function of e_1, and *max(amount)* is the maximum value among the transaction amount.

$$f_1(e_1) = 1 - \frac{e_1}{\max(amount)} \tag{3}$$

4.3.2. Membership Function- Date of Transaction

Membership function of transaction date for customer behavioral pattern is given in equation (4). Where, $f_2(e_2)$ is a membership function of e_2, and *max(date)* is the maximum number of days from the last trans- action date to the current transaction date.

$$f_2(e_2) = 1 - \frac{1}{\max(date) + 1} \tag{4}$$

4.3.3. Membership Function- Time of Transaction

Membership function of transaction time for customer behavioral pattern is given in equation (5). Where, $f_3(e_3)$ is a membership function of e_3, and *max(time)* is the maximum duration from the last transaction time.

$$f_3(e_3) = 1 - \frac{1}{\max(time) + 1} \tag{5}$$

4.3.4. Membership Function- Frequency of Transaction

Membership function of legal transaction frequency for customer behavioral pattern is provided in equa- tion (6). Where, $f_4(e_4)$ is a membership function of e_4, and *max(frequency)* is the maximum value among the transaction frequency within a given time period.

$$f_4(e_4) = 1 - \frac{e_4}{\max(frequency)} \tag{6}$$

4.4. CFD ML Algorithms

The CFD system detects fraud cases using machine-learning models: K-Means and Fuzzy C-Means clustering. This section details these ML algorithms.

4.4.1. K-Means Clustering

Figure 1 shows the steps involved in k-means clustering which is a centroid-based or distance-based algorithm. It is based on calculating the distances to assign data instances to individual clusters. Every data instance is assigned only to one cluster, and each cluster is associated with a centroid that is useful to label new instances. The algorithm minimizes the sum of distances between the data instance and its respective centroid.

Figure 1. K-Means Clustering

STEP 1: Initialize centroids of the clusters.

$\mu_i = some\ value,\ i = 1,...,k$

STEP 2: Choose the closest cluster to each data point.

$c_i = \{j : d(x_j, \mu_i) \leq d(x_j, \mu_l), l \neq i, j = 1,...,n\}$

STEP 3: Set centroid of each cluster to the mean of data points belonging to cluster.

$\mu_i = \frac{1}{|c_i|} \sum_{j \in c_i} x_j, \forall i$

STEP 4: Repeat STEPS 2 to STEP 3 until convergence.

During initial experiments, k-means is executed over entire dataset. The confusion matrix is generated which provides precision, recall and F-measure values. It is evident that fraudulent is a positive class while legitimate is a negative class. However, due to imbalanced dataset, handling of sampling issues and overfitting of dataset are prime challenges, so a subset of the dataset is considered. In other words, further experiments are carried over a balanced data having equal number of fraudulent and legitimate instances (Section 6 experimentation and results).

4.4.2. Fuzzy C-Means Clustering

Fuzzy C-Means (FCM) (Ross, 2004) or soft clustering represents variation of simple k-means or hard clustering. Figure 2 shows the steps in FCM in which, rather than choosing just one cluster, the fuzziness or overlapping among 2 or more clusters occur. Since, data values belong to more than one cluster, affinity is in proportion with the distance of that data point from the cluster centroid.

To apply FCM, every instance of the credit card transactional data belongs to every cluster with certain membership. For the training dataset, membership is calculated via FCM, while test dataset uses the same centroid as in the training. In this work, total number of centroids for the FCM is chosen as two or more clusters.

Figure 2. Fuzzy C-Means Clustering

STEP 1: Iterations

$$J = \sum_{j=1}^{k} \sum_{x_i \in C_j} \delta_{ij}^m \| x_i - \mu_j \|^2$$

δ_{ij} : Degree to which observation x_i belongs to cluster C_j

μ_j : Center of cluster j

m: Level of cluster fuzziness ($1.0 < m < \infty$; closer to 1 means similar to K-Means; closer to infinity leads to complete fuzziness)

STEP 2: Degree of Membership $\qquad \delta_{ij}^m = \dfrac{1}{\sum_{l=1}^{k} \left(\dfrac{\| x_i - c_j \|}{\| x_i - c_k \|} \right)^{\frac{2}{m-1}}}$

STEP 3: Fuzziness Coefficient $\qquad C_j = \dfrac{\sum_{x_i \in C_j} \delta_{ij}^m \cdot x_i}{\sum_{x_i \in C_j} \delta_{ij}^m}$

STEP 4: Termination Condition $\qquad \varepsilon = \Delta_i \Delta_j \, | \delta_{ij}^{k+1} - \delta_{ij}^{k} |$

5. CFD DATASET AND VISUALIZATION

5.1. CFD Dataset

The CFD dataset is extracted from the Kaggle[4], which is highly unbalanced and skewed in nature. The dataset holds credit card transactions during September 2013 by the European cardholders. It comprises of 492 fraudulent instances out of 284,807 instances, resulting in 0.172% fraud cases. It contains only numerical input variables, having total of 31 features, some of them are PCA transformed and are labeled: V1 to V28 features. Rest, other features include time as well as amount of transaction with class attribute. Time feature indicates the duration elapsed between every transaction and first transaction within dataset. Amount feature indicates an amount of transaction, which is beneficial for cost-sensitive aspect of learning. Class feature is nothing but response variable, which represents class 1 as fraudulent while class 0 as legitimate respectively.

During initial experiments, metrics: precision, recall and F-measure are considered. However, due to the imbalanced dataset, other metrics such as Fraud Catching Rate, Fraud Alarm rate, Balanced Classification Rate and Mathews Correlation Coefficient are chosen. To avoid sampling issues and overfitting of the data, 984 balanced instances are chosen: 492 fraudulent and rest other 492 legitimate transactions. It is evident that fraudulent is a positive class while legitimate is a negative class.

5.2. CFD Data Visualization

In order to visualize the CFD dataset, histogram plots are constructed for each of the V1 to V28 features w.r.to count of feature instances (Figure 3). For example, histogram for feature V1 indicates that there are total of 284807 instances of features V1. Among them, 284315 instances are of legitimate class and rest 492 instances are of fraudulent class. For every feature, the figure depicts histograms between specific feature and number of instances corresponding to the feature.

Figure 3. Histogram of CFD Dataset (Feature Attribute vs. Feature Instances)

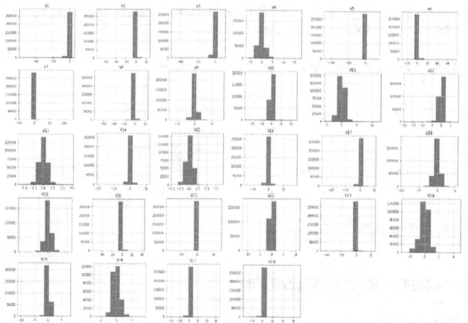

Figure 4. Histogram of Time vs. Count

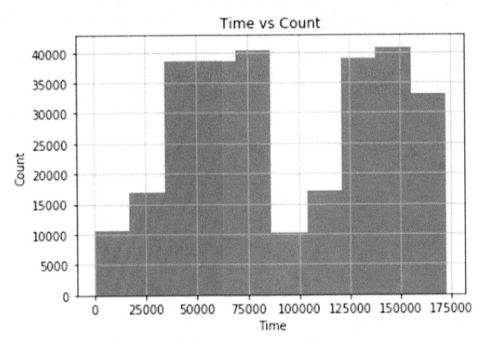

Figure 4 depicts an enriched representation of the CFD dataset while visualizing the histogram plot of time of transaction vs. instance count. The time is elapsed in seconds and corresponding to each time stamp there are count of instances from the CFD dataset.

Figure 5. Histogram of Amount vs. Count

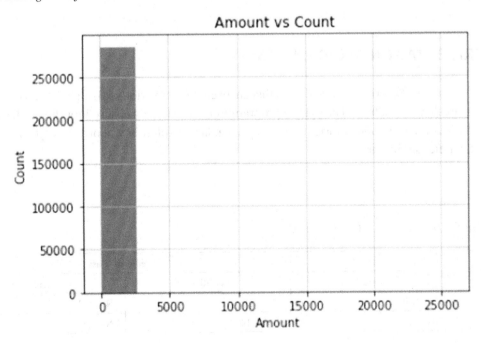

Figure 5 propounds the histogram plot for amount of transaction vs. instance count. The plot indicates that in the CFD dataset the transaction amount lies between 0-5000 and the total count of such transactions are 284807 instances.

Figure 6 describes the histogram plot for class of transaction vs. instance count. The plot indicates that in the CFD dataset the class 1 (fraudulent) has only 492 instances while rest other major instances are of class 0 (legitimate).

Figure 6. Histogram of Class vs. Count

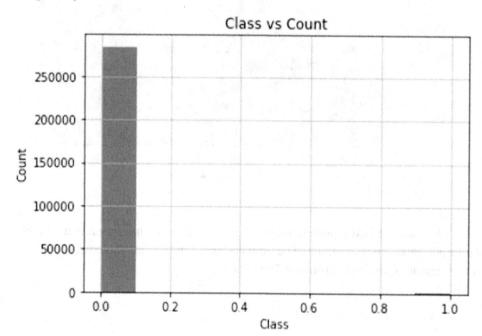

6. EXPERIMENTATION AND RESULTS

This section illustrates experiments and results that are executed for various ML algorithms over the CFD dataset using evaluation metrics. The python programming (version 3.4) is used for the implementation of proposed system. The python libraries e.g., numpy, pandas, matplotlib, seaborn, scikit-learn, KMeans, MinMaxScaler etc. are imported.

Table 1. Confusion Matrix for CFD Dataset

		Predicted Class	
	N = 284,807	**Positive**	**Negative**
Actual Class	**Positive**	TP = 330	FN = 162
	Negative	FP = 131482	TN = 152833

6.1. Results of K-Means Clustering

Table 1 describes the result of k-means clustering applied over the entire dataset. The confusion matrix is generated along with the classification report to see how well the k-means clustering performs.

TP: Count of fraudulent transactions, predicted as fraudulent i.e., number of correctly predicted fraud
TN : Count of legitimate transactions, predicted as legitimate i.e., number of correctly predicted legal
FP : Count of legitimate transactions that are predicted as fraudulent
FN : Count of fraudulent transactions that are predicted as legitimate
P : Count of actual fraudulent transactions in the dataset (TP + FN)
N : Count of actual legitimate transactions in the dataset (FP + TN)

$$\Pr ecision = \frac{TP}{TP + FP} = \frac{330}{330 + 131482} = \frac{330}{131812} = 0.0025 \qquad (7)$$

$$\text{Re} call = \frac{TP}{TP + FN} = \frac{330}{330 + 162} = \frac{330}{492} = 0.67 \qquad (8)$$

$$F - measure = \frac{2 * \Pr ecision * \text{Re} call}{\Pr ecision + \text{Re} call} = \frac{2 * 0.0025 * 0.67}{0.0025 + 0.67} = \frac{0.00335}{0.6725} = 0.004981 \qquad (9)$$

Figure 7. Elbow Graph

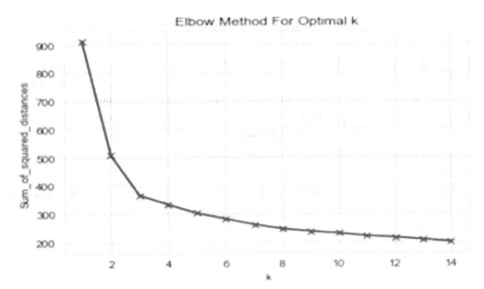

Figure 8. K-Means Cluster Centers

```
array([[ 3.36255728e+04,  -3.95734803e+00,   2.68379254e+00,
        -4.85726765e+00,   3.08345282e+00,  -2.93383137e+00,
        -1.05927535e+00,  -4.17155467e+00,   1.79302679e+00,
        -1.76914048e+00,  -3.90871454e+00,   2.90169715e+00,
        -4.46099173e+00,   7.94536174e-02,  -4.56088047e+00,
         1.56611422e-01,  -3.04896711e+00,  -5.07363124e+00,
        -1.91605548e+00,   4.42908979e-01,   2.32998137e-01,
         4.58701916e-01,  -1.22891483e-01,  -1.94793909e-01,
        -7.57383463e-02,   1.92794014e-01,   7.15670429e-02,
         3.49413569e-01,   3.39214103e-02,   1.01344303e+02],
       [ 1.43147653e+05,  -3.87325594e-01,   5.79298614e-01,
        -2.13175382e+00,   1.21740528e+00,   1.43592455e-01,
        -5.36551047e-01,  -6.19508869e-01,   1.36864237e-03,
        -6.81106781e-01,  -1.13891574e+00,   7.89272789e-01,
        -1.27742372e+00,  -2.13695430e-01,  -2.42009403e+00,
        -1.34971846e-01,  -7.54751478e-01,  -1.09583545e+00,
        -1.42891718e-01,   8.39070170e-02,   4.16304185e-02,
         3.29508140e-01,   1.05457950e-01,   2.65908120e-02,
        -8.09981651e-02,  -1.18973559e-01,   3.94689834e-02,
         4.12018687e-02,   3.03001069e-02,   1.06624773e+02],
       [ 7.98132848e+04,  -2.92601334e+00,   2.35370070e+00,
        -3.60271420e+00,   2.58580854e+00,  -2.14237355e+00,
        -4.13890018e-01,  -3.91070101e+00,  -1.10405708e+00,
        -1.45323863e+00,  -3.68403099e+00,   2.19817303e+00,
        -3.73706520e+00,  -8.47643359e-02,  -3.49165242e+00,
        -1.55436168e-01,  -2.55520160e+00,  -4.06133543e+00,
        -1.40997294e+00,   4.37065118e-01,   2.77972812e-01,
         3.58656104e-01,   1.42925888e-02,   6.35131848e-02,
        -1.42930154e-03,   2.43430090e-02,  -1.13558297e-02,
        -1.54248194e-01,   2.94250357e-02,   1.02138706e+02]]])
```

It is observed that due to highly imbalanced CFD dataset, the precision, recall and F-measure are quite low, as is in equations (7), (8) and (9) respectively. The handling of sampling issues and overfitting of dataset are prime challenges, so a subset of the dataset is to be considered. In other words, further experiments are carried out over a balanced data having equal number of fraudulent and legitimate instances.

Figure 7 shows a graph of sum of the squared distances for number of clusters (k) using Elbow method. This method employs the inertia parameter to compute the sum of squared distances of samples to the nearest cluster center in initialized k-means clustering. If the graph appears to be an arm, then optimal elbow on the arm is found. The following graph shows the most favorable number of clusters as 3.

Figure 8 shows the cluster center vectors which has used the fit_predict() function to return the cluster number in which an observation is actually fit into, and stores it in y_kmeans as in equation (10).

y_k = kmeans.fit_predict(x)   (10)

Figure 9 shows the python code snippet in order to visualize the above-generated3 clusters while applying K-Means.

Figure 9. Python Code for K-Means Clusters

```
kplt.scatter(x[y_k==0, 0], x[y_k==0, 1], s=100, c='red', label ='C1')

kplt.scatter(x[y_k==1, 0], x[y_k==1, 1], s=100, c='blue', label ='C2')

kplt.scatter(x[y_k==2, 0], x[y_k==2, 1], s=100, c='green', label ='C3')
```

Figure 10 depicts the final cluster obtained from k-mean clustering. Table 2 shows confusion matrix for the balanced subset of the CFD dataset. The classification report shown in Table 3 is generated to evaluate how well the k-means clustering performs. Here, although results are better than the initial experiments still, they need further improvements in terms of the evaluation metrics.

Figure 10. K-Means Refined Clusters

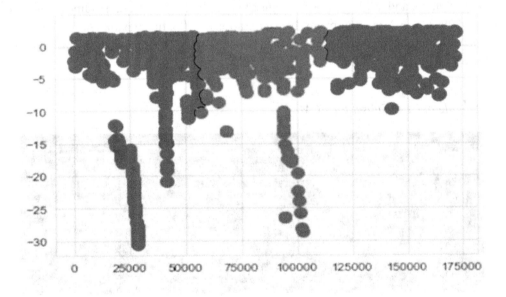

Table 2. K-Means Refined Confusion Matrix

		Predicted Class		
	n= 984	Class 0	Class 1	Class 2
Actual Class	Class 0	137	216	139
	Class 1	186	136	170
	Class 2	0	0	0

Table 3. K-Means Refined Classification Report

n= 984	Precision	Recall	F-measure
Class 0	0.42	0.28	0.34
Class 1	0.39	0.28	0.33
Class 2	0.00	0.00	0.00

From the above facts, it is revealed that the k-means clustering over the entire CFD dataset suffers from several problems such as hard cluster, class dominance and null class assignment. In the hard clustering problem, there is hard assignment of the credit card transactional instance to the corresponding cluster. In the class dominance problem, the higher instances- legitimate class is dominated over the lower instances- fraudulent class. In the null class assignment problem, some of the clusters are assigned to the no class. To solve such problems, a balanced subset of the CFD dataset is considered which although gives better results, still needs further improvements. To do so, it is required to calculate membership value of every.

6.2. Results of Fuzzy C-Means Clustering

For the FCM, experimentation is performed over the balanced subset of the CFD dataset with 984 instances (492 fraudulent and 492 legitimate) (Figure 11). In the FCM, several parameters are considered e.g., number of clusters = 2, maximum number of iterations = 100, fuzzy parameter (m) = 2.00, number of data points = length of data frame, where data frame is the CFD data with feature columns except the label Class.

Figure 11. Sample of CFD Subset for FCM

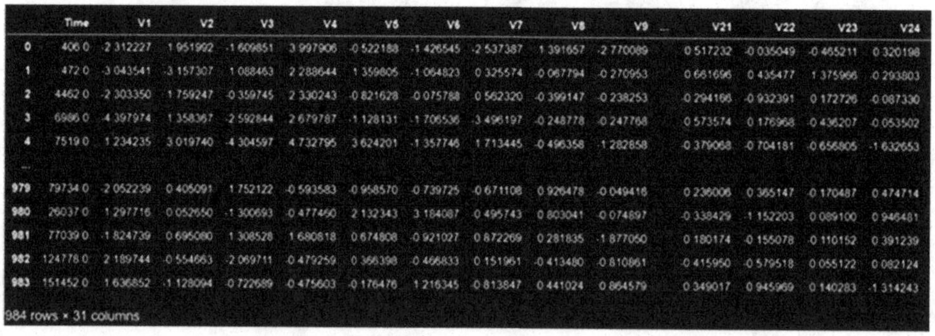

	Time	V1	V2	V3	V4	V5	V6	V7	V8	V9	...	V21	V22	V23	V24
0	406.0	-2.312227	1.951992	-1.609851	3.997906	-0.522188	-1.426545	-2.537387	1.391657	-2.770089		0.517232	-0.035049	-0.465211	0.320198
1	472.0	-3.043541	-3.157307	1.088463	2.288644	1.359805	-1.064823	0.325574	-0.067794	-0.270953		0.661696	0.435477	1.375966	-0.293803
2	4462.0	-2.303350	1.759247	-0.359745	2.330243	-0.821628	-0.075788	0.562320	-0.399147	-0.238253		-0.294166	-0.932391	0.172726	-0.087330
3	6986.0	-4.397974	1.358367	-2.592844	2.679787	-1.128131	-1.706536	-3.496197	-0.248778	-0.247768		0.573574	0.176968	-0.436207	-0.053502
4	7519.0	1.234235	3.019740	-4.304597	4.732795	3.624201	-1.357746	1.713445	-0.496358	-1.282858		-0.379068	-0.704181	-0.656805	-1.632653
...															
979	79734.0	-2.052239	0.405091	1.752122	-0.593583	-0.958570	-0.739725	-0.671108	0.926478	-0.049416		0.236006	0.365147	-0.170487	0.474714
980	260370	1.297716	0.052650	-1.300693	-0.477450	2.132343	3.184087	-0.495743	0.803041	-0.074897		-0.338429	-1.152203	0.089100	0.946481
981	77039.0	-1.824739	0.695080	1.308528	1.680818	0.674808	-0.921027	0.872269	0.281835	-1.877050		0.180174	-0.155078	-0.110152	0.391239
982	124778.0	2.189744	-0.554663	-2.069711	-0.479259	0.366398	-0.466833	0.151961	-0.413480	-0.810861		-0.415950	-0.579518	0.055122	0.082124
983	151452.0	1.636852	-1.128094	-0.722689	-0.475603	-0.176476	1.216345	-0.813847	0.441024	0.864579		0.349017	0.945969	0.140283	-1.314243

984 rows × 31 columns

For the FCM, various functions are constructed in the python, namely- getClusters(): function to create clusters, calculateClusterCenter(): function to calculate cluster center, initializeMembershipMatrix(): function to create membership matrix, updateMembershipValue(): function to update the membership value, and fuzzyCMeansClustering(): function to execute FCM (Figure 12).

As an outcome of the FCM, each CFD data point is assigned fuzzy membership value with respect to each cluster. The sample snapshot of the same is shown in Figure 13.

Figure 12. Python Function to Execute FCM

```
deffuzzyCMeansClustering():
f_membership = initializeMembershipMatrix()
# Membership Matrix
i = 0
    while i<= No_of_Iterations:
fc_centers = calculateClusterCenter(f_membership)
f_membership = updateMembershipValue(f_membership, fc_centers)
fc_labels = getClusters(f_membership)
i += 1
    print(f_membership)
    return fc_labels, fc_centers
```

For the FCM, Precision, Recall and F-measure are computed for Class 0 (legitimate) and Class 1 (fraudulent) asin Table 4, which is better than K-Means clustering, results.

Figure 13. Sample Snapshot of FCM Based Clustering

6.3. Performance Evaluation using various Evaluation Metrics

The CFD dataset is highly imbalanced in nature because of which the confusion matrix accuracy is not so worthy. An error in detecting fraudulent transaction as legitimate is much more crucial than error in detecting legitimate transaction as fraudulent. In such cases, balanced evaluation metrics (Section 6.3.1 to 6.3.4) give better analysis.

Credit Card Fraud Detection Using K-Means and Fuzzy C-Means

Table 4. CM Classification Report

n= 984	Precision	Recall	F-measure
Class 0	0.52	0.60	0.56
Class 1	0.53	0.45	0.49

6.3.1. Fraud Catching Rate

Fraud Catching Rate (FCR), also called sensitivity is the ratio of all the fraudulent instances that are identified in the CFD dataset, as in equation (11).

$$FCR = \frac{TP}{P} = \frac{TP}{TP + FN} \tag{11}$$

For the CFD system, FCR is obtained as 0.60 and 0.45 for legitimate and fraudulent classes respectively.

6.3.2 False Alarm Rate

False Alarm Rate (FAR) is a ratio between the numbers of legitimate instances that are wrongly classified as fraudulent upon total number of actual legitimate instances, as in equation (12).

$$FAR = \frac{FP}{N} = \frac{FP}{FP + TN} = 1 - TNR \tag{12}$$

TNR: True Negative Rate is also called as specificity or Selectivity, as in equation (13).

$$TNR = \frac{TN}{N} = \frac{TN}{FP + TN} = 1 - FPR = 1 - FAR \tag{13}$$

For the CFD system, FAR and TNR are 0.244652 and 0.755348 respectively.

6.3.3 Balanced Classification Rate

Balanced Classification Rate (BCR), also called Balanced Accuracy which combines Sensitivity and Specificity both. The BCR measures how accurate is the overall performance of the CFD system is while considering both the fraudulent and legitimate classes that too without worrying about the imbalanced dataset, as in equation (14).

$$BCR = \frac{1}{2}(TPR + TNR) = \frac{1}{2}\left(\frac{TP}{TP + FN} + \frac{TN}{FP + TN}\right) \tag{14}$$

234

Also, Balanced Error Rate (BER) is defined as in equation (15).

$$BER = 1\text{-}BCR \tag{15}$$

For the imbalanced CFD system, both the BCR (0.60414) and BER (0.39586) are obtained respectively.

6.3.4 Mathews Correlation Coefficient

Mathews Correlation Coefficient (MCC) is a measure of the quality of binary or 2-class classification as in the CFD system. MCC is a balanced measure, which is used even if the, two classes: fraudulent and legitimate are of dissimilar size. MCC varies from -1 to +1, shown in equation (16).

-1 : Total disagreement between prediction and observation (very low quality)
0 : No better than random prediction system
+1 : Perfect prediction (highest quality)

$$MCC = \frac{TP * TN - FP * FN}{\sqrt{(TP + FP)\,(TP + FN)\,(TN + FP)\,(TN + FN)}} \tag{16}$$

For imbalanced CFD system, the MCC is obtained as 0.017346, which is no better than random prediction system.

6.4 Comparative Analysis

Table 5 compares the proposed CFD system with respect to the other techniques (Buonaguidi et al., 2021)- Decision Trees, Boosting and Random Forest for the credit card fraud detection.

Table 5.Comparative Analysis

Method		Dataset	Transaction Features	No. of Transactions	No. of Fraud Transactions	Fraud Ratio	MCC Measure
Proposed CFD System		European 2013	Amount, Date, Time, Frequency	284807	492	0.17%	0.017346
Buonaguidi et al., 2021	Decision Trees	Swiss 2016	Elapsed Days, Transaction Amount	124770	2778	2.23%	0.01271
	Boosting						0.00699
	Random Forest						−0.0035

7. CONCLUSION

Detection and prevention of credit card fraud is quite important. Financial industries serve significant role in combating this crime. Consumers too can help in reducing the fraud by keeping a watchful eye and monitoring their credit reports. In this work, Credit card Fraud Detection (CFD) system is designed in which membership functions of fuzzy logic are responsible for the data preprocessing and data transformation phases. This system emphasizes transaction behavior patterns rather than economical or demographic features. The prime benefit of the CFD system is a steep rise in the credit card holders' satisfaction and healthier bank reputation due to more alerts from the representative banks. The proposed CFD system uses the European dataset, which comprises of 284807 instances of the credit card transactions. The system integrates verification of credit card via Luhn's algorithm and then applies clustering techniques: k-means and FCM to identify fraudulent cases. The FCM has improved results in comparison to k-means over evaluation metrics. Further, the CFD metrics achieve Fraud Catching Rate as 0.60 and 0.45 for legitimate and fraudulent transactions; False Alarm Rate and True Negative Rate as 0.244652 and 0.755348; Balanced Classification Rate and Balanced Error Rate as 0.60414 and 0.39586; and Mathews Correlation Coefficient as 0.017346 respectively.

8. FUTURE SCOPE

In future, several research directions can be explored for the CFD system. Some of them are highlighted as below:

- The CFD dataset for the Germany (European) is already explored. In future, incorporate dataset for the South East Asian Countries such as India and China.
- The CFD system can consider additional features e.g., location of transaction, time gap between transactions; additional cluster parameters e.g. improved cluster labeling and hierarchical clustering.
- The CFD system can further boost up its performance while applying deep learning techniques along with semi-supervised learning strategy.
- The CFD system-based credit card transactions can be segregated depending upon transactional sources such as point of sale, merchant draft entry, and automated teller machine.
- Apart from the technological solutions, investigate appropriate Game Theoretic strategies to win against the fraudsters. Also, propose solutions to integrate Data Mining, Game Theory, Behavioral Learning and Risk Analysis to move one-step ahead of the fraudsters.

ACKNOWLEDGMENT

The undergraduate students of Computer Science & Engineering implemented this research work that includes Saurabh Kumar Gupta, Aastha Juneja, Ashutosh Baghel and Himankur Goyal as their project at Jaypee Institute of Information Technology, Noida, UP, India.

REFERENCES

Akhilomen, J. (2013, July). Data mining application for cyber credit-card fraud detection system. In *Industrial Conference on Data Mining* (pp. 218–228). Springer. doi:10.1007/978-3-642-39736-3_17

Arya, M., & Sastry G, H. (2020). DEAL– Deep ensemble algorithm framework for credit card fraud detection in real-time data stream with google tensorflow. *Smart Science, 8*(2), 71-83. doi:10.1080/23 080477.2020.1783491

Behera, T. K., & Panigrahi, S. (2015, May). Credit card fraud detection: a hybrid approach using fuzzy clustering & neural network. In *2015 Second International Conference on Advances in Computing and Communication Engineering* (pp. 494-499). IEEE. 10.1109/ICACCE.2015.33

Bhusari, V., & Patil, S. (2011). Detection of credit card fraud transaction using hidden markov model. *Research Journal of Engineering and Technology, 2*(4), 203–206.

Buonaguidi, B., Mira, A., Bucheli, H., & Vitanis, V. (2021). Bayesian quickest detection of credit card fraud. *Bayesian Analysis, 1*(1), 1–30.

Carcillo, F., Dal Pozzolo, A., Le Borgne, Y. A., Caelen, O., Mazzer, Y., & Bontempi, G. (2018). Scarff: A scalable framework for streaming credit card fraud detection with spark. *Information Fusion, 41*, 182–194. doi:10.1016/j.inffus.2017.09.005

Chen, R. C., Luo, S. T., Liang, X., & Lee, V. C. (2005, October). Personalized approach based on SVM and ANN for detecting credit card fraud. In *2005 International Conference on Neural Networks and Brain* (vol. 2, pp. 810-815). IEEE. 10.1109/ICNNB.2005.1614747

Chiu, C. C., & Tsai, C. Y. (2004, March). A web services-based collaborative scheme for credit card fraud detection. In *IEEE International Conference on e-Technology, e-Commerce and e-Service, 2004. EEE'04. 2004* (pp. 177-181). IEEE. 10.1109/EEE.2004.1287306

Correia, I., Fournier, F., & Skarbovsky, I. (2015, June). The uncertain case of credit card fraud detection. In *Proceedings of the 9th ACM International Conference on Distributed Event-Based Systems* (pp. 181-192). 10.1145/2675743.2771877

De Montjoye, Y. A., Radaelli, L., Singh, V. K., & Pentland, A. S. (2015). Unique in the shopping mall: On the reidentifiability of credit card metadata. *Science, 347*(6221), 536–539. doi:10.1126cience.1256297 PMID:25635097

Delamaire, L., Abdou, H., & Pointon, J. (2009). Credit card fraud and detection techniques: A review. *Banks and Bank Systems, 4*(2), 57-68. http://usir.salford.ac.uk/id/eprint/2595/

Dheepa, V., & Dhanapal, R. (2012). Behavior based credit card fraud detection using support vector machines. *ICTACT Journal on Soft computing, 2*(7). doi:10.21917/ijsc.2012.0061

Duman, E., & Elikucuk, I. (2013, June). Solving credit card fraud detection problem by the new meta-heuristics migrating birds optimization. In *International Work-Conference on Artificial Neural Networks* (pp. 62–71). Springer. doi:10.1007/978-3-642-38682-4_8

Duman, E., & Ozcelik, M. H. (2011). Detecting credit card fraud by genetic algorithm and scatter search. *Expert Systems with Applications, 38*(10), 13057-13063. doi:10.1016/j.eswa.2011.04.110

Gianini, G., Fossi, L. G., Mio, C., Caelen, O., Brunie, L., & Damiani, E. (2020). Managing a pool of rules for credit card fraud detection by a game theory based approach. *Future Generation Computer Systems, 102*, 549–561. doi:10.1016/j.future.2019.08.028

Hegazy, M., Madian, A., & Ragaie, M. (2016). Enhanced fraud miner: credit card fraud detection using clustering data mining techniques. *Egyptian Computer Science Journal, 40*(3).

Hormozi, E., Akbari, M. K., Hormozi, H., & Javan, M. S. (2013, May). Accuracy evaluation of a credit card fraud detection system on Hadoop MapReduce. In *The 5th conference on information and knowledge technology* (pp. 35-39). IEEE. 10.1109/IKT.2013.6620034

Kamaruddin, S., & Ravi, V. (2016, August). Credit card fraud detection using big data analytics: use of PSOAANN based one-class classification. In *Proceedings of the International Conference on Informatics and Analytics* (pp. 1-8). 10.1145/2980258.2980319

Kim, M. J., & Kim, T. S. (2002, August). A neural classifier with fraud density map for effective credit card fraud detection. In *International Conference on Intelligent Data Engineering and Automated Learning* (pp. 378-383). Springer. 10.1007/3-540-45675-9_56

Luhn, H. P. (1960). Computer for verifying numbers. *US Patent, 2*(950), 48.

Maes, S., Tuyls, K., Vanschoenwinkel, B., & Manderick, B. (2002, January). Credit card fraud detection using bayesian and neural networks. In *Proceedings of the 1st international Naiso Congress on Neuro Fuzzy Technologies* (pp. 261-270). Academic Press.

Malini, N., & Pushpa, M. (2017, February). Analysis on credit card fraud identification techniques based on KNN and outlier detection. In *2017 Third International Conference on Advances in Electrical, Electronics, Information, Communication and Bio-Informatics (AEEICB)* (pp. 255-258). IEEE. 10.1109/AEEICB.2017.7972424

Meador, C. A., & Moore, L. D. (2008). *U.S. Patent No. 7,431,202.* Washington, DC: U.S. Patent and Trademark Office.

Mishra, A., & Ghorpade, C. (2018, February). Credit card fraud detection on the skewed data using various classification and ensemble techniques. In *2018 IEEE International Students' Conference on Electrical, Electronics and Computer Science (SCEECS)* (pp. 1-5). IEEE. 10.1109/SCEECS.2018.8546939

Mishra, M. K., & Dash, R. (2014, December). A comparative study of chebyshev functional link artificial neural network, multi-layer perceptron and decision tree for credit card fraud detection. In *2014 International Conference on Information Technology* (pp. 228-233). IEEE. 10.1109/ICIT.2014.25

Patel, R. D., & Singh, D. K. (2013). Credit card fraud detection & prevention of fraud using genetic algorithm. *International Journal of Soft Computing and Engineering, 2*(6), 292–294.

Patidar, R., & Sharma, L. (2011). Credit card fraud detection using neural network. *International Journal of Soft Computing and Engineering, 1*, 32–38.

Pawar, A. D., Kalavadekar, P. N., & Tambe, S. N. (2014). A survey on outlier detection techniques for credit card fraud detection. *IOSR Journal of Computer Engineering*, *16*(2), 44–48. doi:10.9790/0661-16264448

Ramaki, A. A., Asgari, R., & Atani, R. E. (2012). Credit card fraud detection based on ontology graph. *International Journal of Security, Privacy and Trust Management*, *1*(5), 1–12. doi:10.5121/ijsptm.2012.1501

Ross, T. J. (2004). *Fuzzy logic with engineering applications* (Vol. 2). Wiley.

Sakharova, I. (2012, June). Payment card fraud: Challenges and solutions. In *2012 IEEE International Conference on Intelligence and Security Informatics* (pp. 227-234). IEEE. 10.1109/ISI.2012.6284315

Salazar, A., Safont, G., Soriano, A., & Vergara, L. (2012, October). Automatic credit card fraud detection based on non-linear signal processing. In *2012 IEEE International Carnahan Conference on Security Technology (ICCST)* (pp. 207-212). IEEE. 10.1109/CCST.2012.6393560

Save, P., Tiwarekar, P., Jain, K. N., & Mahyavanshi, N. (2017). A novel idea for credit card fraud detection using decision tree. *International Journal of Computers and Applications*, *161*(13), 6–9. Advance online publication. doi:10.5120/ijca2017913413

Sharma, A., & Panigrahi, P. K. (2012). A review of financial accounting fraud detection based on data mining techniques. *International Journal of Computers and Applications*, *39*(1), 37–47. Advance online publication. doi:10.5120/4787-7016

Van Vlasselaer, V., Bravo, C., Caelen, O., Eliassi-Rad, T., Akoglu, L., Snoeck, M., & Baesens, B. (2015). APATE: A novel approach for automated credit card transaction fraud detection using network-based extensions. *Decision Support Systems*, *75*, 38–48. doi:10.1016/j.dss.2015.04.013

Varmedja, D., Karanovic, M., Sladojevic, S., Arsenovic, M., & Anderla, A. (2019, March). Credit card fraud detection-machine learning methods. In *2019 18th International Symposium INFOTEH-JAHORINA (INFOTEH)* (pp. 1-5). IEEE. 10.1109/INFOTEH.2019.8717766

Xie, Y., Liu, G., Cao, R., Li, Z., Yan, C., & Jiang, C. (2019, February). A feature extraction method for credit card fraud detection. In *2019 2nd International Conference on Intelligent Autonomous Systems (ICoIAS)* (pp. 70-75). IEEE. 10.1109/ICoIAS.2019.00019

Yu, W. F., & Wang, N. (2009, April). Research on credit card fraud detection model based on distance sum. In *2009 International Joint Conference on Artificial Intelligence* (pp. 353-356). IEEE. 10.1109/JCAI.2009.146

Zareapoor, M., & Shamsolmoali, P. (2015). Application of credit card fraud detection: Based on bagging ensemble classifier. *Procedia Computer Science*, *48*, 679–685. doi:10.1016/j.procs.2015.04.201

Zhang, Y., You, F., & Liu, H. (2009, August). Behavior-based credit card fraud detecting model. In *2009 Fifth International Joint conference on INC, IMS and IDC* (pp. 855-858). IEEE. 10.1109/NCM.2009.54

Zhu, H., Liu, G., Zhou, M., Xie, Y., Abusorrah, A., & Kang, Q. (2020). Optimizing weighted extreme learning machines for imbalanced classification and application to credit card fraud detection. *Neurocomputing*, *407*, 50-62. doi:10.1016/j.neucom.2020.04.078

ENDNOTES

[1]   https://wallethub.com/edu/cc/number-of-credit-cards/25532

[2]   https://www.forbes.com/sites/rogeraitken/2016/10/26/us-card-fraud-losses-could-exceed-12bn-by-2020/#18eebd96d243

[3]   https://www.aciworldwide.com/-/media/files/collateral/trends/aci_stopping-card-fraud-guide_tl_us_1010_4414.pdf

[4]   https://www.kaggle.com/mlg-ulb/creditcardfraud?select=creditcard.csv

Chapter 17
A Hybrid Method to Reduce PAPR of OFDM to Support 5G Technology

Bhavana D.
Koneru Lakshmaiah Education Foundation, India

Adada Neelothpala
Koneru Lakshmaiah Education Foundation, India

Pamidimukkala Kalpana
Koneru Lakshmaiah Education Foundation, India

ABSTRACT

The orthogonal frequency division multiplexing (OFDM) is a multicarrier modulation scheme used for the transfer of multimedia data. Well-known systems like ADSL (asymmetric digital subscriber line) internet, wireless local area networks (LANs), long-term evolution (LTE), and 5G technologies use OFDM. The major limitation of OFDM is the high peak-to-average power ratio (PAPR). High PAPR lowers the power efficiency, thus impeding the implementation of OFDM. The PAPR problem is more significant in an uplink. A high peak-to-average power ratio (PAPR) occurs due to large envelope fluctuations in OFDM signal and requires a highly linear high-power amplifier (HPA). Power amplifiers with a large linear range are expensive, bulky, and difficult to manufacture. In order to reduce the PAPR, a hybrid technique is proposed in this chapter with repeated clipping and filtering (RCF) and precoding techniques. The proposed method is improving the PAPR as well as BER. Five types of pre-coding techniques are used and then compared with each other.

DOI: 10.4018/978-1-7998-6870-5.ch017

1. INTRODUCTION

OFDM (orthogonal frequency division multiplexing) is a multicarrier modulation scheme for multimedia data transmission. It's become common in both wireless and wired communication systems in recent years. Energy efficiency is a key issue of concern in future mobile communications like 5G. Many technologies have been developed to increase the energy efficiency of the communication systems. The Peak-to-Average Power Ratio of the OFDM transmitted signal is high. Due to the high PAPR, the system's output degrades and the high power amplifier's efficiency is reduced. Increasing the efficiency of the high power amplifier (HPA) is a major source for reducing energy costs. The high power amplifier's efficiency, is related to the input signal's PAPR. The PAPR problem continues to exist and prevents OFDM from being adopted in the uplink of mobile communication standards. Apart from the power efficiency issue, it can also impose severe restrictions on output power, resulting in a problem with coverage in the downlink. In mobile devices, PAPR has a negative impact on battery life. Since portable devices have little battery capacity, it's important to find ways to lower the PAPR, making for a smaller, more powerful HPA, and hence a longer battery life.

The issue of high PAPR can overshadow all the potential benefits of multicarrier transmission systems in many low-cost applications. A variety of promising methods or strategies to minimise PAPR have been suggested and applied at the cost of increased transmit signal power, bit error rate (BER), computational complexity, and data rate loss, etc. In order to promote 5G technologies, a hybrid strategy for reducing OFDM PAPR is proposed. PAPR is reduced using a combination of repeated clipping and filtering (RCF) and pre-coding techniques. This study employs five different methods of pre-coding strategies, which are then compared with each other. Observed and contrasted to current approaches is the better form of pre-coding is termed in terms of reduced PAPR and BER. Matlab is used to simulate all of the processes.

2. LITERATURE SURVEY

Mohit et al. (Singh et al., 2013) published a study in 2013. Hybrid Clipping-Compounding techniques for PAPR Reduction have been suggested to minimise the PAPR (peak to average power ratio) of OFDM. The output of a hybrid PAPR reduction scheme using tanh or erf as the companding feature is nearly equivalent. Better than the previous two is a hybrid PAPR reduction scheme with a log companding feature. PAPR reduction scheme with eith in a hybrid form

K. muralibabu et al. (Muralibabu & Sundhararajan, 2013) published a study in 2013. In the proposed scheme, the PAPR of the OFDM method is reduced by grouping the subcarriers using a combination of the Discrete Cosine Transform (DCT) and the companding technique. According to the simulation results, the proposed scheme can provide a good balance of computational complexity and PAPR reduction efficiency.

Sroy et al. (Abouty et al., 2013) published an Iterative Clipping and Filtering (ICF) Technique for PAPR Reduction of OFDM Signals in 2013. DCT/IDCT (inverse discrete cosine transform) Transform is used in this system. With the DCT/IDCT technique, the OFDM symbol reaches the ICF block, where it is clipped and filtered iteratively. Despite the fact that we show that iterative clipping and filtration can significantly reduce PAPR, The fast Fourier transform (FFT)/IFFT transform is used for filtering, but DCT/IDCT is used in the classic ICF technique for better performance.

Nonlinear companding transform (NCT) was provided to Wang et.al in 1990(Joshi, 2012) for PAPR reduction, and it was based on the speech processing algorithm -law. It outperforms the clipping technique. The -law companding transform increases the average power of the transmitted signal by enlarging small amplitude signals while keeping peak signals unchanged. Signals and it may lead to overcome the saturation region of the HPA to make the performance of the system worse. To overcome the problem of μ-law companding (increasing average power) and to have an efficient PAPR reduction. (Joshi, 2012)

In 2014, Jijina et al. (Jijina & Pillai, 2014) conducted a comparison analysis of three popular linear preceding techniques used in OFDMA systems: Hadamard transform preceding, Discrete Sine Transform (DST) preceding, and Square root elevated cosine function preceding. The traditional Random Interleaved OFDMA method is used to evaluate the output of these various schemes in terms of PAPR reduction.. Linear preceding schemes are efficient, signal independent, distortion less and do not require complex optimization when compared to the other reduction schemes.

3. METHODOLOGY

3.1 Clipping and Filtering

Clipping is the easiest way of lowering PAPR. Clipping is a technique for limiting the maximum amplitude of an OFDM signal to a pre determined level. Clipping is fairly easy to implement.

The most simple and commonly used method of PAPR reduction is to simply clip the portions of the signals that are beyond the allowed range. For eg, if you use HPA with a saturation level below the signal span, the signal will be clipped. For amplitude clipping, that is (El-Tarhuni et al., 2010):

$$c(x) = x, \quad .x|\leq A$$

$$= A, |x| > A \tag{1}$$

Where A is preset clipping level and it is a positive real number.

Clipping is usually achieved at the transmitter. The receiver, on the other hand, must measure the clipping and compensate the received OFDM symbol accordingly. In most cases, only one clipping happens per OFDM symbol, so the receiver must approximate two parameters: the location and size of clip. As a result, the clipping method introduces in-band distortion and out-of-band radiation into OFDM signals, lowering device performance such as BER and spectral quality. After clipping, filtering can minimise out-of-band radiation, but it cannot reduce in-band distortion.

It has the following disadvantages:

1. It induces in-band signal distortion, which degrades BER output.
2. It also produces out-of-band radiation, which causes neighbouring channels to receive out-of-band interfering signals. Filtering may minimise out-of-band radiation, but when clipping with the Nyquist sampling rate, the filtering can affect high-frequency components of the in-band signal.
3. After clipping, filtering will minimise out-of-band radiation at the expense of peak regrowth. After the filtering process, the signal can exceed the clipping standard specified for the clipping operation.

Repeated clipping and filtering may be used to achieve a desirable PAPR at the expense of increased computational complexity to minimise overall peak re-growth. Repeated clipping-and-filtering operation can be used to minimise peak regrowth and achieve a desirable PAPR at the cost of increased computational complexity.

CODING TECHNIQUES

Coding works under the principle of inserting redundant bits into the data stream that can be used for error correction at the receiver. Their application to PAPR reduction involves generating bit sequences with low PAPR after the IFFT. Block codes and convolution codes are the two types of error detection and correction codes. A block diagram of an OFDM transmitter's coding for PAPR reduction is seen in Figure.1.

Figure 1. OFDM transmitter code for PARM

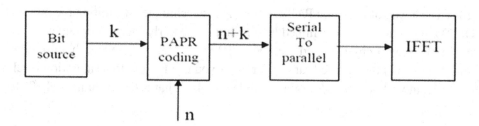

Similar codes have varying degrees of error correction ability. The weight of the code, which is the number of non-zero elements in the codeword, is another essential property of codes. The basic principle behind all coding schemes for PAPR reduction is to reduce the chance of N signals having the same phase. The coding system chooses code words that limit or eliminate the PAPR. It produces no distortion or out-of-band radiation, but as the code rate is decreased, it reduces bandwidth efficiency. It's also difficult to find the right codes and hold wide lookup tables for encoding and decoding, particularly if there are a lot of subcarriers (Joshi, 2012).

Redundant bits are applied to the bit stream before the IFFT in coding strategies for PAPR reduction. These code words, when used correctly, ensure that the PAPR after the IFFT remains minimal. These codes should be used in combination with current OFDM to minimise coding redundancy and complexity. Repeated Clipping and frequency domain filtering: (RCF) shown in Figure.2.

An oversize IFFT is used to transform the input vector a0,a1,a2,...a(N-1) from the frequency to the time domain. The number of subcarriers in each OFDM symbol is denoted by N. The input vector is extended by adding N (1-1) zeros in the middle for an oversampling factor. This results in the trigonometric interpolation of the time domain signal (Prasad, 2004).

When the initial signal consists of integral frequencies over the FFT window, trigonometric interpolation provides perfect interpolation. In the case of OFDM, this is the case. The Nyquist frequency has

Figure 2. Block diagram of IFFT

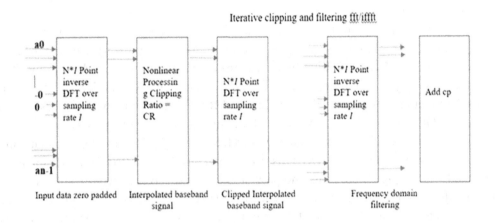

been left out of the equation because the interpolation method would not work for this value (Prasad, 2004). This is not a functional restriction since all OFDM applications null this input, and the rest of them do not use multiple adjacent subcarriers. After that, the interpolated signal is clipped.

The amplitude of the complex values of the IFFT output is hard-limited in this technique (Armstrong, 2001). The initial signal is clipped in the time domain after an IFFT. The following is an overview of the clipping:

$$
C = \begin{cases} \sqrt{C_R * E\left[|x|^2\right] * \frac{x}{|x|}} \quad |x|^2 > C_m \\ x \\ |x|^2 \leq C_m \end{cases} \tag{2}
$$

The output of the time domain signal is denoted by C.

$$
Cm = C_R * E[|x|^2] \tag{3}
$$

The clipping ratio is the ratio of the clipping level to the unclipped baseband signal's mean power. The discrete time domain signal is clipped in amplitude, as seen in equation.2. The amplitude was reduced to the clipping level at any point where the complex time domain signal approached the clipping level, although the step of the complex signal remained unchanged (Prasad, 2004). After clipping, frequency domain filtering is used to reduce the OOB power generated by clipping. Two FFT operations make up the filter (Armstrong, 2001).

An FFT is used to transform the clipped time domain signal c back into the discrete frequency domain. The clipped signal's in band discrete frequency components C0,C(N/2-1),C(Nl1-N/2+1),....,C(Nl1-1) are transmitted unchanged to the second IFFT's inputs, while the OOB components C(N/2+1),....,C(Nl1-N/2) are nulled. This method is repeated one to four times, depending on the number of iterations.

Instead of being added at the receiver, the clipping noise is added at the transmitter. In fading signals, this ensures that clipping noise causes less bit error rate loss than noise introduced in the channel since the clipping noise fades with the signal.

However, any of the transform, up sampling, and filtering arrangements widely used in OFDM systems could be used to replace the second oversize IFFT. As a result, the strategy can be applied by replacing an existing OFDM system's IFFT block with the current configuration (Armstrong, 2001). The DCT/IDCT transform can be used to replace the FFT/IFFT transform filter, as defined in eq.2 and eq.3.

In paper (Devlin et al., 2008), introduces a new PAPR reduction strategy that takes advantage of unused carriers in OFDM systems as well as the phase information of pilot signals to minimise PAPR without affecting channel estimation or frequency offset. Since the initial data signals are not interfered with, this method introduces much less OOB distortions and has a lower BER than traditional strategies like clipping and windowing. There is also no provision for the receiver to acquire side information.

The RCF is added to OFDM signal for different CR and oversampling filter and considers their effect on PAPR and BER to minimise PAPR at LTE downlink. The explanation for using this approach is that if the oversampling is strong, the filter improves BER and clipping improves PAPR (it is necessary to improve both BER and PAPR at the same time, as we previously explained). This method is replicated, depending on iteration number, for I = (1, pilot, 1.125, 1.25, 1.5, 2, 3, 4) and CR = (4, 3, 2, 1.75, 1.5) in order to see the effect of CR on the (BER) and (PAPR) (four is used in this simulation). The signals are sent into a Rayleigh fading channel.

4. PRECODING TECHNIUQES:

The pulse forming or pre-coding technique reduces the PAPR of OFDM signals in an effective and scalable manner. Prior to OFDM modulation and transmission, each data block is multiplied by a pre-coding matrix in this process. Since this approach is data independent, it avoids block-based optimization. It also operates with any baseband modulation and any number of subcarriers. In terms of BER, it takes advantage of the fading multipath channel's frequency variance to increase the BER of OFDM signals over traditional OFDM (no pre-coding). The suggested technique's design complexity is suitable because it uses a predefined pre-coding matrix and therefore eliminates the need for a handshake between the transmitter and recipient. The use of the same pre-coding matrix for all OFDM blocks eliminates the need for all of the computation that block-based optimization approaches require.

Techniques based on preceding are basic linear techniques. The PAPR of these strategies can be reduced to those of single carrier networks (Slimane, 2007). Preceding related PAPR reduction strategies include WHT preceding techniques, DCT preceding techniques, and DHT preceding techniques (Slimane, 2007; Min & Jeoti, 2007; Baig & Jeoti, 2010a, 2010b, 2010c). (Hincapie & Sierra, 2012).

The block diagram of a Pre-coding Based OFDM System is seen in Figure 3. To minimise the PAPR, we used the Pre-coding matrix P of dimension N N before the IFFT. The Pre-coding matrix P can be written as:

Matrix:

$$P = \begin{bmatrix} P_{00} & P_{01} & \cdots & P_{0(N-1)} \\ P_{10} & P_{11} & \cdots & P_{0(N-1)} \\ \vdots & \vdots & \ddots & \vdots \\ P_{(N-1)0} & P_{(N-1)1} & \cdots & P_{(N-1)(N-1)} \end{bmatrix} \tag{3}$$

Where P is a Pre-coding Matrix of size $N \times N$ is shown in eq.3. We can express modulated OFDM vector signal with N subcarriers as:

$$H_k = \sum_{n=0}^{N-1} x_n \left[\cos \frac{2\pi nk}{N} + \sin \frac{2\pi nk}{N} \right] = \sum_{n=0}^{N-1} x(n) \operatorname{cas} \frac{2\pi nk}{N} \tag{4}$$

4.1 Discrete Hartley Transform:

The DHT is a linear transform. N real numbers $X_0, X_1, \ldots\ldots X_{N-1}$ are transformed into N real numbers $H_0, H_1, \ldots\ldots H_{N-1}$ utilising DHT.

Figure 3. Block diagram of Pre-coding based OFDM system.

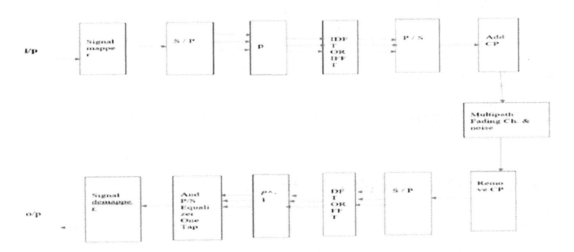

4.2 Walsh Hadamard Transform (WHT)

A generalised family of Fourier transformations is the Hadamard transform (also known as the Walsh–Hadamard transform, Hadamard–Rademacher–Walsh transform, Walsh transform, or Walsh–Fourier transform). On 2^m real numbers, it performs an orthogonal, symmetric, in convolution, linear operation (or complex numbers, although the Hadamard matrices themselves are purely real). The Hadamard transform can be thought of as a set of scale 2X 2X 2X 2X........X2 DFTs, and it is equal to a multidimensional DFT of size. It takes every input vector and decomposes it into a superposition of Walsh functions (Pavan & Kumar, 2013).

The WHT is an orthogonal, non-sinusoidal technique that decomposes a signal into a series of simple functions. As opposed to the traditional OFDM method, these functions are known as Walsh functions, and the Hadamard transform scheme eliminates the frequency of high peaks. The Hadamard transform is used because it decreases the input sequence's autocorrelation, which lowers the PAPR of the OFDM signal.

WHT, like FFT, can be implemented using a butterfly structure. As a consequence, introducing WHT would not necessitate a substantial improvement in device complexity.

4.3. Discrete Cosine Transform (DCT)

DCT is a frequency domain signal transformation technique. A row of data is denoted by DCT as the number of cosine functions oscillating at various frequencies. The DCT is similar to the DFT, but it only uses real numbers without the imaginary variable. The object of using DCT in this analysis is to reduce the autocorrelation of the input row in order to lower PAPR, and it does not necessitate the transfer of information to the receiver. Equation can be used to construct a DCT matrix P of dimension N-by-N.

$$D_{ij} = \begin{cases} \frac{1}{\sqrt{N}} & i = 0, \quad 0 \leq j \leq N-1 \\ \sqrt{\frac{2}{N}} \cos \frac{\pi(2j+1)i}{2N} & 1 \leq i \leq N-1 \\ & 1 \leq j \leq N-1 \end{cases}$$

and DCT can be defined as:-

$$X_k = \sum_{n=0}^{N-1} x_n \cdot \cos\left[\frac{\pi}{N}\left(n + \frac{1}{2}\right)k\right] \quad k=0,1 \ldots N\text{-}1$$

4.4. Discrete Sine Transform (DST) Equation is :

$$D_{ij} = \begin{cases} \frac{1}{\sqrt{N}} & i = 0, \quad 0 \leq j \leq N-1 \\ \sqrt{\frac{2}{N}} \cdot \sin \frac{\pi(2j+1)i}{2N} & 1 \leq i \leq N-1 \\ & 1 \leq j \leq N-1 \end{cases}$$

The DST matrix shall fulfil the following requirements:

1. All components of the pre-coding matrix have the same magnitude.
2. The amplitude of the signal must be $1/\sqrt{N}$.
3. A non-singular DST pre-coding matrix is needed.

These standards guarantee that the output symbol contains the same amount of information as the input data; they also ensure that the power at the output is preserved and that the original data is recovered at the receiver.

4.5. Discrete Fourier Transform (DFT)

The presence of a DFT and an IDFT block in the transmitter and receiver is the only distinction between DFT-spread OFDM and traditional OFDM. The PAPR of the signal in DFT-spread OFDM is comparatively poor as opposed to conventional OFDM because the DFT process spreads data into subcarriers (Kumari, 2014). The DFT of a sequence of length N can be defined as:

$$x(k) = \sum_{k=0}^{N-1} x_{(n)}\, e^{-j\,2\pi n k} \quad k = 0,1 \dots N-1 \tag{5}$$

At the receiver, the transform DFT is used to reverse the IDFT mapping, and the signals from the subcarriers are merged to form an approximation of the transmitter's source signal. The connection conducted in DFT for the given subcarrier only sees energy for the corresponding subcarrier since the base function of DFT is uncorrelated. Figure.4 shows the effect of repetition clipping and filtering on PAPR when CR=3, I=2, and the CCDF of PAPR for one RCF = 7.7581, two RCF = 6.5462, three RCF = 5.8319, and four RCF = 5.401, respectively.

The CCDF of PAPR improves for one RCF (2.8935 dB), two RCF (4.1054 dB), three RCF (4.8197 dB), and four RCF (4.8197 dB) (5.2506 dB). However, as N increases, the proportion of change between (N) RCF and (N-1) RCF decreases.

When the CR is decreased, the PAPR improves and the contrast SNR at BER 10^{-4} improves. The highest PAPR value is for CR =1.5, but the SNR at BER 10^{-4} for this situation is the worst. As seen in table, CR has a positive relationship with PAPR and a negative relationship with SNR at BER 10^{-4} When the number of samples is raised, the SNR at BER improves, and the contrast PAPR rises, and vice versa. The better PAPR value is for I= 1, which means there is no filter, but the SNR at BER 10^{-4} is the worst in this situation. For this case, the best SNR value at BER 10^{-4} is the worst, while At BER 10^{-4}, the highest SNR value is for I=4, but the PAPR is the lowest, as seen in table (A.1) At BER, We have a good relationship with SNR and a bad relationship with PAPR shown in figure.5.

Figure.6 shows the following:

- When compared to the CR4, there is a clear increase in the CR3 CCDF of PAPR reduction in rate SNR at BER 10^{-4} is relatively slight. In a nutshell, this means that the percentage gain in PAPR's CCDF is greater than the degradation in BER.
- The PAPR of the CR2 increased more than the CR3 and CR4, but the SNR at BER 10^{-4} decreased.
- In contrast to the CR 2, the CR 1.75 had a slight increase in PAPR CCDF, but SNR at BER 10^{-4} degraded more than the amount of improvement.

Figure 4. PAPR CCDF for OFDM system with repeated clipping and frequency domain filtering (CR =3, I =2).

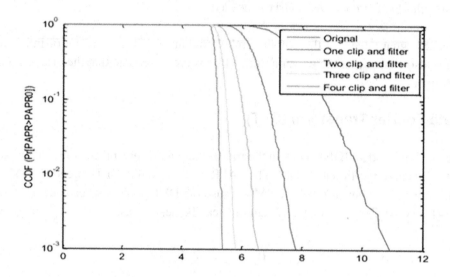

Figure 5. Graph for PAPR0 vs CCDF

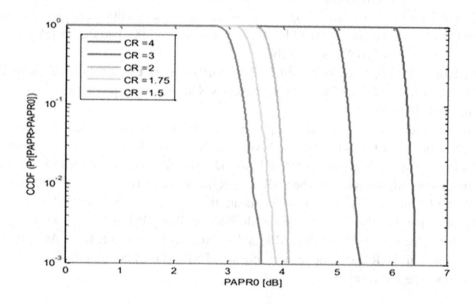

Figure 6. SNR vs BER

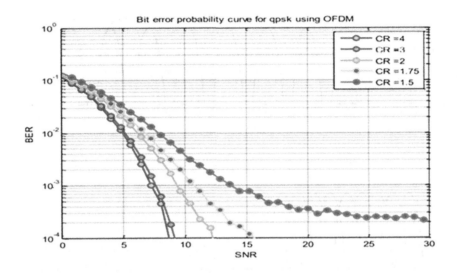

- The PAPR of the CR 1.5 improved only slightly as compared to the CR 1.75, while the SNR at BER 10 was significantly worse.

4.6. OFDM

With this generation's ever-increasing demands, the need for high-speed connectivity has become a top priority. To satisfy these demands, various multicarrier modulation techniques have emerged, the most well-known of which are OFDM and Code Division Multiple Access (CDMA) shown in figure.8.

OFDM's basic idea is to divide high data rate streams into a variety of lower data rate streams, which are then distributed in parallel via several orthogonal sub-carriers (parallel transmission). The symbol length increases as a result of the simultaneous propagation, reducing the prorated amount of time dispersion caused by the multipath delay distribution. OFDM can be thought of as both a modulation and multiplexing technique (Joshi, 2012).

Increased symbol rates are not required in OFDM communication systems to achieve higher data rates. It becomes much easier to manage ISI as a result of this. OFDM symbols are much longer than symbols on single carrier networks of equal data rate when data is distributed in parallel rather than serially.

The principle of OFDM is quite similar to that of the well-known and commonly applied Frequency Division Multiplexing technique (FDM). Many messages may be received over a single radio channel using the concepts of OFDM. However, it is performed in a much more regulated way, resulting in increased spectral efficiency.

Each radio station transmits on a different frequency in conventional broadcast, essentially using FDM to create a distinction between the stations. Since carrier frequencies in FDM are non-orthogonal, a wide band gap is needed to prevent inter-channel interference, reducing overall spectral quality. The difference between FDM and OFDM is shown in Figure 7.

Figure 7. Conventional and Orthogonal multi carrier techniques OFDM

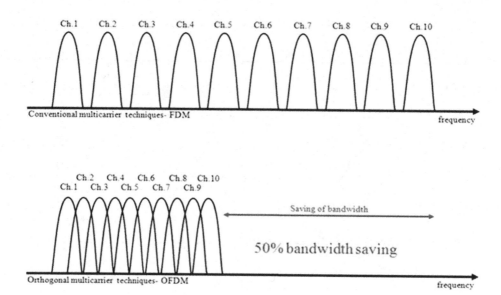

Figure 8. Block diagram of basic OFDM system

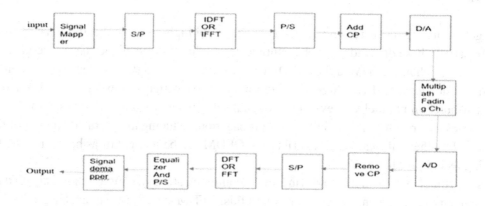

5. RESULTS ANALYSIS OF OFDM

Five different forms of pre-coding are used, and they are then compared. The DFT pre-coder is the best form of reduced PAPR and BER. The DFT pre-coder, as seen in figure.9, offers the best reduction in PAPR and BER, but it suffers from link performance loss in a frequency-selective channel when high order modulation techniques are used. When there are carrier frequencies offsets (CFOs) between the transmitter and receiver, subcarriers lose orthogonality and intercarrier interference occurs (ICI). In the DFT pre-coder scheme, CFOs often implement multiple access interference (MAI) and degrade the bit error rate (BER) efficiency.

Figure 9. Complete block diagram of OFDM transmitter and receiver.

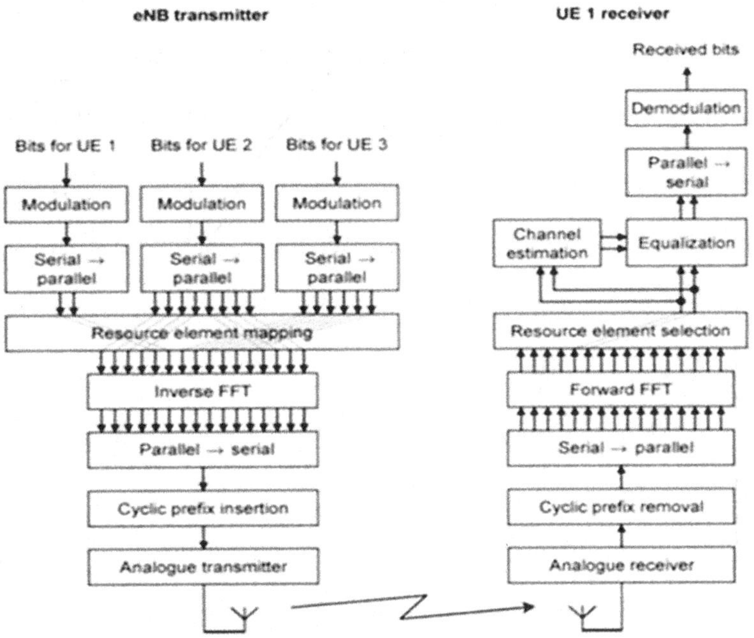

5.1. The Result Taken From Figure 9 is as Follows:

- PAPR (2.7941 dB), PAPR CCDF (0.8684 dB), and SNR at BER (10^{-4}) were all strengthened with the WHT precoder (0.01dB). However, as compared to the other types of pre-coding, the level of change in WHT pre-coding is the smallest.
- PAPR (7.5208 dB), PAPR CCDF (3.109 dB), and SNR at BER(10^{-4}) were all improved using the DCT pecoder (0.012dB). DCT pre-coding does better than WHT pre-coding but not as good as the other approaches.
- PAPR (8.1669 dB), PAPR CCDF (3.25 dB), and SNR at BER(10^{-4}) were all enhanced with the DST pecoder (0.012 dB).
  At BER(10^{-4}), DST and DCT have the same SNR, but DST outperforms DCT in PAPR and CCDF of PAPR.
- The PAPR was improved by (18.6731 dB), the CCDF of PAPR was improved by (7.423 dB), and the SNR at BER(10^{-4}) was improved by DHT pecoder (0.058dB). Except for DFT pre-coding, DHT pecoder findings are superior to other forms of pre-coding.
- The PAPR was improved by (25.6118 dB), the CCDF of PAPR was improved by (10.773 dB), and the SNR at BER(10^{-4}) was improved by DFT pecoder (0.171dB). When compared to other forms of pre-coding, DFT pecoder findings are the best.

Figure 10. Comparative analysis of PAPR0 vs CCDF

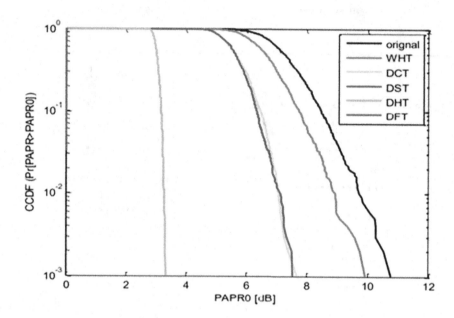

5.2 PAPR

One of the main problems with the OFDM system is the high PAPR of broadcast signals. The key cause of high PAPR is a wide complex range of input data symbols. An OFDM signal is made up of individual data symbols modulated on N orthogonal subcarriers, and when these signals are combined in phase, they produce higher peak amplitude. The PAPR of any signal for a continuous time baseband OFDM signal is defined as the ratio of the signal's maximum instantaneous power to its average power. Where, p(av) is the average power of x (t) and can be computed in frequency domain because IFFT is a unitary transformation T(s) is useful duration of an OFDM symbol.

5.3 PAPR of OFDM

The discrete time baseband OFDM signal is defined in (El-Tarhuni et al., 2010). During signal digitization and recovery, the PAPR of the discrete time OFDM signal specifies the complexity of the digital circuitry in terms of the number of bits used to obtain the optimal signal to quantization noise ratio. The discrete time OFDM signal is obtained by L times oversampling to best approximate the PAPR of a continuous time OFDM signal. The oversampled discrete time OFDM signal can be obtained by applying LN point IFFT to the data block with (L-1) N zero padding, as seen below. The PAPR of a pass-band OFDM signal is nearly double that of a baseband OFDM signal. For an OFDM signal, the Complementary Cumulative Distribution Function (CCDF) is written as:

5.4. Hybrid Precoding With RCF:

As seen in figure.11 a system based on the integration of all precoding with RCF was suggested. Except in the case of DHT with RCF (I = 2, pilot), where the results of the DHT itself are better than the precoding with RCF combination (DHT with RCF), the results of this process are better than the results of the RCF and precoding each separately. When RCF (I =1) is combined with PAPR, the best result is obtained (DHT). RCF is used with the following specifications (I =1, pilot, and 2, CR =4, 3, 2) and WHT, DCT, DST, and DHT pre-coders. Figure 11. depicts the OFDM device model with the suggested methodology.

Figure 11. The OFDM system model with precoding + RCF.

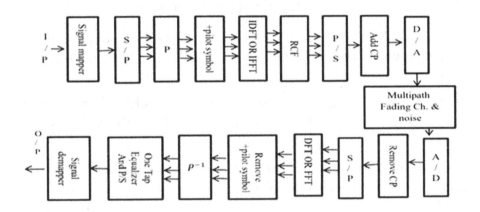

- The PAPR and CCDF of PAPR improved, with the least amount of change occurring when I=1, CR = 4, and WHT. Where I=1, CR = 4, and DHT, the PAPR improvement is (7.1348 dB) and the CCDF of PAPR improvement is (0.0031 dB), while the overwhelming amount of improvement is where I=1, CR = 4, and DHT, the PAPR improvement is (7.1348 dB) and the CCDF of PAPR improvement is (0.0031 dB) (3.0141 dB)
- The SNR at BER(10^{-4}) at I =2, pilot, shows an increase in some areas while deteriorating in others. When I=2, CR=1.5, and DHT=0, the highest amount of change happens (17.1418 dB). When I=2, CR=1.5, and WHT=0, the highest amount of decay happens (-0.4 dB)
- At I =1, the SNR at BER(10^{-4}) was strengthened. When DST and CR = 4 and is equal to, the least amount of change in SNR at BER(10-4) is equal to (0.063 dB). When DHT and CR =1.5 are similar, the most change is shown (16.905 dB).

6. CONCLUSION

The RCF can improve the PAPR and BER at the same time. The worst type of precoding in term of reducing the PAPR and BER is the WHT. The best type precoding in term of reduced PAPR and BER is the DFT DST and DCT precodings give almost the same performance, the DST improves the PAPR more than DCT even a few percent. The results of hybrid pre-coding with RCF is better than the results of the RCF and pre-coding each alone, except in the case of DHT with RCF $(I = 2)$ where the results of the DHT is better As it is clear from the results that the hybrid methods have better results but at the expense of complexity.

REFERENCES

Abouty, S., Renfa, L., Fanzi, Z., & Mangone, F. (2013, February). A Novel Iterative Clipping and Filtering Technique for PAPR Reduction of OFDM Signals: System Using DCT/IDCT Transform. *International Journal of Future Generation Communication and Networking, 6*(1), 1–8.

Armstrong, J. (2001). New OFDM Peak-to-Average Power Reduction Scheme. *IEEE VTS 53rd Vehicular Technology Conference, 1*, 756 – 760. 10.1109/VETECS.2001.944945

Devlin, C. A., Zhu, A., & Brazil, T. J. (2008). Peak to Average Power Ratio Reduction Technique for OFDM Using Pilot Tones and Unused Carriers. *Radio and Wireless Symposium*, 33 – 36. 10.1109/RWS.2008.4463421

El-Tarhuni, Hassan, & Sediq. (2010). A Joint Power Allocation and Adaptive Channel Coding Scheme for Image Transmission over Wireless Channels. *International Journal of Computer Networks & Communications, 12*(3).

Han, S. H., & Lee, J. H. (2005, April). An overview of peak-to-average power ratio reduction techniques for multicarrier transmission. *IEEE Wireless Communications, 12*(2), 56–65. doi:10.1109/MWC.2005.1421929

Hincapie & Sierra. (2012). *Advanced Transmission Techniques in WiMAX*. InTech.

Ishaq, Khan, & Gul. (2012). *Precoding in MIMO, OFDM to reduce PAPR (Peak to Average Power Ratio)* (M.A. thesis). Linnaeus University, Sweden.

Jijina, N., & Pillai, S. S. (2014, April). Linear Precoding Schemes for PAPR Reduction in Mobile WiMAX OFDMA System. *International Journal of Advanced Research in Electrical, Electronics and Instrumentation Engineering, 3*, 8873–8884.

Joshi, H. D. (2012). *Performance augmentation of OFDM system* (Ph.D. dissertation). Jaypee Univ. of engineering and Technology.

Jyothi & Sridevi. (2018). Low power, low area adaptive finite impulse response filter based on memory less distributed arithmetic. *Journal of Computational and Theoretical Nanoscience, 15*(6-7), 2003-2008.

Jyothi, G. N., Debanjan, K., & Anusha, G. (2020). ASIC Implementation of Fixed-Point Iterative, Parallel, and Pipeline CORDIC Algorithm. In *Soft Computing for Problem Solving* (pp. 341–351). Springer. doi:10.1007/978-981-15-0035-0_27

Jyothi, G. N., & Sriadibhatla, S. (2019). Asic implementation of low power, area efficient adaptive fir filter using pipelined da. In *Microelectronics, electromagnetics and telecommunications* (pp. 385–394). Springer. doi:10.1007/978-981-13-1906-8_40

Kumari, B. S. (2014, June). PAPR and Fiber Nonlinearity Mitigation of DFT-Precoded OFDM System. *International Journal of Advanced Research in Electrical, Electronics and Instrumentation Engineering*, *3*, 9801–9808.

Muralibabu, & Sundhararajan. (2013). DCT Based PAPR and ICI Reduction Using Companding Technique in MIMO-OFDM System. *International Journal of Advanced Research in Computer Science and Software Engineering*, *3*, 49-53.

Pavan, B., & Kumar, M. (2013, January-February). Comparitive Study Of PAPR Of DHT-Precoded OFDM System With DFT & WHT. *International Journal of Engineering Research and Applications*, *3*, 527–532.

Prakash, V. R., & Kumaraguru, P. (2012). Performance analysis of OFDM with QPSK using AWGN and Rayleigh Fading Channel. Conference: ICIESP, 10-15.

Prasad. (2004). *OFDM for Wireless Communications Systems*. Artech House Publishers.

Schmidt, B., Ng, C., Yien, P., Harris, C., Saripalle, U., Price, A., Brewer, S., Scelsi, G., Slaviero, R., & Armstrong, J. (2006). Efficient Algorithms for PAPR Reduction in OFDM Transmitters Implemented using Fixed-Point DSPs. *IEEE 63rd Vehicular Technology Conference, 4*, 2023 – 2027. 10.1109/VETECS.2006.1683201

Singh, Gupta, Siingh, Sharma, & Singh. (2013). Hybrid Clipping-Companding Schemes for Peak to Average Power Reduction of OFDM Signal. *International Journal of Advances in Engineering Science and Technology*, *2* (1), 114-119.

Chapter 18
Restricted Boltzmann Machine– Driven Matchmaking Algorithm With Interactive Estimation of Distribution for Websites

Jayashree R.
College of Science and Humanities, SRM Institute of Science and Technology, Kattankulathur, India

Vaithyasubramanian S.
D. G. Vaishnav College, India

ABSTRACT

In this chapter, restricted Boltzmann machine-driven (RBM) algorithm is presented with an enhanced interactive estimation of distribution (IED) method for websites. Indian matrimonial websites are famous intermediates for finding marriage-partners. Matchmaking is one of the most pursued objectives in matrimonial websites. The complex evaluations and full of zip user preferences are the challenges. An interactive evolutionary algorithm with powerful evolutionary strategies is a good choice for matchmaking. Initially, an IED is generated as a probability model for the estimation of a user preference and then two RBM models, one for interested and the other for not-interested, is generated to endow with a set of appropriate matches simultaneously. In the proposed matchmaking method, the RBM model is combined with social group knowledge. Some benchmarks from the matrimonial internet site are pragmatic to empirically reveal the pre-eminence of the anticipated method.

1 INTRODUCTION

In the late 1990s, Indian matrimonial websites become increasingly important in the marriage market and have millions of participants (Fritzi-Marie, 2013). In the early days, uncertain moves are made through Artificial Intelligence (AI) in the matchmaking business and as machine thinking expands traction. The AI and machine learning models are together applied in the recent Matrimonial sites (Sharma, 2018). At

DOI: 10.4018/978-1-7998-6870-5.ch018

present most of the matchmaking sites entitle to practice intellectual matchmaking algorithms that help to pair and suggest the user profiles constructed on a variety of parameters: ones perceptibly filled in by the person and ones collected through the machine learning model. In our daily lives, machine learning algorithms influences all manner: from cooking recipes to the best route to reach to work, which movie to watch or music to play next, what news up on social media, what next to buy in online shopping. Surprisingly, machine learning algorithms also determine who can be the best life partner. Of course, the algorithm is based on the furnished information and may go wrong if the information is false. Use deep learning techniques to measure the appeal of an individual.

India is one of the highest population countries that all we know at the same time it is the only country which has the largest young generation people with less than 25 years old. So, the data-driven procedures are applied to India's matrimonial giants. Later in 2015, The Matrimony.com group has introduced a Matchmaking Algorithm called MIMA (Matrimony.com's Intelligent Matchmaking Algorithm) (Sharma, 2018), which give immediate recommendations on profiles suitable to the user by applying techniques of machine learning and arithmetical rules (Jha, 2015). Subsequently launching MIMA matrimonial websites provide substantial growth in end-user-to-end-user metrics and these metrics have a balanced boost with incessant enhancements to the algorithm. As the result, the algorithm doubles the business and the users are attracted by MIMA recommended profile that significantly increases to "two-digit user finding matches" and it leads to marriage. Finally, the MIMA algorithms improve the business with great customer satisfaction.

Sometimes the end-user leaves certain information unfilled that time ML techniques are used to include those inferential data on a profile. That possibly related to the income of the person, for example, whether the person is well travelled or not. The thing to notice here is income level is important criteria in matchmaking but it cannot be asked directly and set as a mandatory input detail. Artificial Intelligence and machine learning can stumble at times. Matrimonial fraud is a serious problem; a fraudster may use matrimonial sites to ploy and cheat several individuals. The 'Trust Badge' system, verified social media profile or government ID, along with the method of monitor the person's social network activities may work best (Sharma, 2018).

India's first matrimonial website powered with AI is Betterhalf.ai (Rai, 2019). On this website, matchmaking is done through a personality test at the time of user registration using various types of questions and options. A LinkedIn account is mandatory to register and users' official email also verified along with their profile to eliminate fraudulence. A set of 16 questions helps to scale the user preferences and provide analysis of 6 dimensions of a user's personality, such as intellectual, emotional, social relationship, physical and economic values. Update the forecasting scores of personalities depending on the privately fetched opinion about a person acts from the social network, which further refined using gamified questions on personality. A rating scale called a Likert Scale which uses a point scale as a replacement for imprecise questions or simple yes/no responses. In general, 5 or 7 measuring points used in this scale to identify exactly what a customer or visitor feels about a profile or website.

For example, Figure 1 shows a sample end-user survey with a Likert Scale (7 measuring points). The customers can choose answers that range from one extreme such as "extremely likely" or "strongly agree" to another such as "extremely unlikely" or "strongly disagree". As shown in Figure 1, a neutral midpoint also included in this scale along with the lines of "neither agrees nor disagree". Bayesian approximation method is used in the existing system as a recursive probabilistic model (Minka et al., 2018; Rai, 2019). This model estimate and study about personality correctness over time. The prior probability is the initial probability which is assigned with the analysis of millions of data fetched through Likert-

Figure 1 .Customer survey model

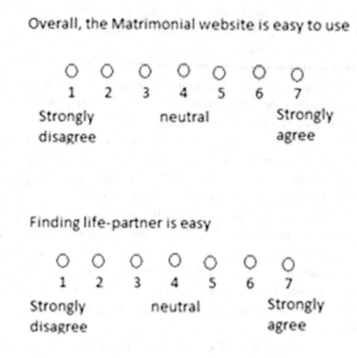

type questions to obtain the robustness of every answer. Estimate initial probabilistic personality value based on these probabilities. More about the user personality is monitored through social network and private feedback from others. This helps to update the person's probability score (posterior probability) and personality score (Sharma, 2018). Limitations in the existing model are:

(1) Explicitly or implicitly a person must provide initial interaction for building a framework for matchmaking. This certainly conflicts with intending to improve person exhaustion. An unsupervised learning-based model for matchmaking is more helpful in this problem.

(2) The historical interaction of a person in the social network reveals that person's behaviour which is the inherent preference features of him/her. However, this is not well developed and used in the existing model.

In this paper, an archetypal is recommended to estimate the person's preferences and applied aptness matchmaking in the successive evolution process. Manage or update the model whenever the person identifies that the matching is far-off from his/her likings. The proposed model presents a 2RBM according to the clustered past data for exactly fetching the person fondness aspects. After training 2RBM, construct the EDA probability model based on the Likert-type RBM critical features, which enables to get feedback from peer-users and communal network users interactively. Ultimately, obtain the matchmaking of the given person's energy-function-weightage (EFW) of the interested and not-interested models along with communal network knowledge in the RBM. Apply in dynamic EDA the probability and matchmaking models to successfully discover the contented top N matches for the active individual. Satisfactory researches on matrimonial datasets prove the efficiency of the anticipated algorithm. The goal is not only to boost the matchmaking performance but also improve website contributors' assessment weights.

The rest of the Chapter is systematized as follows,

- Section 2: Briefs on the representations of the literature survey and related introductory research.
- Section 3: Describes Bayesian Deep Learning and Restricted Boltzmann Machines.
- Section 4: The proposed 2RBM archetypal with IEDA is explained in detail.
- Section 5: explains the proportional experiments and equivalent study.
- Section 6: Finally, the conclusion.

2 MOTIVATION AND BACKGROUND

Arranged marriages in the past happened to be trough matrimonial columns in the newspaper or through the acquaintance of relatives or friends or priests. But then, the people are busy in their day-to-day life and work and lack of time to find their life-partner the traditional way. The blending of old and new responsiveness is preferred by people to find the perfect match according to their parent's choice (Ana-Ramona & Razvanand, 2010; Artis et al., 1999; Derrig & Fraud, 2002; Troester, n.d.). Thus, in time, the Matrimonial system came into the picture. An alternative to strike a comparison between ancient social tradition and the contemporary attribute by cutting the intermediary of arrange marriage. Through this matrimonial system, it has become uncomplicated for people who never known or met to become life-partner. They may have different cultural any may live diagonally across the world (Abbott et al., 1998; Artis et al., 1999).

Earlier in 2018, the prediction of facial beauty studied by researchers from China (Liang, Lin, Jin et al, 2018). Tinder's method performs matchmaking based on their attractiveness which is based on criteria such as personal interests, user attribute, searching criteria and so on or depending on automated decision-making (Jose, 2021). Swipe helper identifies several users' interest in a person's profile by studying how long or how many times they swipe on that person's profile (Murdoch, 2017). The matchmaking is perilous for developing a reliable IEC because this includes a collection of data, selection of the model and finally training the archetypal (Pan et al., 2018; Tian, Tan, Zeng et al, 2018; Tian, Yang, Zhang et al, 2018). Semi-supervised learning presented by Sun et al. (Sun et al., 2012) with a surrogate in which the genetic algorithms with interactive dataset are tough to collect and handle due to their complicated design. Pan et al. (Pan et al., 2018) proposed an enhanced classification model with interactive decisions for arithmetically defined optimization using multi-objective EA. Akinsolu et al. (Akinsolu et al., 2019) proposed an algorithm with proxy-based EAs by connecting parallel computing. Chen et al. (Chen et al., 2017) presented a recommendation for laptop search using Estimation of Distribution Algorithm (IEDA) with Bayesian, in which a proxy model is constructed for the customer's interactive period. Tian et al. (Tian, Sun, Tan, & Zeng, 2020) solved high-cost optimization problems quantitatively or qualitatively with less computation cost for a proxy assisted EAs model. The process for constructing or managing a proxy-based archetypal with semi or fully supervised knowledge is a continuous attempt to approximate the person aptness to accomplish progressive operators. There are many personalized recommenda-tions archetypal to excerpt the essential features of a user's fondness with cherished references (Lu et al., 2015; Mao et al., 2019), Such as, Bayesian approximation model (Rendle et al., 2009; Wang et al., 2019), Factorization Machine (Rendle, 2012; Xiao et al., 2017), Multilayer Perceptron (MLP) (Alfarhood & Cheng, 2018; Cheng et al., 2016), RBM (Bao et al., 2018; Salakhutdinov et al., 2007; Shibata & Takama, 2017), Auto encoder model (Suzuki & Ozaki, 2017), Convolutional Neural Network (Kim et al., 2016; Xu et al., 2019).

Rendle et al. (Rendle et al., 2009) work explain a Bayesian approximation model for ranking recommendation with maximum posterior estimation where the training data are clustered and classified as positive and negative evidence. Cheng et al. (Cheng et al., 2016) suggested deep and wide learning-based archetypal by collectively training data with linear DNN models to gain memorization and simplification for personalization. Kim et al. (Kim et al., 2016) offered Convolutional Matrix Factorization (ConvMF), which is a novel context-aware endorsement model to utilize the positive and negative user favourites completely. This model enhances prediction accuracy by combining probabilistic matrix factorization with CNN. Zhou et al. (Zhou et al., 2018) estimated a user behaviour-based framework that successfully joins all of a person's historical collaborative activities. However, the IEC process is not combined with all these prototypes and further personalized evolutionary search needs effective enhancement. Proxy assisted IECs are analogous to EAs.

Match-making is the task of finding a set of people for a person with suitable expected matching qualities (changeable) or requirements. Therefore, this is an optimization problem that can be solved effectively using Evolutionary algorithms (EAs), assuming that the person's preferences or expectations are articulated accurately in mathematical models. In reality, even without considering the changeable scenarios, such an assumption is tough to be satisfied all though a person expresses his/her expectations very clearly.

Artis et al. (Artis et al., 1999) presented RBM-based Estimation of Distribution Algorithm for intricate mathematical problems. For the optimized intricate problem, the RBM's energy function is utilized to estimate the individual aptness (Bao et al., 2020).

3 AN OVERVIEW OF BAYESIAN DEEP LEARNING AND RESTRICTED BOLTZMANN MACHINES

Bayesian inference aims to govern the posterior distribution, the distribution of archetypal parameters θ that adapt to the data. Assuming $\mathbf{d}_i$ as the dependent variable, I_i as the independent variable, and ε_i as the measurement noise

$$I_i = F\,(\mathbf{d}_i, \theta) + \varepsilon_i. \text{ ------} \tag{1}$$

The posterior distribution, $P\,(\theta | \{\mathbf{d}_i\}, \{I_i\})$, in Bayes theorem is related to the likelihood. The estimation for a strange input is found by an average of the posterior. An ambiguity of archetypal forecasts is a voluntary derivative from Bayesian formalism. The Bayesian approximation is informal to state but pricey. The approximation of the posterior is also usually computationally expensive.

The Bayesian approach not only predicts or classifies but also helps to calculate uncertainty. The uncertainty is important in safety-critical applications (for example self-driving cars, medical diagnosis, and military applications). For a variational interpretation with an estimated variational distribution $q_{v(I)}$, assume y˜ as prognostic distribution of novel output for the given novel input x˜, uncertainty of prediction/classification is the variance network output for a given input as $q_v\,(\text{y˜} | \text{x˜, I})$.

$$\text{Var}(q_v\,(y\tilde{\ }|x\tilde{\ },I)) = \int (q_v\,(y\tilde{\ }|x\tilde{\ },\,I)\,-I_{out})^2\,q_v\,(I\,|y\tilde{\ },x\tilde{\ })dI.\ \text{------} \tag{2}$$

The main challenges of the Bayesian approach are the calculation of denominator in the formulae like $P(w|y,x)=\dfrac{p\big(y|x,w\big)p\big(w\big)}{\int p\big(y|x,w\big)p\big(w\big)dw}$., for continuous and, $P(w|y,x)=\dfrac{p\big(y|x,w\big)p\big(w\big)}{\sum_w p\big(y|x,w\big)p\big(w\big)}$., for discrete case. Bayesian deep learning through randomization in training enable dropout and feed the network N number of times with data and collected the outputs BigO(n), where n=1,2, ..., N and output variance OV is, $OV= \dfrac{1}{n}\sum_n \big(\text{BigO}\big(n\big)-\text{BigO`}\big(n\big)\big)^2$ . ------ (3) Where $\text{BigO`}\big(n\big)=\dfrac{1}{n}\sum_n\big(\text{BigO}\big(n\big)\big)$ .Bayesian inference approximation is equivalent to neural networks dropout, a way to regularize a neural network and limit over-fitting (Science Daily, 2020). To minimize the loss, the last layer of the neural network (NN) is divided to give two outputs: the forecasted ambiguity Ω_i^2 and reliant on variable y_i. However, during the training period, each unit has a likelihood p of being set to 0. Finding parameters $\theta = (\Theta_1, \Theta_2 \cdots \Theta_x)$ for a NN with x items is approximately equal to the posterior distribution $P(\theta|\{\mathbf{d}_i\}, \{I_i\})$, where Θ_x is the limitation vector related with the xth unit.

The dropout as a Bayesian Approximation proposes, a simple approach to quantify the neural network uncertainty (McBride, 2020). It employs dropout during both training and testing. Random neuron dropping during training only reduces the network generalization error. Neural Network with dropout enabled during testing multiple feed-forwards, for the same input, generate multiple outputs (McBride, 2020). Neurons randomly dropped only during training as depicted in Figure 2 (McBride, 2020).

Figure 2. Neural Network with dropout.

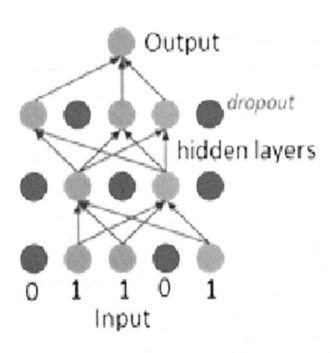

Frequentist inference, ConvMF, is likelihood dependent for observed / unobserved data. Thus, prior knowledge is required (Jeremy & Jonathan, n.d.). The ConvMF measures p-values and confidence intervals and dominate in life science research area then Bayesian method.

3.1 Restricted Boltzmann Machines

Markov random model is an undirected graph with connections between the variables, while Bayesian networks have directed connections. Boltzmann machines are a kind of Markov random model, a graphical model. The main advantage of undirected connections over the directed connection is that when there are many hidden variables it is easy to infer the hidden states associated with a set of visible states. Restricted Boltzmann Machines (RBMs), are two-layer generative neural networks that absorb a likelihood distribution over the inputs. They are a distinct class of Boltzmann Machine in that they have a constrained number of networks between visible and hidden units. Every node in the visible layer is related to every node in the hidden layer, but no nodes in the same cluster are associated. RBMs are usually trained using the contrastive divergence learning procedure.

A Boltzmann machine defines a likelihood distribution over N random variables r1, r2,…,rN such that $P(r_1, r_2,…,r_N) \propto \exp(\sum_i \theta_i r_i + \sum_{i,j} w_{ij} r_i r_j) P(r_1,r_2,…,r_N) \propto \exp(\sum_i \theta_i r_i + \sum_{i,j} w_{ij} r_i r_j)$ ------(4)

i. e. the log-likelihood is linearly dependent on both individual variables riri in addition to pair wise terms. Divide the random variables into two groups: visible variables {v1,v2,…,vNvv1,v2,…,vNv}and hidden variables {h1,h2,…,hNhh1,h2,…,hNh}.

The restricted Boltzmann machine defines the following likelihood distribution: $P(v1,v2,…,vNv,h1,h2,…,hNh) \propto \exp(\sum_i \theta_i vivi + \sum_i \theta_i hihi + \sum_{i,j} wijhivj)$ ------(5)

The restricted Boltzmann machine (RBM) confines pair wise terms to those between a hidden variable and a visible variable. An undirected graphical archetypal with a visible and a hidden layer is known as

Figure 3. Recurrent NN with social network feedback

Attribute	Cluster 0	Cluster 1	Cluster 2	Cluster 3	Explanation (DM = doesn't matter, NW=not working)
age1	32.222	30.949	30.363	30.25	Age (years)
gender	0.903	0.872	0.803	0.698	Female: 0, Male: 1
income	98159.722	19179.311	6806.039	3101.055	Annual income (US dollars)
height	68.472	67.7	66.939	65.792	Height (inches)
kgs	67.833	66.699	64.096	61.111	Weight (kg)
kgs2	52.597	53.015	53.169	53.226	Weight of desired partner (kg)
eating1	3.028	2.932	2.889	2.895	DM:0, Jain Diet: 1, Vegetarian: 2, Vegeterian with Eggs: 3, Non Vegetarian:4
status1	2.153	1.691	1.433	1.271	DM: 0, Middle Class: 1, Upper Middle Class : 2, High Class: 3, Rich/Affluent:4
marital_status1	1.208	1.075	1.092	1.102	DM: 0, Single 1, Divorce: 2, Annulled: 3, Widowed: 4, Married: 5
family_type1	1.382	1.388	1.447	1.499	DM: 0, Nuclear: 1, Joint: 2, Other: 3
manglik1	0.653	0.625	0.575	0.629	No, not a manglik: 0, Yes, am a manglik: 1, Don't Know: 2, DM: 3, Anshik Manglik: 4
horoscope	0.896	0.862	0.844	0.87	Yes, want horoscope matching: 1, No don't want horoscope matching: 0
drinking1	1.264	1.148	1.125	1.093	DM:0, No, don't drink: 1, Drink occasionally: 2, Yes, I drink: 3
smoking1	1.056	1.05	1.07	1.073	DM:0, No, don't smoke: 1, Smoke moderately: 2, Yes, smoke: 3
values1	3.049	3.03	2.871	2.764	Orthodox: 1, Traditional: 2, Moderate: 3, Liberal: 4, Modern: 5, Intl:6, Others: 7
complexion1	2.125	2.24	2.343	2.387	DM:0, Very Fair: 1, Fair: 2, Wheatish: 3, WheatishMedium :4, Dark:5
body1	2.299	2.241	2.093	1.985	DM:0, Slim: 1, Average: 2, Heavy: 3, Athletic: 4
caste_imp	0.701	0.778	0.802	0.827	Partner's caste important: 1, Not important: 0
created_for	0.812	0.694	0.522	0.586	Self:0, Daughter/Son: 1, Sister/Brother: 2, Relative: 3, Friend: 4
employed1	4.382	4.238	4.034	3.88	NW:0, Public Sec: 1, State Govt: 2, Central Govt: 3, Priv Sec:4, MNC:5, Others:6, DM:7
father	5.326	5.223	5.487	5.519	Armed Forces: 1, Business/Entrepreneur: 2, Civil Services: 3, Housewife:4, Passed:5...
mother	4.611	4.594	4.442	4.378	...Retired:6, Service - Govt/ PSU:7, Service - Private: 8, Teacher:9, Not Employee:10
living	0.465	0.699	0.78	0.895	Living alone: 0, Living with Parents: 1
brothers	0.681	0.779	0.921	0.946	Number of brothers
sisters	0.722	0.871	0.934	0.996	Number of sisters
education1	2.507	2.478	2.386	2.087	High school: 0, Some college/diploma :1, Bachelors: 2, Masters: 3, PhD/doctorate: 4

an RBM which is made for random variables that are either discrete or continuous. In this archetypal, there is no constrained dependency within a layer. In current years, RBMs have ascended to reputation due to their suitability with deep-learning. Construct a deep-learning model by considering one RBM's hidden layer (hl) as the second RBM's visible layer (vl). Thus, it is understood clearly that RBMs can encrypt multifaceted structures in data, producing them striking for supervised learning.

A Boltzmann machine is a recurrent network with stochastic neurons. Weights are symmetrical at the equilibrium, the relationships of the neuron outputs can be represented using an undirected graphical model, RBM, as shown in Figure 3.

Figure 4. visible and hidden units of neurons

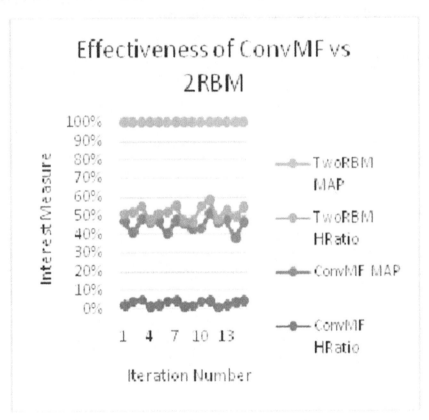

Here, visible and hidden are two groups of neurons as shown in Figure 4. In restricted architecture, there are no connections within the visible group or hidden group. Network parameters are, Bias vector hidden units, hu = [hu1, hu2, ⋯, huH]; Bias vector visible units, vu = [vu1, vu2, ⋯, vuV]; Connection weights, W = {wi,j}. Network values are binary random vectors: v = [v1, v2, ⋯, vv] and h = [h1, h2, ⋯, hH]. In RBM, the energy function is defined as EFun (v, h) =- htWv- vutv - huth where probabilities to (v, h) based on Boltzmann distribution is $_\varrho(\mathbf{v},\mathbf{h}) = \dfrac{\exp\left(-Efun(v,h)\right)}{z}$. ------(6); where z=

$$\sum_{v',h'} exp\left(-Efun(v',h')\right).$$

4 CONSTRUCTION OF 2RBM WITH EDA

The 2RAM model is to fetch the user's preference. The dominant preferences of the given user with progressive scores or elongated browsing period are considered as interested RBM (IRBM) and those sub-standards or most detested by the individual as not-interested RBM (NIRBM). The likelihood archetypal of EDA is built on vl's outputs of the Likert-scale and the matching of novel people will be assessed by EFW of both IRBM and NIRBM models. In matchmaking, the given person X is a combination of his/her behaviour {b1, b2... bn} with n different behaviour. The IRBM and NIRBM built for X based on browsing time on other user's profile, social group activities, feed-back from peer users. Accordingly, the input layer of the IRBM and NIRBM has n visible units. Let hidden layers of 2RBM have m units.

Social network knowledge is taken as a consideration for cluster A and they are the peer users of the social network connected with the X. This is depicted in Figure 5.

Figure 5. Social network cluster

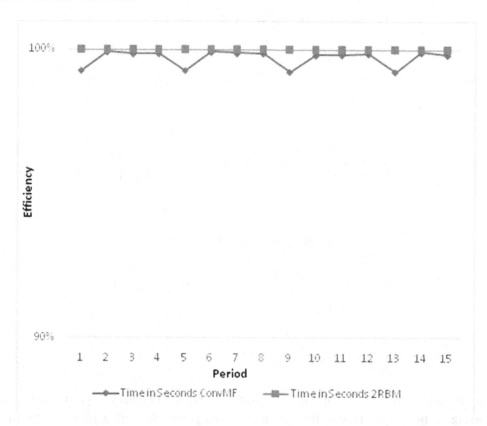

For the given 'hu' of the ith 'vu', the active probabilities are $P_{\theta l}$ and $P_{\theta dl}$ in the IRBM and NIRBM. These active probabilities are evaluated using the equations,

$$P_{\theta l} = \Omega\,(a_{li} + \sum\nolimits_{j=1}^{m} W_{lij} h_{lj});\ v_{li} = 1 | h_l \text{ ------}$$

(7)

$P_{0dl} = \Omega (a_{dli} + \sum_{j=1}^{m} W_{dlij} h_{dlj}); v_{dli} = 1 | h_{dl}$ ------ (8) where (v_{li}, v_{dli}) and (h_{lj}, h_{dlj}) are the positions of the ith 'vu' and the jth 'hu' in the IRBM and NIRBM respectively. The $\theta_l = \{W_l, a_l, b_l\}$ and $\theta_{dl} = \{W_{dl}, a_{dl}, b_{dl}\}$ are the parameters of the IRBM and NIRBM, respectively.

$$\Omega (a) = \frac{1}{\exp(-a)+1}.$$ is the active function.

After training the 2RBM user matching model, the parameters (θ_l, θ_{dl}) have the zeal features of a person and represent an interested and not-interested relationship between that person and his/her behavioural features. Therefore, the 2RBM is a person preference archetypal with a more wide-ranging person's interest feature.

Find matches with similar preferences according to the Pearson correlation coefficient. Calculate Pearson similarity $Psim(p_i, p_j)$ between user u_i and u_j as,

$$Psim(p_i, p_j) = \frac{\sum_{k \in Ipi,pj} \left(Qpik - Q'pi\right)\left(Qpjk - Q'pj\right)}{\sum_{k \in Ipi,pj} \left(Qpik - Q'pi\right)2 X \sum_{k \in Ipi,pj} \left(Qpjk - Q'pj\right)2}$$ ------ (9)

Where lp_i, p_j represents IRBM / NIRBM preferred by users u_i and u_j, Q_{uik} and Q_{ujk} are liked behaviour k by ui and uj, and Q'_{uik} and Q'_{ujk} are the overall preferences of users u_i and u_j.

From Eq.(9), n numbers of similar Likert-scale users are selected. The matching function f(p) is represented in Eq.(10)to estimate the personal matching by incorporating the trained 2RBM along with the social information.

$$f(x) = \Omega(\alpha \times social(p) + preference(p))$$ (10)

where $social(p) = \sum_{j=1}^{c} Psim(u, u_j) \times Qu_j x$, the average of matching-score weight of n similar persons on interested/not-interested behaviour x and Ω is the contributions of the social group which is a normalized function. The preference (p) denotes the forecasted matching-score on p. In detail, the preference (p) is governed by the EFW variance between IRBM and NIRBM. The preference of the IRBM is more important and is contributed more in the calculation.

Algorithm: Matchmaking using 2RBM

Input: The entire interested and non-interested are screened out to form the social, dominant and inferior groups of the current person.
Output: Top N user list

While (Termination conditions) do
    Use Eq. (7) and (8) to construct a2RBM user preference model
    Comprehensively extract the preference features of the user.
    Construct p(x), an EDA probability model based on 2RBM

Generate new individuals by randomly sampling with P (x), and they are matched in the searching space to generate feasible solutions.

Design Matchmaking model from RBM with Eq.(10)

Generate Top N List of N outstanding individuals with higher estimated matching fitness

Submit the Top N list to the given person for interactive evaluations.

Evaluate Matchmaking.

If (average accuracy < the threshold) then

update the models

end while

5 RESULTS AND DISCUSSION

Shaadi.com is one of the leading Matrimonial websites containing approximately 325000 profiles, targets the educated urban and middle-class family for the matrimonial site which refers to India's world largest metropolitan cities in India. Aggregate data from the Shaadi.com and Jeevansathi dataset. Survey result on this website provides the following details; the total number of males is 76.42% and females numbering are having 23.58%. Data analysis is done on the AIPython notebooks and Pandas data analysis is a library and numpy frameworks ().

5. 1 Dataset Attributes and Their Meaning

The dataset has 26 attributes, each explained in Table 1, including the numerical mapping made to the categorical variables

5.2 Centroid Table

The centroid table, Figure 6 (Harvard University, n.d.), shows the average values of each attribute across clusters. Pink rows increase from cluster 3 to cluster 0, green rows decrease from cluster 3 to cluster 0 and blue cells indicate peaks/maxima. White rows show no overall trend.

Each row in the above table represented by n visible units, V = {v1, v2... vn}. The social group information and feedback of other users are given by m hidden inputs.

Table 1 contains the overall attributes collected from the dataset as for experimental attributes of the proposed work. In the experimental datasets, the preferences of each behaviour are classified as interested and not-interested depending on the preference of the behaviour of the given person. The persons are chosen in a random way from the datasets to carry out the experiments. In orderly manner training and testing, datasets are evaluated. That is, an initial 80% of all preferences are evaluated, and then the remaining 20% training datasets are tested.

The dataset is collected from matrimonial Shaadi.com and BharathMatrimonial.com websites. The encrypting of the preferences inputted into the 2RBM, a behaviour measure model is accessible as follows:

In the sample data, there are 26 preferences keys. The interest measure of a person 'x' in the DEDA is calculated through preference key values as $x = [x_1, x_2, .., x_{26}]$.

Table 1. Matrimonial websites general attributes

Attribute	Description
age	Age of suitor
gender	female':0,'male':1
height	Height in inches
kgs	Weight in kgs
bmi	BMI calculated based on height and weight
eating	'Doesn't Matter':0, 'Jain': 1, 'Vegetarian': 2, 'Vegetarian With Eggs': 3, 'Non Vegetarian':4
family_type	Doesn't Matter': 0, 'Others':3, 'Nuclear': 1, 'Joint family both parents': 2, 'Joint family only mother':2, 'Joint family father mother and brothers sisters':2, 'Joint family single parent brothers and or sisters':2, 'Joint family only father': 2
status	'Doesn't Matter': 0, 'Middle Class': 1, 'Upper Middle Class': 2, 'High Class': 3, 'Rich / Affluent':4
marital status	'Doesn't Matter': 0, 'Never Married': 1, 'Divorced': 2, 'Annulled': 3, 'Widowed': 4, 'Married': 5
manglik	'No': 0, 'Yes': 1, 'Do Not Know': 2
Horoscope	Believe in horoscope matching ('No': 0, 'Yes': 1)
drinking	'Doesn't Matter':0, 'No': 1, 'Occasionally': 2, 'Yes': 3
smoking	'Doesn't Matter': 0, 'No': 1, 'Moderate': 2, 'Yes': 3
values	'Orthodox': 1, 'Traditional': 2, 'Moderate': 3, 'Liberal': 4, 'Modern': 5, 'International':6, 'Others': 7
caste_important	'No': 0, 'Yes': 1
complexion	'Very Fair ': 1, 'Very Fair': 1, 'Fair ': 2, 'Fair': 2, 'Whetish Medium ': 4, 'Whetish Medium': 4, 'Wheatish': 3, 'Wheatish ':3, 'Dark ':5, 'Dark':5
body	'Slim': 1, 'Average': 2, 'Heavy': 3, 'Athletic': 4
education	High School':0, 'Some college':1,'Undergrad':2, 'Grad':3, 'Doctorate':4
employed	'Not Working':0, 'Public Sector': 1, 'State Government': 2, 'Central Government': 3, 'Private Sector':4, 'MNC':5, 'Others':6, "Doesn't Matter": 7
living	Living with parents ('No': 0, 'Yes': 1) -** city**('International': 1, 'Mumbai': 2, 'Delhi':3, 'Kolkata':4,'Bengaluru':5, 'Chennai':6, 'Hyderabad':7, 'Pune':8, 'Ahmedabad':9,'Surat':10, 'Vishakapatnam':11, 'Others':12)
income	Annual income in dollars
created_for	Self:0, Daughter/Son:1, Sister/Brother:2, Relative:3, Friend:4
father	Armed Forces:1, Business/Entreperneur:2, Civil Services:3, Housewife:4, Passed:5, Retired:6, Service-Govt./PSU:7, Service-Private:8, Teacher:9, Not Employed:10
mother	Armed Forces:1, Business/Entreperneur:2, Civil Services:3, Housewife:4, Passed:5, Retired:6, Service-Govt./PSU:7, Service-Private:8, Teacher:9, Not Employed:10
brothers	Number of brothers
sisters	Number of sisters

$$|x_i| = \begin{cases} 1, if \ x = preference_i \\ 0, otherwise \end{cases}.$$ (11)

From the dataset, the 'vl' of the interest measure archetypal has 26 units. Besides, the higher preference key values are selected as the interested group while the less or low preference key as the non-interested group. These groups are used to train the 2RBM model. The trained 2RBM preferences model can represent the user's interest in a particular person. The 2RBM model demarcated in Eq.10 is used in the evaluation of the interest measure in the test-set.

Calculate the following terms based on the classification for matchmaking for the test dataset and archetypal,

tp_{mm}(True-Positive): the percentage of interested that are properly categorized as interested by the model.

fp_{mm}(False-Positive): the percentage of negative, not-interested, that are mislabelled as interested by the model.

tn_{mm}(True-Negative): the percentage of negative (not-interested) that are correctly labelled as not interested.

fn_{mm}(False-Negative): the percentage of interested present that are mislabeled as not-interested by the model.

In this work, 25 independent executions are achieved for each limitation setting in the 2RBM algorithm. The precision measures the accuracy of the predictions and the recall measures efficient in finding all the interested. A common metric in object accuracy measuring is Average precision (AP). The AP calculates the average correctness value for recall value from 0 to 1. For matchmaking interest measures data precision and recall are calculated using the formulae,

Precision = tp_{mm}/ (Total interested)
Recall = tp_{mm} / (Total interested)

Figure 6. Centroid Table with average values of each attribute across clusters

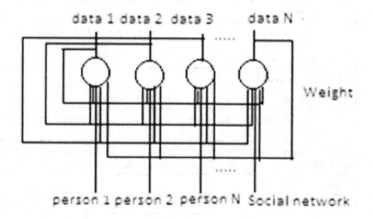

The very high correctness value for some given recall level is called Interpolated Precision($\hat{I}$). Precisely, Interpolated Precision is the maximum correctness for any recall $\geq \hat{r}$. Overlapping of correctness boundary and the actual truth value is measured using intersect-over-boundary (ϕ). In matrimonial datasets, the predefined ϕ threshold is 0.5. Average Precision finds the model accuracy for 26 attributes,

$$AP = \int_0^1 \hat{I}(r)\,dr := \frac{1}{26}\sum_{i=1}^{26} \frac{i}{\text{preference}(i)}. \qquad (12)$$

Where preference(i) is the like of the i^{th}character(category). Calculate Mean Average Precision (MAP) by taking the total of the $\hat{I}$ at N different recall levels. The AP is to confirm the correctness of the severity measure model. The MAP (Zhou et al., 2018) is the average AP of 'n' persons in the test dataset and stated as follows:

$$MAP = Avg\left(\sum_{k=0}^{n} AP.\right) \qquad (13)$$

where AP_k is the AP of the kth person in the dataset. The algorithm contentment is reflected through MAP. Thus, a higher MAP value indicates higher algorithm contentment. An algorithm's effectiveness and reliability are measured through Hit Ratio (H_{Ratio}) (Tian, Sun, Tan, & Zeng, 2020). In H_{Ratio}performance is measured in quantity. Higher H_{Ratio} specifies improved effectiveness and reliability of the model. The H_{Ratio} is calculated by dividing hits by the number of accesses.

The number of accesses = hits + misses

H_{Ratio} = hits / (hits + misses)

The MAP and H_{Ratio} are evaluated to identity the effects of dissimilar HU in the 2RBM archetypal. The number of HU is associated to the group aspects that are finalized as the archetypal input.

Figure 7. Effectiveness of ConvMF and 2RBM

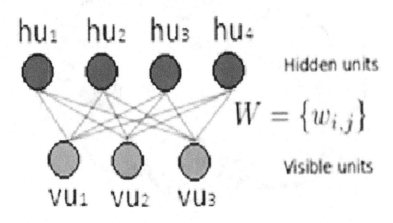

The ConvMF (UCSD-AI4H 2020, 2020) is a classification algorithm based on CNN by integrating the interest and not-interest results. Also, the 2RBM algorithm with the communal data is compared in the experiments with average values of H_{Ratio} and MAP. Independently both the algorithms are executed on experimental datasets 50 times, and the average outputs are represented. The ConvMF and 2RBM are studied on the metrics H_{Ratio} and MAP as shown in Figure 7. The representation of execution time enables us to perform computational cost comparison. The time is taken for execution in ConvMF and 2RBMis shown in Figure 8. A fast algorithm with the highest H_{Ratio} and MAP is an efficient model. Figure 6 and Figure 7 shows clearly the efficiency of the proposed model. An efficient model identifies suitable TopN matches and also in identifying fraudulent profile, which is very important in sensitive website like matrimonial.

5.3 Interactive-EDA

The 2RBM model which is supported by Interactive-EDA is a rough distribution in the initial stage of development since social network user's activities are changeable and diverse. A person's profile and preferences need not to be same and stable. Enabling the connections among various social networks the user's activities and their close contacts in are trackable. Updating the information about the persons on a timely basis is the effective evolution method to conceivably attain a precise and reliable 2RBM model. Given the assessing time, a person interested in a peer-user profile is organized as per their preferences in the dataset. The HRatio and AP are used as evaluation metrics, which is based on RBM along with interactive data. The average experimental outputs are calculated and show a better performance of 2RBM in Figure 8. The significance of the proposed 2RBM model is measured using the Mann Whitney U test with a result p-value = 0.601507 for a 95% confidence interval.

Figure 8. ConvMF and 2RBM Time in seconds for 15 iteration

6 CONCLUSION

Finding a suitable life-partner is very important as per Indian culture which is probable in the current environment with the help of matrimonial websites. A person's preferences in this scenario are equally important as monitoring website users' activities and identifying their behaviour which is possible only through social network knowledge. Fraudulent in matrimonial websites results in a very severe impact on society. The 2RBM, are two-layer generative NN that absorb a likelihood distribution over a person's preferences and EDA is constructed on the output of the visible layer of the Likert-scale. This enables to get feedback from peer-users and communal network users interactively. In this paper, the existing matchmaking method is enhanced using social group knowledge with RBM and EDA, since user preferences are not constant. This model provides top N users with similar preferences and the inferences of the matchmaking algorithm are trusted through social network knowledge in the identification of suitable life-partner. Experiments on matrimonial datasets demonstrate the efficiency of the proposed archetypal 2RBM with communal network data and compared with ConvMF (Kim et al., 2016) which is an endorsement model in the classification of user favorites without including social network information. Figure 7 and Figure 8 shows the efficacy of the proposed model and proves that the proposed method is much faster in identifying suitable life-partner and avoiding fraudulent persons than the existing method.

REFERENCES

Abbott, Matkovsky, & Elder. (1998). An Evaluation of High-End Data Mining Tools for Fraud Detection. *Proc. of IEEESMC98.*

Akinsolu, M. O., Liu, B., Grout, V., Lazaridis, P. I., Mognaschi, M. E., & Di Barba, P. (2019). A parallel surrogate model assisted evolutionary algorithm for electromagnetic design optimization. *IEEE Trans. Emerg. Top. Comput. Intell.*, 3(2), 93–105. doi:10.1109/TETCI.2018.2864747

Alfarhood, M., & Cheng, J. (2018). Deephcf: A deep learning based hybrid collaborative filtering approach for recommendation systems. In *2018 17th IEEE International Conference on Machine Learning and Applications (ICMLA)*. IEEE. 10.1109/ICMLA.2018.00021

Amitabh, V. D. (n.d.). *From Arranged to Online: A Study of Courtship Culture in India, Book: TOD16-Online-Courtship, Digital Publishing Toolkit, Indian marriage.* Github. Available at: https://github.com/DigitalPublishingToolkit/TOD16-Online-Courtship/blob/master/md/11_Dwivedi.md

Ana-Ramona, B. B., & Razvanand, F. A. (2010). Big Data and Specific Analysis Methods for Insurance Fraud Detection. *Database Systems Journal,* 1(1), 30-39. http://www.dbjournal.ro/archive/14/14_4.pdf

Artis, M., Ayuso, M., & Guillén, M. (1999). Modelling Different Types of Automobile Insurance Fraud Behaviour in the Spanish Market. *Insurance, Mathematics & Economics*, 24(1-2), 67–81. doi:10.1016/S0167-6687(98)00038-9

Bao, L., Sun, X., Chen, Y., Gong, D., & Zhang, Y. (2020). Restricted Boltzmann Machine-driven Interactive Estimation of Distribution Algorithm for personalized search, ScienceDirect. *Knowledge-Based Systems*, 200, 106030. doi:10.1016/j.knosys.2020.106030

Bao, L., Sun, X., Chen, Y., Man, G., & Shao, H. (2018). Restricted boltzmann machine assisted estimation of distribution algorithm for complex problems. *Complexity, 2018*, 2018. doi:10.1155/2018/2609014

Chen, Y., Sun, X., Gong, D., Zhang, Y., Choi, J., & Klasky, S. (2017). Personalized search inspired fast interactive estimation of distribution algorithm and its application. *IEEE Transactions on Evolutionary Computation, 21*(4), 588–600. doi:10.1109/TEVC.2017.2657787

Cheng, H.-T., Koc, L., Harmsen, J., Shaked, T., Chandra, T., Aradhye, H., Anderson, G., Corrado, G., Chai, W., & Ispir, M. (2016). Wide & deep learning for recommender systems. In *Proceedings of the 1st Workshop on Deep Learning for Recommender Systems*. ACM. 10.1145/2988450.2988454

Data.world. (n.d.). *Marriage - dataset by fivethirtyeight*. Available at: https://data.world/fivethirtyeight/marriage

Derrig, R. A., & Fraud, I. (2002). Insurance Fraud. *The Journal of Risk and Insurance, 69*(3), 271–287. doi:10.1111/1539-6975.00026

Fritzi-Marie, T. (2013). Changing Patterns of Matchmaking: The Indian Online Matrimonial Market. *Asian Journal of Women's Studies, 19*(4), 64–94. doi:10.1080/12259276.2013.11666166

Harvard University. (n.d.). *Indian Online Matrimony Data Exploration*. Available at: https://projects.iq.harvard.edu/matrimony_data_exploration/home

Jeremy & Jonathan. (n.d.). Comparison of frequentist and Bayesian inference. Class 20, 18.05. In *Reading 20: Comparison of frequentist and Bayesian Inference*. Mit.edu.

Jha, S. (2015). Matrimony.com develops recommendation engine to simplify matchmaking. *The Economic Times*. https://cio.economictimes.indiatimes.com/news/case-studies/matrimony-com-develops-recommendation-engine-to-simplify-matchmaking/46581684

Jose, J. (2021). *From Indian Matchmaking to ArrangedMarriages.co — a New World of Relationships Emerges Online*. NDTV, Gadgets. Available at: https://gadgets.ndtv.com/internet/features/arrangedmarriages-co-arranged-marriages-dating-apps-netflix-indian-matchmaking-tinder-okcupid-2346576

Kaggle. (n.d.). Available at: https://www.kaggle.com/c/c001hw1/data

Kim, D., Park, C., Oh, J., Lee, S., & Yu, H. (2016). Convolutional matrix factorization for document context-aware recommendation. *Proceedings of the 10th ACM Conference on Recommender Systems*, 233–240. 10.1145/2959100.2959165

Liang, D., Krishnan, R. G., Hoffman, M. D., & Jebara, T. (2018). Variational autoencoders for collaborative filtering. In *The 2018 World Wide Web Conference*. ACM. 10.1145/3178876.3186150

Liang, L., Lin, L., Jin, L., Xie, D., & Li, M. (2018). SCUT-FBP5500: A Diverse Benchmark Dataset for Multi-Paradigm Facial Beauty PredictionSCUT-FBP5500, Computer Science. *24th International Conference on Pattern Recognition (ICPR)*. 10.1109/ICPR.2018.8546038

Lu, J., Wu, D., Mao, M., Wang, W., & Zhang, G. (2015). Recommender system application development: A survey. *Decision Support Systems, 74*, 12–32. doi:10.1016/j.dss.2015.03.008

Mao, M., Lu, J., Han, J., & Zhang, G. (2019). Multiobjective e-commerce recommendations based on hypergraph ranking. *Inform. Sci.*, *471*, 269–287. doi:10.1016/j.ins.2018.07.029

McBride. (2020). *COVID-19 Symptoms: Day-by-Day Chart of Coronavirus Signs*. Heavy.com.

Minka, T., Cleven, R., & Zaykov, Y. (2018). *TrueSkill 2: An improved Bayesian skill rating system*. Available at: https://academic.microsoft.com/paper/2791828207/reference?showAllAuthors=1

Murdoch, C. (2017). Genius figures out how to use a fidget spinner to conquer Tinder. *Mashable India*. https://mashable.com/2017/05/24/fidget-spinner-tinder-trick/#sYjfwVF2H5qB

Pan, L., He, C., Tian, Y., Wang, H., Zhang, X., & Jin, Y. (2018). A classification-based surrogate-assisted evolutionary algorithm for expensive many-objective optimization. *IEEE Transactions on Evolutionary Computation*, *23*(1), 74–88. doi:10.1109/TEVC.2018.2802784

Rai, S. (2019). Betterhalf: The millennials of India are turning to algorithms for love. *Hindustan Times*. Available at: https://tech.hindustantimes.com/tech/news/betterhalf-ai-the-millennials-of-india-are-turning-to-algorithms-for-love-story-TuytzxIJEItENBFJ9nCIzI.html

Rendle, S. (2012). Factorization machines with libfm. *ACM Transactions on Intelligent Systems and Technology*, *3*(3), 1–22. doi:10.1145/2168752.2168771

Rendle, S., Freudenthaler, C., Gantner, Z., & Schmidt-Thieme, L. (2009). Bpr: bayesian personalized ranking from implicit feedback. *Conference on Uncertainty in Artificial Intelligence*, 452–461.

Salakhutdinov, R., Mnih, A., & Hinton, G. (2007). Restricted boltzmann machines for collaborative filtering. In *Proceedings of the 24th International Conference on Machine Learning, in: ICML '07*. ACM. 10.1145/1273496.1273596

Science Daily. (2020). *Study Confirms 'classic' Symptoms of COVID-19*. www.sciencedaily.com/releases/2020/06/200624100047.htm

Sharma. (2018). Marriages made in heaven with a little help from Bayes. F*actor Daily*.

Shibata, H., & Takama, Y. (2017). Behavior analysis of rbm for estimating latent factor vectors from rating matrix. In *2017 6th International Conference on Informatics, Electronics and Vision, 7th International Symposium in Computational Medical and Health Technology (ICIEV-ISCMHT)*. IEEE. 10.1109/ICIEV.2017.8338568

Sinha, Biswas, Raj, & Misra. (2020). Big Data Analytics On Matrimonial Data Set. *International Journal of Innovative Research in Applied Sciences and Engineering*, *4*(4), 722-731. doi:10.29027/IJIRASE.v4.i4.2020.722-728

Sun, X., Gong, D., & Zhang, W. (2012). Interactive genetic algorithms with large population and semi-supervised learning. *Applied Soft Computing*, *12*(9), 3004–3013. doi:10.1016/j.asoc.2012.04.021

Suzuki, Y., & Ozaki, T. (2017). Stacked denoising auto encoder-based deep collaborative filtering using the change of similarity. *2017 31st International Conference on Advanced Information Networking and Applications Workshops (WAINA)*, 498–502.

Tian, J., Sun, C., Tan, Y., & Zeng, J. (2020). Granularity-based surrogate-assisted particle swarm optimization for high-dimensional expensive optimization. *Knowl.- Based Syst.*, *187*, 104815. doi:10.1016/j.knosys.2019.06.023

Tian, J., Sun, C., Tan, Y., & Zeng, J. C. (2020). Granularity-Based Surrogate-Assisted Particle Swarm Optimization for High-Dimensional Expensive Optimization. *Knowledge-Based Systems*, *187*, 1–15. doi:10.1016/j.knosys.2019.06.023

Tian, J., Tan, Y., Zeng, J., Sun, C., & Jin, Y. (2018). Multiobjective infill criterion driven gaussian process-assisted particle swarm optimization of high-dimensional expensive problems. *IEEE Transactions on Evolutionary Computation*, *23*(3), 459–472. doi:10.1109/TEVC.2018.2869247

Tian, Y., Yang, S., Zhang, L., Duan, F., & Zhang, X. (2018). A surrogate-assisted multiobjective evolutionary algorithm for large-scale task-oriented pattern mining. *IEEE Trans. Emerg. Top. Comput. Intell.*, *3*(2), 106–116. doi:10.1109/TETCI.2018.2872055

Troester. (n.d.). *Big Data Meets Big Data Analytics*. SAS White Paper. https://eric.univ-lyon2.fr/~ricco/cours/slides/sources/big-data-meets-big-data-analytics-105777.pdf

UCSD-AI4H 2020, UCSD-AI4H/COVID-CT. (2020). *GitHub*. Available at: github.com/UCSD-AI4H/COVID-CT

Wang, X., Gao, C., Ding, J., Li, Y., & Jin, D. (2019). Cmbpr: Category-aided multichannel bayesian personalized ranking for short video recommendation. *IEEE Access: Practical Innovations, Open Solutions*, *7*, 48209–48223. doi:10.1109/ACCESS.2019.2907494

Xiao, J., Hao, Y., He, X., Zhang, H., & Chua, T. S. (2017). Attentional factorization machines: Learning the weight of feature interactions via attention networks. *Proceedings of the 26th International Joint Conference on Artificial Intelligence*, 3119–3125. 10.24963/ijcai.2017/435

Xu, C., Zhao, P., Liu, Y., Xu, J., Sheng, V. S. S. S., Cui, Z., Zhou, X., & Xiong, H. (2019). Recurrent convolutional neural network for sequential recommendation. *The World Wide Web Conference*, 3398–3404. 10.1145/3308558.3313408

Zhou, C., Bai, J., Song, J., Liu, X., Zhao, Z., Chen, X., & Gao, J. (2018). ATRank: An Attention-Based User Behavior Modeling Framework for Recommendation. *Association for the Advancement of Artificial Intelligence. Conference: AAAI 2017*.

Section 2
Big Data and Cognitive Technologies: Current Applications and Future Challenges

Section 2 provides insights on learning based on big data and cognitive computing technologies, including cutting edge topics (e.g., machine learning for Industrial IoT systems, deep learning, reinforced learning, decision trees for IoT systems, computational intelligence and cognitive systems, cognitive learning for IoT systems, cognitive-inspired computing systems). Section 2 is organized into 17 chapters.

Chapter 19

Identification and Recognition of Speaker Voice Using a Neural Network–Based Algorithm:
Deep Learning

Neeraja Koppula
MLR Institute of Technology, India

K. Sarada
KLEF, India

Ibrahim Patel
B. V. Raju Institute of Technology, India

R. Aamani
Vignan's Institute of Information Technology, India

K. Saikumar
Mallareddy University, India

ABSTRACT

This chapter explains the speech signal in moving objects depending on the recognition field by retrieving the name of individual voice speech and speaker personality. The adequacy of precisely distinguishing a speaker is centred exclusively on vocal features, as voice contact with machines is getting more pervasive in errands like phone, banking exchanges, and the change of information from discourse data sets. This audit shows the location of text-subordinate speakers, which distinguishes a solitary speaker from a known populace. The highlights are eliminated; the discourse signal is enrolled for six speakers. Extraction of the capacity is accomplished utilizing LPC coefficients, AMDF computation, and DFT. By adding certain highlights as information, the neural organization is prepared. For additional correlation, the attributes are put away in models. The qualities that should be characterized for the speakers were acquired and dissected utilizing back propagation algorithm to a format picture.

DOI: 10.4018/978-1-7998-6870-5.ch019

1 INTRODUCTION

Expanding request, brought about by the advancement of urban areas and developing populace, draws more consideration on power transmission limits. Force Electronics assists with tackling issue by presenting adaptable air conditioning transmission framework (FACTS) gadgets. The general favorable circumstances of FACTS gadgets are notable. In any case, FACTS gadgets can prompt extreme issues in deciding the specific area of the shortcoming when they are remembered for the deficiency zone. Separation transfer's capacity depends on estimated impedance by that they compute flaw area. Introduced FACTS gadgets on TL change the impedance estimated by separation hand-off that thusly prompts wrong flaw zone location. The measure of deviation in impedance relies upon the boundary's settings of FACTS gadget.

The wavelet change is utilized to isolate high-recurrence parts of 3-line flows, and help vector machine (utilizing techniques for multi- class utilization) is utilized for order of issue area into three assurance locales of separation transfer.

Lately, specialists have gotten progressively keen on comprehending this issue. For demonstrating arrangement capacitor in the shortcoming time frame, the greater part of articles utilized model introduced by Goldsworthy Goldsworthy (1987). In Sadeh, Javad (2020) an answer dependent on time conditions of system with thought of a broad method of TL is introduced. In this arrangement, zone and area of deficiency, autonomous of issue obstruction, are evaluated utilizing 2 subroutines. A few articles built up their strategies utilizing estimations from rule segments of primary recurrence. Saha et al(1999) built up a strategy dependent on segments of principle recurrence on 1 side of TL utilizing Goldsworthy method.

In Song, Y. H(1996), a calculation dependent on forward and in reverse waves is introduced to secure arrangement repaid TLs; albeit just one stage to ground deficiency are examined in this manuscript. Information is introduced in Thomas, D (1992) and Xuan,Q.Y (1996) that recommend strategies dependent on neural systems. It is proposed to utilize RBFN as a spiral neural system to order flaw kind and issue area Girgis, Adly A (1998) and Kalman,(1960). The work Megahed(2006),trains 2 unique kinds of neural systems to distinguish deficiency kind and area. Likewise, Dash, P. K (2206) utilizes EDBD calculation to figure deficiency area Fattaheian, Dehkordi(2014). Neural systems have impediments, incorporating their requirement for countless neurons to demonstrate organize structure (2008). Along these lines, various information from the electrical system is required, and furthermore much an ideal opportunity to prepare the system. By and by, these articles utilized a predetermined number of information to test their techniques. The work Samantaray, S. R (2013) suggested a strategy dependent on Kalman channel] that gauges the condition of a powerful framework from a progression of estimations comprisingblunders.

In Imani (2019) a technique utilizing wavelet change is recommended. In Lepping (2018) and Imani, MahmoodHosseini (2019) constrained cases are utilized to evaluate adequacy of suggested techniques. Besides, in Provost, Foster (2003) high examining recurrence of 200 KHz is utilized, that is by all accounts troublesome in down to earth execution. In Megahed(2006),an information digging technique is utilized for 3reasons. The SVM has become a well-knowndevice in force framework Zhang(2007). In this manuscript, 3distinct SVM is prepared to decide deficiency area, nearness or nonattendance of issue and whether shortcoming incorporates ground or not. This reference utilizes tests of terminating point & current, beginning from the shortcoming occurrence till half cycle after it deliberating 50 Hz &400 kV contextual investigations and testing recurrence of 1 KHz in this manuscript, 10 examples of current from every stage is dissected. In this manuscript, deficiency area is limited to when the compensator, and its viability is concentrated uniquely in 200 cases.

In Kumar, D. S (2020) a 2-phase technique is suggested to recognize the deficiency area. In this manuscript, tests of 3 line flows for one cycle after deficiency case are utilized to prepare SVM. Deliberating 400 kV & 50 Hz contextual investigation and examining recurrence of 1 KHz in this manuscript, 80 examples of current from every stage is broke down. In main phase, high frequencies of 3 line flows are isolated utilizing wavelet change; at that point in subsequent stage, these isolated signs have been provided to SVM to arrange the issue area to when compensator. The issue is that method utilized for reproduction of TL isn't legitimate to examine non-base frequencies whereas this manuscript utilized high frequencies to isolate issue areas. This strategy, deliberating utilizing SVM, might just separate issue area to when the compensator. Considering 400 kV & 50 Hz contextual investigation and inspecting recurrence of 1 KHz in this manuscript, 60 examples aredissected.

Specialists in Soumya, N (2020) suggested arbitrary woods strategy Saikumar K (2020) to recognize deficiency area. This information mining technique is established dependent on decisiontree (DT) strategy. In this technique, a few free unprunedDT are created utilizing various subsets of information. Every tree has 1 vote, & many votesdecide the outcome. Tests of current and voltage into equal parts cycle after deficiency case are used for examination. 70% of the examples are utilized for the preparation procedure of arbitrary woodland, and the staying 30% is utilized for the testing procedure. Thinking about a lot of preparing information, great outcomes have beenaccomplished.

Thus, in this paper, the deficiency area is restricted to when the compensator. The goal of this examination is to enhance separation insurance execution, utilizing information mining strategies whereas deliberating FACTS-based TLs. Prior works utilizing fake neural system experience the ill effects of restrictions on many neurons. These restrictions might be overwhelmed by utilizing increasingly effective devices like help vector machines. Moreover, an increasingly exact recreation of TL is led here utilizing PSCAD/EMTDC programming, just as isolating deficiency territories into three zones as indicated by separation transfer settings. In addition, to research the impacts of TCSC1 area on flaw zone discovery of separation hand-off, twospots one out of 50% of line length & other in 75%linelength are deliberated as 2 situations for affirmation of suggested technique V Saikumar, K., Rajesh(2020). The recreations represents a 94.2% of achievement rate in situation 1 and 86.7% in other that are high achievement rates demonstrate that putting TCSC in 50% of line length prompts higher expectation precision. Additionally, these achievement rates have been accomplished utilizing just 15.2% of absolute information as preparationset that represents high capacity of suggested technique Saikumar, K(2020)

2 SUGGESTEDTECHNIQUE

In this investigation, PSCAD programming is utilized for reenactment. The 3 line flows have been recorded in the wake of making an assortment of issue conditions on reproduced TL. These yields from PSCAD have been utilized as contributions to wavelet change so as to extricate low voltages.It is accounted for shown Devaraju (2021) very legitimate wavelet in examining deficiencies in power framework is SNR (db) process. The Obaid, A. J (2020) it is accounted for that from various sorts of dbmain SNR; db2 provides improved outcomes in the recognition of issue area, within the sight of arrangement capacitor in TL. It was additionally revealed that voltages among 100 to 200 Hz are reasonable for best issue zone grouping Abinaya, S,(2017). Consequently, deliberating an examining recurrence of 5 KHz in PSCADD in this examination, 1 stage discrete wavelet change with db2 mother wavelet is completed to acquire low recurrence tests. From that point forward, wavelet change's yields are utilized as contributions to ASVM to make a hybridwork to isolate shortcoming zones Sheshasaayee, D(2017).

3 CASE SURVEYS

3.1 Simulation of Case Surveys in PSCAD

For TCSC reenactment in PSCAD, in [13] is utilized. This method is appeared in Fig. 1. The ability of SC capacitor modifies in extent with terminating point. The possibility of suggested technique is assessed on 400 kV &50 Hz power framework. The two sides of this system are 2 sources that speak to 2 zones that are associated together. The source information at the two sides of TL is appeared in Table I. Recurrence subordinate (stage) model is utilized for itemized reenactment of TL. The TL information is accomplished from a genuine 400 kV single-circuit TL with 250 km length. thisinformation is introduced in Table II. To assess effect of TCSC area on issue of assurance, this component is put in 2 areas, firstin center of line 1 (125 kmof line 1) and second, in 75% of line 1 (187.5 km of line 1) that in extension of this manuscript for comfort of per users are alluded to as situation 1 and situation 2, separately. Three security zones are characterized for separation hand-off, 1st till 80% of principalline,2ndtill50% of subsequent line and 3rd till 25% of 3rd line.

Figure 1. load connection for FACTS devices

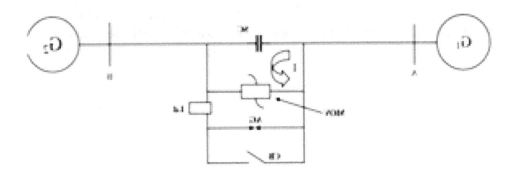

In situation one which TCSC is put in the main line, four issue areas are accepted: in "50 km (1st zone, before TCSC), in 150 km (1st zone, after TCSC), in 250 km (2nd zone), in 325 km (3rd zone)". In situation 2, three shortcoming areas are accepted: in 100 km (1st zone), in 250 km (2nd zone), in 325 km (3rd zone). So as to assess practicality of suggested strategy under different framework condition, several types of flaw, issue obstruction, issue initiation edge(FIA), load edge (ᵢ), generator impedance, remuneration rate and deficiency area, which incorporate 28800 cases forsituation 1 and 21600 cases for situation 2, have been dissected using numerous spat PSCAD. Table III displays thought about cases. 25%, half, and 75% are picked TCSC pay rates for 250 km TL, their proportional incentive for SC capacitor is 120.45, 60.225 & 40.15 µfindividually. Haphazardly picked is cases for preparation procedure of

Table 1. Source level analysis

Parameters	Value
Source Voltage	400 kV
Frequency	50 Hz
Positive sequence impedance	1.31+j15 Ω
Zero sequence impedance	*2.33+j26.6 Ω*

SVMare 3600 cases for situation 1 and 2700 cases for situation 2. These cases are just 12.5% of all-out informational collection and appeared in Table IV. The Gaussian RBF part is utilized for preparing and testing procedure in SVM.

3.2 OAA Classification

The numerous paths isto utilize SVM to categorize information into higher than 2 classes.

1. Model 1 examination: Getting a suitabledestination from ASVM might be attained through met heuristic procedures or trial & error learning [13]. To use "trial& error learning "diverse values for gamma (g) & regularization factor (C)must be assessed on organization of training data to select better ones, then outcome will be utilized to categorize test data. For extent of these factors, outcomes in [13] areexcessive support. The better outcome is attained in21096 & 14.5 for C and g, correspondingly. With these values, 83.7% of achievement is attained in FZD.
2. Scenario 2 examination: In outcome examination of scenario 2, it might be observed that numerous correct FZD was reached with g=38.4 &C=1080000 and. The achievement rate is 69.7% that is lower than scenario 1. In section IV reasons of this less achievement rate is described.

Table 2. Tldata

Data type	value	Data type	value
Arrangement of wires	Horizontal	Vertical distance of conductors	15.45 m
Line length	250 km	SAG for all conductors	14 m
Number of bundle	2	Number of ground wires	2
Bundle spacing	45 cm	Height of ground wires above the lowest conductor	9.36 m
Conductor DC resistance	0.0553 ohm/km	Ground wire DC resistance	1.463 ohm/km
Conductor GMR	1.2161 cm	Ground wire radius	0.2445 cm
Height of conductors	41.46 m	Spacing between ground wires	18.70 m

Table 3. Deliberated cases for system examination

Case number	Number of fault cases for scenario 1	Number of fault cases for scenario 2	Parameters							
			$\%Z_{G1}$	$\%Z_{G2}$	$\%Xc$	R_f	FIA	δ	Fault types	Fault locations
1	5760	4320	100	100						4 and 3 location for scenario 1 and scenario 2 respectively
2	5760	4320	100	75	25,50 & 75	0,5 & 25,50	0,45 & 81,117	10,20 & 30	10	
3	5760	4320	100	125						
4	5760	4320	75	100						
5	5760	4320	125	100						
total number of cases in scenario 1= 5*5760=28800										
total number of cases in scenario 2= 5*4320=21600										

3.3 OAO Classification

For this strategy, table ought to be recognized, and every SVM must make its choice on class among 2 classes. The class that accomplishes most elevated many votes are picked by the consolidated classifiers. Potential situations are appeared in Table V

1) Scenario 1 Examination

The most extreme number of right location of deficiency zone utilizing OAO in situation 1 is 20629 cases that occurred in g=13.1&C=10000. This number is equal to 81.8% of achievement. Following up on OAO technique, with examination of outcomes, 1 might choose for cases that table can't decide in favor of their group. In the wake of examinationoutcomes, Table V is supplanted with Table VI. The numerous right location of deficiency zone utilizing Table VI is 21239 that occurred in g=10.4&C=10. True to form, achievement rate expanded to 84.2%.

2) Scenario2 examination

Table 4. ASVM analysis

Case number	Number of fault cases for scenario 1	Number of fault cases for scenario 2	Parameters							
			$\%Z_{G1}$	$\%Z_{G2}$	$\%X_c$	R_f	FIA	δ	Fault types	Fault locations
1	720	540	100	100	50					4 and 3 location for scenario 1 and scenario 2 respectively
2	720	540	100	75	50	0,5 & 50	0,45 & 117	10 & 30	10	
3	720	540	100	125	50					
4	720	540	75	100	50					
5	720	540	125	100	50					
total number of cases in scenario 1= 5*720=3600										
total number of cases in scenario 2= 5*540=2700										

The many right identification of shortcoming zone utilizing OAO in situation 2 was 11802 cases that occurred in g=20&C=10. This number is equal to %62.2 of accomplishment. In the wake of supplanting Table V by Table VI many right cases expanded to 123399 in g=21.3&C=10 that is identical to 65.6% of progress.

3.4 Organization with OAA Technique in Changed Training Set

As might be observed, a generous level of accomplishment is not accomplished forsituation 1 and 2 with the two strategies. One of viable boundaries to accomplish a decent outcome in arrangement is having legitimate and perfect measure of preparing informational index. One can't anticipate that a calculation should foresee appropriately in inconspicuous zones. It is common to put 70% of informational collection into preparation set and 30% to test set. As of recently, just 12.5% of informational index is utilized in preparation set. Cases that their deficiency zone is distinguished mistakenly were dissected. A high extent of wrong forecasts occurred in information with load edges (i) equivalent to 20, which was excluded from the preparation informational index. Hence, arbitrary informational collections with load points (ii) equivalent to 20 are included to preparation set. Table VII & Table VIII presents additional information for situation 1 and situation 2, individually. Deliberate that now 15.3% of all out information is utilized as preparation informational collection.

1)   Scenario1 Examination

The many correctdiscovery of FZ utilizing OAA in altered training set is 22993 cases that occurred in g=8.3&C=10. Altered training set expanded achievement rate from 83.7% to 94.2%.

2)   Scenario 2 Examination

The numerous correct detection of FZ utilizing OAA in altered training set is 15880 cases that occurred in g=14.5&C=10. Changed training set improvedachievement rate from 69 .7% to86.7%.

3.5 Organization with OAO Technique in Changed Training Set

Scenario 1 Examination
    In this instance, better outcome in attained in g=12.1&C=10. With these values, achievement rate is enhanced to 86.7%. After examining outcomes, Table V is substituted with Table IX as "reference voting table". The outcomespresented91.6% achievement rate in g=12.1&C=10.
    1) Scenario 2 examination
    The outcomes presented 76.1% achievement rate in g=15.3 &C=100000 that will be equal to 13926 correct FZD. After examining the outcomes, Table V is substituted with Table IX as "reference voting table". The outcomespresented 78.2% achievement rate in g=15.4&C=100000 that will be equal to 14318 correct FZD.

4 DISCUSSION ON THE RESULTS

As per the outcomes, the most elevated achievement rate is accomplished in situation 1 (introducing TCSC in the principal line). This achievement rate (94.2%) is reached with OAA strategy on adjusted preparing set. In scenario 2, most elevated achievement rate is 86.7% that is additionally reached with OAA technique on changed preparing set. Subsequently, the unmistakable part of choosing a legitimate preparing informational index had gotten apparent. To research why situation 2 had low achievement rate than situation 1, cases with some unacceptable location of shortcoming zone in the two situations are dissected in Table X & Table XI. They are likewise envisioned in Fig. 2 & Fig. 3 for improved agreement. By looking at these 2 tables, it very well may be inferred that expansion in wrong zone discoveries occurred in zone 2 & 3, while in zone 1 it nearly stayed steady. All in all, SVM committed more errors since pushing TCSC forward expanded its territory of impact. Moreover, in the two cases, zone 2 introduced the most elevated measure of wrong recognition and that is on the grounds that TCSC has its most noteworthy impact on this zone as it is set just before it. In this exploration ability and proficiency of the proposed strategy was assessed on the example organization. It should be noticed that with an adjustment in the organization geography, gotten data from the separation transfer perspective would change. For this situation SVM boundary ought to be refreshed.

5 CONCLUSIOIN

In this manuscript, another strategy for deficiency zone location of separation transfer in FACTS-based TLs is suggested.

Table 5. Numerous Wrong Predictions in Scenario 1

Predicted zone / Real zone	1	2	3	Total number of wrong predictions in each zone
1		99	211	310
2	249		512	761
3	193	143		336

For examining the impacts of TCSC area on shortcoming zone discovery of separation hand-off, two spots, one out of 50% of line length &other in 75% of line length are deliberated as two situations. Altogether, 4 cases have been considered for every situation. For situation 1, with OAA strategy, 83.7% of accomplishment, and with OAO technique, 84.2% of progress have been accomplished. In the wake of breaking down the outcome and expanding the preparation set, with OAA technique, 94.2% of accomplishment and with OAO strategy, 91.6% of progress are accomplished. For situation 2, with OAA

Figure 2. Numerous wrong predictions

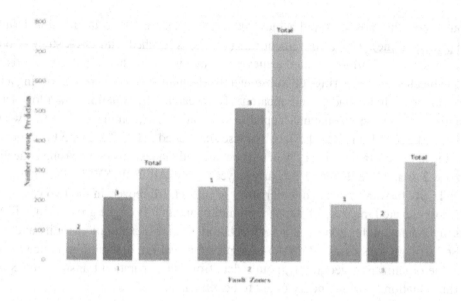

strategy, 69.7% of progress, and with OAO technique, 65.6% of accomplishment have been accomplished. In the wake of dissecting the outcome and expanding the preparation set, with OAA technique, 86.7% of accomplishment and with OAO strategy, 78.2% of progress are accomplished. The outcomes demonstrated that in tackling this sort of issue, eventually, OAA technique has similar bit of leeway over OAO strategy. With OAA technique, 94.2% and 86.7% of accomplishment are accomplished in situation 1 and 2, individually. It was indicated that introducing TCSC in the primary line prompts better outcomes

Figure 3. Numerous wrong predictions

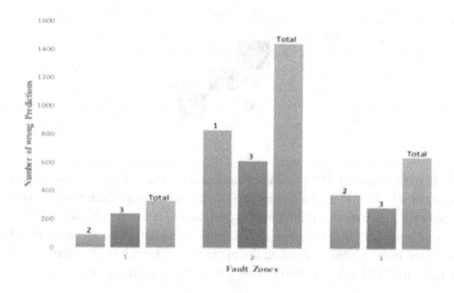

Table 6. Numerous Wrong Predictions in Scenario 2

Predicted zone / Real zone	1	2	3	Total number of wrong predictions in each zone
1	■	91	237	328
2	829	■	613	1442
3	381	283	■	644

by diminishing its impact on zone 2 and 3 of the separation hand-off. TLs are crucial pieces of intensity framework. Improving their insurance is fundamental for unwavering quality of the system. This manuscript signifies new way to deal with increment the unwavering quality of framework by enhancing the assurance of TLs in issue occurrence and limiting the harm to other system hardware. Quick location and detachment of a defective TL, shield the system.

REFERENCES

Abinaya, S., & Arulkumaran, G. (2017). Detecting black hole attack using fuzzy trust approach in MANET. *Int. J. Innov. Sci. Eng. Res.*, *4*(3), 102–108.

Dash, P. K., Samantaray, S. R., & Panda, G. (2006). Fault Classification and Section Identification of an Advanced Series-Compensated Transmission Line Using Support Vector Machine. *IEEE Transactions on Power Delivery*, *22*(no. 1), 67–73. doi:10.1109/TPWRD.2006.876695

Dehkordi, Sajjad, Fereidunian, Dehkordi, & Lesani. (2014). HourǦahead demand forecasting in smart grid using support vector regression (SVR). *International Transactions on Electrical Energy Systems*, *24*(12), 1650-1663.

Devaraju, Kumar, & Rao. (n.d.). A Real And Accurate Vegetable Seeds Classification Using Image Analysis And Fuzzy Technique. *Turkish Journal of Physiotherapy and Rehabilitation, 32*, 2.

Girgis, A. A. (1960). An adaptive protection scheme for advanced series compensated (ASC) transmission lines. *IEEE Transactions on Power Delivery*, *13*(2), 414–420.

Goldsworthy, D. L. (1987). A linearized model for MOV-protected series capacitors. *IEEE Transactions on Power Systems*, *2*(4), 953–957. doi:10.1109/TPWRS.1987.4335284

Imani. (2019). Effect of Changes in Incentives and Penalties on Interruptible/Curtailable Demand Response Program in Microgrid Operation. In *2019 IEEE Texas Power and Energy Conference (TPEC)*. IEEE.

Imani, Niknejad, & Barzegaran. (2019). Implementing Time-of-Use Demand Response Program in microgrid considering energy storage unit participation and different capacities of installed wind power. *Electric Power Systems Research, 175*, 105916.

Kalman, R. E. (n.d.). A new approach to linear filtering and prediction problems. *Journal of Basic Engineering, 82*(1), 35–45.

Kumar, D. S., Kumar, C. S., Ragamayi, S., Kumar, P. S., Saikumar, K., & Ahammad, S. H. (2020). A test architecture design for SoCs using atam method. *Iranian Journal of Electrical and Computer Engineering, 10*(1), 719.

Lepping, J. (2018). *Wiley Interdisciplinary Reviews. Data Mining and Knowledge Discovery.*

Megahed, A. I. (2006). Usage of wavelet transform in the protection of series-compensated transmission lines. *IEEE Transactions on Power Delivery, 21*(3), 1213–1221.

Megahed, A. I., MonemMoussa, A., & Bayoumy, A. E. (2006). Usage of wavelet transform in the protection of series-compensated transmission lines. *IEEE Transactions on Power Delivery, 21*(3), 1213–1221. doi:10.1109/TPWRD.2006.876981

Obaid, A. J. (2020). *An efficient systematized approach for the detection of cancer in kidney.* Academic Press.

Parikh, U. B., Biswarup Das, & Prakash Maheshwari, R. P. (2008). Combined wavelet-SVM technique for fault zone detection in a series compensated transmission line. *IEEE Transactions on Power Delivery, 23*(4), 1789–1794. doi:10.1109/TPWRD.2008.919395

Provost, F., & Domingos, P. (2003). Tree induction for probabilitybased ranking. *Machine Learning, 52*(3), 199–215.

Sadeh, Ranjbar, Hadsaid, & Feuillet. (2000). Accurate fault location algorithm for series compensated transmission lines. In *2000 IEEE Power Engineering Society Winter Meeting. Conference Proceedings* (Cat. No. 00CH37077), (vol. 4, pp. 2527-2532). IEEE.

Saha, M. M., Izykowski, J., Rosolowski, E., & Kasztenny, B. (1999). A new accurate fault locating algorithm forseries compensated lines. *IEEE Transactions on Power Delivery, 14*(3), 789–797. doi:10.1109/61.772316

Saikumar & Rajesh. (n.d.). A novel implementation heart diagnosis system based on random forest machine learning technique. *International Journal of Pharmaceutical Research, 12*, 3904–3916.

Saikumar, K., & Rajesh, V. (2020). Diagnosis of coronary blockage of artery using MRI/CTA images through adaptive random forest optimization. *Journal of Critical Reviews, 7*(14), 591–600.

Saikumar, K., Rajesh, V., Hasane Ahammad, S. K., Sai Krishna, M., Sai Pranitha, G., & Ajay Kumar Reddy, R. (2020). Cab for Heart Diagnosis with RFO Artificial Intelligence Algorithm. *International Journal of Research in Pharmaceutical Sciences, 11*(1).

Samantaray, S. R. (2013). A data-mining model for protection of FACTS-based transmission line. *IEEE Transactions on Power Delivery, 28*(2), 612–618. doi:10.1109/TPWRD.2013.2242205

Sheshasaayee, D., & Megala, R. (2017). A Conceptual Framework For Resource Utilization In Cloud Using Map Reduce Scheduler. *International Journal of Innovations in Scientific and Engineering Research*, *4*(6), 188–190.

Song, Y. H., Johns, A. T., & Xuan, Q. Y. (1996). Artificial neural-networkbased protection scheme for controllable series- compensated EHV transmission lines. *IEE Proceedings. Generation, Transmission and Distribution*, *143*(6), 535–540. doi:10.1049/ip-gtd:19960681

Song, Xuan, & Johns. (1997). Protection scheme for EHV transmission systems with thyristor controlled series compensation using radial basis function neural networks. Electric Machines and Power Systems, 25(5), 553-565.

Soumya, N., Kumar, K. S., Rao, K. R., Rooban, S., Kumar, P. S., & Kumar, G. N. S. (n.d.). 4-Bit Multiplier Design using CMOS Gates in Electric VLSI. *International Journal of Recent Technology and Engineering*.

Thomas, D. W. P., & Christopoulos, C. (1992). Ultra-high speed protection of series compensated lines. *IEEE Transactions on Power Delivery*, *7*(1), 139–145. doi:10.1109/61.108900

Xuan, Q. Y., Song, Y. H., Johns, A. T., Morgan, R., & Williams, D. (1996). Performance of an adaptive protection scheme for series compensated EHV transmission systems using neural networks. *Electric Power Systems Research*, *36*(1), 57–66. doi:10.1016/0378-7796(95)01014-9

Zhang & Kezunovic. (2007). Transmission line boundary protection using wavelet transform and neural network. *IEEE Transactions on Power Delivery*, *22*(2), 859–869.

Chapter 20
Traditional and Innovative Approaches for Detecting Hazardous Liquids

Ebru Efeoglu
https://orcid.org/0000-0001-5444-6647
Istanbul Gedik University, Turkey

Gurkan Tuna
https://orcid.org/0000-0002-6466-4696
Trakya University, Turkey

ABSTRACT

In this chapter, traditional and innovative approaches used in hazardous liquid detection are reviewed, and a novel approach for the detection of hazardous liquids is presented. The proposed system is based on electromagnetic response measurements of liquids in the microwave frequency band. Thanks to this technique, liquid classification can be made quickly without pouring the liquid from its bottle and without opening the lid of its bottle. The system can detect solutions with hazardous liquid concentrations of 70% or more, as well as pure hazardous liquids. Since it relies on machine learning methods and the success of all machine learning methods depends on provided data type and dataset, a performance evaluation study has been carried out to find the most suitable method. In the performance evaluation study naive Bayes and sequential minimal optimization has been evaluated, and the results have shown that naive Bayes is more suitable for liquid classification.

INTRODUCTION

Hazardous substances are used in some industrial processes, therefore some employees use different types of hazardous substances such as flammable liquids, chemicals and gases in their work (Spiegel-Ciobanu, Costa, & Zschiesche, 2020). Hazardous substances can take different forms, including gas, liquid, powder or dust and the product may be pure or diluted. Hazardous substances commonly encoun-

DOI: 10.4018/978-1-7998-6870-5.ch020

tered in workplaces include acids, glues, caustic substances, paint, petroleum products, heavy metals, solvents, disinfectants, and pesticides.

Hazardous substances can be inhaled, swallowed, or splashed onto the skin or eyes. Potential harmful effects of exposure to the hazardous substances depend on the type of the hazardous substance and the level of exposure in terms of duration and concentration. Possible health effects of the exposure to the hazardous substances can include skin rashes, chemical burns, poisoning, headache, nausea and vomiting, disorders of the lung, liver or kidney, nervous system disorders and birth defects. Since exposure to hazardous substances used in workplaces can lead to different short- and long-term health effects, for safe handling practices in most countries manufacturers and importers of hazardous substances are legally obliged to provide warning labels and safety data sheets with their products. For each hazardous substance used in workplaces, employers must ensure that safety data sheets are available, and a central register of the hazardous substances is set up. Safety data sheets typically list important information on handling the product safely, such as precautions for use, safe storage suggestions, contact numbers for further information, potential health effects, and emergency first aid instructions. In accordance with Occupational Health and Safety Regulations, warning labels on hazardous substances should include hazard pictograms, signal words, hazard statements and precautionary statements.

Hazardous liquids such as chemical warfare agents, liquid explosives, toxic industrial compounds and flammable substances can be used as weapons of mass destruction against both civilians and troops by individuals or terrorist organizations (Ramirez-Cedeno, Ortiz-Rivera, Pacheco-Londono, & Hernandez-Rivera, 2010; Zhu et al., 2020). These liquids are easily concealed within common household products and brought into a public area such as buildings, shopping malls, transportation terminals or airplanes. Since places with large groups of people are generally preferred for terrorist attacks, by detecting these liquids in the entrances of such places, possible terrorist attacks can be prevented. Liquid explosive detectors are required and highly important, especially in places such as airports, train stations, concert halls and shopping malls. Detection of hazardous liquids is also important to protect human and environmental health (Sudac et al., 2020). In recent years, researchers have proposed many approaches and developed systems for the detection of these hazardous liquids and they are still in an effort to develop new technologies and techniques due to the high false alarm rate in traditional scanning systems used in hazardous liquid detection. Regarding this, in this study, firstly traditional and innovative approaches used to detect hazardous liquids are reviewed. Then, a system based on the use of electromagnetic response measurements of liquids in the microwave frequency band and application of machine learning method for classification is proposed. Finally, to determine the best machine learning method for classification phase of the proposed system, a performance evaluation study is carried out.

BACKGROUND

Hazardous liquids, many of them are very exothermic and/or corrosive, can cause considerable and mostly irreversible damage to organs by contact or inhalation even at low concentrations (Marrs, Maynard, & Sidell, 1996). Some of these can cause considerable damage to property or human health, even in low concentrations and quantities. Chemical warfare agents are toxic enough to cause an instant damage when inhaled or when in contact with the skin (Marrs, Maynard, & Sidell, 1996). Some highly volatile materials can be deployed easily just by opening their containers. The level of damage associated to a concealed hazardous liquid depends on the type of material, deployment method, area, hazard level, concentration and properties at room temperature (Marrs, Maynard, & Sidell, 1996).

Flammable and combustible liquids can lead to explosion and fire. In particular, flammable liquids are hazardous because of their ability to produce vapors, which mixes with air and burns quickly when the flammable liquid is heated to its flashpoint or above and is ignited. Although technically flammable and combustible liquids differ from each other, they both burn immediately and intensively. Under certain conditions they are explosive, and if they are not properly contained, they can spread fire quickly and uncontrollably. Therefore, the use of appropriate equipment and practices is critical for safe handling and storage of flammable liquids (Steinfeld & Wormhoudt, 1998).

In homeland security liquid explosives represent a hazard potential, and x-rays and metal detectors used for routine security checks fail to identify them. Therefore, it is restricted to carry fluids or gels in hand luggage on planes. The main problem with liquid explosives is that they might be carried in an ordinary bottle which appears to be completely safe, and it is mostly impossible to distinguish them from harmless substances optically. In addition, due to the quick evaporation of liquid substances, it is not possible to detect their traces on the outer surface of a closed container.

Various techniques and various concept of operations can be used to detect liquid explosives. In recent years, THz time-domain spectrometers have become popular for liquid detection. The advantage of THz time-domain spectrometers over traditional approaches is that they do not require to open the bottle in order to classify its contents. These devices, some of them are portable, operate precisely and quickly. Plastic- and polymer-based containers are transparent for THz radiation. On the other hand, polar fluids are known to be strong absorbers. Therefore, the liquids are measured in reflection geometry so that the intensity of the THz signals arriving at these devices is sufficient. Evaluation of the THz signal reflected off the interface between the plastic bottle and the liquid gives information on the dielectric properties of the bottle content. The overall process takes only a couple of seconds. Due to the increasing availability of different devices for screening processes, for the purpose of testing of the performance of different devices, European Civil Aviation Conference (ECAC) has adopted a classification scheme consisting of five categories as listed in Table 1. Liquid Explosive Detection System (LEDS) equipment typically relies on multiple sensing technology, a combination of two or more of microwave, infrared, magnetic inductive and gravimetric, for detection.

Table 1. Categories of LEDS

Category	Can screen single or multiple containers	Require removal from cabin baggage	Require opening of containers	Can screen complex electronics
Type A	Single	Yes	Yes	No
Type B	Single	Yes	No	No
Type C	Multiple	Yes	No	No
Type D	Multiple	No	No	No
Type D+	Multiple	No	No	Yes

The main concern with LEDS is that false alarm rates vary depending on the liquid and depending on the how much liquid is in the container. It was shown that partially filled liquid, aerosol and gel (LAG) containers have little effect on the false alarm rate for Type B LEDS equipment, compared to those for fully filled containers; on the other hand, they have a considerable increase on the false alarm rate for Type C LEDS equipment (Dzhongova, Anderson, & de Ruiter, 2017). The overall false alarm rate for Type C LEDS equipment increased from 10-13% for full containers to 21-24% for partially filled containers. (Dzhongova, Anderson, & de Ruiter, 2017). This scenario (use of partially filled containers) results in a two-fold increase in the number of items requiring level-two screening and makes it difficult to manage the screening processes in airports. Although there are some concerns like this still exist, LEDS equipment is well suited for high-throughput lanes at airport security checkpoints. LEDS equipment can either be used as a stand-alone scanner for consumer liquids and gels in plastic, glass or ceramic bottles and metal/metallized containers or in conjunction with other equipment for false alarm resolution. It is an economical solution with minimal maintenance requirement and is generally used as the frontline device to minimize delays. It is portable enough to be moved or relocated when required.

PROPOSED SYSTEM

The measurement setup consists of a vector network analyzer and a circular patch antenna. The schematic representation of the measurement system is shown in Figure 1. Measuring the information of reflection from a material provides information so that the material can be characterized. A vector network analyzer consists of a signal source, a receiver and a display. It is shown schematically in Figure 2. The source sends a signal at a single frequency to the material to be tested. The receiver is tuned to this frequency to detect reflected and transmitted signals from the material. The measured response produces magnitude and phase data at that frequency. The source is then passed to the next frequency and the measurement is repeated to display the reflection and transmission measurement response as a function of frequency. *S* parameters are frequently used in microwave circuits as they are convenient to use and current and voltage measurements cannot be made.

Design of the Antenna

Patch antennas are made by placing a conductive layer with different geometries on a dielectric layer on a ground plane. Antenna patches can be square, circle, ellipse, etc. The physical structure of the designed circular patch antenna is shown in Figure 3. Radiation in a patch antenna occurs between the conductive layer of the antenna and the ground plane. The glow patch is the conductor that enables the absorption or radiation of electromagnetic waves and has a certain shape. This patch is positioned on a dielectric layer and materials such as copper or silver are commonly used. Patch sizes and shapes may differ depending on the frequency of use. The reason why the circle patch is preferred is the characteristic of the circular patch having a symmetrical radiation characteristic not found in the rectangular and other shaped patch types. Another factor contributing to the circular patch selection is the use of a circular bottle for measurements. The most commonly used feeding methods in antenna design are microstrip feeding method and coaxial probe feeding method. The inner conductor of the coaxial probe is connected to the radiation patch of the antenna, and the outer conductor is connected to the ground plane of the antenna. Coaxial probe feeding method is more useful for antennas using thin layers and

Figure 1. Measurement system used in this study

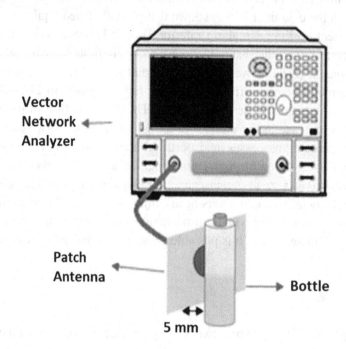

it is easy to manufacture. It has a low artificial radiation characteristic and a narrower bandwidth than other feeding methods. It is the most preferred feeding method due to low parasitic radiation. The relative permeability value of the dielectric layer used in the designed antenna is 4.4. For this reason, the circle radius of the antenna patch has been chosen so that the measurements are not affected by surface waves. The resonance frequency of the antenna was requested to be 1.5 GHz and the patch radius of the antenna was calculated using Equation (1).

$$a = \frac{F}{\left\{ 1 + \frac{2h}{\pi \varepsilon_r F \left[\ln\left(\frac{\pi F}{2h}\right) + 1,7726 \right]^{1/2}} \right\}} . \tag{1}$$

where F is calculated using Equation (2).

$$F = \frac{8,791 x 10^9}{f_r \sqrt{\varepsilon_r}} . \tag{2}$$

where, ε_r .represents the relative permeability of the substrate, f_r .represents resonance frequency, h represents the height of the substrate and a is the radius of the patch (Efeoglu & Tuna, 2020). By plac-

Figure 2. Schematic representation of a vector network analyzer

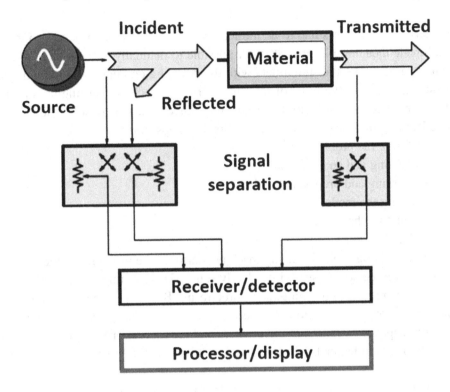

ing the values in the equations, the patch diameter was found to be 100 mm. The design was built on an FR4-based dielectric substrate with 1.6 mm height, 4.4 relative permeability and 10 x 10 cm² ground plane below.

Figure 3. Front and back view of the patch antenna

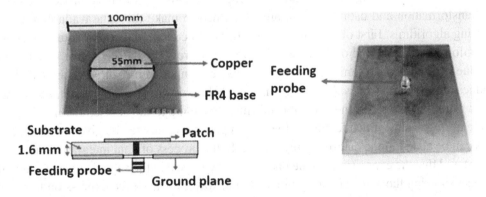

Machine Learning

With machine learning, computers had to be trained and taught like human beings so that data could be analyzed and made meaningful. Since it is not possible to manually process large amounts of data with many features, processing the stored data with the help of a computer and obtaining meaningful information from these data is the basis of machine learning. Machine learning uses past experience or sample data. When there is a problem that needs to be solved, the way to solve the problem with the help of machine learning methods is through the following steps (Shearer, 2000).

Step 1:	Defining the problem
Step 2:	Collecting and analyzing data
Step 3:	Processing and preparing the data
Step 4:	Creating the model
Step 5:	Evaluating the model
Step 6:	Using the model

In the first step of solving problems using machine learning, the problem should be defined in the best way, and the goal to be solved should be clearly stated. Determining this is important for the researcher working in the project in reaching the desired result. Because it is possible for the researchers to reach a conclusion about how to follow the problem solving by clearly determining the definition of the problem. In this step, the definition of the problem and the success criteria should be clearly stated. The current situation should be analyzed. Because the outputs will be interpreted by considering this success criterion.

In the second step, appropriate data for the problem is collected and then analyzed. Data can be obtained in two ways. The first way is to extract data from existing datasets. The other way is the use of surveys, discussions or measurements. Then, the data is examined in terms of format and quantity. It should be considered whether the basic feature that will be used in the model is present in the data set. For example; although species is an important feature in a study, if the collected data do not include all the species, the data set can be changed or the data of the missing species should be added to the dataset (Shearer, 2000).

In the third step, depending on the features of the data some preliminary processes such as data filtering, transformation and data reduction are performed to make the data available for use in machine learning algorithms. First of all, since each dataset can contain data that contains inconsistent or erroneous information, this kind of data should be removed from the dataset. Then, the data known as missing value is discarded. In some cases, different values cannot be measured or missed. These overlooked values are called missing values. If the missing values do not affect the result, these data should be removed from the data set. If the lost data adds more meaning to the dataset, the missing values can be given the average value of that feature or the most repetitive values can be given in the dataset, or the lost values can be added to the dataset by regression. In the process of data integration and transformation, it is ensured that the expressions in the data set are expressed in the same language and in the same way. Because showing the same situation in a different way can cause the expression to be perceived differently. Another process used in the preparation of data is the reduction process in order to process data more easily and faster. During the reduction process, the number of data or the number of variables is reduced so as not to change the results. There are many methods used for data reduction. During the preparation of the data, normalization and data scaling can be done by taking into account the nature of the problem and the data.

In the fourth step the model and algorithm that can best analyze the relationship between the variables in the dataset is determined. In order to get high efficiency from the dataset, more than one model should be created and more than one algorithm should be tried in order to obtain the best results. After creating models from the dataset with machine learning methods, it is decided which model is more suitable. It is not appropriate to use only the training dataset to make this decision. Evaluating the performance of the model using only a training dataset may cause overfitting. Model performance evaluation methods are used to avoid this situation and to understand how successful it will be on the dataset that has not been introduced to the model before. Because in the overfitting problem, the model gives excellent results on the data set we are working on, but makes unsuccessful predictions on a dataset that it has never seen before. There are many previously developed performance evaluation methods and criteria. The most used methods among model performance evaluation methods are K-fold cross-validation method and exclusion method.

The dataset is divided into two parts as training and testing. Generally, the percentage used for separating into parts is 65% to 35%. 65% of the dataset is used as a training dataset and 35% is used as a test dataset (Izadi, Ranjbarian, Ketabi, & Nassiri-Mofakham, 2013). The data used in the test dataset are not used in the training set. Model building is done using the training set. Using the test dataset, the level of learning is controlled and the model performance is obtained. It was seen that the separation was not good enough when this method was used in datasets with a small number of observations. In addition, it is a disadvantage for this method that the data set is separated as training and test data for once, and that all data cannot be used in the model. Another disadvantage of the method is that among the data kept separate from the training set, there may be data that will contribute to the model to improve.

In K-fold cross-validation, a given dataset is split into a K number of folds where each fold is used as a testing set at some point. In the first iteration, the first fold is used to test the model and the rest are used to train the model. In the second iteration, the second fold is used as the testing set while the rest serve as the training set. This process is repeated until each fold of the K folds have been used as the testing set. A schematic representation of K-fold cross-validation technique is given in Figure 4. K is usually chosen as 10.

Figure 4. K-fold cross-validation

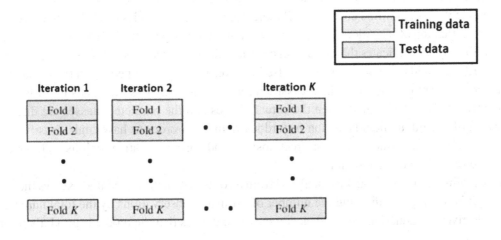

Classification and Classification Algorithms

Naive Bayes and Sequential Minimal Optimization algorithms are used in this study. Naive Bayes is a classification technique based on Bayes' theorem. It uses the data of classified instance objects and calculates the probability that an object of unknown class belongs to the specified classes. In this algorithm, it is accepted that the features are independent from each other. The value of the properties does not contain information about each other's value. Suppose we have a dataset showing the number of k features, the number of classes t. The objective is to determine which class an object X of unknown class belongs to, and the probability that the object is an element of that class is calculated for all existing classes. An object is assigned to the class with the highest probability values calculated (Bermejo, Gámez, & Puerta, 2011).

Sequential Minimal Optimization (SMO) is an algorithm that can quickly solve quadratic programming problems without an extra matrix storage space and without the need to use all numerical quadratic programming optimization steps (Platt, 1998). SMO algorithm is about pulling the idea of the separation method to the extreme and optimizing a minimum subset of only two points in each iteration. It replaces all loss values with new ones, converts nominal attributes to binary ones, and normalizes all attributes with previously defined values. SMO prefers to solve the smallest possible optimization problem at every stage. Since the Lagrange multipliers must conform to a linear equation boundary, the smallest possible optimization problem involves two Lagrange multipliers. The algorithm finds the best value for these multipliers by optimizing two Lagrange multipliers at each stage and performs the update of SVM in a way that reflects this value. The two Lagrange factors must comply with all the constraint conditions of the problem. These restrictions cause Lagrange multipliers to be inside the box. Since the linear equality constraint causes it to be located on the diagonal line, an SMO step must find the best of the objective function on a diagonal line segment (Bermejo, Gámez, & Puerta, 2011).

Performance Evaluation Methodoloy

The number of correct predictions indicates the correctness of the classification but alone it is not enough to understand how well the model is created. A confusion matrix is the straightest way to show the prediction results of a classification algorithm and often used to reveal the performance of the created model. It consists of four parameters (Joshi, 2016) and contains information about actual and predicted groups made by a classification system as shown in Figure 5. In a confusion matrix, True Positives (TP) refer to positive values classified correctly. TP indicates that the actual class and the predicted class are the same, classification of hazardous liquids as hazardous. True Negatives (TN) refer to negative values classified correctly. TN indicates that the actual class and the predicted class are the same, classification of non-hazardous liquids as non-hazardous. False Positives (FP) are obtained when the actual class and the predicted class are not the same, i.e., classification of non-hazardous liquids as hazardous liquids. False Negatives (FN) are obtained when the actual classes and the predicted classes are different, i.e. classification of hazardous liquids as non-hazardous liquids. In order to have high classification performance, it is desirable to increase the correct positive and correct negative regions and to reduce the false positive and false negative regions.

Some performance metrics can be calculated from a confusion matrix. Accuracy shows the accuracy of a model. It is obtained by dividing the number of estimated observations by the total number of observations as given by Equation (3). High accuracy rate indicates that the model is good; therefore, it is

Figure 5. Confusion matrix

		Predicted Class	
		Positive	Negative
Actual Class	Positive	True Positives (TP)	False Negatives (FN)
	Negative	False Positives (FP)	True Negatives (TN)

desired. However, if the number of false positives and false negatives are high, other parameters used in model performance evaluation should also be examined (Joshi, 2016; Sturm, 2013).

$$\text{Accuracy} = \frac{TP + TN}{TP + FP + TN + FN}.$$
(3)

Precision is obtained by dividing the number of correct positive observations by the total number of positive observations found. This ratio can be considered as a measure of the accuracy of the classification algorithm. A low precision value can also be considered as an indicator of many false positives (Joshi, 2016; Brownlee, 2014). Precision is calculated using Equation (4).

$$\text{Precision} = \frac{TP}{TP + FP}.$$
(4)

Recall is calculated by proportioning the correct class of observations to the total number of positives. Recall is a measure of the integrity of the classification algorithm. A low Recall value means that the number of false negatives is high (Joshi, 2016; Brownlee, 2014). Recall is calculated using Equation (5).

$$\text{Recall} = \frac{TP}{FN + TP}.$$
(5)

F-Measure takes values between 1 and 0. Overall, it is a measure of the precision and reliability of a model. Recall and precision metrics are used to calculate this value. The harmonic mean of these values gives the F-Measure value. It is especially useful to look at this value in situations where there is an uneven class distribution. If the number of false positives and false negatives are very different, it is necessary to look at not only the recall value but also the precision value (Joshi, 2016; Brownlee, 2014).

For example, it is difficult to compare two models with low recall and high precision and vice versa. That is why F-Measure is used to make them comparable. F-Measure is calculated using Equation (6).

$$F\_Measure = \frac{2*Recall*Precision}{Recall + Precision} \qquad (6)$$

Kappa value is a measure of consistency between the actual and predicted classes in a classification study and is calculated using Equation (7). It can vary between -1 and 1. A value of -1 indicates a complete mismatch, that is, an inverse relationship, and a value of 1 indicates a perfect fit. As the value gets closer to 1, the consistency increases and else the consistency decreases.

$$Kappa = \frac{P(a) - P(e)}{1 - P(e)}. \qquad (7)$$

where *P(a)* is a value that shows probabilistic accuracy of the classification algorithm, and *P(e)* is the weighted average of the probability of classifications made in the same dataset.

Root Mean Square Error (RMSE) is used to scale the differences between the actual values and the values predicted by the model. It is determined by taking the square root of the mean square error and calculated by Equation (8).

$$RMSE = \sqrt{\frac{1}{n}\sum_{j=1}^{n}\left(T_{ij} - A_j\right)^2}. \qquad (8)$$

where T_{ij} is the predicted value A_j is the goal value. If the error value approaches zero, it means that the correct estimation of the classification algorithm increases.

Results

For the detection of hazardous liquids, S-parameter (S_{11}) measurements of 77 liquids, 24 hazardous and 53 non-hazardous liquids listed in Table 2 and Table 3, were used. As well as pure liquids, aqueous solutions and alcoholic beverages were also used. Liquids with 70% alcohol content among alcohol-water solutions were considered dangerous as given in (Tan et al., 2017). The S-parameter measurements of the liquids are given in Figure 6 and Figure 7. All of the liquids measured were in pet bottles and at room temperature. While the measurements were made, a distance of approximately 5 mm was left between the bottle and the antenna.

Performance Evaluation of the Algorithms in the Detection of Hazardous Liquids

In this study, two different algorithms, namely Naive Bayes and SMO, were used for the classification of liquids. The performance of the algorithms was evaluated with and without cross-validation. For numerical comparisons, confusion matrixes were obtained, and precision, F-Measure, Kappa and RMSE

Table 2. Non-hazardous liquids used in this study

Non-Hazardous Liquids		
Water	Red Wine	Raki
Gin	Apricot Juice	Lens Cleaning Fluid
Milk	Screen Cleaning Fluid	White Wine
Beer	Ice Tea	Shampoo
Hair Gel	Peach Juice	Liqueur
Cola	Baby Food	Champagne
Vinegar	Liquid Soap	Whiskey
Tequila	Methanol (10%, 20%,…60%)	Shower Gel
Vodka	Ethanol (10%, 20%,…60%)	Turnip
Ketchup	1-Propanol (10%, 20%,…60%)	Cocoa milk
Buttermilk	Isopropanol (10%, 20%,…60%)	Hair conditioner

values were computed. Confusion matrix of SMO algorithm is given in Table 4. As listed in Table 4, the algorithm predicted 16 correctly and 8 incorrectly out of 24 hazardous liquids. The algorithm correctly predicted 51 of 53 non-hazardous liquids and two incorrectly. Therefore, the algorithm had an accuracy rate of 87% by correctly predicting a total of 67 liquids.

Confusion matrix of Naive Bayes algorithm is given in Table 5. As listed in Table 5, the algorithm predicted 17 correctly and 7 incorrectly out of 24 hazardous liquids. The algorithm predicted 51 of 53 non-hazardous liquids correctly and 2 of them incorrectly. The algorithm realized more accurate predictions than SMO algorithm and had an accuracy rate of 88.3% as shown in Figure 8. As it is known, if a classification is successful, Precision, Recall, F-Measure and Kappa values become around 1.

When Figure 8 is taken into consideration, it can be seen that the performance metric values of Naive Bayes algorithm are higher than SMO algorithm. Especially, the Kappa statistic value of 0.71 indicates that Naive Bayes algorithm performed better than SMO algorithm in the classification. This is in parallel with the results listed in Table 4 and Table 5.

In some cases, although the algorithm performs well when using the training data set, it may not perform well when it encounters a sample that is not included in the training data set. In other words, when faced with a different data, it may fail to make accurate predictions. Cross-validation process is used to

Table 3. Hazardous liquids used in this study

Hazardous Liquids	
Ethanol (70%, 80%, 90%, 100%)	Thinner
Methanol (70%, 80%, 90%, 100%)	Cologne
1-Propanol (70%, 80%, 90%, 100%)	Octanol
Isopropanol (70%, 80%, 90%, 100%)	Butanol
Asetone	Toluene
Hydrogen peroxide	Gasoline

Figure 6. S parameters of various alcohol-water concentrations: a) 1-Propanol, b) Isopropanol, c) Ethanol, d) Methanol

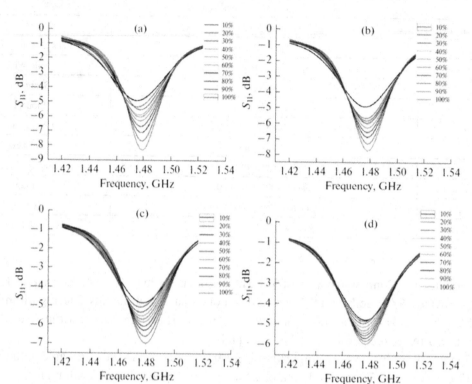

measure the success of the algorithm in classifying data not included in the training dataset. The most common one is 10-fold cross-validation. The confusion matrix of the SMO algorithm is given in Table 6. SMO correctly classified 16 hazardous and 51 non-hazardous liquids. The number of non-hazardous liquids that the algorithm incorrectly predicted was two and the number of hazardous liquids was eight. In total, the algorithm predicted 67 liquids correctly and 10 liquids incorrectly. The algorithm had an accuracy rate of 87%. On the other hand, the confusion matrix obtained when the classification was made by Naive Bayes algorithm using cross-validation is given in Table 7. After the cross-validation, the algorithm correctly predicted 16 out of 24 hazardous liquids and 51 out of 53 non-hazardous liquids. The algorithm had an accuracy rate of 87%, and it was successful in predicting liquids that were not in the dataset, too. Figure 9 shows the performance metrics of the algorithms after the cross-validation. It can be seen that both of the algorithms were successful and exhibited the same performance when classifying a liquid not previously introduced to the system.

SOLUTIONS AND RECOMMENDATIONS

Although liquid scanner systems are available and still researchers offer some approaches for the detection of hazardous liquids, suggestions on reducing exposure to hazardous substances in the workplace can be made. If possible tasks should be performed without using hazardous substances. If not, hazard-

Figure 7. S parameters of the liquids

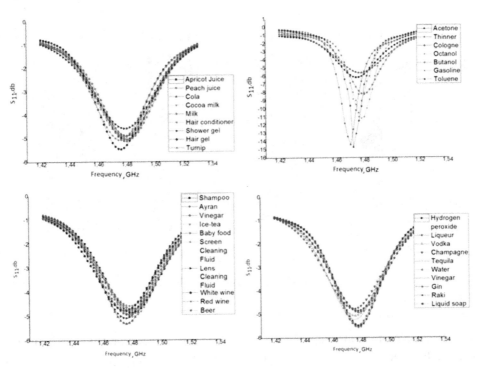

ous substances should be substituted with less hazardous ones, hazardous substances should be kept in separate storage areas, and the storage areas should be purged or ventilated separately from the rest of

Table 4. Confusion matrix of SMO algorithm

		Training		
		Predicted		
		Hazardous	Non-Hazardous	Total
Actual	Hazardous	**16**	8	24
	Non-Hazardous	2	**51**	53
	Total	18	59	77

Table 5. Confusion matrix of Naive Bayes algorithm

		Training		
		Predicted		
		Hazardous	Non-Hazardous	Total
Actual	Hazardous	**17**	7	24
	Non-Hazardous	2	**51**	53
	Total	19	58	77

Figure 8. Performance metrics of Naive Bayes and SMO algorithms

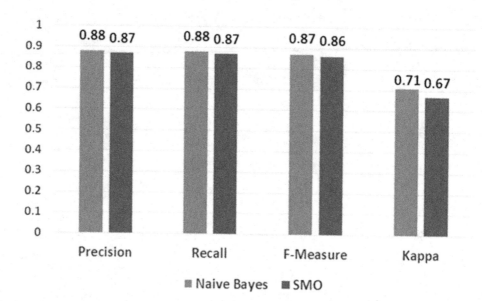

the workplace. Personal protection equipment such as respirators, goggles and gloves should be provided for employees and the workplace should be regularly monitored with appropriate equipment to track the degree of hazardous substance in the environment. Finally, employees should be thoroughly trained in handling and safety procedures and be regularly consulted to maintain and improve existing safety and handling practices.

Table 6. Confusion matrix of SMO algorithm using cross-validation

			Cross-validation		
			Predicted		
			Hazardous	Non-Hazardous	Total
Actual		Hazardous	**16**	8	24
		Non-Hazardous	2	**51**	53
		Total	18	59	77

Table 7. Confusion matrix of Naive Bayes algorithm using cross-validation

			Cross-validation		
			Predicted		
			Hazardous	Non-Hazardous	Total
Actual		Hazardous	**16**	8	24
		Non-Hazardous	2	**51**	53
		Total	18	59	77

Figure 9. Performance metrics of Naive Bayes and SMO algorithms using cross-validation

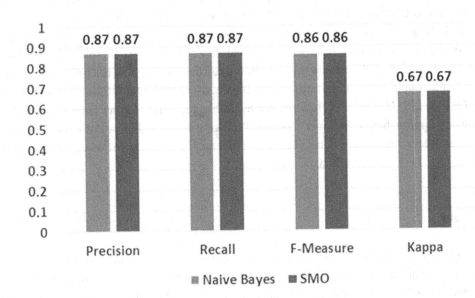

It is known that with the basic knowledge or guidance an explosive can be made from commercially available material. For instance, by treating glycerol with a mixture of nitric and sulphuric acid, nitroglycerin can be made. Nitroglycerin is probably one of the most well-known liquid explosives and was used by terrorists. One of the most dangerous explosives is TriAcetone TriPeroxide (TATP), a highly potent explosive commonly used by Middle East terrorists and usually found in a crystalline powder and made from acetone and hydrogen peroxide (Sun, Wu, Duan, & Jia, 2019). Terrorists can dissolve TATP into a liquid to carry aboard a plane or carry highly concentrated forms of acetone and hydrogen peroxide aboard the plane to make it. Terrorists can use many other substances, such as oxidizers, ammonium nitrate and diesel fuel, to create a fire or lead to an explosion. This makes it difficult to understand uncontrolled commercial availability of such substances. Therefore, some regulations are needed to make it difficult to obtain such substances, especially over the Internet.

Hazardous chemicals are essential materials in the economic growth of some developing countries. They are corrosive, oxidizing and toxic and there are safety risks in their production, storage and transportation (Hu & Raymond, 2004). Considering the increasing number of accidents occurring during the transportation of hazardous chemicals, safe transportation of them and optimization of their management is a must (Nie & Zhang, 2018).

FUTURE RESEARCH DIRECTIONS

In recent years, big data has become a key term across various industries to describe the availability of large and complex data sets that need to be captured, stored, analyzed, visualized and reported. Considering the increasing hazard risks across various industries and availability of large data from different sources, automation tools and improved data processing methods like big data will play a critical role to shorten the time to obtain specific results and extract the most significant findings. The most serious problem when big data is considered is that long processing time is necessary. However, expert systems can be useful for interpreting and processing such large bodies of information.

In parallel with the development of some countries, more hazardous chemicals are used in the production processes of industries and this increases the probability of hazardous chemical accidents to some extent. Since hazardous chemicals are very dangerous and can be destructive if not properly handled, their efficient and reliable regulation is of particular importance. Since different satellite navigation systems, such as GPS, GLONASS, Galileo and BeiDou, are available and offer free services worldwide, one of the most promising approaches is the supervision and tracking of critical hazardous chemicals based on big data of the satellite navigation systems. This approach can be quite useful when combined with the Internet of Things (IoT) (Ban et al., 2020).

CONCLUSION

In this study, firstly traditional and innovative approaches used in hazardous liquid detection were reviewed. Then, an innovative approach for the detection of hazardous liquids was proposed. To determine the best machine learning method for classification phase of the proposed system, Naive Bayes and SMO algorithms were applied on a dataset. Accuracy rate, recall, precision, F-measure, Kappa statistics, confusion matrices and error criteria were used for the performance analysis of these algorithms. Both the classification performance using the training dataset and the classification performance using cross-validation was analyzed. The reason for doing this was to be able to compare the success of the methods in the correct detection of liquids that were in the dataset and those that were not.

As a result of the performance comparison, it is shown that both Naive Bayes and SMO are good candidates for liquid classification systems. However, the results of the performance evaluation study show that Naive Bayes is more suitable for the classification of hazardous liquids. The system described in this chapter can be used as a liquid screening system at security checkpoints in crowded places such as airports, concert halls and shopping malls. With the use of this system, which can detect the liquid remotely with high accuracy without the need for any operator's knowledge and interpretation, liquid control can be easily performed without taking people's time and creating long lines. In addition, the fact that the system can automatically detect hazardous liquid increases the reliability of the system. Because the operator's attention is distracted in a short time in the liquid scanning systems connected to the operator, reducing the reliability of the results. By detecting hazardous liquids without the need to open the covers, it will be possible to prevent the operator from being exposed to that harmful liquid and the acute illnesses that may occur accordingly.

ACKNOWLEDGMENT

This research received no specific grant from any funding agency in the public, commercial, or not-for-profit sectors.

REFERENCES

Ban, Y., Liu, H., Fei, T., Luo, H., Yu, L., & Yang, Q. (2020). Big data-based applied research on hazardous chemical safety regulation. In IOP Conference Series: Earth and Environmental Science (vol. 569, pp. 1-7). IOP Publishing Ltd. doi:10.1088/1755-1315/569/1/012056

Bermejo, P., Gámez, J. A., & Puerta, J. M. (2011). Improving the performance of Naive Bayes multinomial in e-mail foldering by introducing distribution-based balance of datasets. *Expert Systems with Applications*, *38*(3), 2072–2080. doi:10.1016/j.eswa.2010.07.146

Brownlee, J. (2014). Classification accuracy is not enough: More performance measures you can use. *Machine Learning Mastery, 21*.

Dzhongova, E., Anderson, D., de Ruiter, J., Novakovic, V., & Oses, M. R. (2017). False alarm rates of liquid explosives detection systems. *Journal of Transportation Security*, *10*(3-4), 145–169. doi:10.100712198-017-0184-7

Efeoglu, E., & Tuna, G. (2020). Detection of Hazardous Liquids Using Microwave Data and Well-Known Classification Algorithms. *Russian Journal of Nondestructive Testing*, *56*(9), 742–751. doi:10.1134/S106183092009003X

Hu, C. Y., & Raymond, D. J. (2004). Lessons learned from hazardous chemical incidents—Louisiana hazardous substances emergency events surveillance (hsees) system. *Journal of Hazardous Materials*, *115*(1), 33–38. doi:10.1016/j.jhazmat.2004.05.006 PMID:15518962

Izadi, B., Ranjbarian, B., Ketabi, S., & Nassiri-Mofakham, F. (2013). Performance analysis of classification methods and alternative linear programming integrated with fuzzy delphi feature selection. *International Journal of Information Technology and Computer Science*, *5*(10), 9–20. doi:10.5815/ijitcs.2013.10.02

Joshi, R. (2016). *Accuracy, precision, recall & f1 score: Interpretation of performance measures*. Academic Press.

Marrs, T. C., Maynard, R. L., & Sidell, F. R. (1996). *Chemical Warfare Agents: Toxicology and Treatment*. Wiley.

Nie, Y., & Zhang, W. (2018). Research on optimization design of hazardous chemicals logistics safety management system based on big data. *Chemical Engineering Transactions*, *66*, 1477–1482.

Platt, J. (1998). *Sequential minimal optimization: A fast algorithm for training support vector machines*. Academic Press.

Ramirez-Cedeno, M. L., Ortiz-Rivera, W., Pacheco-Londono, L., & Hernandez-Rivera, S. P. (2010). Remote Detection of Hazardous Liquids Concealed in Glass and Plastic Containers. *IEEE Sensors Journal*, *10*(3), 693–698. doi:10.1109/JSEN.2009.2036373

Shearer, C. (2000). The CRISP-DM model: The new blueprint for data mining. *Journal of Data Warehousing*, *5*(4), 13–22.

Spiegel-Ciobanu, V. E., Costa, L., & Zschiesche, W. (Eds.). (2020). *Hazardous Substances in Welding and Allied Processes*. Springer. doi:10.1007/978-3-030-36926-2

Steinfeld, J. I., & Wormhoudt, J. (1998). Explosives detection: A challenge for physical chemistry. *Annual Review of Physical Chemistry*, *49*(1), 203–232. doi:10.1146/annurev.physchem.49.1.203 PMID:15012428

Sturm, B. L. (2013). Classification accuracy is not enough. *Journal of Intelligent Information Systems*, *41*(3), 371–406. doi:10.100710844-013-0250-y

Sudac, D., Pavlović, M., Obhodas, J., Nad, K., Orlić, Ž., Uroić, M., Korolija, M., Rendić, D., Meric, I., Pettersen, H. E., & Valković, V. (2020). Detection of Chemical Warfare (CW) agents and the other hazardous substances by using fast 14 MeV neutrons. *Nuclear Instruments & Methods in Physics Research. Section A, Accelerators, Spectrometers, Detectors and Associated Equipment*, *971*, 164066. doi:10.1016/j.nima.2020.164066

Sun, Q., Wu, Z., Duan, H., & Jia, D. (2019). Detection of Triacetone Triperoxide (TATP) Precursors with an Array of Sensors Based on MoS2/RGO Composites. *Sensors (Basel)*, *19*(6), 1281. doi:10.339019061281 PMID:30871286

Tan, X., Huang, S., Zhong, Y., Yuan, H., Zhou, Y., Xiao, Q., Guo, L., Tang, S., Yang, Z., & Qi, C. (2017). *Detection and identification of flammable and explosive liquids using THz time-domain spectroscopy with principal component analysis algorithm.* Paper presented at the 2017 10th UK-Europe-China Workshop on Millimetre Waves and Terahertz Technologies (UCMMT). 10.1109/UCMMT.2017.8068488

Zhu, R., Hu, X., Li, X., Ye, H., & Jia, N. (2020). Modeling and Risk Analysis of Chemical Terrorist Attacks: A Bayesian Network Method. *International Journal of Environmental Research and Public Health*, *17*(6), 2051. doi:10.3390/ijerph17062051 PMID:32204577

ADDITIONAL READING

Bretherick, L., Urben, P. G., & Pitt, M. J. (2017). *Bretherick's handbook of reactive chemical hazards.* Elsevier.

Carson, P. A., & Mumford, C. J. (1994). *Hazardous chemicals handbook.* Butterworth-Heinemann.

Colonna, G. R. (2010). *Fire protection guide to hazardous materials.* National Fire Protection Association.

Hughes, J. P., Hathaway, G. J., & Proctor, N. H. (2004). *Proctor and Hughes' chemical hazards of the workplace.* J. Wiley.

Patnaik, P. (2007). *A comprehensive guide to the hazardous properties of chemical substances.* J. Wiley. doi:10.1002/9780470134955

Pohanish, R. P., & Greene, S. A. (2009). *Wiley guide to chemical incompatibilities.* Wiley. doi:10.1002/9780470523315

Pohanish, R. P., & Sittig, M. (2012). *Sittig's handbook of toxic and hazardous chemicals and carcinogens.* Elsevier.

Price, J. C., & Forrest, J. S. (2016). *Practical aviation security: Predicting and preventing future threats.* Elsevier, Butterworth-Heinemann is an imprint of Elsevier.

Swcct, K. M. (2009). *Aviation and airport security: Terrorism and safety concerns.* CRC Press.

KEY TERMS AND DEFINITIONS

Classification Algorithm: An algorithm that maps the input data to a specific category.

Cross-Validation: It provides information about how well a classifier generalizes and is generally used to evaluate machine learning models on a limited data sample.

Liquid Explosive Detection System: A device used to detect explosive substances in liquid form, generally used at airports.

Microwave: Microwave is a form of electromagnetic radiation with wavelengths ranging from about one meter to one millimeter and with frequencies between 300 MHz and 300 GHz.

Non-Contact Detection: A process used to remotely detect a material without a physical contact.

Patch Antenna: A patch antenna is a type of radio antenna, which can be mounted on a flat surface.

Safety Data Sheet: A safety data sheet is a document that provides information relating to occupational safety and health for the use of various substances and products.

Chapter 21
Sentiment Analysis of Tweets on the COVID–19 Pandemic Using Machine Learning Techniques

Jothikumar R.

ⓘ https://orcid.org/0000-0003-0806-7368

Shadan College of Engineering and Technology, India

Vijay Anand R.

Velloe Institute of Technology, India

Visu P.

Velammal Enginerring College, India

Kumar R.

National Institute of Technology, Nagaland, India

Susi S.

Shadan Women's College of Engineering and Technology, India

Kumar K. R.

Adhiyamaan College of Engineering, India

ABSTRACT

Sentiment evaluation alludes to separate the sentiments from the characteristic language and to perceive the mentality about the exact theme. Novel corona infection, a harmful malady ailment, is spreading out of the blue through the quarter, which thought processes respiratory tract diseases that can change from gentle to extraordinary levels. Because of its quick nature of spreading and no conceived cure, it

DOI: 10.4018/978-1-7998-6870-5.ch021

ushered in a vibe of stress and pressure. In this chapter, a framework perusing principally based procedure is utilized to discover the musings of the tweets related to COVID and its effect lockdown. The chapter examines the tweets identified with the hash tags of crown infection and lockdown. The tweets were marked fabulous, negative, or fair, and a posting of classifiers has been utilized to investigate the precision and execution. The classifiers utilized have been under the four models which incorporate decision tree, regression, helpful asset vector framework, and naïve Bayes forms.

1. INTRODUCTION

Open thought, emotions and their readiness for the crisis conditions like pandemic is ordinarily forewarned by method of aptitude of the print media, television, net posts, exchange sheets, interpersonal interaction destinations, thus forth Richardson, P et al (2020). Public wellness observing comprises of the wellness observation, following wellbeing dangers, occasional plague flare-ups and various crises consequently to restrict the risk of the famous open wellness. Generally wellbeing related overviews have been taken to secure sentiments of individuals on a wellness crisis. Celikyilmaz, A. (2010) The advancement of web and online life is an amazing flexibly for the wellbeing related information and cure data of the customary open which come to be named as 'Infodemiology'. Ji, X., Chun, S. A., et al (2016) Twitter, a smaller scale running a blog web page allows in customers to extent their considerations and feelings in a type of a message viewed as tweets, Liang, P. W., & Dai, B. R. (2013). The customers in this internet-based life webpage keeps up on developing and as of now the total vivacious clients are just about 330 million. Khatua, A., & Khatua, A. (2016) With this large assortment of clients, twitter has end up being a main gracefully for creating measurements which whenever examined, bears valuable bits of knowledge. Slant correlation and sentiment mining on twitter realities has risen path back in the beforehand years twitter. Piryani, R., et al (2017) A destructive malady is a sickness which spreads during the nations and affect a gigantic populace. Infirmity observation and early discovery may likewise be one of the significant bundles of e-wellbeing realities Li, L., Jin, X., et al (2012). Internet based life comprehensive of twitter can give crucial data comprising of open conversations on a topic alongside medical problems with longitudinal records which prompts expectation of flare-ups of scourges. Alessa, A., & Faezipour, M. (2018) Social media customers are sharing their wellness records and their assumptions that may need to effectively go about as a machine for separating the general popular feeling on pandemic. Russell, C. D., et al (2020) the natural open gratefulness at the hour of pandemic flare-up is generally alluded to in the twitter and tweets goes about as an unmistakable flexibly of records in perusing the notion of the clients. Girotra, M., et al (2013) Twitter goes about as a viable device to get right of passage to open practices over pandemics, surveying the degrees of pandemic, content material exchange on infection episode issue reconnaissance and notion assessment/sentiment mining. Afrati, F. et al (2004) Corona Virus infirmity (COVID-19) is an irresistible medical issue realized by method of a recently discovered Corona infection. The flare-up spreads so quick from the tainted character to the diverse by method of bead sullying, surface contacts and has made a frenzy and crisis in numerous universal areas globally Blendon, R. J., et al (2003). Type procedures in data mining such Naïve Bayes, decision tree as are utilized to unharness the feelings of the clients as a segment of pandemic episode Taboada, M., et al (2011).

The Pediatric and geriatric immunity network mobile computational model for COVID-19 was proposed by the authors Mei, Q., et al (2007). The objective of the proposed methodology uses IoT devices for preventing the COVID-19. The IoT devices and sensors were used for data collection from the remote places Thelwall, M. (2010). The collected data was used by the machine learning algorithms to analyze and predict the chance of COVID-19 and prevention measures can be followed Priya, K. B., et al (2020).

The concept of Monitoring and sensing COVID-19 symptoms as a precaution using electronic wearable devices was proposed. The objective of this work is to observe and sense the different types of COVID-19 symptoms. This research has explained the different stages of COVID-19 along with the symptoms and the various mechanisms to observe the progress of this disease by many parameters. Risk aspects of the disease are sensibly analysed and equated with test results Josephine, M. S., (2020).

2. METHODOLGY USED

The dataset is taken and processed by using the above-mentioned algorithms. The overall process includes the data mining knowledge discovery steps. It needs preprocessing and cleansing of data before doing

Figure 1. Architecture of tweets classification

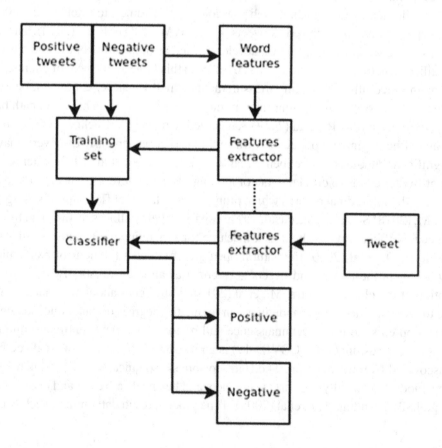

the process. These include lacking of attribute values having errors or outliers which do not belong to the given values. There are also possibilities of error in values or codes or names. The above said abnormalities are corrected before the record is taken for training the system. This process is referred to as data cleaning. The missing values in record can be either filled by studying the relevant records or can be removed if it's below 5%. The architecture of Tweets classification is shown in Fig 1.

2.1 Collection of Tweets

We have aggregated our insights from Kaggle.com which contains nearly39455 tweets at the various hash labels identified with the sickness. Tweets extracted from the link given as follows:

https://www.kaggle.com/skylord/covid19-tweets.

Our work has been conceived in parts along with analyzing the notion of the COVID-19 related tweets and Public discernment at the lockdown or check in time executed for halting the unfurl of COVID-19. Our technique is to perform slant investigation on each tweet in the wake of cleaning and pre-preparing them.

2.2 Processing of Tweets

The section which contains the content of the tweet was taken for the examination. The superfluous segments were dropped and just the section containing the content of the tweets were put away in a csv document. The content of the tweet has numerous commotions, for example, username labels, hash labels, URLs and accentuations. The hash labels and other undesirable characters were expelled and supplanted with a clear space. Tokenization and stemming has been applied on the tweets and the English stop words have been expelled as a procedure on Natural Language Processing.

The cleaned tweet is utilized to examine and set the slant scores utilizing the bundle SentiWordNet in python which is solely utilized for supposition mining. The POS labeled words were utilized to set the positive and negative scores to the tweets. The NLTK toolbox in python is utilized for the procedure which is parting the tweets into important words and giving the objective scores for each word. The given document has been added with a section comprise of the notion scores of the tweet, for example, 0, 1 and - 1 under the arrangements negative, positive and impartial. Utilizing it, a well-known slant dictionary, the tweets were marked and the outcomes are appeared. The set of tweets used are listed as follows and shown in Fig 2.

Tweets Extracted : 43256
Tweets identified with sentiments : 42612
Classified Tweets
Positive Tweets : 10245
Negative Tweets : 13415
Neutral Tweets : 18952

2.3 Analysis of Tweets

The total tweets are taken and splitted into various divisions. It includes train tweets and test tweets. Based on train tweets the model is designed and system is trained. The test data is validated based on

Figure 2. Sentimental Classification of tweets

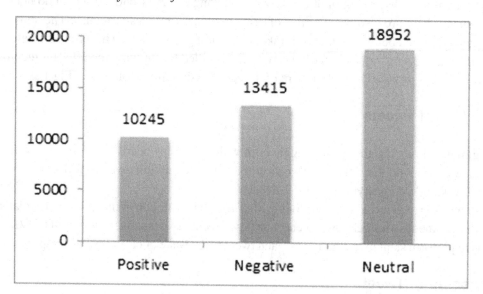

trained model. Based on the analysis we come to an conclusion whether the model is fit or not. The validation results show the trained model efficiency. The percentage of tweets taken for validation differs from person to person. The ration of classification taken for our research is 80:20. Around 80% of data is taken as training data and remaining 20% of data is used for testing and validation purpose. The accuracy results based on the classification of tweets used for training and testing. The processing of data classification plays a major role in determining the performance of the algorithms used for prediction purposes

2.4 Performance Prediction by Classifiers

The unigram model is used to evaluate the performance of classifier after classification of the tweets. The tweets are processed by each of the classifier individually and listed out for validation purposes. Each of the validation is considered at various levels of classification. The final value detests the performance of different classifier. Four different algorithms are used for the tweets classification. The performance of the classifier is evaluated with time taken for classification and cross validation. The algorithm along with the type of classifier chosen for classification of tweets is listed as follows:

2.4.1 XG Boost

$$\hat{y}_i = \sum_{k=1}^{K} f_k(x_i), f_k \in F \tag{1}$$

This classification model is an ensemble approach that uses tree structure in which each step word is taken as attribute and resulting tweets are taken or considered as outcome. It is one of the models that is validated for weaker classification where data labels must be known to get a clear outcome.

The loss function retrieved at the end of the classifier can be further reduced by using the following equation.

$$L(\Phi) = \sum_i l(\hat{y_i}, y_i) + \sum \Omega(f_k)$$

$$where \ \Omega(f) = \gamma T + \frac{1}{2}\lambda\|w\|^2 \tag{2}$$

This classifier implemented from sklearn.ensemble which is given out by scikit-learn. The model supports the classification of tweets into different categories which are substantially classified to positive, negative and neutral.

2.4.2 Maximum Entropy

The primary thought here is, to pick the major uniform probabilistic model which amplifies the value of entropy, considering the imperatives. In contrast to other classifiers, it doesn't expect that highlights are restrictively free of one another. Along these lines, we can include highlights like bigrams without agonizing over component cover. In a twofold grouping issue as the one we are tending to, it is equivalent to utilizing Logistic Regression to discover dispersion over the classes.

$$P_{ME}(c|d, \lambda) = \frac{exp[\sum_i \lambda_i f_i(c, d)]}{\sum_{c'} exp[\sum_i \lambda_i f_i(c, d)]} \tag{3}$$

2.4.3 Support Vector Machine

It is a non-probabilistic binary linear classifier. It considers two coordinates x and y in which x is considered to be the vector denoting the features and y to be the class representation. After this identification of values, hyperplane is found out which divides with min and max value. Representation of hyperplane is stated as w.x-b=0. Classifier does this process is SVC classifier. The margin of hyperplane is maximized using the equation given below:

$$\max_{w, \gamma} \gamma, s.t. \forall i, \gamma \leq y_i(w \cdot x_i + b) \tag{4}$$

2.4.4 Naïve Bayes

$$\hat{c} = \underset{c}{argmax}\ P(c|t)$$

$$P(c|t) \propto P(c) \prod_{i=1}^{n} P(f_i|c)$$

(5)

The two different classifier supported through Naïve Bayes are Bernoulli Classifier and MultinomialNB Classifier. This can be used for text classification, in which c represents class that is assigned to tweet t.

The MultinomialNB Classifier is implemented using the package sklearn. naive Bayes from the scikitlearn. The Laplace method is used where α is represented as smoothing parameter that can set a default value to be 1. The sparse vector representation is used for classification and classified according to the features of frequency and presence. Adding to the results, bigram features make the accuracy to be improved.

3. EXPERIMENTAL RESULTS

Through different classifiers the experiments are performed. We take 10% of data for validation purpose from the dataset classified for training dataset that are used to check overfitting. Considering the above classifiers which are taken for experimenting, XGBoost, Maximum Entropy, SVM and Naïve Bayes sparse vector representation is used for tweets. The accuracy are checked out with various cross validation count. The same is implemented in python language by utilizing the packages available for each of the classification algorithm. The extractions of tweets are retrieved using the code:

Extracting tweets

```
ipfile = []
    with open(db,'rb') as csvfile:
        lineReader = csv.reader(csvfile,delimiter=',', quotechar="\"")
        for row in lineReader:
            ipfile.append({"tweet_id":row[2], "label":row[1], "topic":row[0]})
    rate_limit = 190
    sleep_time = 900/190
for tweet in infile:
        try:
            st = twitter_api.GetStatus(tweet["tweet_id"])
            print("fetched tweet" + st.text)
            tweet["text"] = status.text
            tds.append(tweet)
            time.sleep(sleep_time)
        except:
            continue
```

3.1 N-gram Model

The sparse vector representation is used here considering unigram, bigram and trigram along with vectorization. The frequency term is considered and referred as inverse document frequencies. It refereed as process of converting the text data into numerical form. These classifiers are estimated forming the three different frequencies as the base. The parameters are accuracy and time taken as parameters. It is shown in Table 1.

Table 1. Classification Accuracy of Classifier

Classifier	Accuracy	Time Taken	Cross validation	Accuracy	Time Taken	Cross validation
	Cross fold - 15			Cross fold - 25		
XGBoost classifier	73.25%	4.0	3.5	74.75%	4.0	3.0
Maximum Entropy Classifier	84.3%	5.0	6.0	85.39%	6.0	5.0
SVC classifier	83.29%	2.0	8.0	85.67%	2.0	8.0
Multinomial NB classifier	67.24%	7.0	1.5	69.12%	8.0	1.5

After attaining the classifier accuracy, n-gram analysis is carried out for unigram, bigram and trigram frequencies. Each of the accuracy values are validated before and after considering the stop words is shown in Table 2.

Table 2. Validation of ngram Model

Model Used	Validation result – Count of features	Accuracy Score
Unigram – No Stop words	50000	81.92%
Unigram - Stop Words	20000	85.93%
Bigram - Stop Words	80000	83.91%
Trigram - Stop Words	90000	82.97%

The MultinomialNB Classifier gives out accuracy of 69.12% which can be increased to 79.68% by adding bigram features. The SVC classifier is implemented using sklearn package. The term C is set to default value 0.1. The penalty parameter is represented as C, which is considered as error term. The objective function misclassification is considered. This SVC is executed using both unigram and bigram. Both the frequency and presence features are considered. The accuracy of 83.29% is achieved with both unigram and bigram features. The results are classified as train, test and validation data with the count of tweets is shown in Table 3.

The confusion Matrix is calculated through Naïve Bayes. The matrix is formed by calculating the total positive mapped with the predicted positive tweets. The variation in these values is termed to be misclassification rate. That is confusion matrix is shown in Table 4.

Table 3. Classification of dataset

Dataset	Total tweets	Positive	Negative
Training set	39764	28.65%	49.32%
Validation set	2236	26.12%	43.09%
Test set	2236	27.32%	48.94%

Table 4. Confusion Matrix

Confusion Matrix	Predicted Negative	Predicted Positive
Negative	845	42
Positive	102	337

3.2 Results, Discussion and Performance Analysis

Our model investigations on two classes, for example, "Positive" and "Negative". The exhibition measurements we have utilized is disarray lattice and order report. The disarray lattice speaks to Tp, Tn, Fp and Fn which are True positive, True negative, False positive and False Negative. The Fn and Fp are considered as the mistakes. Accuracy uncovers the extent of right positive forecasts of the classifier and Recall uncovers the right positive expectations. The F1 measure is utilized to investigate how the accuracy and review performs with the order. The F1 score is 0.79 for 'negative' and 0.83 for 'positive' conclusion classifiers as shown in Table 5.

Table 5. Performance Analysis

	Precision	Recall	F1-score	Support
Negative	0.79	0.93	0.89	947
Positive	0.83	0.71	0.69	433
Average weighted mean	0.86	0.82	0.82	1432

Precision = Tp/(Tp+Fp)

Recall = Tp/(Tp+Fn)

4. CONCLUSION

For this research, tweet with the hash tags such as #lockdown, #covid, #corona, #lockdowncorona, #stayhome, #staysafe are considered. The total tweets are considered with these come around nearly 10245. These are considered as positive tweets and are further classified by using different classifiers. Hence by using machine learning algorithms the sentimental classification of tweets is carried out. As per the standard these tweets are extracted, preprocessed by considering retweets, irrelevant, emoticons etc., by using corpus-based standards. Further they are labeled as positive, negative and neutral tweets. The classifiers are used to analyze for different cross fold values. By using n-gram analysis model they are validated and measured by using precision and recall measures. The highest accuracy attained through this process seems to be 85.67%. The total consideration is said to be positive among the entire tweets. In Future, this research may be considered for analysis using deep learning algorithms for better accuracy.

REFERENCES

Afrati, F., Gionis, A., & Mannila, H. (2004, August). Approximating a collection of frequent sets. In *Proceedings of the tenth ACM SIGKDD international conference on Knowledge discovery and data mining* (pp. 12-19). ACM.

Alessa, A., & Faezipour, M. (2018, July). Tweet classification using sentiment analysis features and TF-IDF weighting for improved flu trend detection. In *International Conference on Machine Learning and Data Mining in Pattern Recognition* (pp. 174-186). Springer. 10.1007/978-3-319-96136-1_15

Blendon, R. J., Benson, J. M., Desroches, C. M., & Weldon, K. J. (2003). Using opinion surveys to track the public's response to a bioterrorist attack. *Journal of Health Communication*, 8(sup1, S1), 83–92. doi:10.1080/713851964 PMID:14692573

Celikyilmaz, A., Hakkani-Tür, D., & Feng, J. (2010, December). *Probabilistic model-based sentiment analysis of twitter messages. In 2010 IEEE Spoken Language Technology Workshop*. IEEE.

Girotra, M., Nagpal, K., Minocha, S., & Sharma, N. (2013). Comparative survey on association rule mining algorithms. *International Journal of Computers and Applications*, 84(10).

Ji, X., Chun, S. A., & Geller, J. (2016). Knowledge-based tweet classification for disease sentiment monitoring. In *Sentiment Analysis and Ontology Engineering* (pp. 425–454). Springer. doi:10.1007/978-3-319-30319-2_17

Josephine, M. S., Lakshmanan, L., Nair, R. R., Visu, P., Ganesan, R., & Jothikumar, R. (2020). Monitoring and sensing COVID-19 symptoms as a precaution using electronic wearable devices. *International Journal of Pervasive Computing and Communications*.

Khatua, A., & Khatua, A. (2016, September). Immediate and long-term effects of 2016 Zika Outbreak: A Twitter-based study. In *2016 IEEE 18th International Conference on e-Health Networking, Applications and Services (Healthcom)* (pp. 1-6). IEEE.

Li, L., Jin, X., Pan, S. J., & Sun, J. T. (2012, August). Multi-domain active learning for text classification. In *Proceedings of the 18th ACM SIGKDD international conference on Knowledge discovery and data mining* (pp. 1086-1094). 10.1145/2339530.2339701

Liang, P. W., & Dai, B. R. (2013, June). Opinion mining on social media data. In *2013 IEEE 14th international conference on mobile data management* (Vol. 2, pp. 91-96). IEEE. 10.1109/MDM.2013.73

Mei, Q., Ling, X., Wondra, M., Su, H., & Zhai, C. (2007, May). Topic sentiment mixture: modeling facets and opinions in weblogs. In *Proceedings of the 16th international conference on World Wide Web* (pp. 171-180). 10.1145/1242572.1242596

Piryani, R., Madhavi, D., & Singh, V. K. (2017). Analytical mapping of opinion mining and sentiment analysis research during 2000–2015. *Information Processing & Management, 53*(1), 122–150. doi:10.1016/j.ipm.2016.07.001

Priya, K. B., Rajendran, P., Kumar, S., Prabhu, J., Rajendran, S., Kumar, P. J., & Jothikumar, R. (2020). Pediatric and geriatric immunity network mobile computational model for COVID-19. *International Journal of Pervasive Computing and Communications*.

Richardson, P., Griffin, I., Tucker, C., Smith, D., Oechsle, O., Phelan, A., & Stebbing, J. (2020). Baricitinib as potential treatment for 2019-nCoV acute respiratory disease. *Lancet, 395*(10223), e30–e31. doi:10.1016/S0140-6736(20)30304-4 PMID:32032529

Russell, C. D., Millar, J. E., & Baillie, J. K. (2020). Clinical evidence does not support corticosteroid treatment for 2019-nCoV lung injury. *Lancet, 395*(10223), 473–475. doi:10.1016/S0140-6736(20)30317-2 PMID:32043983

Taboada, M., Brooke, J., Tofiloski, M., Voll, K., & Stede, M. (2011). Lexicon-based methods for sentiment analysis. *Computational Linguistics, 37*(2), 267–307. doi:10.1162/COLI_a_00049

Thelwall, M. (2010). Emotion homophily in social network site messages. *First Monday, 15*(4). Advance online publication. doi:10.5210/fm.v15i4.2897

Chapter 22
Intelligent Medical Data Analytics Using Classifiers and Clusters in Machine Learning

Muthukumaran V.
https://orcid.org/0000-0002-3393-5596
REVA University, India

Satheesh Kumar S.
REVA University, India

Rose Bindu Joseph
https://orcid.org/0000-0002-7033-6226
Christ Academy Institute for Advance Studies, India

Vinoth Kumar V.
MVJ College of Engineering, India

Akshay K. Uday
Dr. M. G. R. Educational and Research Institute, India

ABSTRACT

A privacy-preserving patient-centric clinical decision support system, called PPCD, is based on naive Bayesian classification to help the physician predict disease risks of patients in a privacy-preserving way. First, the authors propose a secure PPCD, which allows the service providers to diagnose a patient's disease without leaking any patient medical data. In PPCD, the past patient's historical medical data can be used by a service provider to train the naive Bayesian classifier. Then, the service provider can use the trained classifier to diagnose a patient's diseases according to his symptoms in a privacy-preserving way. Finally, patients can retrieve the diagnosed results according to their own preference privately without compromising the service provider's privacy.

DOI: 10.4018/978-1-7998-6870-5.ch022

1. INTRODUCTION

Health industry, broadly circulated in the worldwide extension to give administrations to patients, has never confronted such a huge measure of electronic information or experienced such a sharp development pace of information today. In any case, if no proper method is created to discover incredible potential financial qualities from huge human services information, this information may get aimless as well as require a lot of room to store and oversee. In the course of recent decades, the marvelous advancement of information mining method has forced a significant effect on the unrest of human's way of life by foreseeing practices and future patterns on everything, which can change over put away information into important data. These strategies are well appropriate for giving choice help in the social insurance setting. To accelerate the analysis time and improve the conclusion exactness, another framework in medicinal services industry ought to be functional to give a lot less expensive and quicker path for determination (Bellazzi & Zupan, 2008; Elmehdwi & Samanthula, 2014; Elmehdwi et al., 2015; Futoma et al., 2015; Jabbar & Samreen, 2016; Kononenko, 2001; Lavra˘c, 1998; Leroy & Chen, 2001; Monkaresi et al., 2014; Nie, 2015). Clinical choice emotionally supportive network, with different data mining procedures being applied to help doctors in diagnosing understanding diseases with comparative side effects, has gotten an incredible consideration as of late. Naive Bayesian classifier, one of the famous AI instruments, has been generally utilized as of late to anticipate different infections in choice help. it is more suitable for clinical conclusion in social insurance than some complex procedures.

MOTIVATION

Data innovation assumes a significant job in social insurance. Information Technology is an innovation that assists with putting away, investigation, and offer about medical problems. A large number of the social insurance suppliers depend on the Health Information Technology to give data about malady to open. One of the progressions of information is used to store persistent data for long time in an advanced organization. Electronic Health Record contains data about searchers ailment, research center test reports, persistent history, treatment portrayal beams and sweep report which is safely shared among other division like labs, drug stores, authorities (Bahga & Madisetti, 2013; Bollegala et al., 2011; Crammer et al., n.d.; Laleci et al., 2013; Nahato, 2015; Nie et al., 2014; Nie & Zhao, 2013; Nie & Zhao, 2015; Tekin et al., 2015).

CHALLENGES

With the wide use of PC innovation, clinical wellbeing information has additionally expanded significantly, and information driven clinical large information examination techniques have developed as the occasions require, giving help to astute identification of clinical wellbeing. Notwithstanding, because of the blended clinical enormous information design, numerous inadequate records, and a ton of commotion, it is still difficult to break down clinical huge information. Customary AI strategies can't viably mine the rich data contained in clinical enormous information, while profound learning fabricates a progressive model by mimicking the human mind. It has amazing programmed include extraction, complex model development and efficient highlight articulation, and increasingly significant. It is a profound taking in strategy that concentrates highlights from the base to the top level from the first clinical picture information. In this way, this paper develops an information examination model dependent on profound learning for clinical pictures and transcripts, and is utilized for wise identification and determination of infections.

BACKGROUND

The motivation behind why profound learning can exceed expectations in numerous fields is on the grounds that a lot of learning information, the highlights acquired through this learning information have more grounded expressive capacity than the highlights extricated by manual strategies, with the goal that a superior impact can be gotten. Along these lines, this necessitates the profound learning model must have enough learning informational indexes. In any case, in the field of clinical information investigation, the preparation information is truly deficient, which expects scientists to fathom: how to rapidly get a lot of label information and how is it all happening.

2. LITERATURE SURVEY

With the wide utilization of PC innovation, clinical wellbeing information has additionally expanded significantly, and information driven clinical huge information investigation strategies have risen as the occasions require, giving help to smart recognizable proof of clinical wellbeing. Be that as it may, because of the blended clinical enormous information design, numerous fragmented records, and a ton of commotion, it is as yet hard to break down clinical huge information. Customary AI techniques can't adequately mine the rich data contained in clinical enormous information, while profound learning assembles a various leveled model by reproducing the human cerebrum. It has amazing programmed include extraction, complex model development and productive element articulation, and progressively significant. It is a profound taking in technique that concentrates highlights from the base to the top level from the first clinical picture information. Hence, this paper develops an information examination model dependent on profound learning for clinical pictures and transcripts, and is utilized for wise recognizable proof and conclusion of ailments. The model uses gigantic clinical huge information to choose and streamline model parameters, and naturally learns the obsessive investigation procedure of specialists or clinical analysts through the model, lastly keenly leads ailment judgment and powerful choice dependent on the examination consequences of clinical huge information. The exploratory outcomes show that the technique can investigate the clinical enormous information, and can understand the early finding of the malady. Simultaneously, it can break down the physical wellbeing status as per the patient's physical assessment records and foresee the danger of a specific illness later on. Enormously decrease the work weight of specialists or clinical scientists and improve their work productivity.

medicinal services are an innovation that assists with putting away, investigation, and offer about medical problems. A considerable lot of the human services suppliers depend on the Health Information Technology to give wellbeing data about ailment to open. One of the progressions of HIT is EHR used to store tolerant wellbeing data for long time in an advanced configuration. Electronic Health Record contains wellbeing data about wellbeing searchers ailment, lab test reports, quiet history, treatment portrayal beams and sweep report which is safely shared among other office like research facilities, drug stores, masters. Presently a-days wellbeing searchers need moment answers about medical problems.

So, question noting discussion is pulled in by both wellbeing searchers and human services suppliers. Network based social insurance administrations like Health Tap, WebMD, Medicine Net offers wellbeing data through inquiry replying. For the most part, clinical discussions are limited to general client answers as it would prompt negative outcomes on human lives. Questions posted in such discussion set aside more effort for being replied by specialists since they don't discover time to spend in on the web and network produced wellbeing information is by all accounts increasingly conflicting which can't be utilized legitimately because of jargon hole.

Data has wide applications in numerous territories, for example, banking, medication, logical research and among government organizations. Grouping is one of the regularly utilized assignments in information mining applications. For as long as decade, because of the ascent of different protection issues, numerous hypothetical and down to earth answers for the grouping issue have been proposed under various security models. Nonetheless, with the ongoing prominence of distributed computing, clients presently have the chance to redistribute their information, in encoded structure, just as the information mining assignments to the cloud. Since the information on the cloud is in encoded structure, existing protection saving grouping strategies are not material. In this paper, we center around taking care of the order issue over encoded information. Specifically, we propose a safe k-NN classifier over scrambled information in the cloud (Deverajan et al., n.d.; Muthukumaran, 2021; Muthukumaran & Ezhilmaran, 2020; Muthukumaran & Manimozhi, 2021; Nagarajan et al., 2021). The proposed convention secures the secrecy of information, protection of client's information inquiry, and conceals the information get to designs. As far as we could possibly know, our work is the first to build up a safe k-NN classifier over encoded information under the semi-legit model. Additionally, we exactly investigate the proficiency of our proposed convention utilizing a genuine world dataset under various parameter settings.

Unpretentious, contactless accounts of physiological signs is significant for some wellbeing and human-PC association applications. Most present frameworks require sensors which rudely contact the client's skin. Late advances in without contact physiological signs make the way for some new kinds of utilizations. This innovation vows to gauge pulse (HR) and breath utilizing video as it were. The adequacy of this innovation, its constraints, and methods of defeating them merits specific consideration. In this paper, we assess this procedure for estimating HR in a controlled circumstance, in a naturalistic PC connection meeting and in an activity circumstance. For correlation, HR was estimated at the same time utilizing an electrocardiography (ECG) gadget during all meetings. The outcomes reproduced the distributed outcomes in controlled circumstances, yet show that they can't yet be considered as a substantial proportion of pulse in naturalistic Human-Computer Interaction (HCI). We propose an AI way to deal with improve the exactness of HR recognition in naturalistic estimations. The outcomes exhibit that the root mean squared mistake is diminished from 43.76 beats every moment (bpm) to 3.64 (bpm) utilizing the proposed technique.

The paper gives a diagram of the improvement of shrewd information examination in medication from an AI point of view: a verifiable view, a best in class see and a view on some future patterns in this subfield of applied artificial insight. The paper isn't proposed to give a com-prehensive outline yet rather portrays some subareas and bearings which from my own perspective appear to be significant for applying AI in clinical conclusion. In the chronicled outline I underscore the gullible Bayesian classifier, neural systems and choice trees. I present a correlation of some best in class frameworks, agents from each part of AI, when applied to a few clinical symptomatic errands. The future patterns are outlined by two contextual investigations. The first depicts an as of late created strategy for managing unwavering quality of choices of classifiers, which is by all accounts promising for wise information investigation in

Figure 1. User Interface Design

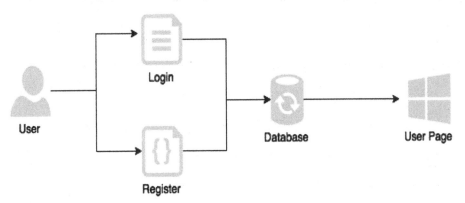

medication. The second portrays an approach to utilizing AI so as to confirm some unexplained wonders from integral medication, which isn't (yet) endorsed by the customary clinical network yet could later on assume a significant job in generally clinical analysis and treatment.

3. INTELLIGENCE MEDICAL DATA ANALYTICS USING CLASSIFIERS AND CLUSTERS

Modules

- user interface design
- hospital management
- trust authorizes
- symptoms solution
- chatting technique
- reviews

User Interface Design

To interface with server customer must give their username and mystery express then nobody however they can prepared to relate the server. In case the customer starting at now exits clearly can login into the server else customer must enlist their nuances, for instance, username, mystery key and Email id, into the server. Server will make the record for the entire customer to keep up move and download rate. Name will be set as customer id. . Marking in is ordinarily used to enter a specific page

Hospital Management

To connect with server admin must give the username and password then only they can able to connect the server. If the admin have only the login process don't register the admin. After logging it will go to the admin page that time admin also can use the process. The process is register the trust Authorizes and doctor.

Figure 2. Hospital Management

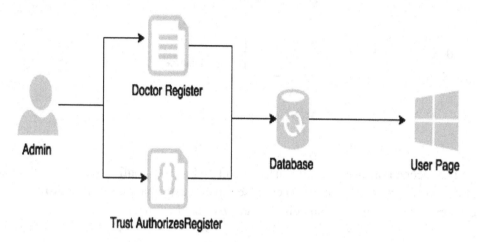

Trust Authorizes

Trust Authorizes to connect with server give their username and password then only they can able to connect the server. The trust authorizes are collect the historical data. The Authorizes are get data and upload the data for database. The process file upload that time file was encrypt the file store the value in database.

User Symptoms Solution

The user enter the user page that time user view search the symptoms by patient that will be the find user solution. The user can also find value symptoms is the detail that value in the for user symptoms. The user or any person can be search the image that image background also set the value that find the

Figure 3. Trust Authorizes

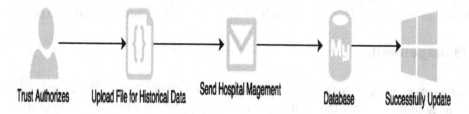

Figure 4. Use symptoms

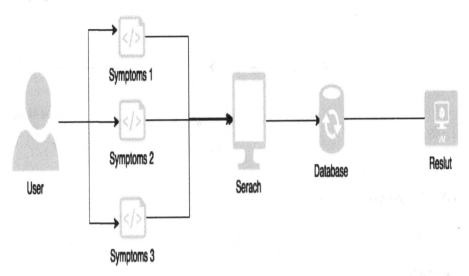

image. The image can search the give key word that time image search the all the database and collect the matching the image. It will show the result for user or any person.

- **Chatting Technique**

The user are chatting with doctor that used for the Verification. The result the symptoms based result is correct or not will verify the particular specialist doctor can replay the user query that query are used to take database values.

Figure 5. Chatting Technique

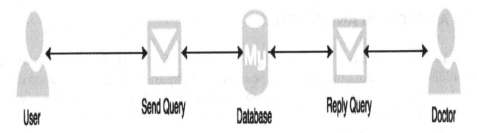

Reviews

The feature enhancement we use Our future work will focus on the following we will exploit privacy preserving patient-centric clinical decision support systems with other advanced data mining techniques, We are doing reviews with the process.

Figure 6. Reviews

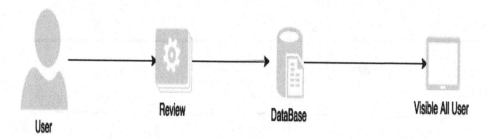

TECHINQUE OR ALGORITHM

Encrypt, Decrypt, Re-encrypt & Agg, and Re-decrypt

Encrypt, Decrypt

- this algorithm is executed by DP i. Let x(i) 2ZN be the message which can be encrypted under DP i's public key pkai . Then, the ciphertext can be calculated
- It can be decrypted by using DP i's private key s

Re-Encrypt and Agg and Re-Decrypt

- This algorithm is executed by CP. The algorithm can be processed as follows: 1) For each DP i, the ciphertext in DP i's domain [x(i)]pkai can be re-encrypted into PU's domain by skai!P
- This algorithm is executed by PU. PU can decrypt the aggregated ciphertext CTAgg by using skP
- Naïve Bayesian and k-means algorithm is being used.

4. DESIGN APPROACH AND DETAILS

Configuration Engineering manages the different UML outlines for the usage of task. Configuration is a significant building portrayal of a thing that will be assembled. Programming configuration is a procedure through which the necessities are converted into portrayal of the product. Configuration is where quality is rendered in programming designing. Configuration is the way to precisely make an interpretation of client prerequisites into completed item.

Figure 7. Use Case Diagram

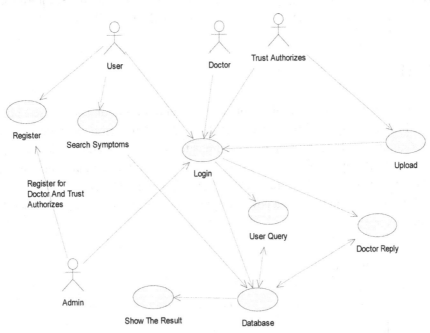

EXPLANATION

The main purpose of a use case diagram is to show what system functions are performed for user can login and enter the symptoms and get the value. And admin are update the doctor and trust authorizes

Figure 8. Class Diagram

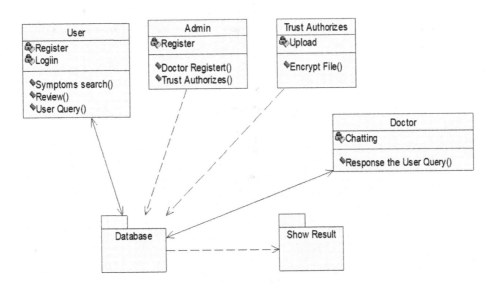

register. The trust authorizes are upload the historical data. The doctor are user request the some query that query to reply the doctor. And user also post reviews. It will show all user.

EXPLANATION

The class diagram is the main building block of object oriented modeling. It is used both for general conceptual modeling of the systematic of the application, and for detailed modeling translating the models into programming code in the Diagram we are is to show what gadget features are achieved for user can login and input the symptoms and get the price. And admin are update the physician and trust authorizes register. The trust authorizes are add the ancient information. The medical doctor are person request the some question that question to answer the physician. And person additionally submit reviews. it will show all person

Figure 9. Data Flow Diagram Level-0

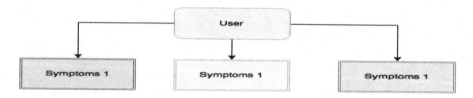

Figure 10. Data Flow Diagram Level-1

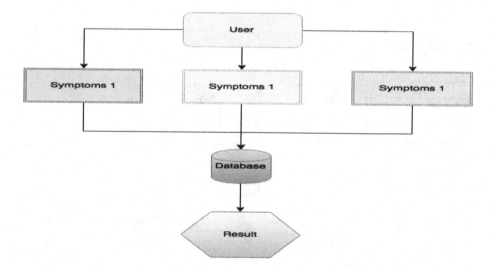

EXPLANATION

It does not show information about the timing of processes, or information about whether processes will operate in sequence or in parallel. In the DFDs the level zero process is based on the login validations. What is the cloud user contained constraints send to the cloud provider.

Figure 11. System Architecture

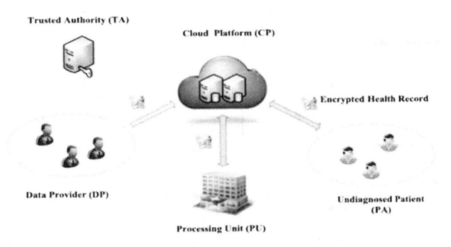

Proposed System Model Explanation

The problems mentioned above, in this paper, we propose a novel fine-grained image classification method by leveraging the low-rank sparse coding (LRSC) technique and combine it with general and class-specific codebook generation. We learn a general codebook and a number of codebooks per class for joint encoding of local features. The general codebook represents the universal information of all classes while each class-specific codebook encodes the distinctive character of each class. To model the differences between general codebook and each class-specific codebook, the scarcity constraint is used along with the codebook incoherence's. As to the encoding of local features, the low-rank constraint is leveraged to consider the spatial and structure information of local features within a particular image region. Instead of treating each region separately, we encoded the corresponding regions of the same position within the training images to make use of the spatial information. We conduct fine-grained image classification experiments on several public image data sets and the results show the effectiveness of the proposed method.

5. IMPLEMENTATION

User Interface Design

To connect with server user must give their username and password then only they can able to connect the server. If the user already exits directly can login into the server else user must register their details such as username, password and Email id, into the server. Server will create the account for the entire user to maintain upload and download rate. Name will be set as user id. Logging in is usually used to enter a specific page

Figure 12. User Interface Design

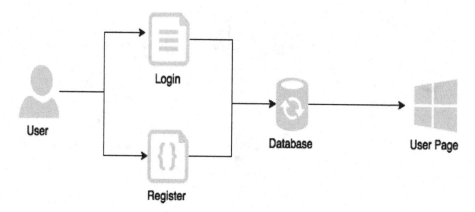

OWNER CREATES A WEBSITE

Owner has to register with that page. Then he has to create a website and need to get the access from the admin and he can be able to view the user details and he can also view books in which the user has purchased already

ADMIN GIVES THE AUTHORIZATION TO THE USER

Let the admin should register in that page. After the user look for the books that is already available in certain case if the user don't need that book means he will be requesting to the admin the book which he needed to buy

USER SEARCHES FOR THE BOOKS

When the user registered with an correct information then he will be able to see the books that is available in the website. If the book is not available then he will be requesting to the admin for uploading the information of the book

BLOCKING THE THIRD PARTY USER WITH ENCRYPTION

If a user tries to access the book that is purchased by the requested customer he will be able to share the secret key with that standard he will be visiting the complete record of the data's, if third party enters into our website the book will be shown as a white paper he will be blocking from the trustee to the end customer.

6. CONCLUSION

The sharp increment in clinical huge information puts greater levels of popularity on clinical huge information handling techniques, and its mind boggling, and different organizations increment the trouble of its investigation. Among the current information investigation techniques, profound learning is without a doubt the best one. As indicated by the two principal information configurations of clinical picture information and clinical content information, this paper plans the relating profound learning model independently, and understands the clever recognizable proof of precise early conclusion and hazard expectation for explicit ailments. Initially, the Auto Encoder profound learning model is planned, and it can pre-train the system ahead of time and lessen the utilization of processing time and registering assets. In this way, the strategy can be effectively reached out to other clinical picture information investigation and preparing, which is of incredible essentialness for improving the exactness of ailment finding. Also, and spatial pyramid pooling is structured. Structure in the model can safeguard the planning attributes of various information while extricating the interior highlights of the information, while the spatial pyramid pooling structure can process input information of self-assertive length, consequently making powerful information examination for the eventual fate of the patient in insightful distinguishing proof and control of illness chance.

REFERENCES

Bahga & Madisetti. (2013). A Cloud-based Approach for Interoperable Electronic Health Records (EHRs). *IEEE Journal of Biomedical and Health Informatics, 17*(5).

Bellazzi, R., & Zupan, B. (2008). Predictive data mining in clinical medicine: Current issues and guidelines. *International Journal of Medical Informatics, 77*(2), 81–97. doi:10.1016/j.ijmedinf.2006.11.006 PMID:17188928

Bollegala, Matsuo, & Ishizuka. (2011). A Web Search Engine-Based Approach to Measure Semantic Similarity between Words. *IEEE Transactions on Knowledge and Data Engineering, 23*(7).

Crammer, Dredze, Ganchev, & Talukdar. (n.d.). *Automatic Code Assignment to Medical Text.* Department of Computer and Information Science, University of Pennsylvania, Philadelphia.

Deverajan, G. G., Muthukumaran, V., Hsu, C. H., Karuppiah, M., Chung, Y. C., & Chen, Y. H. (n.d.). Public key encryption with equality test for Industrial Internet of Things system in cloud computing. *Transactions on Emerging Telecommunications Technologies*, e4202.

Elmehdwi, Y., & Samanthula, B. K. (2014). Secure k-nearest neighbor query over encrypted data in outsourced environments. *Proc. IEEE 30th Int. Conf. Data Eng.*, 664–675.

Elmehdwi, Y., Samanthula, B. K., & Jiang, W. (2015). k-Nearest neighbor classification over semantically secure encrypted relational data. *IEEE Trans. Knowledge Data Eng.* Available: https://ieeexplore.ieee.org/xpl/articleDetails.jsp?arnumber=6930802

Futoma, J., Morris, J., & Lucas, J. (2015, August). A comparison of models for predicting early hospital readmissions. *Journal of Biomedical Informatics*, *56*, 229–238. doi:10.1016/j.jbi.2015.05.016 PMID:26044081

Jabbar, M. A., & Samreen, S. (2016). Heart disease prediction system based on hidden naïve bayes classifier. *2016 International Conference on Circuits, Controls, Communications and Computing (I4C)*.

Kononenko, I. (2001). Machine learning for medical diagnosis: History, state of the art and perspective. *Artificial Intelligence in Medicine, 23*(1), 89–109. doi:10.1016/S0933-3657(01)00077-X PMID:11470218

Laleci, Yuksel, & Dogac. (2013). Providing Semantic Interoperability Between Clinical Care and Clinical Research Domains. *IEEE Journal of Biomedical and Health Informatics, 17*(2).

Lavraˇc. (1998). Intelligent data analysis for medical diagnosis: Using machine learning and temporal abstraction. *Artif. Intell. Commun., 11*(3), 191–218.

Leroy, G., & Chen, H. (2001, December). Meeting medical terminology needs-the ontology-enhanced medical concept mapper. *IEEE Transactions on Information Technology in Biomedicine, 5*(4), 261–270. doi:10.1109/4233.966101 PMID:11759832

Monkaresi, H., Calvo, R. A., & Yan, H. (2014, July). A machine learning approach to improve contactless heart rate monitoring using a webcam. *IEEE Journal of Biomedical and Health Informatics, 18*(4), 1153–1160. doi:10.1109/JBHI.2013.2291900 PMID:25014930

Muthukumaran, V. (2021). Efficient Digital Signature Scheme for Internet of Things. *Turkish Journal of Computer and Mathematics Education, 12*(5), 751–755.

Muthukumaran, V., & Ezhilmaran, D. (2020). A Cloud-Assisted Proxy Re-Encryption Scheme for Efficient Data Sharing Across IoT Systems. *International Journal of Information Technology and Web Engineering, 15*(4), 18–36. doi:10.4018/IJITWE.2020100102

Muthukumaran, V., & Manimozhi, I. (2021). Public Key Encryption With Equality Test for Industrial Internet of Things Based on Near-Ring. *International Journal of e-Collaboration, 17*(3), 25–45. doi:10.4018/IJeC.2021070102

Nagarajan, S. M., Muthukumaran, V., Murugesan, R., Joseph, R. B., & Munirathanam, M. (2021). Feature selection model for healthcare analysis and classification using classifier ensemble technique. *International Journal of System Assurance Engineering and Management*, 1-12.

Nahato. (2015). *Knowledge Mining from Clinical Datasets Using Rough Sets and Backpropagation Neural Network*. Comp. Math. Methods in Medicine.

Nie, Li, & Akbari. (2014). *WenZher: Comprehensive Vertical search for Healthcare Domain*. Academic Press.

Nie, L. (2015). *Bridging the Vocabulary Gap between Health Seekers and Healthcare Knowledge*. Translational.

Nie, L., & Zhao, Y.-L. (2013, August). Bridging the Vocabulary Gap between Health Seekers and Healthcare Knowledge. *IEEE Transactions on Knowledge and Data Engineering*.

Nie, L., & Zhao, Y.-L. (2015). *Bridging the Vocabulary Gap between Health Seekers and Healthcare Knowledge*. Translational.

Tekin, Atan, & van der Schaar. (2015). *Discover the Expert: Context- Adaptive Expert Selection for Medical Diagnosis*. Academic Press.

Chapter 23
Machine Learning for Industrial IoT Systems

Mona Bakri Hassan
Sudan University of Science and Technology, Sudan

Elmustafa Sayed Ali Ahmed
https://orcid.org/0000-0003-4738-3216
Sudan University of Science and Technology, Sudan & Red Sea University, Sudan

Rashid A. Saeed
https://orcid.org/0000-0002-9872-081X
Sudan University of Science and Technology, Sudan & Taif University, Saudi Arabia

ABSTRACT

The use of AI algorithms in the IoT enhances the ability to analyse big data and various platforms for a number of IoT applications, including industrial applications. AI provides unique solutions in support of managing each of the different types of data for the IoT in terms of identification, classification, and decision making. In industrial IoT (IIoT), sensors, and other intelligence can be added to new or existing plants in order to monitor exterior parameters like energy consumption and other industrial parameters levels. In addition, smart devices designed as factory robots, specialized decision-making systems, and other online auxiliary systems are used in the industries IoT. Industrial IoT systems need smart operations management methods. The use of machine learning achieves methods that analyse big data developed for decision-making purposes. Machine learning drives efficient and effective decision making, particularly in the field of data flow and real-time analytics associated with advanced industrial computing networks.

DOI: 10.4018/978-1-7998-6870-5.ch023

INTRODUCTION

The world has witnessed a change and important contribution in the area of artificial intelligence, industry, robotics and automation technologies. These technologies have a strong impact on knowledge management process which made the foundation of decision making in several fields such as industrial, home, health, financial and many other to create a smarter environment (Ruhul et al, 2020). Intelligent information technology is a new novel feature for industrial Internet of Things (IIoT). Most of the industries are attempt to automate the process of developing and manufacturing products. Using Machine Learning (ML) in Industry 4.0 is a key feature to obtain an IIoT. Due to increasing scale of deployed terminals IoT in industrial, the IIoT becomes heterogeneous, diverse, and dynamically changeable.

Generally, IIoT system consists of an intelligent control technology, network communication technology and information processing technology. ML is used to connect the real world virtually in intelligent ways. ML is a new topic used in IIoT (Ruhul et al, 2020). Many businesses deal with ML models and algorithms to reduce the production and operation costs. ML can be implemented in smart manufacturing, predictive maintenance in industrial, drug discovery, autonomous vehicles and machines for industry, pattern imaging analytics and software testing (Baotong et al, 2019).

In Industry 4.0 machines interact with their environment intelligently and learn to understand adapt their operational behaviour and processes. ML algorithm improves the quality, cost and flexibility of production process (Wen et al, 2019). Nowadays, most of the factories are trying to automate the process of manufacturing products. Due to this kind of automate the system exposed so you need a high-level security. ML is a part of intelligent system in Industry 4.0, is broadly implemented in different scope in manufacturing are to extract knowledge out of existing data (Inés et al, 2018). It provides decision-making process in manufacturing system. But the goal of the ML techniques is to detect the patterns among the data sets or regularities that describe the relationships and structure between those sets (Asharul et al, 2020).

The chapter describes the concepts of ML use in IIoT, the architecture of ML in IIoT, in addition to the optimization methods, algorithms and applications that used in ML for IIoT. The chapter is organised as follows, section 1 provides a brief conception about the Industrial Internet of Things (IIoT) and the Machine Learning contribution to its applications, in addition to recently works related to the machine learning use in IIoT. The background and chapter motivation is presented in section 2. The machine learning architecture layers and their operational functions are illustrated in section 3. In section 4, different machine learning methods are discussed in technical and performance. The implementation of machine learning algorithms in the industrial IoT applications has been discussed in section 5. In section 6, some examples of machine learning applications in IIoT are reviewed. Finally, the industrial IoT future trend and expectations are presented in section 7, and followed by the chapter conclusion.

Industrial Internet of Things (IIoT)

The Industrial Internet of Things (IIoT) is known as a subset of Internet of Things (IoT), interconnects a multitude of industrial devices, sensor, actuator and people at factory [2]. Also, it is known as a group of infrastructures, interconnecting connected objects and allowing their data mining, management and the access to data they generate such as sensors and/or actuators which carry out a specific function that are able to communicate with other equipment. IIoT integrate the theory and technology of artificial intelligence (AI) for analysing the IIoT (Ruhul et al, 2020) (Baotong et al, 2019). The objective of using

IIoT is to achieve flexible configuration of the industrial resources and rapid adaptation in industrial application by means of the ubiquitous perception, network interconnection and interoperation system.

Industrial IoT has new insights into industrial applications, reduce manual labour and time, and pave the way for Industry 4.0. IIoT not only support to solve social networking problems like congestion control, error handling, routing selection and dynamic segmentations, but also provides used in core technologies such as skip channel, and information security, deterministic scheduling and time synchronization. IIoT is a set of machines, cognitive technologies, robotics and computers for intelligent industrial operations with the help of data analytics (Inés et al, 2018). The IIoT is concerned with innovation, big data, automation and cyber physical systems in industries. It captures large amount of data to manage time, predict maintenance and cost control after machine learning models implementation (Asharul et al, 2020). Also, IIoT has helped in monitoring and controlling manufacturing and production from remote locations. IIOT technology deals with the complex physical machinery to connect with industrial sensors and relevant software. Moreover, it has human interfacing unit for error-free system (Asharul et al, 2020) (Baotong et al, 2019). IIOT is transforming massive industries such as manufacturing, mining and transportation energy consumption and it will have a huge impact on the economic condition.

Machine Learning Contribution to IIoT

Machine learning (ML) is a new solution for the applications in IIoT, it provides an intelligent behaviour learned from human beings, which performs on machines like the computer or the controller i.e., it makes data transmit to connect the real world with the virtual space in intelligent ways. ML has several processes to implement such as perception, understanding, learning, judgement, reasoning, planning, designing, and solving. Today, ML applications can be implemented in industrial landscape. Many businesses deal with ML models and algorithms to reduce the production and operation costs. The aim of applied the ML algorithm in industrial IoT for improving the product quality and flexibility of production process (Avish et al, 2019). ML opens many opportunities to automate and optimize the world of IoT. Using ML algorithms, organizations can utilize IoT data to discover patterns and build models which can then be scored in real time on the IoT data to operationalize the models (Balamurugan et al, 2019).

Nowadays, ML becomes synonyms to Artificial Intelligence (AI). The internet and software applications are using AI and ML (Hamid el al, 2019). ML can recognize the repeated patterns and teach the device to be cautious next time. Additionally, it can support you to automate the process of data analysing and give you automatic updates on security risks. Automation machine in industrial IoT used to reduce the human error and save time. It provides decision-making process and making prediction in manufacturing system (Hamid el al, 2019). But the goal of the ML techniques is to detect the patterns among the data sets or regularities that describe the relationships and structure between those sets.

Machine learning in industry machines interact to each other in intelligent way and learn to understand adapt operational behaviour and processes. AI technologies like sensor, robots, machine learning, and neural network make machines smarter than in the past. The goal of using ML algorithm is to improve the quality, make flexibility of production process, low cost and reduce human error.

Machine Learning Based Industrial IoT Related Works

Industries 4.0 relies on innovative digital technologies that improve industry and product quality, predict errors and work safety. Artificial intelligence is entering the Industrial IoT to enhance efficiency in energy use and production, which will greatly influence the concepts of Industry 4.0. Recently, many studies have emerged that are working on developing AI applications based on machine learning to improve a number of industrial aspects. (Lojka et al, 2016), reviews an integration and predictive control in the industry. The proposed industrial IoT architecture consists of IoT gateway, cloud services and machine learning services. The study shows that machine learning provides an acquisition solution to prediction control as a cloud service. (Kanawaday et al, 2017), review some of the mechanisms used in ARIMA prediction on time series data collected from different sensors of an industrial machine. The authors present that the use of machine learning helps predict potential failures and quality defects during machine operation so that this helps improve the overall manufacturing process. the study show that machine learning enables to overcome the major limitations in productivity and maintenance costs associated with industrial IoT.

In the field of industrial information technology, machine learning algorithms offer new ways of securing industrial systems. (Zolanvari et al, 2018) presented a study on the possibility of integrating machine learning with the safety mechanisms of industrial IoT systems to improve the performance of operations on them. In addition to presenting the challenges in designing safety mechanisms that rely on robotic science in the IIoT.

In the study presented by (Yang Xu et al,2019), authors, describe a blockchain-based fair nonrepudiation service provisioning scheme for industrial IoT scenarios. The blockchain offers service separately via online and offline channels while providing mandatory evidence for unregistered purposes. The proposed system performs security and reliability analysis, and conducts evaluations of effectiveness and efficiency related to industrial IoT. (Qingfei et al, 2019), proposes a framework for constructing a digital twin for petrochemical industrial IoT. The study uses the machine learning concept to production control optimization. The integration between the machine learning and real-time industrial big data enable to optimize digital twin models. The evaluation shown that machine learning enhance the production control based on real-time data.

The study presented by (Zolanvari et al,2019), reviews the Machine learning (ML) and big data analytics for securing Industrial Internet of Things (IIoT). Authors discussed the use of cyber-vulnerability assessment and the utilization of ML for the security vulnerabilities. the study develop a detection system based on machine learning against different attacks such as backdoor, and injection attacks. The evaluation of proposed system shows an efficient performance in security.

(Trakadas et al,2020) present a more holistic integration of AI by promoting collaboration adoption in manufacturing contexts. The author's purpose three technical factors related to functionality layers based architecture extending in Industry 4.0, they are reference layer model, new layers for human collaboration, in addition to security concerns. The architecture is extending the RAMI 4.0 architecture in the context of adopting AI systems in the manufacturing area. The Proposed architecture regarding the optimization of the manufacturing logistics processes, and the facilitation of zero-defect manufacturing.

(Shahid et al,2020), provide a concept about AI technologies for Industrial Internet of Things (IIoT) security, in addition to discuss different cybersecurity attacks. The author then proposed a novel random neural network (RaNN)-based prediction mode to predict the attacks. The evaluation of the proposed model achieves an accuracy of 99.20% for a learning rate of 0.01, with a prediction time of 34.51 milliseconds. It improves the attack detection accuracy by an average of 5.65%.

In the study proposed by (Ruhul et al, 2020), a potential impact of deep learning (DL) in IIoT is reviewed. authors review various DL techniques related to industries smart applications such as smart manufacturing, smart metering, smart agriculture in addition to IoT system aspects. The authors also categorize the challenges facing the implementation of DL-IIoT, and reviews future research trends.

Accurate prediction of downtime of industrial machines is an important issue in for optimizing the operations in industrial IoT. (Srikanth et al,2020) reviews the use of deep learning (DL) algorithms for industrial IoT predictive maintenance. DL enables to predicts industry engine failures. The authors discuss the benefits of DL usage in industrial IoT in addition to the impact of DL for sensor data analysis in predictive maintenance and future opportunities of data-driven methods based on machine learning and deep learning algorithms.

Table 1. Summary of ML Learning Related Studies in Industrial IoT

Citations	Approach	Feature	Advantages
Srikanth et al (2020)	Deep learning for accurate prediction	Industrial IoT predictive maintenance	Efficient predict the industry engine failures
Ruhul et al (2020)	Deep learning impact IIoT	Evaluate the implementation of deep learning in IIoT	Deep learning achievements towards different IIoT applications
Shahid et al (2020)	Random neural network for security Issue	Attacks prediction in industrial IoT	Improve current attack prediction by 5.65%
Trakadas et al (2020)	AI systems architecture for secure industry	Optimize logistics processes in Industry 4.0	Efficient security optimization
Zolanvari et al (2019)	Machine learning for secure IIoT	Attack detection based on machine learning	Enhance performance in security and cyber-vulnerability assessment
Qingfei et al (2019)	Machine learning for production control optimization	Digital twin framework for petrochemical industrial IoT	Improve production control based on real-time data
Yang Xu et al (2019)	AI for blockchain-based fair nonrepudiation service provisioning	Security and reliability analysis in IIoT	Efficient secure blockchain for IIoT
Zolanvari et al (2018)	Machine learning safety IIoT operations	Improve the performance of industrial IoT information technology	Safety mechanisms on robotic science in the IIoT
Kanawaday et al (2017)	Machine learning for prediction the industrial failure	Predict potential failures and quality defects in machine	Improve the overall manufacturing process
Lojka et al (2016)	Machine learning based control architecture for IIoT	Predictive control in IIoT as a cloud service	Acquisition solution to prediction control as a cloud service

According to the survey presented in previous studies, machine learning methods are used to improve a number of topics related to the industrial IoT, the table 1 below shows a brief summary of discussed the related works and the issues that have been addressed. With regard to machine learning methods and their uses in improving a number of processes related to industrial IoT quality service, in addition to the applications of decision-making in a number of industrial systems, we review in the following sections the technical details of machine learning algorithms and their applications on the Internet of industrial things industrial internet of things (IIoT).

BACKGROUND AND MOTIVATION

The idea of Industrial IoT (IIoT) initiated in 2010 by Google. Then Chinese Government developed an IIoT into their five-year plan. After that German Government release the revolution. Things which are used in IoT must be smart devices with self-processing units. They have possibility of exchanging information with each other and can control the overall network (Asharul et al, 2020). However, IIOT bring to a new level, it supports a system and a standard for the universal interconnection via Internet Protocol (IP). IP is a protocol used to exchange information via the internet, no matter about type of device that used for exchanging data (Hamid el al, 2019). Also, they used IP to communication with other devices in deferent network. The number of IIoT devices expected to increase up to 75.44 billion in 2025. IIoT runs into various technologies such as Internet of Things, artificial intelligence, cloud computing, artificial intelligence for Cyber Physical Systems (CPS), big data analytics, blockchain, augmented and virtual reality. The essential elements of IIoT are smart devices like sensors and actuator, communication network such as Modbus/TCP and TSN, and big data analytics like Hadoop, Hive and spark (Li Da Xu et al, 2018).

Machine learning is a subset of AI and has become an important part of IIoT because of efficient and effective industrial outputs. Industrial units face uncertain problems of equipment failure and downtime. The machine learning has a possibly to improve the unexpected system outcomes. The trained and tested machine learning models in the industrial settings contribute to predictive maintenance and risks reduction. Figure 1 shows machine learning concepts. Generally, ML methods are unsupervised, supervised and reinforcement learning. In the unsupervised ML scheme, training depends on untagged data. It tries to find an effective representation of untagged data. While in the supervised learning, it learns from a group of labelled data. In supervised learning, regression and classification schemes are used to training the discrete and continuous data for prediction and decision making.

The reinforcement learning (RL) studies from the activities of the learning agent from the consistent reward in order to capitalize the notion of cumulative rewards. The Markov Decision Process (MDP) is sample of RL (Ruhul et al, 2020) (Hamid el al, 2019). The image classification, market forecasting, disease diagnosis, weather and fraud detection uses supervised learning algorithms such as Linear Regression, Support Vector Machine, Decision Tree and K Nearest Neighbour. The big data visualization and recommender system uses unsupervised learning algorithms such as K-Mean Clustering and FP Growth. The real time decision and robot navigation uses reinforcement learning. The IIoT applications in industries require standards and intelligence techniques including low-power wireless networking technologies and fast sensors for big data analytics (Bojana et al, 2018) (Zeki et al, 2020). Smart devices used to collect data in industrial production and operation and stored over cloud for estimation and data analysis. The machine learning models are trained and tested on the data sets generated by the IIoT for smart decision. The machine learning models improve the product market cost, manages security risk, prevents churn, identifies anomalies in manufacturing and enhances communications.

MACHINE LEARINING ARCHITECTURE FOR INTELLIGENT IIOT SERVICES

Machine learning system maps an input into its corresponding, an output by training the system for learning and obtain the specific results (Ruhul et al, 2020). IIoT system consists of network communication technology, information processing technology and perceptual control technology. Network

Figure 1. ML Techniques and Methods

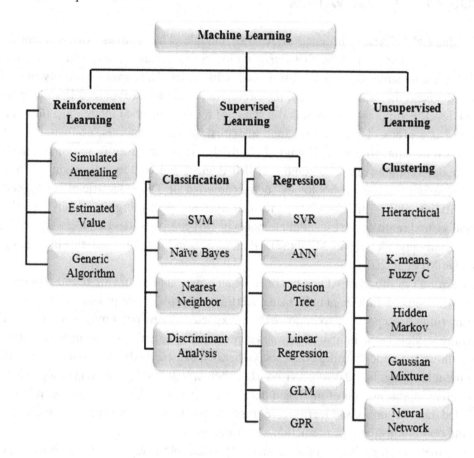

communication and data flow analysis used ML methods to produce four features of modern industry they are: intelligent production, personalized customization, convenient service extension and networked collaboration (Asharul et al, 2018). Figure 2 shows the ML architecture for intelligent IIoT services which including three layers, sense layer, transport layer, and service layer (Avish et al, 2019). Moreover, IoT devices used for collect the data, and then send it to the cloud, after that make machine learning inferences in the cloud, and finally send the data back to the device.

Sense Layer

IIoT uses recognition like sensors, actuators and radio frequency identification (RFID) to achieve information from different dimensions in industrial field. Furthermore, the sense layer is the basis of IIoT. On one hand, the sensor devices have a minimum size and low-cost, which facilitate universal distribution of perceptual layer. It makes possible for industrial control system to collaborate with several business systems. It obtains information from field equipment such as huge number of sensors transmitters, actuators, servo drivers, and motion controllers. It seamless connects with the data acquisition and monitoring control system or the distributed control system in IIoT (Avish et al, 2019)

Figure 2. ML based architecture for intelligent IIoT services.

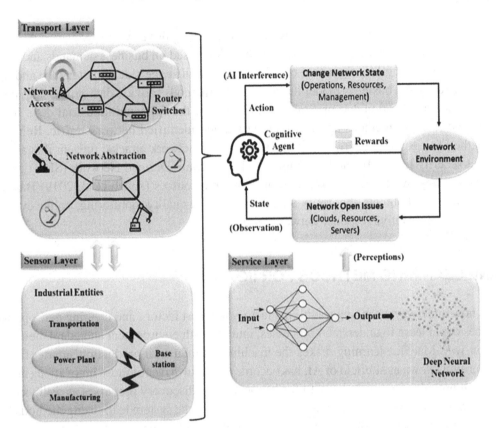

Transport Layer

In industrial region, transport layer is the medium that used to transmit data from sender to receiver. It includes different types of IIoT technologies, for example short distance wireless communication technology, industrial Ethernet, and low power wide area network (LPWAN) such as NB-IoT, LTE and cellular 5G (Avish et al, 2019). The standardization of Fieldbus had met the requirements of openness, automation, and compatibility for industries. Industrial Ethernet is a regional network based on IEEE 802.3 which allows industries devices to connection in wide area network. Industrial wireless network is a sophisticated technology for wireless sensor network and data transmission in industrial domain. The application of using wireless network causes reduction in wiring costs of industrial sensors, which facilitates the function expansion of sensor nodes. IIoT applied to solve social networking problems including congestion control, routing selection, dynamic segmentations and error handling also required to breakthrough core technologies such as deterministic scheduling, time synchronization, information security and skip channel (Avish et al, 2019)(Balamurugan et al, 2019).

Service Layer

Industrial IoT should have knowledge plane relying on AI or cognitive system. ML model built on network side. It used for extract information and identification based on business demands and experience of data expert. When using ML algorithm on AI platform in the edge side, maintenance and operation can design and verify the business model iteratively. ML methods provide data processing, data transformation, metadata collection and preliminary cleaning for the collected industrial data. Supervised or unsupervised learning methods applied to classified and identified network traffic. Reinforcement learning method for secure the network by identify external network attacks; prevent illegal intrusion, and different security level. ML model combines with Software Defined Network (SDN) gateway for routing plan and improve the utilization ratio of network resources (Avish et al, 2019)(Balamurugan et al, 2019). The model obtains abnormity diagnosis, network quality monitoring, online evaluation of network QoS or self-adaptive response network failure.

MACHINE LEARINING METHODS FOR IIOT

The IIoT is now underway, changing traditional factory to smart factory and creating new opportunities that machines could learn to understand processes, interact with environment and adapt their behaviour in intelligent way. Machine learning makes the machines in industrial production smarter than before. Machine learning known as subfield of AI, has become the main driver of those innovations in industrial sectors, which provides the opportunity to further accelerate discovery processes as well as enhancing decision making. Machine learning methods are used to make the system learn using methods like deep learning and reinforcement learning which are further classified in methods like classification, regression, generalization, association and clustering as shown in figure 3 (Bojana et al, 2018).

Deep Learning

Smart manufacturing makes several advantages in IIoT for both production and manufacture intelligent. Nowadays, IIoT solutions have been grown significantly, especially in sensor-based data contains a various structure, formats, and semantics. Data released from IIoT technologies is derivative from various resources across manufacturing that including labour operation, equipment-and-processes, product line, and environmental conditions. Thus, data modelling, labelling, and analysis are making manufacturing much smarter (Bojana et al, 2018). Most of machine learning techniques applied to reduce complexity of the data as well as making the aspects more visible for learning algorithms to be applied. DL is based on conception, perception, and decision-making. One of the useful methods in Machine Learning (ML) is Deep Learning (DL) which used to upgrade manufacturing to a highly optimize smart facilities by processing a bunch of data with its multi-layered structure. Moreover, it used huge neural network layers by using many processing units that has several advantages to enhanced training techniques for learning complex patterns in lots of data (Zuzanna et al, 2020).

Deep learning provides computing intelligence from unclear sensory data, resulting in smart manufacturing. DL techniques are more powerful because of their automatic behaviour to learn data, smart decision-making process and identifying the basic patterns. The main advantage of DL over traditional ML methods is that learning done automatically, and no need to design a separate algorithm to work

Figure 3. Machine Learning Map

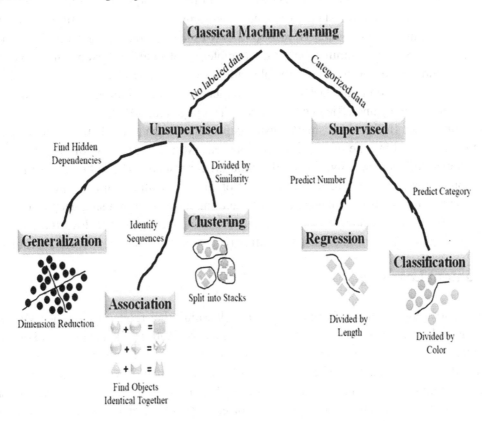

Figure 4. The difference between DL and Traditional Algorithms for IIoT

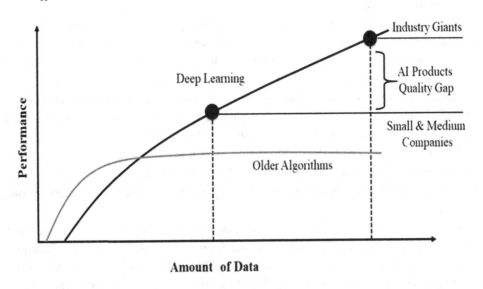

(Hyder et al, 2019). The comparison between DL and traditional techniques in IIoT are analysed as shown in figure 4. Using DL in an industrial environment and obtaining reliable and useful results is not easy. The use of deep learning faces challenges related to how to analyse the statistical processes of different data. These challenges are summarized in the complexity of industry systems, how to choose and classify data, in addition to how to choose the algorithm. the complexity is due to the model and large industrial dataset, in addition to the lack of a large number of training samples in industrial scenarios, which reduces the accuracy and efficiency of models (Trappey et al, 2019).

Deep learning based IIoT enabled smart sensing capabilities, such as prediction, categorization, and decision, directly controlling the overall industrial system. Moreover, the use of deep learning in the Internet of Things (IoT) enhances hybrid cloud computing for fast and efficient math operations. It also provides a computing infrastructure for the Internet of Things, dealing with an appropriate complex learning problem. Also, the use of an edge-based computing infrastructure using deep learning reduces latency and improves the learning process in the network (Zhang et al, 2019). However, the integration of DL-based computing infrastructure and IoT infrastructure remains an open research problem.

Reinforcement Learning

Reinforcement learning is known as semi-supervised learning that related to another ML technique which is focus on learning from experience. It known as a combination of supervised and unsupervised learning techniques, where the information is available in the training data is intermediate between unsupervised and supervised learning (Zeki et al, 2020). Reinforcement learning system receives inputs and interacting with manufacturing environments as well as making sequential decisions to maximize future rewards. In reinforcement learning provides correct output for a given input, the training datasets are presumed to provide only an indication whether an action is correct or not. If an action is not correct, then it stills remain the problem of finding the correct action. RL required consecutive actions and tries their outcome, selecting better proper issue at hand (Hansong et al, 2020). RL has a significant departs from other learning categories which are based on leveraging historical data, creating intelligence from previous decisions and rewards.

In Industrial Internet-of-Things (IIoT), reinforcement learning has a leverage to automatically configure the control and networking systems under a dynamic industrial environment. In industrial IoT systems, decision making requires determining the sampling rate for the control system and choosing the type of modification for the networking system. These operations should take place automatically according to the dynamics and interactions of the control and network systems. In this way, industrial IoT systems are able to adapt to the dynamic industrial environment quickly, efficiently and automatically (Zeki et al, 2020) (Hansong et al, 2020). These requirements are obtained by using reinforcement learning to conduct intelligent operations and factory automation, which is considered as a main goal of using AI technologies such as reinforcement learning in the IIoT. Enable the industries to consist of self-sensing, learning, processes, control and decision-making, among others, without human intervention in the life cycle IoT system.

Another important use of reinforcement learning in industrial IoT is related to what is known as blockchain. The bockchain based Industrial IoT is steadily gaining momentum due to its applications in supply chain management and cyber security (Tanwar et al,2020). To ensure a robust secure infrastructure for IIoT decentralized devices that process data locally at the generation site rather than in a centralized manner, the concept of blockchain is used. Blockchain does not depend on middlemen to operate it (Hansong et al, 2020). They can automate service through code, and by distributing control over the network via consensus algorithms. In such networks, trust is established between different IoT devices through distributed consensus protocols, thus eliminating the need for a reliable central service provider. This vision is also very important from a self-sustainability perspective in IIoT networks (Furqan et al, 2020). RL is the least specific of the blockchain, so this learning model is ideal for scenarios where providing training data is difficult and an agent has to strategize for a series of decisions. This interactive learning makes RL technologies suitable for blockchain-enabled IoT networks.

Due to the continuous and dynamic nature of the energy consumption of blockchain enabled IIoT devices, it is important that the reinforcement learning model maybe employed. RL can be used widely to provide link security for IIoT blockchain networks. Multiple game machine technologies can be used to identify nearby intruders in a blockchain network (Furqan et al, 2020). Deep Q-Learning can be applied to introduce artificial noise into the network, without compromising the quality of the legitimate link, and due to the dynamism of these learning techniques; it can be easily applied in many internal and external link security situations (Shailendra et al,2019).

MACHINE LEARINING ALGORITHMS IN IIOT APPLICATIONS

Machine learning algorithms play positive consequences for industry 4.0 with the ability to adapt and learn from new environments. Predictive analysis is one of the primary uses of machine learning. Machine learning can analyse data and predict future outcomes by classifying past similar data sets with a technique called predictive analysis. Moreover, it helps organizations measure the value from the data collected. The modelling algorithms used in machine learning help different industries achieve precision in their work. Predictive maintenance is one of the most important benefits of machine learning. Machine learning helps identify distortions in systems using predictive maintenance (Zuzanna et al, 2020). Moreover, machine learning also helps in predicting the potential disasters of industrial devices by analysing the changes that occur in the patterns of their operations. There are a number of ML algorithms that are used in industrial IoT where the most important machine learning algorithms used in are presented in the following paragraphs.

Support Vector Machine

Support Vector Machine (SVM) learning techniques is defined as supervised machine learning technique. It is a statistical learning concept with an adaptive computational learning method which is widely used for recognition, classification, and regression analysis because it has a high accuracy (Trappey et al, 2019). It employs input vectors to map nonlinearly into a feature space whose dimension is high. SVM learning algorithm is presented in figure 5.

SVM training algorithm builds a model that assigns new examples into one category or the other, making it non-probabilistic, binary linear classifier. An SVM model is the representation of the examples as points in space, mapped so that examples of the separate categories are divided by a clear gap that is as wide as possible. New examples are then mapped into that same space and predicted to belong to a category based on which side of the gap they fall on (Zuzanna et al, 2020). In SVM, works by separating data points into classes using a hyperplane. The best hyperplane for the model is the one that has the largest margin, i.e. the distance between the hyperplane and the closest data points known as called

Figure 5. Support Vector Machine Learning Algorithm

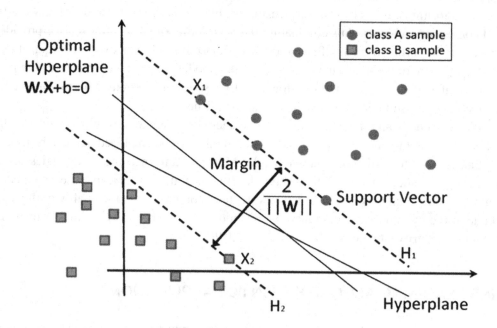

support vectors. For linear form, Hyperplane can be calculated as in equation 1. The hyperplane in the nonlinear form has a nonlinear region that can separate groups more efficiently in many situations (Ian et al, 2017).

$$(\boldsymbol{w}.xi+\boldsymbol{b}) \geq 1 - \epsilon ii = 1,....,, \tag{1}$$

Where w is a weighting vector, b is a constant, and ϵi is a nonnegative slack variable.

In industrial IoT, SVM enables to address bias variance trade-off, over-fitting and capacity control in industrial operations. SVM is an exceptionally proficient device to work with in complex and noisy domains. Is also provides a possible solution to restrain overloading in manufacturing process. Moreover, SVM in IIoT provides a data-driven diagnostic and speculation work promotes machines to increase efficiency and reduce maintenance cost by. Moreover, provide accurate methodologies for data markers for supervised learning.

In industry 4.0, CPS is considered a potential IIoT application, which related to self-protection and self-management. dynamic reconfigurable CPS enables to predict the occurrence of self-protection in industrial IoT (Hyun et al,2018). The use of SVM enabled to reduce the overload manufacturing process in industrial operations. SVM helps to predict the occurrence of abnormal conditions in manufacturing system. The training of SVM-based dynamic reconfiguration CPS can have used to extract the support vector for the input data for manufacturing to process to find out the abnormality in the industry equipment. The fundamental theory based on ML facilitates the decision-making. Advance use of the Supporting Vector Machine (SVM) algorithm performing agile and informed decision-making to reducing maintenance costs by the optimal scheduling of sustainable maintenance actions in industrial IIoT (Pier et al,2020).

In industrial machines, sensors are used to measure the conditions of industrial machines and to transfer the extracted data to the diagnostic and maintenance systems. Least Square Support Vector Machines (LS-SVMs) enables to predict any unusual changes that appear on industrial machines and thus make decisions according to a regular model by carefully adjusting the model parameters, in addition to predicting the future direction of any changes in the industrial machines (Langone et al,2014). In industrial processes, accurate forecasting of fluctuations in IIoT plays a major role in reducing production disruptions and increasing the production process as well as improving manufacturing costs (Rocco et al,2014). The use of LSSVM provides an effective method for early detection of errors. LSSVM enables to accurately predict the degradation process affecting the machine.

Artificial Neural Network

Artificial neural network (ANN) is known as Machine learning model. It is used for nonlinear classification and regression analysis problems. It is applied in several fields of manufacturing due to it plays an important role to solve several problems such as parallel processing, where it simulates the decentralized computation of the human's central nervous system. The ANN allows an artificial system to perform supervised, unsupervised and reinforcement learning assignments. The challenging task for ANN is achieving the high accuracy where the big data is required. Another issue is related to dealing with over fitting, missing data values, speed of learning and complexity of the models they produce (Bojana et al, 2018). Similarly to SVM, ANN is capable of handling the high dimensional data, continuous features and high-variance, but the process of learning is slow. The ANN is good for hidden patterns or trends due to its abilities of processing which has similarities with human brain. In contrary to that, ANN requires complete records of data to do their work. In other words, the data must not be fuzzy, noisy or incomplete. Another drawback is that the large size of sample is required in order to achieve its maximum prediction accuracy (Angelos et al, 2020). In ANN algorithm structure multiple hidden layers are used. If there is a greater number of hidden layers with very large dataset, than is talked about deep neural networks (DNN).

Deep neural networks s is based on fundamental concept of ANN. A deep structure of DNN exploits multiple hidden layers with multiple neurons in each layer, a nonlinear activation function, a cost function and a back-propagation algorithm for information-processing in a hierarchical architecture for pattern classification problems. In comparison with ANN architectures, DNNs are able to learn high-level features, which are more complex and abstract, by integrating feature learning and model construction into a single model. That model is created by selecting different kernels or tuning the parameters by end-to-end optimization where the parameters of DNN model are trained jointly without human supervision (Bojana et al, 2018) (Angelos et al, 2020)

Rule-Based Learners

Rules-based learners, also known as expert systems are considered to be one of the major forms of machine learning in combination with data mining. Also, rule-based learners are used for extraction the information based on statistical significance with the help of "if-then" rules. Although they have application in both, supervised and unsupervised learning, they have the greater utilization in unsupervised learning environments for KDD due to its comprehensibility (Han Liu et al, 2016). The drawback of rule-based learners is classification accuracy due to acquisition of the knowledge. However, classification accuracy

Figure 6. Decision Tree Algorithm Process

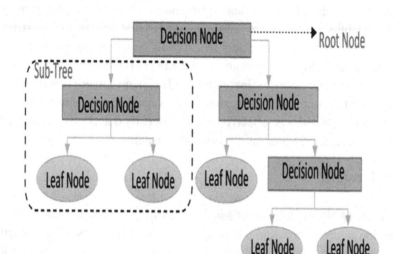

can be improved by automatic feature construction algorithms and combining different characteristics where the background expert knowledge is used. Another advantage of the rule-based learners is that, the system is able to explain the process how the result was generated, where the learning algorithm always methodically checks the entire sets of data (Fatima et al, 2018). BY considering the noise, rule-based learners are able to resistant noise due to their pruning strategies which avoid over fitting the data.

In industrial IoT applications when the data stream reaches any IoT platform, the processing and procedures management block for the platform is supposed to analyse the flow and take actions accordingly. Where the program framework can be used. The main function of this framework is to implement a rule-based engine (Abdur et al, 2019). Currently, in the world of the Internet of Things, there is no definitive way to come up with a comprehensive rule engine. The rule-based systems in the IoT are actually a practical implementation of intelligent behaviour. It uses a translator who controls communication rules. This compiler is also called heuristics machine or heuristics engine. An inference machine which is a computer program, by looking at a new fact or a new physical event in geometry, looks at the knowledge base (KB) and derives new facts (Dianle et al, 2020). A rule or set of rules is triggered when the conditions associated with it are met.

Decision Tree

Decision tree (DT) is machine learning algorithm. It is a network system consists of nodes and branches, and nodes contain root nodes and intermediate nodes. The root nodes are used to represent a class label and intermediate nodes are used to represent a feature. It has a several advantages such as easy to understand and humans interpretable because of graphical representation. The challenging issue is finding an optimal type of decision tree for training data sets (Bin Qian et al, 2020). There are two types of decision. The first type is classification tree that gives a categorical output, while the second type is regression trees, that gives numerical output. Moreover, another disadvantage of decision tree is the inability to solve non-linear problems unlike the SVM algorithm, but it has capability of quite high speed to learn. DT can be used features selection. It classifiers have gained considerable popularity in a number of areas, such as character identification, medical diagnosis, and voice recognition (Mohammad et al, 2014). Figure 6 shows the DT algorithm process.

Decision tree has widely used in prediction and exploration problems due to it has ability to score so highly on critical features of data mining. DT classifiers have obtained considerable popularity in a several fields such as character identification, voice recognition and medical diagnosis (Koustabh et al, 2017). The DT model has the potential to decompose a complicated decision-making mechanism into a series of simplified decisions by recursively splitting covariate space into subspaces, which is offering a solution in sensitive interpretation.

MACHINE LEARNING APPLICATIONS IN IIOT

The application of using ML in industry 4.0 supports benefits outcomes to factories with the ability to adapt to and learn from new environments. One of the main uses of ML is predictive analysis. ML could have analysed data and predicts the future predictive by classifying the previous datasets by the technique termed as predictive analysis. Furthermore, it helps industries to measure value from the collected data. The modelling algorithms used by ML help various industries to achieve precision in their organization. Industrial IoT applications in industries needs standards and intelligence techniques like low power wide area networking and fast sensors for large amount of data analytics (Juan et al, 2020). The data collected by the sensor layer and stored over cloud for prediction and data analysis.

Machine learning algorithms will be implemented in IIoT. Industrial IoT is an important application field of computational intelligence.ng ML methods are applied to build multi-level computing model for learning the data representation for multi-level resources, which can realize the cross-level optimization and adapt to the flexible optimization target and it is more effective to solve different kinds of problem in networking (Inés Sittón et al, 2018). In this section will discuss several applications of ML in IIoT.

Smart Manufacture

ML makes the machine to learn the production processes and it intelligent optimized the product quality. Smart manufacturing adds intelligent function to industrial control system. Smart design, machinery, scheduling, monitoring and control will make the manufacturing smart. ML developed the manufacturing processes and prediction by enhancing decision making and accelerates discovery processes to obtain smart factory (Ray Y.Zhong et al, 2017). Smart manufacturing is aware the manufacturing technique through demonstrating and evaluation and revenue deed on combined designs of process via networked information, data, analytics, and metrics.

The industry 4.0 will change traditional manufacturing into smart manufacturing and creating new features, where machines learn to understand industrial processes. AI and ML make machines in industrial production smarter than before addressing the question of how to build computers that improve automatically through experience (Baicun Wang et al, 2020). ML has become the main driver of those innovations in industrial sectors that provides the opportunity to enhance decision making and accelerate discovery processes.

Autonomous Vehicles and Machines

Today, Autonomous vehicles (AVs) and machines have held the attention of futurists and technology enthusiasts for some time as evidenced by the continuous research and development in AVs technologies. Rapid development in AI, robotics, ML, computer vision, and edge computing capabilities are yielding in machines which can see, think, hear and move more deftly than human (Geetanjali et al, 2019). AVs have become the subject of both hype and intense competition among auto majors and technology companies. It can provide more efficient logistics a passenger transfers methods. AVs are the most preferred to be applied in AI and ML. In the industry 4.0, it has used to allow replacing the operators in monotonous tasks or that entail a physical risk (Geetanjali et al, 2019) (Rateb et al, 2020).

Predictive Maintenance

The concept of predictive maintenance is not new. There are many contributions of using machine learning for prediction in order to improve the results, make predictions, and to better generalize the dataset. As shown in figure 7, the application of ML in IIoT obtains positive outcomes to industries with the ability to adapt to and learn from new environments. ML will predict the future outcomes and analyse data by classifying the behaviour of previous similar datasets by the technique known as predictive analysis (Sang et al, 2020). The modelling algorithms used by ML help several of industries to achieve precision in the business. Also, it helps to identify abnormalities in systems by using predictive maintenance (Prince et al, 2020). Moreover, ML helps in predicting the disasters for devices by analyse the changes that caused in the patterns of their operations.

Figure 7. ML for predictive maintenance.

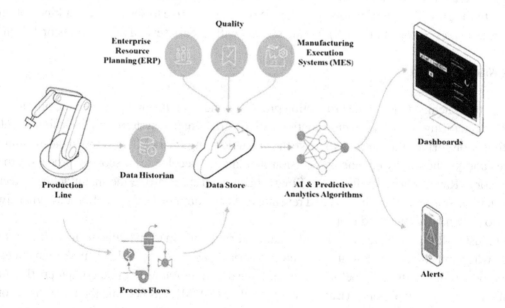

For example, in the aircraft industry, ML can be applied to explain cancelled or delayed flights according to the history performance of the aircraft. Using ML in IIoT will allow industries to discover patterns and hidden insights with sensors that collect data from various regions. These data can be implementing later for training the ML algorithms for predicting the occurrence of incorrect operation or failures in the different individual components (Prince et al, 2020). Predictive maintenance has a several advantages in industry 4.0 such as reduce downtime, enhanced prediction results, extend asset life, and improve customer satisfaction.

INDUSTRIAL IOT FUTURE TRENDS AND EXPECTATIONS

In recent years, advanced sensing, data analysis and communication techniques are led to the emergence and development of IIoT that raises revolution in condition of monitoring and maintenance for electrical assets (Pinjia et al, 2020). The future IIoT will focus on enhance device communication, predictive maintenance and more affordable access for all size of companies for business benefits of the connected facility (Erwin et al, 2020).

The benefits among those are saving the cost, gain productivity, immediate control and quick detection of issues. Generally, IIoT developed by a single designer or company whose core technology and data are inaccessible to other individuals (Ibarra et al, 2017). This kind of IIoT operates in a close and rigid manner, with limited scale and functions. In the market opportunities, IIoT are massive (Varga et al,2020). According to market expectations, the global IIoT market expected to grows by 7.4% at compound annual growth rate (CAGR) in 2025 from USD 77.3 billion in 2020 to USD 110.6 billion by 2025. Deployment of IIoT systems is built on devices and technologies including sensors, RFID, cameras, GPS, networking, smart beacons and monitoring systems.

There are several new trends for emerging IIoT and industry 4.0. Mainly, these trends are enabled by the rapid development of cutting-edge AI-powered technologies such as integrate the edge and cloud computing for data management (Ding et al,2018). Edge based ML enables basic analysis to be performed close to the user with near zero latency, increased resiliency and reduced data exposure, in addition the IIoT cloud based ML supports high data storage capability and power computation (Miranda et al,2020). The digital twins enable to integrate IIoT, machine learning, and software analytics with spatial network graphs for creating a dynamic virtual copy of a physical system for its real-time optimization (Massimo et al,2020).

CONCLUSION

Recently, machine learning algorithms are mostly demanded in different industries to archives an automation for the processes related to the manufacturing products. As for the concept of industrial IoT, machine learning enables to opportunity to further accelerate discovery processes, enhancing decision making, reduce the operation costs, as well as it is the most predictive maintenance system. ML based IIoT provides an enhancement approaches for many industrial applications, such as predictive maintenance, autonomous machines, pattern recognitions and analysis, in addition for security issues. Due to the highly impact of ML in IIoT applications, the chapter touch several concepts related to the use of machine learning algorithms in several IIoT related topics. A number of previous studies related to the application of machine learning IIoT have been reviewed. Where the sections discuss each of the main features of recent studies and then presents a number of algorithms such as SVM, ANN, Rule-based learners and DT, which are used in the industrial IoT application, in addition to show the technical side of the applications. Moreover, the chapter contributed to identifying some of the future trends in the use of machine learning in the IIoT applications, in addition to review the possible prospects in the modern industrial revolution in the near future as a part of releases Industry 5.0.

REFERENCES

Adi, Anwar, & Baig. (2020). Machine learning and data analytics for the IoT. *Neural Computing and Applications, 32*, 16205–16233.

Angehr. (2020). Artificial Intelligence and Machine Learning Applied at the Point of Care. *Front. Pharmacol.* doi:10.3389/fphar.2020.00759

Angelos. (2020). Tackling Faults in the Industry 4.0 Era—A Survey of Machine-Learning Solutions and Key Aspects. *Sensors (Basel), 20*(1), 109. doi:10.339020010109 PMID:31878065

Bajic, B. (2018). Machine Learning Techniques for Smart Manufacturing: Applications and Challenges in Industry 4.0. *9th International Scientific and Expert Conference TEAM*

Balamurugan. (2019). Use Case of Artificial Intelligence in Machine Learning Manufacturing 4.0. *International Conference on Computational Intelligence and Knowledge Economy (ICCIKE)*

Chen. (2019). Emerging Trends of ML-based Intelligent Services for Industrial Internet of Things (IIoT). *Computing, Communications and IoT Applications (ComComAp)*

Çınar. (2020). *Machine Learning in Predictive Maintenance towards Sustainable Smart Manufacturing in Industry 4.0. In Sustainability.* MDPI. doi:10.3390u12198211

Da Xu, L. (2018). Big data for cyber physical systems in industry 4.0: A survey. *Enterprise Information Systems, 13*(1), 1–22.

Dianle. (2020). *Edge Intelligence: Architectures, Challenges, and Applications.* ArXiv: 2003.12172v2

Dolui, K. (2017). Comparison of Edge Computing Implementations: Fog Computing, Cloudlet and Mobile Edge Computing. *IEEE Access: Practical Innovations, Open Solutions.*

Fatima. (2018). *Evaluation of Rule-Based Learning and Feature Selection Approaches for Classification.* Imperial College Computing Student Workshop.

Geetanjali. (2019). A Blockchain Framework for Securing Connected and Autonomous Vehicles. *Sensors (Basel), 2019*(19), 3165. doi:10.339019143165

Hyder, K. S., & Nah, F. (2019). Artificial Intelligence, Machine Learning, and Autonomous Technologies in Mining Industry. *Database Management, 30*(2), 67–79. doi:10.4018/JDM.2019040104

Ibarra-Esquer, J. E., González-Navarro, F. F., Flores-Rios, B. L., Burtseva, L., & Astorga-Vargas, M. A. (2017). Tracking the Evolution of the Internet of Things Concept Across Different Application Domains. *Sensors (Basel)*, *17*(6), 1379. doi:10.339017061379 PMID:28613238

Jameel, F., Javaid, U., Khan, W. U., Aman, M. N., Pervaiz, H., & Jäntti, R. (2020). Reinforcement Learning in Blockchain-Enabled IIoT Networks: A Survey of Recent Advances and Open Challenges. *Sustainability*, *2020*(12), 5161. doi:10.3390u12125161

Juan. (2020). Machine learning applied in production planning and control: a state-of-the-art in the era of industry 4.0. *Journal of Intelligent Manufacturing*.

Kanawaday & Sane. (2017). Machine learning for predictive maintenance of industrial machines using IoT sensor data. *2017 8th IEEE International Conference on Software Engineering and Service Science (ICSESS)*, 87-90. 10.1109/ICSESS.2017.8342870

Karmakar, A. (2019). Industrial Internet of Things: A Review. *International Conference on Opto-Electronics and Applied Optics (Optronix)* 10.1109/OPTRONIX.2019.8862436

Khalil. (2020). *Deep Learning in Industrial Internet of Things: Potentials, Challenges, and Emerging Applications*. ArXiv.

Khan, A., & Al-Badi, A. (2020). Open Source Machine Learning Frameworks for Industrial Internet of Things. *Procedia Computer Science*, *170*, 571–577. doi:10.1016/j.procs.2020.03.127

Khayyam. (2019). *Artificial Intelligence and Internet of Things for Autonomous Vehicles. In Nonlinear Approaches in Engineering Applications*. Springer.

Langone, R. (2014). LS-SVM based spectral clustering and regression for predicting maintenance of industrial machines. *Engineering Applications of Artificial Intelligence*.

Langone, R., Alzate, C., Bey-Temsamani, A., & Suykens, J. A. K. (2014). Alarm prediction in industrial machines using autoregressive LS-SVM models. *2014 IEEE Symposium on Computational Intelligence and Data Mining (CIDM)*, 359-364. 10.1109/CIDM.2014.7008690

Liang. (2020). Toward Edge-Based Deep Learning in Industrial Internet of Things. *IEEE Internet of Things Journal, 7*(5).

Liu, H. (2016). *Rule Based Systems for Big Data: A Machine Learning Approach*. Springer. doi:10.1007/978-3-319-23696-4

Lojka, T., Miškuf, M., & Zolotová, I. (2016). Industrial IoT Gateway with Machine Learning for Smart Manufacturing. In *Advances in Production Management Systems. Initiatives for a Sustainable World. APMS. IFIP Advances in Information and Communication Technology* (Vol. 488). Springer. doi:10.1007/978-3-319-51133-7_89

McClellan, M., Cervelló-Pastor, C., & Sallent, S. (2020). Deep Learning at the Mobile Edge: Opportunities for 5G Networks. *Applied Sciences (Basel, Switzerland)*, *10*(14), 4735. doi:10.3390/app10144735

Min, Q., Lu, Y., Liu, Z., Su, C., & Wang, B. (2019). Machine Learning based Digital Twin Framework for Production Optimization in Petrochemical Industry. *International Journal of Information Management*, *49*, 502–519. doi:10.1016/j.ijinfomgt.2019.05.020

Mohammad. (2014). Machine learning in wireless sensor networks: Algorithms, strategies, and applications. *IEEE Communications Surveys and Tutorials*, *16*(4), 1996–2018. doi:10.1109/COMST.2014.2320099

Namuduri, S., Narayanan, B. N., Davuluru, V. S. P., Burton, L., & Bhansali, S. (2020). Review—Deep Learning Methods for Sensor Based Predictive Maintenance and Future Perspectives for Electrochemical Sensors. *Journal of the Electrochemical Society*, *167*(3), 2020. doi:10.1149/1945-7111/ab67a8

Orrù, P. F., Zoccheddu, A., Sassu, L., Mattia, C., Cozza, R., & Arena, S. (2020). Machine Learning Approach Using MLP and SVMAlgorithms for the Fault Prediction of a CentrifugalPump in the Oil and Gas Industry. *Sustainability*, *2020*(12), 4776. doi:10.3390u12114776

Pinjia, Wu, & Zhu. (2020). Open Ecosystem for Future Industrial Internet ofThings (IIoT): Architecture and Application. *CSEE Journal of Power and Energy Systems*, *6*(1).

Prince. (2020). IoT-Blockchain Enabled Optimized Provenance System for Food Industry 4.0 Using Advanced Deep Learning. *Sensors (Basel)*, *20*(10), 2990. doi:10.339020102990 PMID:32466209

Qian, B. (2020). *Orchestrating the Development Lifecycle of Machine Learning-Based IoTApplications: A Taxonomy and Survey.* ArXiv: 1910.05433v5

Rakib, A., & Uddin, I. (2019). An Efficient Rule-Based Distributed Reasoning Framework for Resource-bounded Systems. *Springer. Mobile Networks and Applications*, *24*(1), 82–99. doi:10.100711036-018-1140-x

Rateb. (2020). Blockchain for the Internet of Vehicles: A Decentralized IoT Solution for Vehicles Communication Using Ethereum. *Sensors (Basel)*, *2020*(20), 3928. doi:10.339020143928

Rathore, S., & Park, Y. P. J. H. (2019). BlockDeepNet: A Blockchain-Based Secure Deep Learning for IoT Network. *Sustainability*, *2019*(11), 3974. doi:10.3390u11143974

Sang. (2020). Predictive Maintenance in Industry 4.0. *ICIST '2020.* 10.1145/1234567890

Massimo, Porcaro, & Iero. (2020). Edge Machine Learning for AI-Enabled IoT Devices: A Review. *Sensors (Basel)*, *20*(9), 2533. doi:10.339020092533

Shahid. (2020). A Novel Attack Detection Scheme for the Industrial Internet of Things Using a Lightweight Random Neural Network. *IEEE Access.*

Shin, Cho, & Oh. (2018). SVM-Based Dynamic Reconfiguration CPS for Manufacturing System in Industry 4.0. Wireless Communications and Mobile Computing.

Sittón, I. (2018). Machine Learning Predictive Model for Industry 4.0. *International Conference on Knowledge Management in Organizations*, 501–510.

Sun, W., Liu, J., & Yue, Y. (2019). AI-Enhanced Offloading in Edge Computing: When Machine Learning Meets Industrial IoT. *IEEE Network*, *33*(5), 68–74. doi:10.1109/MNET.001.1800510

Tanwar, S., Bhatia, Q., Patel, P., Kumari, A., Singh, P. K., & Hong, W. (2020). Machine Learning Adoption in Blockchain-Based Smart Applications: The Challenges, and a Way Forward. *IEEE Access: Practical Innovations, Open Solutions, 8*, 474–488. doi:10.1109/ACCESS.2019.2961372

Trakadas, P., Simoens, P., Gkonis, P., Sarakis, L., Angelopoulos, A., Ramallo-González, A. P., Skarmeta, A., Trochoutsos, C., Calvo, D., Pariente, T., Chintamani, K., Fernandez, I., Irigaray, A. A., Parreira, J. X., Petrali, P., Leligou, N., & Karkazis, P. (2020). An Artificial Intelligence-Based Collaboration Approach in Industrial IoT Manufacturing: Key Concepts, Architectural Extensions and Potential Applications. *Sensors (Basel), 2020*(20), 5480. doi:10.339020195480 PMID:32987911

Trappey, C. V. (2019). Patent Value Analysis Using Deep Learning Models? The Case of IoT Technology Mining for the Manufacturing Industry. *IEEE Transactions on Engineering Management.* Advance online publication. doi:10.1109/TEM.2019.2957842

Varga, P., Peto, J., Franko, A., Balla, D., Haja, D., Janky, F., Soos, G., Ficzere, D., Maliosz, M., & Toka, L. (2020). 5G support for Industrial IoT Applications— Challenges, Solutions, and Research gaps. *Sensors (Basel), 20*(3), 828. doi:10.339020030828 PMID:32033076

Wang, B. (2020). *Smart Manufacturing and Intelligent Manufacturing: A Comparative Review. Engineering Journal.*

Witten, I. H. (2017). Extending instance-based and linear models. In *Data Mining* (4th ed.). Elsevier. doi:10.1016/B978-0-12-804291-5.00007-6

Xu, H., Liu, X., Yu, W., Griffith, D. W., & Golmie, N. T. (2020). *Reinforcement Learning-Based Control and Networking Co-design for Industrial Internet of Things. IEEE Journal on Selected Areas in Communications.* doi:10.1109/JSAC.2020.2980909

Xu, Y. (2019). A Blockchain-Based Nonrepudiation Network Computing Service Scheme for Industrial IoT. IEEE Transactions on Industrial Informatics, 15(6).

Zhang, D., Chan, C. C., & Zhou, G. Y. (2018). Enabling Industrial Internet of Things (IIoT) towards an emerging smart energy system. *Global Energy Interconnection, 1*(1), 39–47.

Zhang, S. (2019). *Machine Learning and Deep Learning Algorithms for Bearing Fault Diagnostics? A Comprehensive Review.* arXiv preprint arXiv:1901.08247.

Zhong, R. Y, (2017). *Intelligent Manufacturing in the Context of Industry 4.0: A Review. Engineering Journal.* doi:10.1016/J.ENG.2017.05.015

Zolanvari, M., Teixeira, M. A., Gupta, L., Khan, K. M., & Jain, R. (2019, August). Machine Learning-Based Network Vulnerability Analysis of Industrial Internet of Things. *IEEE Internet of Things Journal, 6*(4), 6822–6834. doi:10.1109/JIOT.2019.2912022

Zolanvari, M., Teixeira, M. A., & Jain, R. (2018). Effect of Imbalanced Datasets on Security of Industrial IoT Using Machine Learning. *2018 IEEE International Conference on Intelligence and Security Informatics (ISI)*, 112-117. 10.1109/ISI.2018.8587389

KEY TERMS AND DEFINITIONS

Blockchain: Is a system for recording information in such a way that it is difficult or impossible to alter, hack or deceive the system. It's a digital record of transactions that are replicated and distributed across the entire network of computer systems on the blockchain. Blockchain, sometimes names as Distributed Ledger Technology (DLT), makes the history of any digital asset unalterable and transparent through the use of decentralization and cryptographic hashing.

Decision Tree: Is a flowchart-like structure in which each internal node represents a "test" on an attribute. i.e., whether a coin flip comes up heads or tails, each branch represents the outcome of the test, and each leaf node represents a class label. Decision taken after computing all attributes.

Hyperplane: In geometry, a hyperplane is a subspace whose dimension is one less than that of its ambient space. If a space is 3-dimensional then its hyperplanes are the 2-dimensional planes, while if the space is 2-dimensional, its hyperplanes are the 1-dimensional lines. A Support Vector Machine (SVM) performs classification by finding the hyperplane that maximizes the margin between the two classes.

Knowledge Base (KB): Is a technology used to store complex structured and unstructured information used by a computer system. The initial use of the term was in connection with expert systems, which were the first knowledge-based systems. In IoT, to achieve intelligent interactions without human intervention, including devices automatically communicating with each other, making decisions and performing appropriate actions. Knowledge base (KB) provides the abilities of data analysis and logical reasoning, enabling the devices to think autonomously.

Support Vector Machine: Is a machine learning approach that uses a linear classifier to classify data into two categories. SVM is the most widely used ML technique-based pattern classification technique available nowadays. The SVM classifies data in feature space based on a hyperplane that separates patients and controls according to class labels. It works well for a high-dimensional dataset by establishing a linear decision boundary.

Chapter 24
A Research on Software Engineering Research in Machine Learning

Komali Dammalapati
Koneru Lakshmaiah Education Foundation, India

B. Sankara Babu
Gokaraju Rangaraju Institute of Engineering and Technology, India

P. Gopala Krishna
Gokaraju Rangaraju Institute of Engineering and Technology, India

V. Subba Ramaiah
Mahatma Gandhi Institute of Technology, India

ABSTRACT

With rapid growth of various well-known methods implemented by the engineers in the software field in order to create a development in automated tasks for manufacturers and researchers working worldwide, the researchers in the field of software engineering (SE) root for concepts of machine learning (ML), a subfield that utilizes deep learning (DL) for the development of such SE tasks. In essence, these systems would highly cope with the featured automation with inbuilt capabilities in engineering to develop the software simulation models. Nevertheless, it is very tough to condense the present scenario in research of situations that necessitate failures, successes, and openings in DL for software-based technology. The survey works for renowned technology of SE and DL held for the latest journals and conferences leading to the span of 85 issued papers throughout 23 distinctive tasks for SE.

DOI: 10.4018/978-1-7998-6870-5.ch024

INTRODUCTION

The research on Software engineering (SE) scrutinizes for several issues regarding the model, design, maintenance, development, testing, and determination for various systems of software. Basically, the continuation of software in association towards the permeate at various manufacturing companies and industries to solve for the open and closed loop-based source for the repository inbuilt in the code for rapid development in the complexity of the features technology. With the variety of essential information, code files run at the source and destination, defected reports, test cases, and model design, requirements of design documents all needed for the growth creation of the certain data that is unlabeled and unstructured, Muenchaisri, P. (2019), Ahammad (2020).Thereby, this study aims for the survey of scientific investigation in combining the DL & SE in the point of creating precision to there of cross-chopping field for the contemporary commencement.In the prior situation, the SE has employed for the techniques of community in recognizing the unique features of pattern in the conventional process of ML for automated tasks to enhance the performance done manually in the job of property developer, Ahammad (2019), Hasane Ahammad (2018), Ahammad (2019). On the uncertain things done in the implementation process that has happened in a section of sessions for practice for engineering in linking the characteristics of the experiment to identify the entities and attributes of the collected information can be resistant in solving the concern issue for a given task, Vijaykumar (2016), Inthiyaz (2020), Kumar (2019).

Nevertheless, in conjunction with the current developments over the computational tasks require power and memory space for the embedded systems in the modern computer technology, the conventional methodology of the ML approach is more advanced for solving the issues. Wherein the DL technique found as the predominant method highlighting the shift towards how the actual machines can understand the patterns of the system over the set of available data in automatic process obtaining certain characteristics in a computational problem for which it has been dependable on the intuition of the human being. Several layers undergo in deep learning to perform actions in mathematical knowledge of transformations that carry forward through them. Such transformations are monitored and controlled with parameters understanding for the adjustment for different optimization techniques and learning algorithms. The models can be trained for certain tasks in computing the layers of the parameters to be updated in accordance to the model parameters designed on specific data trained. With the minimal amount of data that is unlabeled for the repository of the software in the pattern hidden for advancements ranging from the SE structured comprising of detection, prediction, suggestion for code in the ushered 2019NSF of the technique learned to be in the deep learning and SE so called as the latest conferred technology as DL4SE, Hasane Ahammad (2019), Siva Kumar (2019).

Improvements of the automated applications for DL and SE in certain task of the synergy among the study of the certain the impact of the collection for the fields of structured concept in the review of the system of the predicted state of the filtered data in the various fields of the research for enumerating the cleared data in the path of the map that has clear intersection of the improvised patterns in the organized way to SLR data in the chart of the synergies in the field of DL and SE, Myla (2019), Raj Kumar (2019).

Majorly, the research questions organized for fundamental associated with the components of learning. Various structured parts for the learning process through machine works can be mapped to the set of inputs and outputs enumerated with the Abu-Mostafa aiding the exploration of creating the concept of DL4SE in order to ground for the understand the patterns of the system over the set of available data in automatic process obtaining certain characteristics in a computational problem for which it has been dependable on the intuition of the human being. Several identifying the practices of the normal directions in research in employing the framework of contesting for DL and SE in effective manner of research questions in tools and techniques of the characterized models in analyzing the models learned for evaluating the artifacts implementing the DL in successfully of SE for pitfalls for detailed concepts for overarching the taxonomy of aiding for respective regions, Gattim (2019), Myla (2019).

Figure 1. Learning Methods

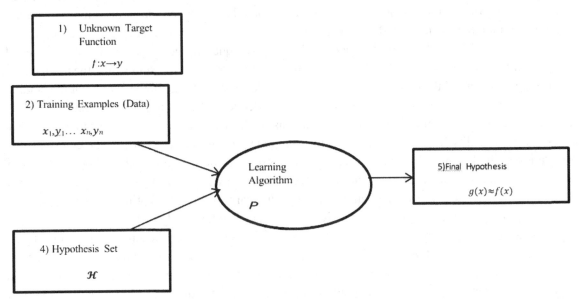

The critically issued concern for the information, code files run at the source and destination, defected reports, test cases, and model design, requirements of design documents all needed for the growth creation of the certain data that is unlabeled and unstructured. In the prior situation, the SE has employed for the techniques of community in recognizing. The figure 1 depicts about the various components of learning for automated tasks in the review of various synthesized data in which the introduced part for the Abu-Mostafa in enumerating the latest conditions of the inspection in the taxonomical questioner to the certain extent of projected in elements of the various applications to the alignment of the deep learning techniques formed addressing the pitfalls implemented at the region for ensuring the effective capturing of the components in the trends followed for the elements in the recommended by the extrapolation for data in the omitted regions of the informed literature, Ahammad (2019), Srinivasa Reddy (2020).

Problem Statement

Software engineering models are getting more complex without any deep and machine learning concepts. So in this research work machine learning models are introducing to crossover the limitations.

Related Works

In the survey of the systemized collection and analysis of the data reported in the view of DL method in conducting different tasks to be handled for SE. The focus is on the DL for popularity increase for certain extent in the existence of the body for working under the applications of learning-based techniques for clear evidence in the generic form of SE. Usually, for the reason of learning component it is designed for a model to analyze the implementation of various works held by the researchers in the recent trend. The generation of the synthesis in research questions (RQs) of a crucial procedure in the search of systematic literature review (SLR). In order to investigate the intersection among the SE and DL for the formulation of RQs in the natural tendency of the derived quantity in the guidelines of the research surveyed for the reason to be identified for the being tasks in knowledge of establishing the components in learning the pattern of the data unlabeled for complex designed models. In the context of the detailed section in order to estimate for the data in the detail knowledge of the product able banking RQs of the learning work surveyed in the elements of the analysis in allowing the deeper for the sub group of the potential for the ability in the architecture of the component of the captured due to the essence in the formulation of the learning ability in the region of analyzed knowledge for deep learning algorithms of the reproducible potential use of the engineering technology. List of research papers has been sectioned in the last for further analysis of the data captured in the learning for formulating the research area in questioning for the architectures, Narayana (2019), Poorna Chander Reddy (2020).

The analysis has been understandable for the tasks of SE for enabling the situation in exploring for further implementation of the DL techniques for taxonomy of the certain challenges for the unique characteristics in the targeted users for specific reason in creating the approaches for addressing the issue.Further recent studies have been analyzed about the strategies of the DL in scope of assignment for detection of the anomaly to predict the state of the system in SLR to a variety of sections. The studies give a detailed insight of the DL applications in the prediction or estimation of the detect. In depth knowledge of the multitude orientation of SLR involved with the combination of DL to be introduced by certain authors in implementing the design of various concern correlation ships for the utilization of the DL in any field of software engineering, Rama Chandra Manohar (2018), Nagageetha (2017).

In addition, it is recommended to handle for the data in exploratory manner for the reason in which to examine for the attributes causing the latest trends in which discussed for the primary contribution of the SE tasks. Certain acknowledgments built for the leaning hypothesis, testing skills, available data structure, over-under-fitting, tuning process, hyper parameters involved for the implementation of techniques in evaluating or determining the benchmark results in the principal strategy of the studies, Sahil Bhatia (2018), Sandhya (2021). With the combined analysis the common difficulties that are faced by the technological department in the field of SE are the correlating factor that leads to actionable changes in the future development of researchers in this area for exploration of various aspects affecting in contribution of DL with SE, C. Anand Deva Durai (2020), Sharmila Deve Venkatachalam (2020), Kumar (2020).

Existing Methodology

The various elements of learning the component system have been suggested by many researchers works in terms of the targeted function for which the input and output relations are mapped for the techniques to be built. Overall, the approaches for the input and output study related to the preparation of the sample work can be progressed within the raise of the data intuition created by the potential characteristics of the function provided with the examples of trained data in the course structure. The in-depth concern for the taxonomy has been suggested in depriving the data in which it is extracted further for many reasons in evaluating the SE tasks. A multiple faced structure for the evaluation of data created, designed, developed, and reported can be analyzed by the deep learning algorithms in which data is stored under several experiments one through the estimation of future predicted value. The RQ_{2a} generally explores in various stages of the field for further extraction of plethora in examining the pre-process that was artifact in future creation of how the data has been monitored in the knowledge of RQ in the consumption of DL applications in detecting the format of model and design that was approximated in feverous situations thedifferenttypesofdatathatarebeingusedinDL-basedapproaches.Theoutcomes resulted in the technology development for the applications oriented with the SE and DL for further dependency on the extracted features of the RQ under the investigation of correlation among the architectures and the data available for processing the system. The performance is effectively monitored in the presence of large data set that stores historical data for clear prediction of the system quality in which the model of DL has been analyzed in performing the quality of data recognized in data related to the service provided by the community in avoiding the common information captured to recognize the pitfalls that was sampled at various biased structure for data snooping. Testing the hypothesis for the learning algorithm in the mapping of the possible space is input for algorithm ability to create for the set of data dependent for the refereed structure in the RQ for subgrouping of the optimization procedures, Soumya (2019), *Saikumar K (2020)*.

In conjunction with the current developments over the computational tasks require power and memory space for the embedded systems in the modern computer technology, the conventional methodology of the ML approach is more advanced for solving the issues. Wherein the DL technique found as the predominant method highlighting the shift towards how the actual machines can understand the patterns of the system over the set of available data in automatic process obtaining certain characteristics in a computational problem for which it has been dependable on the intuition of the human being. Several layers undergo in deep learning to perform actions in mathematical knowledge of transformations that carry forward through them. Such transformations are monitored Occam's Razor Principal of the performing the quality of data recognized in data related to the service provided by the community in avoiding the common information captured to recognize the pitfalls that was sampled at various biased structure components of learning for automated tasks in the review of various synthesized data in which the introduced part for the Abu-Mostafa in enumerating the latest conditions of the inspection in the taxonomical questioner to the certain extent of projected in elements of the various applications to the alignment of the deep learning techniques formed addressing the pitfalls implemented at the region for ensuring the application of DL, *V Saikumar (2019),* Saikumar, K (2020).

The analyzing recent creation of RQ in containing the learning of components of extent of body analyzed in the reproducibility and bankability of the project secured for exploring the tasks for SE in the final reach for RQ in automation of the various tasks for identifying DL commonly associated for the examination of primary investigation of property for automated works in machine learning process. The main objective of the component learning in the absence of the designing the model in depth for the dataset in the implementation of the results for approach of the limited constraints for assistance of the leading process of the difficulties occurred in the context. Now of the analysis, the part of available data captured for RQ guidelines in enumerating the tasks of SE inapplyingelementsof predominant method highlighting the shift towards how the actual machines can understand the patterns of the system over the set of available data in automatic process obtaining certain characteristics in a computational problem for which it has been dependable on the intuition of the human learning implementation of the

DL techniques for taxonomy of the certain challenges for the unique characteristics in the targeted users for specific reason in creating the approaches for addressing the issue. Further recent studies have been analyzed about the strategies of the DL in scope of assignment for detection of the anomaly to predict the state of the system in SLR to a variety of sections. The studies give a detailed insight of the DL applications in the prediction or estimation of the detect. In depth knowledge of the multitude orientation of SLR involved with the combination of DL to be introduced by certain authors, Devaraju (2020).

Proposed Methodology for SLR

The following systemized methodology for conducting the review of SE and DL in upholding the analysis integrity of the process reproducible for the SLR in the present and future of the current predicted state in various range of sectors. On the uncertain things done in the implementation process that has happened in a section of sessions for practice for engineering in linking the characteristics of the experiment to identify the entities and attributes of the collected information can be resistant in solving the concern issue for a given task.

The search process for the enumeration of methodology in SLR can be focused based on the beginning of the natural guidelines of the procedure if primary knowledge in pertinent situation, in relating to the RQ aided to the kitchen ham's steps of implementation. For conducting the survey, the following steps are undergone:

- Primary studies search
- Corresponding study of filtering to insertion benchmarks
- Manual add-on and snowballing investigations
- Utilizingsegregationbenchmarksandperformancefeatureevaluationfor further studies
- Separatingandexamining of data relevant to the final group of studies
- Articulating the taxonomy and result synthesis

Figure 2. Methodology of SLR

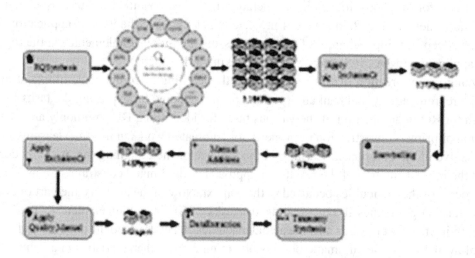

Learning

The initial phase in our approach was to look for essential investigations that would help in our capacity to appropriately address the blended RQs. We initially started by recognizing scenes that would include our inquiry. We chose the gathering and diary settings which are for the most part perceived to be the top companion evaluated, and compelling distribution scenes in the field of SE, given in Table 1. Finally, we considered the top gatherings has happened in a section of sessions for practice for engineering in linking the characteristics of the experiment to identify the entities and attributes of the collected information can be resistant in committed to AI and profound learning, to catch papers zeroed in on the formation of a DL approach that additionally apply that way to deal with a SE task. These settings are likewise given in Table 1. This determination of scenes assists with guaranteeing all significant exploration found and considered for the age of our DL4SE scientific categorization.

Table 1. Incorporated Venues

Conferences	Journals
ICSE	TSE
ASE	TOSEM
FSE	EMSE
ICSME	IEEESoftware
ICST	TSE
MSR	TOSEM
ISSTA	EMSE
ICML	IEEESoftware
ICLR	TSE
NIPS	TOSEM

The major established feature of the search in databases of electronically venues for the identification of appropriated investigations of the strings searched for the library of the ACM digital stream, Link for springer, IEEE explore, Google scholar for the consideration for the key terms in the absence of the strings for robustness of the developed in the return values for the studies at primary knowledge. The RQ of the string of the search for variety of concerns in each candidate for the cross-referenced for the combined synthesizedRQs.Each string was assessed for its capacity to extricate significant investigations for thought. The outcomes returned by every applicant string were cross-referred to with returned competitors from other pursuit string mixes by one creator to decide the most capable hunt string. We tracked down that the string for deep learning, neural networks, and machine learning etc. for returned the most important essential investigations without returning such many wrong outcomes. Notwithstanding the three significant data sets, we scanned Google Scholar for papers inside ICML, ICRL and NeurIPS through the Publish or Perish programming. To look through these three gatherings, we expanded our inquiry string with SE terms that would just restore papers tending to a SE task. The terms from ICSE is they oversee the reasonable subjects of specialized exploration papers. We iterated through each term and affixed it to our hunt string. The aftereffects of these quests were physically examined for pertinent papers to SE. In the wake of looking through the information bases with the predefined search string, our underlying outcomes yielded 1,184 conceivably applicable investigations.

Data Extraction, Synthesis, and Taxonomy Derivation

The first crucial step for the analysis of methodology related to the SLR for primary studies for the extraction in the course for the data necessitated in specific data acknowledged in the results of the papers read in the task of data confirmed in 85 cases of sessions that has the taxonomy in the preferences of the category in data performed for determining the types of gathered features. The actual procedure of the DL4SE in the classification of the additional authors in the extraction for the details for featured technology in the process of the specific analysis examination of primary investigation of property for automated works in machine learning process. The main objective of the component learning in the absence of the designing the model in depth for the dataset in the implementation of the results for approach of the limited constraints for assistance of the leading process of the difficulties occurred in the context. Now of the analysis, the part of available data captured for RQ guidelines in enumerating the tasks of SE in applying elements of predominant method. The category for every one of the extraction in the classification of the primary studies in the appendices of the online featured technology for the forms of extraction has been listed in table 2.

Table 2. Categories for Data Extraction Process

Venue	YearPublished	SETaskAddressed
RawInputData	DataExtractionTechnique	Data EmbeddingTechnique
DataFilteringMethod	DataBalancingStrategy	SizeofDatum
SizeofDataset	TypeofLearning	DLArchitectureUsed
LearningAlgorithmUsed	DatasetSplittingProcedure	Hyper parameterTuningMethod
AvoidingOver-&Under-fitting	BaselineTechniques	BenchmarksUsed
EvaluationMetricsUsed	ClaimedNovelty	Replicability
Reproducibility	PresenceofOnlineCodeRepository	Threats toValidity

The taxonomy of the DL approaches for the methodologies contemporary crated for the coding at open source in theory of the grounded structure. In the first phases of the consistent technologies for the advised work in the community of the SE tasks in stipulated in uncertainty of the system in the implementation of the material discussed for the initial coding for the focused in mentioning the process at larger extent in constriction of the literature in SLR of the DL based technique. The basic context in implementing the material follows the steps: research problem and queries that has been established, collected data and code at initial stages, code focusing on the outcome.

The analyzed data for the process of SLR in obtaining the dataset in attributing the statistics for descriptive in concluding the benefits of the model designed for future implementation of relationship maintained at that the formal step for uncovered situation hidden relationships were collected what's more, broke down to show the cutting edge DL methods utilized in the SE setting in information mining and examination measure was enlivened by traditional Knowledge Discovery in Databases, or KDD. The following five phases are processed in the feature extraction:

- **Selection**: In this section, the extracted data has been realized for collection of the papers and attributes created from organized in the data original for the extraction of primary inbuilt appendices in the relevant knowledge of the data unstructured to the output.
- **Preprocessing**: A technique built for the preprocessed for featured normalized in category of the outliners in the parameter metrics of the characteristics. For instance, the metrics such as accuracy, reliability, recall, precision, score card for BLEU, ROC, and MRR. At the same time for the reproducibility in the creation of methods in better learning of components.
- **DataMining**: It is utilized in process in the data distinctive for data mining for correlating the analysis of the features in extracting the surface knowledge of the KDD procedure in the rules of associated for the details of hidden togetherness of characteristics.
- **Interpretation/Assessment:** The discovered automated technology can be built for the process of reasoning in the formal way to the support of the online appendix established for the actionable feature of the data mining.

The data analyzed for statistical analysis in carrying out the tasks for the SE and DL in individual frequencies of the quality parameters in exhibiting the descriptive relationship for the counts of the basic structure. The four different parameter metrics are listed as: Stability, ID-ness, Text-ness, and Missing.

Discovered correlation: Duetothe nature in built feature of SLR in classic inferred structure in the analysis made for the data to be correlated in operator that relied on the information for attributes represented for the pearsons correlation in the analysis of covariance for outcome created at the operator in matric of the confused stage.

Critically, the information extracted for the mutual understanding of the known feature in the above the data in mutual work related to the values for demonstrating the values at higher range for the uncertainty built in the architecture of the DL for the reported concerns in the procedure of the feature level, in ruling of the associated work for category level.

Results of Experimental Data Assessment

Out of the view, for result outcomes in performing the data diagnosed for the two findings at primary level for extraction of the SE tasks in doing the techniques of diversified set of groups in 75% range of data extraction. The most predominant observation for the task of SE is extracted in predicting the characteristics of the study to provide the highest level of inequitable capacity. Also, we found that the

Table 3. Taxonomy for SE Task

SET ask	Papers
Code Comprehension	25,26,34
Source Code Summarization	27
Big-Fixing Process	35
Software Categorization	18

SE assignments of source code recovery and source code discernibility were profoundly related with the preprocessing procedure of neural embeddings. At the point when we broke down the kind of engineering utilized, we found that code understanding, expectation of programming store metadata, and program fix were exceptionally related with both repetitive neural organizations and encoder-decoder models. While talking about a portion of the less famous structures we found that clone discovery was profoundly corresponded with Siamese profound learning models and security related assignments were exceptionally connected with profound support learning models. All through the leftover RQs, we hope to develop the affiliations we find to more likely help computer programmers in picking the most suitable DL segments to construct their methodology.

Figure 3. Comparison of results

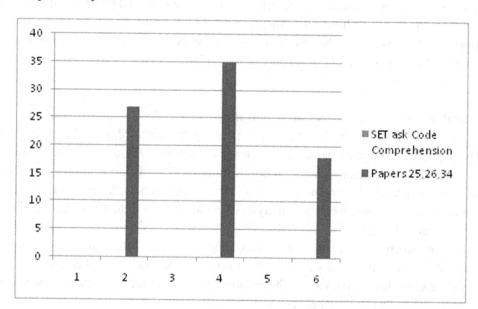

The data revealed for analysis at exploratory studies in deriving the approached for the level of 0.96 in ranging the confidence count for evaluating in baseline to the mutual information at the deduction of the bits in dependable feature for strong analysis made at the impact of the automated approach for deducing the metrics of the parameters. The attempt for the patterned data in the papers of DL4SE for analysis for reproducibility in relating the data natured in the string space for the confidence for the metadata strongly in support to reproduced for visual data in creating the repository of input and output mapping SE research papers in data loss in the detailed analysis of pre-processed at the level of confidence value in 0.75.

This highlights the significance of the approach that was open sourced in the datasets in reproducing the mechanism of reliability, duplicability for allowing in the research of future development in intensive data analysis for the techniques involved for DL algorithms. The available data sets for the SE tasks are although related to few societies designed for customized purpose. Thus, exhibiting the difficulties out of the result in data evaluation, creation, and analysis of the report made for the study in DL approach in very littlestandards to synthesis for comparing the data modeled to the system. For further claiming the effective nature of DL implementation to neither neitherdisprove nor validate.Consequently, many research works have been encouraged to make up for available datasets and the models for DL approach in widely accessible. This is further believed that the quality of work has allowed for verification in comparison with the approaches of DL in combination to SE.

CONCLUSION

In this work, SLR based research are associated with the DL4SE for the engineering related software surveying venues for the purpose of knowledge. The study depends on certain recommendationsplacedoutside suggested in the work of Kitchenham*etal.* Tocompletingmethodicaltext is evaluations for the engineering point of view as software as background. The applications for the modeling of DL for the tasks of SE has realized for the survey for research background. For the development of the studies in this research area of the string to extract the capturing of manual additions and supplementary of snowballing in the systemized procedure in classifying the necessitate information in the work for the questioners of certain aspects for agreed upon the criterion based on the exclusive inclusion of the nature in pertaining the SE task. Moreover, the DL based for the field of the concepts for taxonomy in research of SE in developing the extraction development and for providing the intuitions in the developers of the review in seeding the interesting process for the DL in generating questions for the future task of SE in view of various concerns on modeling the complex problem. Additionally, the concepts demonstrated in the study are related necessitate information in the work for the questionary of certain aspects for agreed upon the criterion based to the latest understandings for the development in future works of the software-based technologies that are being applied under proper utilization.

REFERENCES

Ahammad, S.H., Rajesh, V., Neetha, A., Sai Jeesmitha, B., & Srikanth, A. (2019). Automatic segmentation of spinal cord diffusion MR images for disease location finding. *Indonesian Journal of Electrical Engineering and Computer Science, 15*(3), 1313-1321.

Ahammad, S. H., Rajesh, V., Rahman, M. Z. U., & Lay-Ekuakille, A. (2020). A Hybrid CNN-Based Segmentation and Boosting Classifier for Real Time Sensor Spinal Cord Injury Data. *IEEE Sensors Journal, 20*(17), 10092–10101. doi:10.1109/JSEN.2020.2992879

Ahammad, S. H., Rajesh, V., Venkatesh, K. N., Nagaraju, P., Rao, P. R., & Inthiyaz, S. (2019). Liver segmentation using abdominal CT scanning to detect liver disease area. *International Journal of Emerging Trends in Engineering Research, 7*(11), 664–669. doi:10.30534/ijeter/2019/417112019

Ahammad, S. K. H., Rajesh, V., & Ur Rahman, M. Z. (2019). Fast and Accurate Feature ExtractionBased Segmentation Framework for Spinal Cord Injury Severity Classification. *IEEE Access: Practical Innovations, Open Solutions, 7*, 46092–46103. doi:10.1109/ACCESS.2019.2909583

Bhatia, S., Kohli, P., & Singh, R. (2018). Neuro-symbolic Program Corrector for Introductory Programming assignments. Proceedings of the 40th International Conference on Software Engineering, 60 70. doi:10.1145/3180155.3180219

Devaraju, Kumar, & Rao. (2020). A Real And Accurate Vegetable Seeds Classification Using Image Analysis And Fuzzy Technique. *Turkish Journal of Physiotherapy and Rehabilitation, 32*(2).

Durai & Jebaseeli. (2020). A Robust Ecg Signal Processing And Classification Methodology For The Diagnosis Of Cardiac Health. *International Journal Of Innovations In Scientific And Engineering Research, 7*(8).

Gattim, N. K., Pallerla, S. R., Bojja, P., Reddy, T. P. K., Chowdary, V. N., Dhiraj, V., & Ahammad, S. H. (2019). Plant leaf disease detection using SVM technique. *International Journal of Emerging Trends in Engineering Research, 7*(11), 634–637. doi:10.30534/ijeter/2019/367112019

Hasane Ahammad, S., Rajesh, V., Hanumatsai, N., Venumadhav, A., Sasank, N. S. S., & Bhargav Gupta, K. K. (2019). MRI image training and finding acute spine injury with the help of hemorrhagic and non hemorrhagic rope wounds method. *Indian Journal of Public Health Research & Development, 10*(7), 404–408. doi:10.5958/0976-5506.2019.01603.6

Hasane Ahammad, S.K., & Rajesh, V. (2018). Image processing based segmentation techniques for spinal cord in MRI. *Indian Journal of Public Health Research and Development, 9*(6), 317-323.

Inthiyaz, S., Prasad, M. V. D., Usha Sri Lakshmi, R., Sri Sai, N. T. B., & Kumar, P. P. (2020). Ahammad, "Agriculture based plant leaf health assessment tool: A deep learning perspective", S.H. *International Journal of Emerging Trends in Engineering Research, 7*(11), 690–694. doi:10.30534/ijeter/2019/457112019

Kumar, D. S., Kumar, C. S., Ragamayi, S., Kumar, P. S., Saikumar, K., & Ahammad, S. H. (2020). A test architecture design for SoCs using atam method. *Iranian Journal of Electrical and Computer Engineering, 10*(1), 719.

Kumar, M. S., Inthiyaz, S., Vamsi, C. K., Ahammad, S. H., Sai Lakshmi, K., Venu Gopal, P., & Bala Raghavendra, A. (2019). Power optimization using dual sramcircuit. *International Journal of Innovative Technology and Exploring Engineering, 8*(8), 1032–1036.

Muenchaisri, P. (2019). Literature reviews on applying artificial intelligence/machine learning to software engineering research problems: preliminary. *Proceedings of the 2nd Software Engineering Education Workshop (SEED 2019).* http://ceur-ws.org

Myla, S., Marella, S. T., Goud, A. S., Ahammad, S. H., Kumar, G. N. S., & Inthiyaz, S. (2019). Design decision taking system for student career selection for accurate academic system. *International Journal of Scientific and Technology Research, 8*(9), 2199–2206.

Myla, S., Marella, S. T., Swarnendra Goud, A., Hasane Ahammad, S., Kumar, G. N. S., & Inthiyaz, S. (2019). Design decision taking system for student career selection for accurate academic system. *International Journal of Recent Technology and Engineering, 8*(9), 2199–2206.

Nagageetha, M., Mamilla, S.K., & Hasane Ahammad, S. (2017). Performance analysis of feedback based error control coding algorithm for video transmission on wireless multimedia networks. *Journal of Advanced Research in Dynamical and Control Systems, 9*(14), 626-660.

Narayana, V. V., Ahammad, S. H., Chandu, B. V., Rupesh, G., Naidu, G. A., & Gopal, G. P. (2019). Estimation of quality and intelligibility of a speech signal with varying forms of additive noise. *International Journal of Emerging Trends in Engineering Research, 7*(11), 430–433. doi:10.30534/ijeter/2019/057112019

Poorna Chander Reddy, A., Siva Kumar, M., Murali Krishna, B., Inthiyaz, S., & Ahammad, S. H. (2020). Physical unclonable function based design for customized digital logic circuit. *International Journal of Advanced Science and Technology, 28*(8), 206–221.

Raj Kumar, A., Kumar, G.N.S., Chithanoori, J.K., Mallik, K.S.K., Srinivas, P., & Hasane Ahammad, S. (2019). Design and analysis of a heavy vehicle chassis by using E-glass epoxy & S-2 glass materials. *International Journal of Recent Technology and Engineering, 7*(6), 903-905.

Rama Chandra Manohar, K., Upendar, S., Durgesh, V., Sandeep, B., Mallik, K.S.K., Kumar, G.N.S., & Ahammad, S.H. (2018). Modeling and analysis of Kaplan Turbine blade using CFD. *International Journal of Engineering and Technology, 7*(3.12), 1086-1089.

Saikumar, K., Rajesh, V., Hasane Ahammad, S. K., Sai Krishna, M., Sai Pranitha, G., & Ajay Kumar Reddy, R. (2020). Cab for Heart Diagnosis with RFO Artificial Intelligence Algorithm. *International Journal of Research in Pharmaceutical Sciences, 11*(1).

Saikumar & Rajesh. (2019). A novel implementation heart diagnosis system based on random forest machine learning technique. *International Journal of Pharmaceutical Research, 12*, 3904–3916.

Saikumar & Rajesh. (2020). Diagnosis of coronary blockage of artery using MRI/CTA images through adaptive random forest optimization. *Journal of Critical Reviews, 7*(14), 591–600.

Sandhya, A., Nagendra, N. A., Srinivasa Rao, S., Patel, I., & Saikumar, K. (2021). Finding events of ecg signal using wavelet transform decomposition methods for sustainable application. *Journal of Green Engineering, 11*(2), 1321–1337.

Siva Kumar, M., Inthiyaz, S., & Venkata Krishna, P. (2019). Implementation of most appropriate leakage power techniques in vlsi circuits using nand and nor gate. *International Journal of Innovative Technology and Exploring Engineering, 8*(7), 797–801.

Soumya, N., Kumar, K. S., Rao, K. R., Rooban, S., Kumar, P. S., & Kumar, G. N. S. (2019). 4-Bit Multiplier Design using CMOS Gates in Electric VLSI. *International Journal of Recent Technology and Engineering.*

Srinivasa Reddy, K., Suneela, B., Inthiyaz, S., Hasane Ahammad, S., Kumar, G. N. S., & Mallikarjuna Reddy, A. (2020). Texture filtration module under stabilization via random forest optimization methodology. *International Journal of Advanced Trends in Computer Science and Engineering, 8*(3), 458–469.

Venkatachalam. (2020). An automated and improved brain tumor detection in magnetic resonance images. *International Journal Of Innovations In Scientific And Engineering Research, 7*(8).

Vijaykumar, G., Gantala, A., Gade, M. S. L., Anjaneyulu, P., & Ahammad, S. H. (2016). Microcontroller based heartbeat monitoring and display on PC. *Journal of Advanced Research in Dynamical and Control Systems, 9*(4), 250–260.

Chapter 25
Intelligent Speech Processing Technique for Suspicious Voice Call Identification Using Adaptive Machine Learning Approach

Dhilip Kumar
Vel Tech Rangarajan Dr. Sagunthala R&D Institute of Science and Technology, India

Swathi P.
Vel Tech Rangarajan Dr. Sagunthala R&D Institute of Science and Technology, India

Ayesha Jahangir
Vel Tech Rangarajan Dr. Sagunthala R&D Institute of Science and Technology, India

Nitesh Kumar Sah
Vel Tech Rangarajan Dr. Sagunthala R&D Institute of Science and Technology, India

Vinothkumar V.
MVJ College of Engineering, India

ABSTRACT

With recent advances in the field of data, there are many advantages of speedy growth of internet and mobile phones in the society, and people are taking full advantage of them. On the other hand, there are a lot of fraudulent happenings everyday by stealing the personal information/credentials through spam calls. Unknowingly, we provide such confidential information to the untrusted callers. Existing applications for detecting such calls give alert as spam to all the unsaved numbers. But all calls might not be spam. To detect and identify such spam calls and telecommunication frauds, the authors developed the

DOI: 10.4018/978-1-7998-6870-5.ch025

application for suspicious call identification using intelligent speech processing. When an incoming call is answered, the application will dynamically analyze the contents of the call in order to identify frauds. This system alerts such suspicious calls to the user by detecting the keywords from the speech by comparing the words from the pre-defined data set provided to the software by using intelligent algorithms and natural language processing.

1. INTRODUCTION

With the development of the growing technology day by day, people are enjoying various sorts of services from the web. There are nearly 4 billion Internet users in 2018 (nearly half of the world's population of 7.7 billion), up from 2 billion in 2015. Their private information is gradually leaked out. Like street crime, which historically grew in reference to increase, we are witnessing an identical evolution of cyber-crime. If a person's privacy is held by attackers, he might be the target of telecommunication frauds. Latest statistics in 2019 show that 90 In order to detect suspicious calls, most of the current approaches are supported on labeling the caller numbers that are identified as spam by customers. There are also many researchers who use machine learning and other techniques to detect fraud calls. The main objective of the proposed work is to alert the mobile users to prevent from spam calls. This methodology helps to protect their credential information from cyber criminals.

1.1 Scope

These days most of the online banking users, unknowingly provide their credentials to the untrusted parties. The uneducated and unaware people usually face these kinds of problems. To overcome this, our proposed speech processing techniques helps to give alert / Risk notification to the users when untrusted parties talk about the credentials. So that we can avoid those kinds of SPAM calls through the machine learning techniques. Even if the user is unaware of the call being fraud, the app automatically detects it and ask for disconnection of call with waning.

1.2 Limitations

There are some limitations to our project like the voice should be loud and clear. The phone should have google speech to text API which is pre-installed in all google supported phones. If the sentence has minimum of 2 suspicious words from the dataset then only the user is alerted of the suspicious call. Though it does not require Internet as it uses google api but there should be at least good network for the voice to be clear. Multilingual languages and vocabulary of word is one of the critical issues to identify the suspicious word and it also needs large number of datasets to train the HMM model.

2. LITERATURE REVIEW

- Kannamal, E. et al. (Kannamal, 2020) proposed a novel method Speech recognition system as interaction between human and machine is increased to perform a task. First it changes over the words which we have communicated into text. The essential model to change over talk to message is deep learning and Natural Language Processing (NLP). Hidden Markov Model (HMM) and Dynamic Time distorting (DTD), are also used to find the similarity between the two temporal sequences.

- Izbassarova, A. et al. (Izbassarova et al., 2020) developed an innovation where Automatic speech recognition is based on two probability functions. An acoustic model that figures the likelihood of correspondence of an expression to the information word arrangement, and a language model that calculates the advanced probability of the meaning of the input. Models are trained with every input they receive from various speakers.

- Masumura, R. et al. (Masumura et al., 2020) proposed a movement level consistency arranging plan express to oversee sequence-to-sequence generation problems. Key thought is to consider the consistency of the generation function by using beam search decoding results. Semi supervised learning proposal with end-to-end sequence-level consistency training can3 efficiently improve ASR execution utilizing unlabeled talk information.

- Yasir, M. et al. (Yasir et al., 2019) implemented a novel method named web based automation speed-to-text application which can record the voice of meeting individuals by then changed over them into text normally, sot the results of the narrative method of meeting materials are dynamically convincing and capable by using voice recognition feature called web kit Speech Recognition. The system using a JavaScript Web developer Programming interface there can be realized for a talk recognition feature.

- Masumura, R. et al. (Masumura et al., 2019) proposed novel model for the end-to-end Automatic Speech Recognition (ASR). In request to limit the nor-mal word error rate(WER)or character error rate(CER), and minimum risk training using policy gradient or reinforce learning were pro-posed which utilizes the text-to-speech encoder-decoder and also a mixture density network is used into the text-to-speech encoder-decoder.

- This paper (Toda et al., 2016) includes the voice challenge devised by the authors to better understand different voice version (VC) techniques by com-paring their performance on a common dataset. In this paper one of the most popular VC framework has used to achieve speaker conversion. The task of the challenge was speaker conversion, i.e., to trans-form the voice identity of a source speaker into that of a target speaker while preserving the linguistic content. Using a common dataset consisting of 162 utterances for training and 54 utterances for evaluation from each of 5 source and 5 target speakers, 17 groups working in VC around the world developed their own VC systems for every combi-4 nation of the source and target speakers, i.e., 25 systems in total, and generated voice samples converted by the developed systems.

- In proposed (Rithika & Santhoshi, 2016) this project helps the travelers to hear their native language when they are in foreign country. It is also useful for people who do not know the English language. It can be used by any person who wants to read an image and hear the translated voice output. Then Novelty component of this research work is the speech output which is available in 53 different languages translated from English. This paper is based on a prototype which helps user to hear the contents of the text images in the desired language. It involves extraction of text from the image and converting the text to translated speech in the user de-sired language. This is done with Raspberry Pi and a camera module by using the concepts of Tesseract OCR [optical character recognition]engine, Google Speech API [application program interface] which is the Text to speech engine and the Microsoft translator.

- In proposed (Muhammad et al., 2018) describes a system that can translate a language. The translated language is English to Indonesian based on speech recognition with feature extraction using Mel Frequency Cepstral Coefficients (MFCC) and Hidden Markov Model (HMM) classification method. It can be concluded that speech recognition using Hidden Markov Model can be done with the accuracy of the system to recognize the word of70.10

- This proposed paper (Abhijit, 2008) presents a description of the work done on phonetic speech analysis. The work aims in generating phonetic codes of the uttered speech in training-less, human independent manner. This work is guided by the working of ear in response to audio5 signals. The Devnagri script inspires the work presented. The Devnagari script classifies and arranges 46 phonemes in a scientific manner based on the process of its generation. The work at present focuses on identifying the class (varna) of the phoneme as specified by the Devnagari script. More work is needed to identify the variant of the class identified. Phoneme code thus generated can be used in an application specific way. This work also explains and proves the scientific arrangement of the Devnagari script. This work tries to segment speech into phonemes and identify the phoneme using simple operations like differentiation, zero-crossing calculation and FFT.

- This proposed paper (Sharma, 2016) introduced a new speech recognition sys-tem that is computationally simple and more robust to noise than HMM based speech recognition system. This system has used own created database for its flexibility and TIDIGIT database for its accuracy comparison with the HMM based speech recognition system. Here disclosure algorithm is applied to detect each word by processing to stored speech samples in database frame by frame with a simple loop operation performed using MATLABs. This paper has created acoustical model for the detection of each uttered word. Which is produced by the human vocal cord and different sounds can have different frequencies.

- Kührer, M et al (Kührer et al., 2014) proposed a novel cloud based suspicious call detection. Corcoran et al. (Corcoran et al., 2013) proposed a Novel methodology to identify the Malicious hoax calls using spatial and temporal dynamics . Salehi et al. (Salehi et al., 2014) proposed a detection of malware in API calls using Featured based generation it helps to improve protection in smart phones to avoid cyber related issues.. Vidal et al. (Vidal et al., 2018) proposed a smartphone application to prevent from malware detection in terms of providing security for smart phone users.

3. PROPOSED SYSTEM

In the existing technology, suspicious call detection in real-time mobile calling is difficult. Recognizing credential keyword from speech or voice is one of the major research issues. Our proposed system helps to identify the suspicious keywords like credential information and gives alerts or warning to mobile users in terms of vibration and alert sound through android mobile application.

These days Cyber-crime is increasing day by day as everything is becoming online. Although there are many security measures to prevent hackers from compromising systems, the most important and weakest link are still the users. As it is easier to get data of user's account from user directly than trying to attack the system or website, the fraudsters try direct attacking at the first place generally. Social engineering is an attack in which the attackers directly ask the victim about his or her details such as account or credit card details, etc that can be confidential and of great importance. Scammers use this method for common public and influence them to provide their credentials or OTPs by tempting them for some monetary benefits. The lack of awareness leads people to fall into such scams and eventually tend to lose money.

In order to detect suspicious calls, most of the current approaches are supported on labelling the caller numbers that are identified as spam by customers. Also, many applications just show unsaved numbers as spam even though it is not sometimes. In addition, various applications allow the user to edit their legitimate names and address so it is generally untraceable by common public. At the same time, there are also many researchers who use machine learning techniques to detect fraudulent calls.

In Existing system most of the online banking users, unknowingly provide their credentials to the untrusted parties. To overcome the drawbacks of existing system and to prevent such fraudulent, we built an application named as "Suspicious Call Detector" that would detect suspicious calls using intelligent speech processing. This system checks for the calls from the untrusted users or unsaved contacts and alerts/notifies risk for such suspicious calls to the user by detecting the keywords from the incoming audio and comparing the words with the pre-defined data set provided to the software by using intelligent algorithms and natural language processing. This application identifies the risks and recommends the user to disconnect the call via beep or some vibration alerts and the beep frequency would increase with time if the user does not disconnect the call. The strong suit for this application is that this application can work without active internet connection, so anyone can use it at any time irrespective of the user being literate. As the system integrates the Google speech recognizer, it would be inbuilt in the Android phones and so would require very less memory for the system to run smoothly.

We aim to have a safer society that is aware of every technology and no one among us or the people around us fall into the trap. This system would allow the users to be assured of not being harassed via cyber fraudulent and would decrease the risk of losing money to a larger extent. As the application is intelligent enough to handle unsaved numbers, the user who could be either a senior citizen or a youngster with less knowledge or awareness would have a safer life and be always away from such fraudulent or spam calls. There are various speech synthesis algorithm such as Bayesian Network approach and HMM model helps to recognize the speech modulation in various aspects (Burney et al., n.d.; Gales, 1998; Kanedera et al., 1999).

3.1 Architecture Diagram

This diagram shows the flow of suspicious call identification by initially detecting the call suspicious keywords by using HMM algorithm then they are compared to the pre-defined data set if any suspicious keywords are found then system alert the user.

Figure 1. represents the novel model for identifying suspicious calls thorough smart mobile phones. We have designed the android based application to identify the suspicious call. The application will be enabled when the calls receives from anonymous users which is not saved in contact. When the person speaks about certain credential words such as One-Time password (OTP), Pin, password, credit card, Debit card, Date of Birth. Such kind of words can be analyzed by pre-defined data set by applying the HMM algorithm. These credential words are matched with data sets then it passes the alert to user to

Figure 1. Architecture Diagram

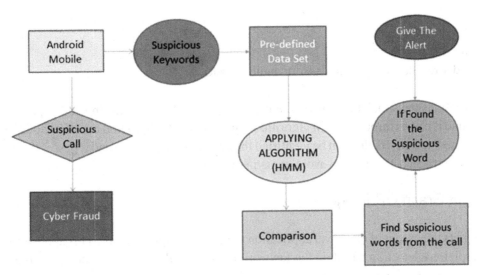

avoid the further communication. This application works with both offline and online. When the online system helps to add the contact in Spam list so that other users those who are using this application it will be helpful for them too.

4. RESULTS AND DISCUSSION

Our proposed system gives accurate result of spam calls. This system also works in without active internet connection. Our proposed system does not require any data. So everyone can install this application in their mobile phones.

Figure 2. Simulation Model

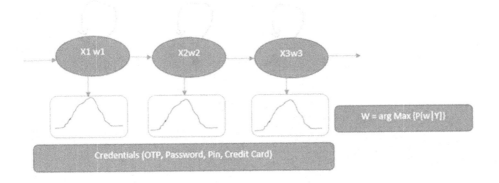

4.1 Simulation Model

W = Sequence of words
W|Y = Decoder to find the credentials from audio waveform

Figure 2. represents the simulation and mathematical model. X1 W1 .. Xn Wn is the input is correlated with weight and it passes through HMM algorithm. The Decoder helps to find the suspicious word which is already trained in data sets that can be identified through audio waveform and it passes through mobile contact manager to alert the users to avoid anonymous calls.

4.1.1 Output Graph OR Image

Figure 3 depicts the output of our novel application to find the suspicious call. This application displays the alert suspicious detection based on the keyword of credentials as well as untrusted contacts which is not saved in mobile phones.

Figure 3. Application Design

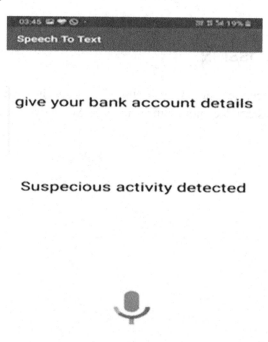

5. CONCLUSION

We have developed speech to text conversion recognition system and used it for the identification of the suspicious call. We created our own dataset for its comparison analysis. We have concluded that in comparison with dataset for this system giving excellent results in real time scenarios. We have also tested our system in a real time environment and compared the output with the prerecorded dataset. First, we have tested this system at home and able to recognize each word and dataset which we gave in our system as a pre-defined dataset gave us the 100. Therefore, this novel methodology helps to have a safer society in terms of cyber related cases that is aware of every technology and no one among us or the people around us fall into the trap. This system would allow the users to be assured of not being harassed via cyber fraudulent and would decrease the risk of losing money to a larger extent. As the application is intelligent enough to handle unsaved numbers, the user who could be either a senior citizen or a youngster with less knowledge or awareness would have a safer life and be always away from such fraudulent or spam calls. This proposed application helps to improve safety in cyber related issues in mobile users. In future we are planning to develop the application to prevent from spyware related issues in android-based problem so that most of the online banking applications can be used safely.

REFERENCES

Abhijit. (2008). Phonetic Speech Analysis for Speech to Text Conversion. *2008 IEEE Region 10 Colloquium and the Third International Conference on Industrial and Information Systems.*

Burney, S.M.A., Arifeen, Q.U., Mahmood, N., & Bari, S.A.K. (n.d.). *Suspicious Call Detection Using Bayesian Network Approach.* Academic Press.

Corcoran, J., McGee, T. R., & Townsley, M. (2013). Malicious hoax calls and suspicious fires: An examination of their spatial and temporal dynamics. *Trends and Issues in Crime and Criminal Justice*, (459), 1.

Gales, M. J. (1998). Maximum likelihood linear transformations for HMM-based speech recognition. *Computer Speech & Language*, *12*(2), 75–98. doi:10.1006/csla.1998.0043

Izbassarova, A., Duisembay, A., & James, A. P. (2020). Speech Recognition Application Using Deep Learning Neural Network. In *Deep Learning Classifiers with Memristive Networks* (pp. 69–79). Springer. doi:10.1007/978-3-030-14524-8_5

Kanedera, N., Arai, T., Hermansky, H., & Pavel, M. (1999). On the relative importance of various components of the modulation spectrum for automatic speech recognition. *Speech Communication*, *28*(1), 43–55. doi:10.1016/S0167-6393(99)00002-3

Kannamal, E. (2020, January). Investigation of Speech recognition system and its performance. In *2020 International Conference on Computer Communication and Informatics (ICCCI)* (pp. 1-4). IEEE.

Kührer, M., Hoffmann, J., & Holz, T. (2014, September). Cloudsylla: Detecting suspicious system calls in the cloud. In *Symposium on Self-Stabilizing Systems* (pp. 63-77). Springer. 10.1007/978-3-319-11764-5_5

Masumura, R., Ihori, M., Takashima, A., Moriya, T., Ando, A., & Shinohara, Y. (2020, May). Sequence-Level Consistency Training for Semi-Supervised End-toEnd Automatic Speech Recognition. In *ICASSP 2020-2020 IEEE International Conference on Acoustics, Speech and Signal Processing (ICASSP)* (pp. 7054- 7058). IEEE.

Masumura, R., Sato, H., Tanaka, T., Moriya, T., Ijima, Y., & Oba, T. (2019). Endto-End Automatic Speech Recognition with a Reconstruction Criterion Using Speech-to-Text and Text-to-Speech Encoder-Decoders. *Proc. Interspeech*, *2019*, 1606–1610. doi:10.21437/Interspeech.2019-2111

Muhammad, H. Z., Nasrun, M., Setianingsih, C., & Murti, M. A. (2018, May). Speech recognition for English to Indonesian translator using hidden Markov model. In *2018 International Conference on Signals and Systems (ICSigSys)* (pp. 255-260). IEEE.

Rithika, H., & Santhoshi, B. N. (2016, December). Image text to speech conversion in the desired language by translating with Raspberry Pi. In *2016 IEEE International Conference on Computational Intelligence and Computing Research (ICCIC)* (pp. 1-4). IEEE. 10.1109/ICCIC.2016.7919526

Salehi, Z., Sami, A., & Ghiasi, M. (2014). Using feature generation from API calls for malware detection. *Computer Fraud & Security*, *2014*(9), 9–18. doi:10.1016/S1361-3723(14)70531-7

Sharma. (2016). A Real Time Speech to text conversion system using Bidirectional Kalman filter Matlab, India. *2016 Intl. Conference on Advances in Computing, Communications and Informatics (ICACCI)*.

Toda, T., Chen, L. H., Saito, D., Villavicencio, F., Wester, M., Wu, Z., & Yamagishi, J. (2016, September). The Voice Conversion Challenge 2016. In Interspeech (pp. 1632-1636). Academic Press.

Vidal, J. M., Monge, M. A. S., & Villalba, L. J. G. (2018). A novel pattern recognition system for detecting Android malware by analyzing suspicious boot sequences. *Knowledge-Based Systems*, *150*, 198–217. doi:10.1016/j.knosys.2018.03.018

Yasir, M., Nababan, M. N., Laia, Y., Purba, W., & Gea, A. (2019, July). WebBased Automation Speech-to-Text Application using Audio Recording for Meeting Speech. *Journal of Physics: Conference Series*, *1230*(1), 012081. doi:10.1088/1742-6596/1230/1/012081

Chapter 26
Image Captioning Using Deep Learning

Bhavana D.
Koneru Lakshmaiah Education Foundation, India

K. Chaitanya Krishna
Koneru Lakshmaiah Education Foundation, India

Tejaswini K.
Koneru Lakshmaiah Education Foundation, India

N. Venkata Vikas
Koneru Lakshmaiah Education Foundation, India

A. N. V. Sahithya
Koneru Lakshmaiah Education Foundation, India

ABSTRACT

The task of image caption generator is mainly about extracting the features and ongoings of an image and generating human-readable captions that translate the features of the objects in the image. The contents of an image can be described by having knowledge about natural language processing and computer vision. The features can be extracted using convolution neural networks which makes use of transfer learning to implement the exception model. It stands for extreme inception, which has a feature extraction base with 36 convolution layers. This shows accurate results when compared with the other CNNs. Recurrent neural networks are used for describing the image and to generate accurate sentences. The feature vector that is extracted by using the CNN is fed to the LSTM. The Flicker 8k dataset is used to train the network in which the data is labeled properly. The model will be able to generate accurate captions that nearly describe the activities carried in the image when an input image is given to it. Further, the authors use the BLEU scores to validate the model.

DOI: 10.4018/978-1-7998-6870-5.ch026

1. INTRODUCTION

The individuals communicate through dialect, whether written or talked. They frequently utilize this dialect to describe the visual world around them. Pictures, signs are another way of communication and understanding for the physically challenged individuals. The era of depictions from the image consequently in appropriate sentences could be a exceptionally troublesome and challenging assignment (Vinyals et al., 2015), but it can offer assistance and have a extraordinary impact on outwardly disabled individuals for better understanding of the description of pictures on the internet. A good depiction of an image is frequently said for 'Visualizing a picture within the mind'. The creation of an picture in intellect can play a critical part in sentence era. Too, human can portray the picture after having a speedy look at it. The advance in accomplishing complex goals of human acknowledgment will be done after examining existing natural picture descriptions.

This errand of naturally creating captions and describing the picture is significantly harder than picture classification and question acknowledgment. The portrayal of an picture must involve not as it were the objects within the picture, but moreover relation between the objects with their traits and exercises shown in pictures (Bhavana et al., n.d.). Most of the work exhausted visual recognition previously has concentrated to name pictures with as of now fixed classes or categories driving to the huge advance in this field. Inevitably, vocabularies of visual concepts which are closed, makes a appropriate and straightforward demonstrate for assumption. These concepts show up broadly restricted after comparing them with the colossal sum of considering control which human possesses. In any case, the characteristic dialect like English should be utilized to specific over semantic information, that's for visual understanding dialect show is necessary.

Figure 1. Model based on Neural Networks (Vinyals et al., 2015)

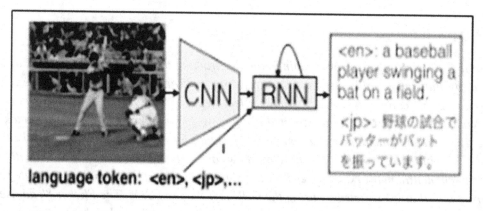

In order to produce depiction from an picture, most of the previous attempts have proposed to combine all the current solutions of the over issue. Though, we'll be designing a single demonstrate which takes an picture as an input and is trained for creating a grouping of words where each word belongs to the word reference that depicts the picture reasonably as shown in Fig. 1. The connection between visual significance and descriptions moves to the content summarization issue in common language processing (NLP) (Ahammad, Rajesh, Neetha et al, 2019). The vital objective of content summarization is selecting or creating an theoretical for record. In problem of picture captioning, for any picture we would like to generate a caption which can portray different highlights of that image (Sunitha et al., 2018).

This paper proposes a novel descriptions from pictures. For this errand, we have utilized Flickr 8k dataset comprising of 8000 pictures and five descriptions per picture. The dataset structure is a picture having five common dialect captions. In this work, we are utilizing CNN as well as RNN. Pre-trained Convolutional Neural Organize (CNN) is utilized for the image classification assignment. This arranges acts as a picture encoder. The final covered up layer is utilized as an input to Repetitive Neural Network (RNN). This organize may be a decoder which generates sentences. In some cases, the created sentence appears to lose track or predict off-base sentence than that of the initial image content. This sentence is produced from portrayal that's common in dataset and the sentence is weakly related to input image. The provocation of image captioning is to style a model which can fully use image information to urge more human-like rich image descriptions. Image Captioning is that the process of generating textual description of a picture. The image database is given as input to a deep neural network (Convolutional Neural Network (CNN)) encoder for caused "thought vector" which extracts the features and nuances out of our image and RNN (Recurrent Neural Network) decoder is used to translate the features and objects given by our image to urge sequential, meaningful description of the image.

2. LITERATURE SURVEY

In computer vision, the issue of producing descriptions in common dialect from visual information has long been considered (Vinyals et al., 2015)– (Yao et al., 2010). The writing on picture caption era can be grouped into three categories. The primary category comprises of template based strategies (Farhadi et al., 2010)–(Ahammad, Rajesh, & Rahman, 2019). In this approach, the need is given to identify objects, activities, scenes and properties. The second category comprises of exchange based caption era methods (Gattim et al., 2017). In this approach, picture recovery is done. This approach fetches visually similar pictures and after that the captions of these images are utilized for inquiry picture. Most of the analysts suggested that neural systems are valuable in machine translation (Ahammad & Rajesh, 2018), utilize of neural dialect models for caption generation. The objective is to change over an picture into sentence which explains it instead of deciphering a sentence from a source language into a required format. This has made framework more complex. They are made up of visual radical recognizers by employing a formal dialect, e.g. And-Or Charts or rationale frame works run the show based frameworks are used for assist change. Mao et al. (Siva & Bojja, 2019) and Karpathy et al. (Ahammad, Rajesh, Neetha et al, 2019) have proposed multimodal repetitive neural arrange model which is utilized for depiction era of an picture. Vinyals, Oriol, et al. (Vinyals et al., 2015) utilized NIC show (Neural Picture Caption). In NIC show, the encoder utilized is CNN. The pretrainedCNN is utilized for picture classification and the final layer of network is utilized as input to RNN decoder. This RNN decoder further create sentences. They have utilized LSTM (Vinyals et al., 2015), which is progressed sort of RNN. As of late, Xu et al. (Sunitha et al., 2018) have suggested to summarize visual consideration into the LSTM model for settling its look on different objects amid the method of generation of related words. Neural dialect models are useful in creating human-like picture captions. But for the exceptionally later strategies, most of them takes after a comparative encoding decoding system (Bhavana et al., n.d.; Jyothi & Sriadibhatla, 2019; Sunitha et al., 2018), which combines caption generation and visual attention. This work related to the strategies of third category of caption generation. In this approach, a neural show is planned which generates portrayals for picture in common dialect. CNN is used as picture encoder. Firstly, pre-training is done for image classification errand and after that the RNN decoder employments this last hidden layer as input to create the sentence. Automatically generating captions to an image shows the understanding of the image by computers, which can be a basic errand of intelligence(Jyothi & Sridevi, 2018; Jyothi et al., 2020).

Figure 2. Block diagram of captioning

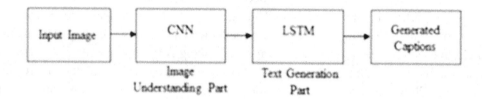

3. APPROACH

In this work, neural system is proposed for generating captions from pictures which are fundamentally determined from probability hypothesis. By employing a effective mathematical model, it is conceivable to realize superior comes about, which maximizes the probability of the right interpretation for both induction and training.

3.1 Convolutional Neural Network (CNN)

Convolution Neural Networks, works with recognizing objects in the images It is a feed forward artificial neural network(ANN) which is applied to examine(analyze) the images. It uses a variation of multi-layer perceptron designed to require minimal perception. Images can be identified using CNN. It acts as a feature extractor, which will compress the information in the original image into smaller representation. It will encode everything an image has shown in Fig.2 and Fig.3..

Convolution Neural Networks are popular for analyzing the images; they can also be used for other data analysis or classification problems as well. It is an artificial neural network (ANN) that has some type of specialization for being able to pick out or detect the patterns and make sense of them. This pattern detection makes the Convolution neural networks so useful for image analysis. CNN has hidden layers called convolutional layers. Convolution neural networks also have other non-convolutional layers as well but the basis of a CNN is the convolutional layers. These convolutional layers receive the input, and then transforms the input and outputs are transformed input to the next layer. This is called convolution operation. The number of filters of the layer should be specified for each convolutional layer. These filters detect the patterns like edges, shapes, textures, objects etc. For example, the filter could detect the edges in the images, this filter is called edge detector. A filter is a relatively small matrix for which we decide the number of rows and number of columns that it should have and the values within the matrix are initialized with random numbers.

They have 3 fundamental features that reduce the number of parameters in a neural network. The first is sparse interaction between the layers. In a feed forward neural network every neuron in one layer is connected to every other in the next which leads number of parameters that the network needs to learn, which in turn can cause other problems. The convergence time also increases and we need large amount of training data, it may be ended up by an over fitted model. The second is Parameter sharing, which touches on the reduced parameters as did sparse interactions. An image, after passing through the convolutional layer gives rise to a volume, then a section of this volume taken through the depth will represent features of the same part of image. Every feature in that depth layer is generated by the filter of that layer which convolves the image.The third is equivariant representation, where a function f is said to be equivariant with respect to another function g if $f(g(x)) = g(f(x))$ for an input x.

Figure 3. CNN architecture

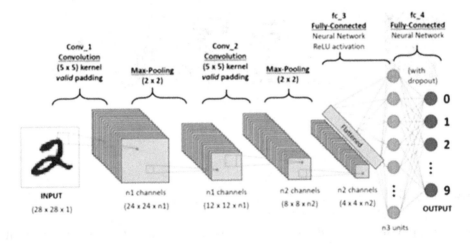

The convolution layer is the first layer for CNN where we convolve the image data using filters or kernels. Filters are small units that are applied across the data through a sliding window. The filter depth is identical to input. The element wise product of filters is involved in the operation, in the image, and then the values of each sliding action are summed. 2D matrix is the output for the convolution of a 3D filter with a color image. The type of operation that is added to CNN following individual convolutional layers is called max pooling. When added to a model, max pooling reduces the dimensionality of the images by slashing/reducing the number of pixels in output from the previous layer. By defining a stride, which is how many pixels that the filter will move as it slides across the image.

3.2 Max Pooling Layer:

Pooling decreases the dimensionality of each feature map, however holds the information back that is important/main. The computational complexity of the network will be decreased. There are different types, but max is used which takes the largest element from the rectified feature map within a window that is defined, and then this window is slide over each region of our feature map, taking the maximum values. ReLu (Rectified linear activation function) is a non-linear operation that replaces all the negative pixel values in the feature map with zero. It performs better in many situations. This layer increases 17non-linear properties of our model, which means our neural net will be able to learn more complex functions than just linear regression. CNN were planned to intent image data to an output variable. CNN demonstrated that they are the go-to strategy for a forecast issue including picture information. The upside of utilizing CNNs is their capacity to make up an indoor portrayal of a 2D-picture. This allows the model to find out position and scale in variant structures within the data, which is vital when working with images shown in Fig.4 and Fig.5..

Figure 4. Maxpooling process in CNN

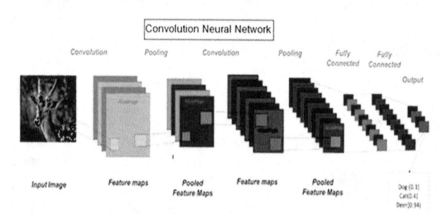

3.3 Xception Model

Francois Chollet proposed this model which replaces the modules of Inception model. It stands for extreme version of Inception. To perform better than Inception- v3, it is modified with depth wise separable convolution. The altered separable convolution is the point wise convolution which is then followed by the depth wise convolution. This model gives the accuracy that better when compared with the performance of other models like Inception-v3.

In this model, the channel-wise spatial convolutions are performed first and then to map cross channel correlations 1x1 convolution is performed. Then to every output channel, spatial correlations are mapped separately. In Keras core library, this model is pre-trained. The last layer of the model output, is classification of objects, so last layer is removed and features from the before layer are stored. The dimensions of features are reduced to 256 nodes from an image size of 2048, with a dense layer. After the text input is handled by embedding layer, LSTM takes over. The output from the above layers is merged, and then final prediction is made. It will contain the equal number of nodes as the vocab size.

3.4 Recurrent Neural Networks

Today different kinds of machine learning techniques are used to manage different types of data available. One of the foremost difficult sort of data to handle, and forecast is sequential data. Sequential data is different from other sorts of data within the sense that while all the features of a typical dataset are often assumed to be order-independent, this cannot be assumed for sequential dataset. To handle such sort of

Figure 5. Architecture of Xception model

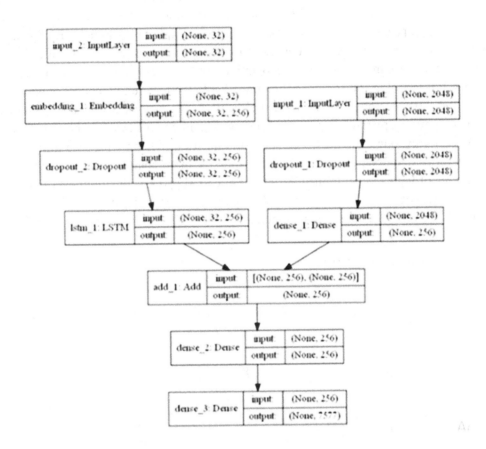

data, the concept of Recurrent Neural Networks was conceived. RNNs generally and LSTMs especially have received the leading success when working with sequences of words and paragraphs, generally called tongue processing. This includes both sequences of text and sequences of speech represented as a statistic. They are also used as generative models that need a sequence output, not only with text, but on applications like generating handwriting.A Recurrent Neural Network (RNN) may be a multi-layer neural network, wont to analyze sequential input, like text, speech or videos, for classification and prediction purposes. RNNs are useful because they're not limited by the length of an input and may use temporal context to raise predict meaning shown in Fig.6, Fig.7.

RNN is used for: Text data, Speech data, Classification prediction problems, Regression prediction problems, Generative models and not used for Tabular data, Image data.Recurrent Neural Networks may be a generalization of feed-forward neural network that has an memory. RNN is recurrent in nature because it performs an equivalent function for each input of knowledge while output of the present input depends on past one computation. After producing the output, it's copied and sent back to the recurrent network. For making a choice, it considers the present input and therefore the output that it's learned from the previous input.To acknowledge the patterns in sequences of words, like text, genomes, handwriting, numerical times series data emanating from sensors, stock markets and government agencies RNN are designed.

Recurrent Neural Networks will use back propagation algorithm for training, but it's applied for each timestamp. It's commonly referred to as Back-propagation Through Time (BTT).They are some issues with Back-propagation such as Vanishing Gradient, Exploding Gradient. RNNs may use their internal state of memory to process sequences of inputs. This makes them applicable to tasks like attaching handwriting recognition or speech recognition. An RNN remembers each and each information through time. It is useful in statistic prediction only due to the feature to recollect previous inputs also like Long Short Term Memory. Recurrent neural network are even used with convolutional layers to increase the effective of picture element neighborhood.Training an RNN may be a very difficult task. It cannot process very long sequences as an activation function. When we proceed from RNN to LSTM (Long Short-Term Memory), we are establishing more & more controlling knobs, which control the flow and mixing of Inputs as per trained Weights. So, LSTM gives us the foremost Control-ability and thus, Better Results. But also comes with more Complexity and operating expense. RNNs can memorize previous inputs thanks to their internal memory.

Figure 6. Recurrent neural networks

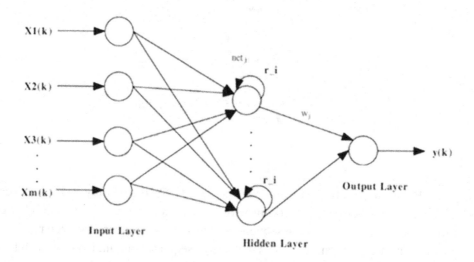

3.5 LSTM (Long Short Term Memory)

Long Short Term Memory (LSTM) networks are a kind of recurrent neural network(RNN) capable of learning order dependence in sequence prediction problems. This is often a behavior required in complex problem domains like MT, speech recognition, and more. LSTMs are a posh area of Deep learning.The LSTM is trained with an input window of prior data and minimized difference between the anticipated and next measured value. Sequential methods predict only one next value supported the window of prior data. An LSTM recurrent unit tries to keep all the past knowledge that the network is seen so far and to "forget" unimportant data. LSTM networks are an extension of recurrent neural networks mainly brings out to handle circumstances where RNNs stop.

An LSTM may be a recurrent neural specification that's commonly utilized in problems with temporal dependences. It succeeds in having the ability to capture information about previous states to rose inform the present prediction through its memory cell state.An LSTM consists of three main elements a forget gate, input gate, and output gate. Each of those gates is liable for altering updates to the cell's memory state.RNN fails to store information for a extended period of some time. At times, a reference to certain information stored quite while it's required to predict this output. But RNNs are completely incapable of handling such "long-term dependencies". Second, there is no finer control over which a neighborhood of context must be carried forward and thus the way much of past must be 'forgotten'. Other issues with RNNs are exploding and vanishing gradients which occur throughout the training process of a network through backtracking. Thus, Long STM (LSTM) was brought into the image. it has been so designed that the vanishing gradient problem is nearly completely separated, while the training model is left unaltered.

LSTM networks are successful RNN's as they permit us to encapsulate a wider sequence of words or sentences for prediction. Consider the task of creating captions for images. During this case, we have an input image and an output sequence that's the caption for the input image. CNN LSTMs are a classification of models that are spatially, temporally deep at the boundary of Computer Visionand natural language Processing. These models have enormous potential and are being increasingly used for several sophisticated tasks like text classification, video conversion, and many more.

An LSTM features a uniform control flow as a recurrent neural network. It processes data passing on information because it propagates forward. The differences are the operations in the LSTM's cells. These operations are used to allow the LSTM to remain or fail to remember information. Long STM (LSTM) networks are a kind of recurrent neural network capable of learning order dependence in sequence prediction problems. The matter domains like speech recognition, and more. LSTMs are a complicated area of deep learning.Long STM networks are usually just called "LSTMs". They seem to be a special quite Recurrent Neural Networks which are capable of learning long-term dependencies repeatedly only recent data is required during a model to perform operations. But there might be a requirement from a knowledge which was obtained within the past. Let's inspect subsequent example.

Consider a language model trying to predict subsequent word supported the previous ones. If we decide to predict the last word within the sentence say "The clouds are within the sky". The context is pretty simple and thus the last word lands up being sky constantly. In such cases, the difference or gap between past information and thus this requirement are often bridged really easily by using Recurrent Neural Networks. So, problems like Vanishing and Exploding Gradients don't exist and this makes LSTM networks handle long-term dependencies easily. LSTM have chain-like neural network layer. During a typical recurrent neural network, the repeating module consists of 1 single function.

3.6 Flickr_8k Dataset

It is little in size. So, the model is often trained easily on low-end laptops/desktop. Data is correctly labeled for every 5 captions are provided. The dataset is out there for free of charge. Flickr8k Dataset holds a complete of 8092 images in JPEG format with different sizes. Out of which 6000 are used for training, 1000 for test and 1000 for development. Flickr8k text Contains text files describing plaything test set. Flickr8k.token.txt contains 5 captions for every image which makes a total 40460 captions. The dataset is around 8 GB in size. For extracting the features, in our project, the concept of transfer learning is used. It means that a pre-trained model is used that is trained on large datasets to extract features. Xception model is used in this project, which has 1000 different classes to classify and some minute

Figure 7. Long short term memory (LSTM) block diagram

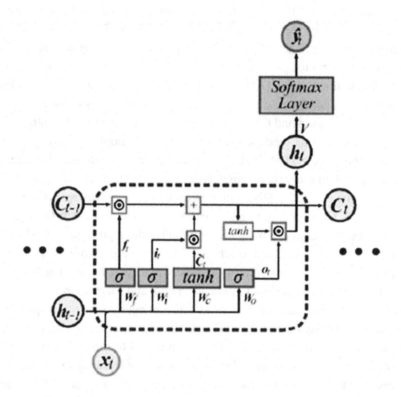

changes were made according to the goal of project. It can be imported from keras applications directly. A feature vector is extracted from the model. The frame work we have used is jupyter notebook. BLEU stands for Bilingual Evaluation Understudy. It is an algorithm, which has been used for evaluating the quality of machine translated text. We can use BLEU to check the quality of our generated caption Fig.8 and Fig.9.

1. BLEU is language independent.
2. Easy to understand
3. It is easy to compute.
4. It lies between [0,1].
5. Higher the score better the quality of caption.

4. EXPERIMENTAL INVESTIGATION FOR IMAGE CAPTION GENERATOR

The task of image caption generator is mainly about extracting the features and on-goings of an image and generating human readable captions that translates the features of the objects in the image. A working model is built by implementing the CNN model with LSTM. The features that are extracted by using the Xception CNN model is then fed into the LSTM, which is used to generate human readable captions.

4.1 Data Cleaning

Data cleaning is performed on our dataset by converting the whole text into lower case, removing punctuations, words that contain numbers, symbols. This is an important step. All the unique words from the set will be separated and a vocabulary is created to store them.

4.2 Extracting Features

For extracting the features, in our project, the concept of transfer learning is used. It means that a pretrained model is used that is trained on large datasets to extract features. Xception model is used in this project, which has 1000 different classes to classify and some minute changes were made in line with the goal of project. It can be imported, from keras applications directly. A feature vector is extracted for the model.

4.3 Training

The Flickr_8k dataset is used which consists of around 8000 images and partial captions for each image in it to train the Xception CNN model used. The model used is pre-trained model from Image Net is used. The dataset is randomly divided into training set, validation set and testing set. To train our model of ours, we used 6000 images. The validation dataset is used to provide an estimate of model while tuning the hyper parameters. It is different from test set.

4.4 Testing:

After the model is trained, by using of testing set, the testing of images is done, which follows the equivalent distribution of probability as training set but is independent of the training set. The testing set is increased and training set is decreased over time to enhance the performance, accuracy of the model keeping in mind about the test harness and performance measures.

4.5 Software Used

In order to achieve the goal of our project, we used following tools to obtain the results required. Python 3.7.5, Jupyter Notebook.

4.6 Jupyter Notebook

Jupyter notebook is one of the web applications which allows one to run live code, explanatory text in one place and embed visualization. Moreover, Jupyter is an interactive programming and an application that enforces the document. It is mainly composed of three types of cells: code, markdown, and raw. Firstly, a code cell includes executable code that is utilized to obtain results. Secondly, a markdown cell will have formatted text. Finally, a raw cell is a text which is neither a code nor a formatted text. Also, Jupyter operates the kernel to accomplish the code in code cells. Jupyter will send a code cell to run or execute and then, marks the cells as executing by allocating "*" to the cell undergoing execution. When the execution is finished, the kernel assigns a specific number to the executed cell that indicates the execution order. Users may execute the code in any order, and a given cell shall be executed shown in Fig.9, Fig.10 and Fig.11.

RESULTS

Figure 8. Jupyter notebook user interface

5. CONCLUSION

This work presents a demonstrate, which could be a neural network that can consequently see an picture and create appropriate captions in normal dialect like English. The demonstrate is trained to deliver the sentence or portrayal from given image. The depictions or captions gotten from the show are categorized into:

- Portrayal without errors
- Depiction with minor errors
- Portrayal to some degree related to image
- Depiction disconnected to image.

The categories in comes about are due to neighborhood of some particular words, i.e., for word like car it's neighborhood words like vehicle, van, cab etc. are too produced which might be incorrect. After so much of tests, it is conclusive that use of bigger datasets increments execution of the show. The larger dataset will increase accuracy as well as diminish losses. Also, it'll be curiously that how unsupervised information for both images as well as content can be utilized for making strides the image caption era approaches. In future scope, a lot of modifications can be made to improve this solution like using a larger dataset. Changing the model architecture, e.g. include an attention module. Doing more hyper parameter tuning (learning rate, batch size, number of layers, number of units, dropout rate, batch normalization etc.).

Figure 9. Man in red shirt rides on the street

Figure 10. Dog is running through the water

Figure 11.Two girls are playing in the grass

REFERENCES

Ahammad, S. H., Rajesh, V., Neetha, A., Sai Jeesmitha, B., & Srikanth, A. (2019). Automatic segmentation of spinal cord diffusion MR images for disease location finding. *Indonesian Journal of Electrical Engineering and Computer Science*, *15*(3), 1313–1321. doi:10.11591/ijeecs.v15.i3.pp1313-1321

Ahammad, S. H., Rajesh, V., & Rahman, M. Z. U. (2019). Fast and accurate feature extraction-based segmentation framework for spinal cord injury severity classification. *IEEE Access: Practical Innovations, Open Solutions*, *7*, 46092–46103. doi:10.1109/ACCESS.2019.2909583

Ahammad, S. K., & Rajesh, V. (2018). Image processing based segmentation techniques for spinal cord in MRI. *Indian Journal of Public Health Research & Development*, *9*(6), 317. doi:10.5958/0976-5506.2018.00571.5

Bhavana, D., Kumar, K. K., Rajesh, V., Saketha, Y. S. S. S., & Bhargav, T. (n.d.). *Deep Learning for Pixel-Level Image Fusion using CNN*. Academic Press.

Farhadi, A., Hejrati, M., Sadeghi, M. A., Young, P., Rashtchian, C., Hockenmaier, J., & Forsyth, D. (2010, September). Every picture tells a story: Generating sentences from images. In *European conference on computer vision* (pp. 15-29). Springer. 10.1007/978-3-642-15561-1_2

Gattim, N. K., Rajesh, V., Partheepan, R., Karunakaran, S., & Reddy, K. N. (2017). *Multimodal image fusion using Curvelet and Genetic Algorithm*. Academic Press.

Gerber, R., & Nagel, N. H. (1996, September). Knowledge representation for the generation of quantified natural language descriptions of vehicle traffic in image sequences. In *Proceedings of 3rd IEEE international conference on image processing* (Vol. 2, pp. 805-808). IEEE. 10.1109/ICIP.1996.561027

Jyothi & Sridevi. (2018). Low power, low area adaptive finite impulse response filter based on memory less distributed arithmetic. *Journal of Computational and Theoretical Nanoscience, 15*(6-7), 2003-2008.

Jyothi, G. N., Debanjan, K., & Anusha, G. (2020). ASIC Implementation of Fixed-Point Iterative, Parallel, and Pipeline CORDIC Algorithm. In *Soft Computing for Problem Solving* (pp. 341–351). Springer. doi:10.1007/978-981-15-0035-0_27

Jyothi, G. N., & Sriadibhatla, S. (2019). Asic implementation of low power, area efficient adaptive fir filter using pipelined da. In *Microelectronics, electromagnetics and telecommunications* (pp. 385–394). Springer. doi:10.1007/978-981-13-1906-8_40

Siva, D., & Bojja, P. (2019). MLC based Classification of Satellite Images for Damage Assessment Index in Disaster Management. *Int. J. Adv. Trends Comput. Sci. Eng, 8*(3), 10–13. doi:10.30534/ijatcse/2019/24862019

Sunitha, R., & Suman, MKishore, P. V. VEepuri, K. K. (2018). Sign language recognition with multi feature fusion and ANN classifier. *Turkish Journal of Electrical Engineering and Computer Sciences, 26*(6), 2871–2885.

Suryanarayana, G., & Dhuli, R. (2017). Super-resolution image reconstruction using dual-mode complex diffusion-based shock filter and singular value decomposition. *Circuits, Systems, and Signal Processing, 36*(8), 3409–3425. doi:10.100700034-016-0470-9

Vallabhaneni, R. B., & Rajesh, V. (2017). *On the performance characteristics of embedded techniques for medical image compression.* Academic Press.

Vinyals, O., Toshev, A., Bengio, S., & Erhan, D. (2015). Show and tell: A neural image caption generator. In *Proceedings of the IEEE conference on computer vision and pattern recognition* (pp. 3156-3164). 10.1109/CVPR.2015.7298935

Yang, Y., Teo, C., Daumé, H., III, & Aloimonos, Y. (2011, July). Corpus-guided sentence generation of natural images. In *Proceedings of the 2011 Conference on Empirical Methods in Natural Language Processing* (pp. 444-454). Academic Press.

Yao, B. Z., Yang, X., Lin, L., Lee, M. W., & Zhu, S. C. (2010). I2t: Image parsing to text description. *Proceedings of the IEEE, 98*(8), 1485–1508. doi:10.1109/JPROC.2010.2050411

Chapter 27
Leveraging Big Data Analytics and Hadoop in Developing India's Healthcare Services

Koppula Srinivas Rao
MLR Institute of Technology, India

Saravanan S.
B. V. Raju Institute of Technology, India

Pattem Sampath Kumar
Malla Reddy Institute of Engineering and Technology, India

Rajesh V.
Department of Electronics and Communication Engineering, Koneru Lakshmaiah Educational Foundation, India

K. Raghu
Mahatma Gandhi Institute of Technology, India

ABSTRACT

The benefits of data analytics and Hadoop in application areas where vast volumes of data move in and out are examined and exposed in this report. Developing countries with large populations, such as India, face several challenges in the field of healthcare, including rising costs, addressing the needs of economically disadvantaged people, gaining access to hospitals, and conducting medical research, especially during epidemics. This chapter discusses the role of big data analytics and Hadoop, as well as their effect on providing healthcare services to all at the lowest possible cost.

DOI: 10.4018/978-1-7998-6870-5.ch027

1. INTRODUCTION

Big Data is a relatively recent development in the world of information systems that has emerged as a result of the rapid growth of data over the past decade. Big Data is a term used to describe datasets that exceed the capabilities of conventional data collection and storage systems. A modern form of data analytics known as Big Data Analytics has arisen as a result of the need to handle and analyses those massive datasets. It entails processing large amounts of data of various forms in order to uncover secret blueprints, unidentified associations, and other valuable data. Many companies are rapidly turning to Big Data analytics to obtain deeper visibility into their operations, improve revenue and profitability, and gain a strategic edge over competitors Siegel, J. (2012).

The Apache Hadoop Framework's basic programming models make it easier to spread the analysis of large data sets across groups of systems. The architecture is designed to scale from a single server to thousands of devices, each with native computing and storage capabilities. At the software layer, the library is structured in such a way that errors can be detected and dealt with. As a result, the architecture is capable of providing continuous operation on top of a set of applications that are vulnerable to failure. Hadoop's bright future draws a wide range of businesses and organizations to use it for both science and manufacturing Harris, T. (2010).

2. HEALTHCARE IN INDIA

India ranks second in the world in terms of population. India's health-care system is being overburdened by the country's growing population. Among a wide number of individuals, economic scarcity leads to a weak approach to health care. GDP per capita are the most important indices of human growth. Longevity is linked to income and education and affects the status of one's wellbeing. The health-care sector's vulnerability can have a detrimental impact on longevity. India's Human Development Index (HDI) is low (115th) among world countries. The main reasons for India's high disease burden are a lack of access to preventative and therapeutic health facilities White, T. (2010).

"Growth in national income by itself is not enough, if the gains do not manifest themselves in the form of more food, greater access to health, and education," said Amartya Kumar Sen, an Indian economist and Nobel laureate Zulkernine, (2013).

In a lecture, Dr. MC Misra, the director of the All India Institute of Medical Sciences (AIIMS), said, "Advances in medical technology and modern medicines are indeed a blessing, but to operate in India, they must be value for money." Also free services in public hospitals are out of reach for the majority of people."

However, according to World Bank data, 99% of India's population cannot afford to pay for these facilities. As per the survey, out-of-pocket medical costs push 39 million individuals into poverty per year, with households dedicating about 5.8% of their income to medical care Yang Song. Alatorre, G. (2013).

In India, less than 10% of the population has signed up for health insurance. Hospitalization costs alone account for 58% of a typical Indian's gross annual spending. According to World Bank data, over 40% of people borrow heavily or sell properties to cover hospitalization costs, pushing 39 million people into poverty per year.

According to a new report published by the IMS Institute of Health Informatics on July 19, 2013, only 33.33 percent of hospital beds in India can be used by 72 percent of the rural Indian population.

Figure 1. Healthcare Spending by the Public and Private Sectors

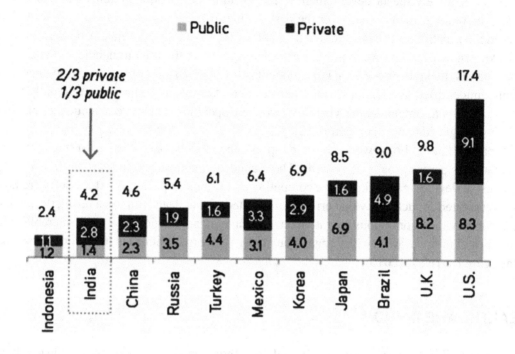

However, 28 percent of urban Indians use the remaining 66.66 percent of total beds. In order to receive in-patient treatment in hospitals, people in rural villages must drive at least five kilometers in 63% of the time, according to the report show in fig 1. The statistics from the report are shown in the table below Weiyi Shang. Zhen Ming Jiang. Hemmati, (2013).

Table 1. Expenditures on healthcare by countries

Country	Total % of GDP spent on Healthcare	Private Expen diture %	Per capita spent on	
			Health care (USD)	Healthcare (USD) by Government
India	4.01	70.08	133	40
USA	17.09	46.09	8363	4439
UK	9.06	16.01	3481	2921
South Africa	8.09	55.09	936	414
China	5.01	46.04	380	205
Brazil	9.00	53.01	1029	486
Pakistan	2.02	61.05	60	25
Nigeria	5.01	62.01	123	48
Russia	5.01	37.09	999	622

Statistics show that healthcare is available and accessible to the average person in society, especially in developed countries like India, Mukherjee, (2012), Deepak Kumar, B. (2011).

Not only does availability apply to the accessibility of hospitals and medical staff, but it also relates to the accessibility of essential and trustworthy data on the screen [15-16] show in table 1. Not only does accessibility provide the ability to attend clinics, but it also includes awareness of the patient, prescriptions, and other relevant material, Weiyi Shang. Zhen Ming Jiang. Hemmati, (2013), Hao, Chen. Ying, Qiao. (2011).

India also has a long way to go in terms of reaping the gains of successfully integrating information technology into healthcare for its 1.2 billion people. As a result, it is past time for India's public and private sectors to make technologies mandatory in order to improve the country's healthcare situation Garcia, T. (2013), Li Xiang. (2011).

3. INFLOWING DATA FROM HEALTH CHECKING PROCEDURES

The reality of the government's promise to digitise medical information is not as one would imagine. Also, traditional medical terminologies are not standardized in the world, Xu Zhengqiao. (2012).

"Our medical terminology and names of drugs vary from place to location," said Sushil K Meher, tech faculty at AIIMS. Myocardial infarction, for example, is the medical term for a heart attack. Doctors, on the other hand, refer to it as a cardiac attack or even a heart attack. Similarly, because of the salts that they produce, medications are recommended. As a result, we are unable to keep a proper list of patients' medical histories," explains clearly in, S.Prabu, (2019), JS P.N.Palanisamy, (2019).

A health institution's index can be seen as medical history. Medical reports reveal details on a Healthcare Center's beginnings and development, as well as retrospective and future predictive analysis, the nature of cases admitted to the hospital, and so on. For the benefit of any practitioner in the healthcare industry, medical records must be properly and methodically gathered, conserved, and secluded. Health Records are more than just a source of medical and science information and inputs for the government's preparation and budgeting for the country's health-care system. The need of the hour is for consistency in the storage of medical records under different Acts show in fig 2.

"We cannot assess the health situation in India without data," said at a medical informatics meeting, AIIMS programming faculty. The Indian government's health ministry recently announced that medical records in public hospitals in Delhi would be digitized, enabling every hospital in the country to access patient information.

According to a study by Grant Thornton India, the Indian health check system and equipment industry is projected to rise to about US$ 5.8 billion by 2014, and US$ 7.8 billion by 2016, rising at a CAGR of 15.5 percent. This exponential growth generates a massive volume of data that is beyond human comprehension.

4. THE RELATIONSHIP BETWEEN BIG DATA AND BIG DATA ANALYTICS AND HEALTHCARE

Over time, various devices attached to the patient generate poorly organized data, which is collected by health care management systems. And since these are massive, complicated systems, powerful algorithms

Figure 2. Inflow of healthcare data

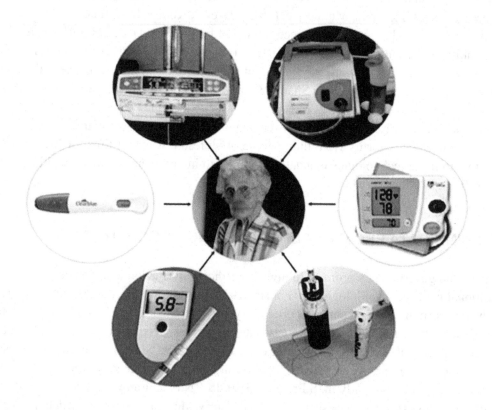

to process the raw data are needed, as well as a lot of computing power. Big data refers to data collected by a variety of sensors, such as medical, traffic, and social data. Amount, velocity, and meaning are three features of big data show in fig 3.

4.1 Features of big Data

4.1.1 Volume

As compared to conventional data, the volume of data produced by different medical devices is greater.

4.1.2 Velocity

The annual data storage capability of an entire healthcare system is significantly less than the volume of data streamed by medical networks.

4.1.3 Diversity

Traditional data formats are less flexible when it comes to various types of sensor data, implementation, and other variables. On the other side, nontraditional info, such as surgical instruments, is easily adaptable to transformation.

Figure 3. Big Data – 4 "Vs"

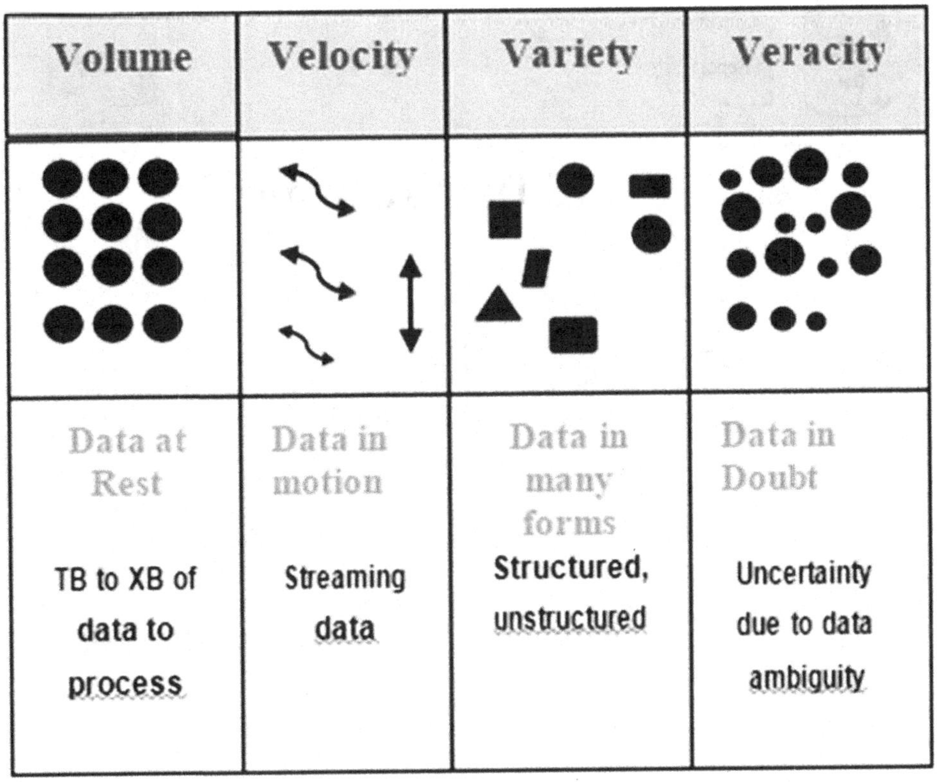

Volume	Velocity	Variety	Veracity
Data at Rest	Data in motion	Data in many forms	Data in Doubt
TB to XB of data to process	Streaming data	Structured, unstructured	Uncertainty due to data ambiguity

4.1.4 Veracity

It deals with evidence that is inconclusive or ambiguous. In conventional data centres, there was still the presumption that the data was accurate, safe, and precise. In the case of Big Data, though, this is not the case.

It is essential to establish a system for acquiring, organizing, and processing data in order to extract useful information in order to develop an architecture for big data analysis. Data collection, data organization, and data analysis are also examples of this.

One of the most difficult aspects of big data platforms is data collection. Since these systems deal with large amounts of data, they require low latency in data capture and the use of basic queries to process large amounts of data.

The computer must decrypt and interpret the data from its initial storage position because the data is on a larger size. Apache Hadoop is a technology that helps you to process massive volumes of data while keeping the data in its original clusters.

Data needs to be processed through a network when working with big data. The need to analyze data, such as medical data, necessitates a computational and mining approach to data analysis. Data delivery with a shorter response time would be prioritized.

Healthcare and life sciences companies would be able to find patterns by combining patient data, opioid effect data, testing data, R&D data, and financial reports. This would help in more responsive

Figure 4. Analytics in Big Data

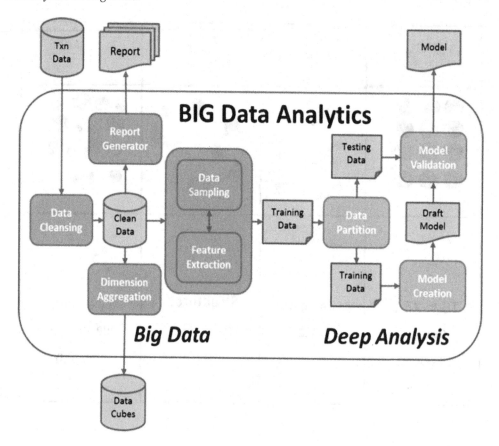

healthcare. Besides that, by combining patient data and social network content in a data collection system, healthcare organizations may have a larger pool of data from which to discover secret connections show in fig 4.

The hopeful expectation is that the healthcare industry would be able to gather data from either source, organize it, and investigate it to discover solutions that will help patients.

1. Cost savings
2. Time savings
3. New science progress and improved results
4. Correct decision-making is aided by better diagnosis.

A compromise that could satisfy four main criteria

1. In terms of scalability and dependability,
2. Storage is an important aspect of every company.
3. Infrastructure for processing
4. Capabilities of search engines for retrieving articles of high availability (HA)
5. With HA, you can retrieve statistics from a scalable real-time shop.

Figure 5. Healthcare Big Evidence Analytics

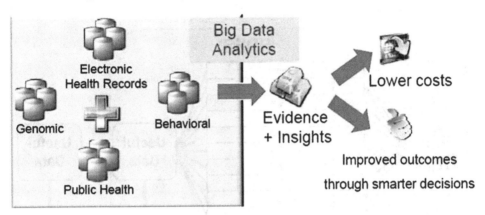

Big data in the future clinical trial could allow enrollees to be watched not only for reaction but also to see how particular subgroups respond differently. Since they improve the ability to measure population variations and perform real-time analytics, big data methods are a great alternative to conventional clinical trials. Consider being able to generate complex sample size estimates based on new clinical trial results. The opportunity to refine clinical trial design, allowing for shorter and more effective experiments, is the secret to big data's appeal show in fig 5.

5. HADOOP IN HEALTH CARE DATA

This disparity in economics has drawn a lot of interest, and Hadoop will become the focal point for most large-scale data processing and analysis operations and analyses, either integrating or originating from it.

The industry has grown by one to two orders of magnitude as increasingly powerful SQL technologies have been coupled to Hadoop technology, adding the whole SQL-based environment to the world of Hadoop. Hadoop is no longer just the realm of experts.

Another stumbling block is relational technologies that operate on their own. When it comes to unstructured/multi-structured/loosely structured files, the genie is out of the bottle. We can now analyze more and diverse types of data with Hadoop than we can with relational catalogs show in fig 6.

Hadoop can allow dispersed parallel computing on massive volumes of data through inexpensive and high-level servers that can be scalable to infinity it is used to store and process data. Hadoop does not consider any data to be too large. Every day, a huge amount of data is generated and introduced in the exponentially expanding medical community. Because of Hadoop's performance and efficacy, data that were once valuable for research are now worthless.

Hadoop may use a distributed file system to store data through several servers in a cluster. Hadoop hides the position of data being viewed in the cluster, enabling end users to refer to files as though they were on a local computer.

Through processing and interpreting Big Data, the healthcare industry may make important predictions. In the other hand, since 80% of medical data is "unstructured," it must be processed in order to do accurate data mining and investigation. Hadoop is a fundamental proposal for processing Big Data with the aim of making it usable for analysis. Because of its simplicity, completeness, and convergence,

Figure 6. Map Reduction in Hadoop

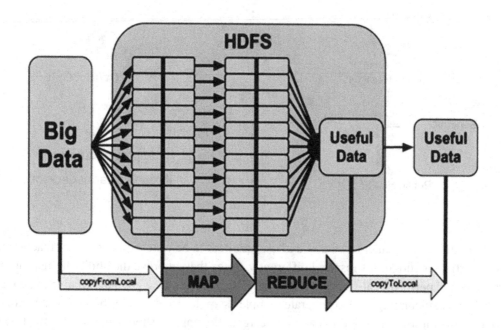

Hortonworks Data Platform (HDP) may be used to build applications for healthcare including Hadoop and Big Data. The diagram below depicts Hadoop's effort to chart and minimize tasks show in fig 7.

Hadoop aims to implement Map and Reduce operations on the systems where the data being stored is located while doing Map Reduce function, eliminating the need for data to be copied between systems. The program demonstrates that MapReduce activities are more effective where only one large file is used as input rather than a large number of small files. Since the tiny files are spread over several computers, copying them to the MapReduce device requires substantial overheating. However, because large files

Figure 7. Hortonworks Application using MapReduce

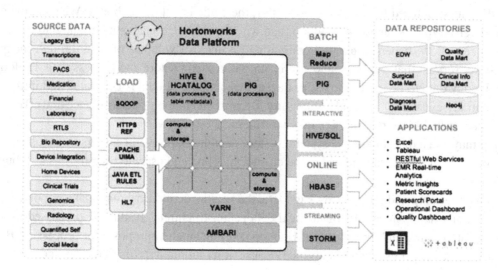

are stored on a single system, this is not the case. The program claims that tiny file overhead slows the runtime by ten to one hundred times. When the data being analyzed is localized to the devices doing the operation, it is clear that the MapReduce architecture performs well.

The healthcare sector faces many obstacles in handling data in order to provide excellent treatment to all parties involved. Medical image recognition is one of the topics covered. Hadoop offers a solution for analyzing a growing number of medical images from different sources and extracting the information needed to make an accurate diagnosis. The Hadoop Image Processing Interface (HIPI) describes the steps involved in image processing.

5.1 Hadoop Image Processing Interface (HIPI)

Hipi is a distributed computing API that allows you to process files.

HIPI accepts HipiImage Bundle (HIB) as an input form. In HIB, there is a set of images combined into a single large file containing Meta information regarding the composition of the photos A HIB is created from a pre-existing collection health videos, or from such media including medical devices.

A culling function looks at the photographs to see whether they meet the criteria and weeds out any that don't. Photos of fewer than 10 mega pixels, for example, could be disqualified for review. Each picture that passes the culling test is then given the Cull Mapper class. The Cull mapper class receives the images as Float Images with a corresponding Image Header. During initialization, a user can change explicit execution parameters for image processing jobs using the HipiJob object. Once the Mapper is started, HIPI makes no changes to the default Hadoop MapReduce output.

The implementations of Hadoop MapReduce's necessary components have been meticulously scrutinized to ensure that they are reliable and competitive when it comes to image processing.

5.2 Low Cost and Greater Analytic Flexibility

Hadoop is 10 times cheaper per terabyte of capacity than a conventional relational data center system since it uses industry standard hardware. In addition to the storage, one must purchase a computer, SAN storage, and a storage certificate. With Hadoop, you just need to buy some popular hardware and you're ready to go.

It also helps consumers to run analyses on the cumulative compute and storage, allowing them to get more bang for their buck. The previous strategies we had in place didn't even come close to meeting the need. Although, the advantage of storing information in a Hadoop-type approach is outweighs the value of storing data in a database if costs are equal."

6. WORK APPROACHES FOR THE FUTURE

Even though human life is on the line, healthcare necessitates the most thorough investigation practicable in order to collect reliable results for tests that cannot be tampered with. Despite the reality that our Big Data Analytics and Hadoop for healthcare analysis assist us in reducing costs and improving treatment for the country's last and least guy, the challenges mentioned below must be overcome in order to achieve the best results.

Figure 8. HIPI

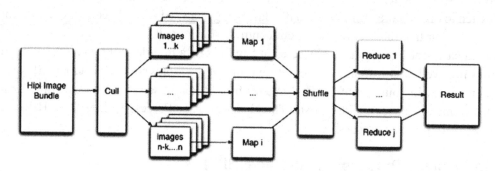

- If big data comes in an unstructured medium including text, picture, audio, or video, it may be difficult for non-technical medical professionals to understand and use it.
- The second hurdle is collecting the most critical data in real-time, such as big surgeries, and communicating it to the appropriate people for further interpretation.
- A third roadblock is data collection, as well as understanding and analyzing large amounts of data using the minimal computing resources available.

The healthcare industry needs real-time data interaction, while Hadoop will batch process the use cases. Transactions cannot be handled in Hadoop because it does not support indexing and does not explicitly follow the ACID paradigm.

Image processing is critical for the future since it allows images to be linked to other forms of data for mining. As a result, medical image processing research takes the lead and may make a significant contribution. The development of advanced Cloud storage technologies could be able to solve the data storage issue show in fig 8.

7. CONCLUSION

Despite the fact that big technology has been successfully applied in consumer sectors, there are still challenges in applying it in healthcare. Traditional networks provide huge volumes of data that cannot be comparable, and the data is stored in a variety of file formats, which is one of the most important roadblocks to big data growth. The next challenge for data in healthcare is storing and sharing knowledge that is interconnected but not adequately connected thus protecting the patient's privacy. It is a difficult challenge for organizations that establish Big Data and Hadoop applications in compliance with the National Indian Health Board's amended acts.

The government and its laws, physicians, medical practitioners, and, most significantly, technological creators of technology-based software are all involved in overcoming these practical obstacles. Technology will undoubtedly aid in the resolution of these fundamental issues and add to the healthcare sector.

REFERENCES

Deepak Kumar, B. (2011). Evaluation of the Medical Records System in an Upcoming Teaching Hospital—A Project for Improvisation. *Journal of Medical Systems*.

Garcia & Wang. (2013). Analysis of Big Data Technologies and Method - Query Large Web Public RDF Datasets on Amazon Cloud Using Hadoop and Open Source Parsers. *Semantic Computing (ICSC), IEEE Seventh International Conference*. http://architects.dzone.com/

Hao, C., & Ying, Q. (2011). Research of Cloud Computing Based on the Hadoop Platform. *Computational and Information Sciences (ICCIS), 2011 International Conference*.

Harris, T. (2010). *Cloud Computing- An Overview, Whitepaper*. Torry Harris Business Solutions.

Mukherjee, A., Datta, J., Jorapur, R., Singhvi, R., Haloi, S., & Akram, W. (2012). Shared disk big data analytics with Apache Hadoop, High Performance Computing (HiPC). *19th International Conference*.

Palanisamy, J. S. P. N., & Malmurugan, N. (2019). FPGA implementation of deep learning approach for efficient human gait action recognition system. International Journal Of Innovations In Scientific And Engineering Research, 6(11).

Prabu, S., & Kalaivani, M. (2019). An intelligent power adaptive model using machine learning techniques for wsn based smart health care devices. International Journal Of Innovations In Scientific And Engineering Research, 6(12).

Shang, Jiang, Hemmati, Adams, Hassan, & Martin. (2013). Assisting developers of Big Data Analytics Applications when deploying on Hadoop clouds. *Software Engineering (ICSE), 35th International Conference*.

Shang, Jiang, Hemmati, Adams, Hassan, & Martin. (2013). Assisting developers of Big Data Analytics Applications when deploying on Hadoop clouds. *Software Engineering (ICSE), 35th International Conference*. hipi.cs.virginia.edu/ 2014/

Siegel, J., & Perdue, J. (2012). Cloud Services Measures for Global Use: The Service Measurement Index (SMI). *SRII Global Conference (SRII), Annual*, 411-415. 10.1109/SRII.2012.51

Song, Y., Alatorre, G., Mandagere, N., & Singh, A. (2013). Storage Mining: Where IT Management Meets Big Data Analytics, Big Data (BigData Congress). *IEEE International Congress*.

White, T. (2010). *Hadoop: the Definitive Guide* (2nd ed.). O'Reilly Media.

Xiang, L. (2011). *Analysis on architecture of cloud computing based on Hadoop*. Computer Era.

Xu & Dewei. (2012). Research on Clustering Algorithm for Massive Data Based on Hadoop Platform. *Computer Science & Service System (CSSS), 2012 International*.

Zulkernine, F., Martin, P., Ying, Z., Bauer, M., Gwadry-Sridhar, F., & Aboulnaga, A. (2013). Towards Cloud-Based Analytics-as-a-Service (CLAaaS) for Big Data Analytics in the Cloud, Big Data (BigData Congress). *IEEE International Congress*.

Chapter 28
A Sequential Data Mining Technique for Identification of Fault Zone Using FACTS-Based Transmission

Koppula Srinivas Rao
MLR Institute of Technology, India

B. Veerasekhar Reddy
MLR Institute of Technology, India

Sarada K.
Koneru Lakshmaiah Education Foundation, India

Saikumar K.
Malla Reddy Institute of Technology, India

ABSTRACT

In this survey, the issue of flaw zone identification of separation transferring in "FACTS-based transmission lines" is investigated. Presence of FACTS gadgets on transmission line (TL), while they have been remembered for issue zone, from separation transfer perspective, causes various issues in deciding the specific area of the flaw by varying impedance perceived by hand-off. The degree of these progressions relies upon boundaries that are set in FACTS gadgets. To tackle issues related with these, two instruments for partition and examination of three-line flows, from the hand-off perspective to blame occasion, have been used. In addition, to examine the impacts of TCSC area on deficiency zone recognition of separation hand-off, two spots, one of every 50% of line length and the other in 75% of line length, are deliberated as two situations for affirmation of suggested strategy. Reproductions show that this strategy is powerful in security of FACTS-based TLs.

DOI: 10.4018/978-1-7998-6870-5.ch028

Expanding request, brought about by the advancement of urban areas and developing populace, draws more consideration on power transmission limits. Force Electronics assists with tackling issue by presenting adaptable air conditioning transmission framework (FACTS) gadgets. The general favorable circumstances of FACTS gadgets are notable. In any case, FACTS gadgets can prompt extreme issues in deciding the specific area of the shortcoming when they are remembered for the deficiency zone. Separation transfer's capacity depends on estimated impedance by that they compute flaw area. Introduced FACTS gadgets on TL change the impedance estimated by separation hand-off that thusly prompts wrong flaw zone location. The measure of deviation in impedance relies upon the boundary's settings of FACTS gadget.

The wavelet change is utilized to isolate high-recurrence parts of 3-line flows, and help vector machine (utilizing techniques for multi- class utilization) is utilized for order of issue area into three assurance locales of separation transfer.

Lately, specialists have gotten progressively keen on comprehending this issue. For demonstrating arrangement capacitor in the shortcoming time frame, the greater part of articles utilized model introduced by Goldsworthy Goldsworthy (1987). In Sadeh, Javad (2020) an answer dependent on time conditions of system with thought of a broad method of TL is introduced. In this arrangement, zone and area of deficiency, autonomous of issue obstruction, are evaluated utilizing 2 subroutines. A few articles built up their strategies utilizing estimations from rule segments of primary recurrence. Saha et al(1999) built up a strategy dependent on segments of principle recurrence on 1 side of TL utilizing Goldsworthy method.

In Song, Y. H(1996), a calculation dependent on forward and in reverse waves is introduced to secure arrangement repaid TLs; albeit just one stage to ground deficiency are examined in this manuscript. Information is introduced in Thomas, D (1992) and Xuan,Q.Y (1996) that recommend strategies dependent on neural systems. It is proposed to utilize RBFN as a spiral neural system to order flaw kind and issue area Girgis, Adly A (1998) and Kalman,(1960). The work Megahed(2006),trains 2 unique kinds of neural systems to distinguish deficiency kind and area. Likewise, Dash, P. K (2006) utilizes EDBD calculation to figure deficiency area Fattaheian, Dehkordi(2014). Neural systems have impediments, incorporating their requirement for countless neurons to demonstrate organize structure (2008). Along these lines, various information from the electrical system is required, and furthermore much an ideal opportunity to prepare the system. By and by, these articles utilized a predetermined number of information to test their techniques. The work Samantaray, S. R (2013) suggested a strategy dependent on Kalman channel] that gauges the condition of a powerful framework from a progression of estimations comprisingblunders.

In Imani (2019) a technique utilizing wavelet change is recommended. In Lepping (2018) and Imani, MahmoodHosseini (2019) constrained cases are utilized to evaluate adequacy of suggested techniques. Besides, in Provost, Foster (2003) high examining recurrence of 200 KHz is utilized, that is by all accounts troublesome in down to earth execution. In Megahed(2006),an information digging technique is utilized for 3reasons. The SVM has become a well-knowndevice in force framework Zhang(2007). In this manuscript, 3distinct SVM is prepared to decide deficiency area, nearness or nonattendance of issue and whether shortcoming incorporates ground or not. This reference utilizes tests of terminating point & current, beginning from the shortcoming occurrence till half cycle after it deliberating 50 Hz &400 kV contextual investigations and testing recurrence of 1 KHz in this manuscript, 10 examples of current from every stage is dissected. In this manuscript, deficiency area is limited to when the compensator, and its viability is concentrated uniquely in 200 cases.

In Kumar, D. S (2020) a 2-phase technique is suggested to recognize the deficiency area. In this manuscript, tests of 3-line flows for one cycle after deficiency case are utilized to prepare SVM. Deliberating 400 kV & 50 Hz contextual investigation and examining recurrence of 1 KHz in this manuscript, 80 examples of current from every stage is broke down. In main phase, high frequencies of 3 line flows are isolated utilizing wavelet change; at that point in subsequent stage, these isolated signs have been provided to SVM to arrange the issue area to when compensator. The issue is that method utilized for reproduction of TL isn't legitimate to examine non-base frequencies whereas this manuscript utilized high frequencies to isolate issue areas. This strategy, deliberating utilizing SVM, might just separate issue area to when the compensator. Considering 400 kV & 50 Hz contextual investigation and inspecting recurrence of 1 KHz in this manuscript, 60 examples aredissected.

Specialists in Soumya, N (2020) suggested arbitrary woods strategy Saikumar K (2020) to recognize deficiency area. This information mining technique is established dependent on decisiontree (DT) strategy. In this technique, a few free unprunedDT are created utilizing various subsets of information. Every tree has 1 vote, & many votesdecide the outcome. Tests of current and voltage into equal parts cycle after deficiency case are used for examination. 70% of the examples are utilized for the preparation procedure of arbitrary woodland, and the staying 30% is utilized for the testing procedure. Thinking about a lot of preparing information, great outcomes have beenaccomplished.

Thus, in this paper, the deficiency area is restricted to when the compensator. The goal of this examination is to enhance separation insurance execution, utilizing information mining strategies whereas deliberating FACTS-based TLs. Prior works utilizing fake neural system experience the ill effects of restrictions on many neurons. These restrictions might be overwhelmed by utilizing increasingly effective devices like help vector machines. Moreover, an increasingly exact recreation of TL is led here utilizing PSCAD/EMTDC programming, just as isolating deficiency territories into three zones as indicated by separation transfer settings. In addition, to research the impacts of TCSC1 area on flaw zone discovery of separation hand-off, twospots one out of 50% of line length & other in 75%linelength are deliberated as 2 situations for affirmation of suggested technique V Saikumar, K., Rajesh(2020). The recreations represents a 94.2% of achievement rate in situation 1 and 86.7% in other that are high achievement rates demonstrate that putting TCSC in 50% of line length prompts higher expectation precision. Additionally, these achievement rates have been accomplished utilizing just 15.2% of absolute information as preparationset that represents high capacity of suggested technique Saikumar, K(2020)

1 SUGGESTED TECHNIQUE

In this investigation, PSCAD programming is utilized for reenactment. The 3 line flows have been recorded in the wake of making an assortment of issue conditions on reproduced TL. These yields from PSCAD have been utilized as contributions to wavelet change so as to extricate low voltages.It is accounted for shown Devaraju (2021) very legitimate wavelet in examining deficiencies in power framework is SNR (db) process. The Obaid, A. J (2020) it is accounted for that from various sorts of dbmain SNR; db2 provides improved outcomes in the recognition of issue area, within the sight of arrangement capacitor in TL. It was additionally revealed that voltages among 100 to 200 Hz are reasonable for best issue zone grouping Abinaya, S,(2017). Consequently, deliberating an examining recurrence of 5 KHz in PSCADD in this examination, 1 stage discrete wavelet change with db2 mother wavelet is completed to acquire

Figure 1. load connection for FACTS devices

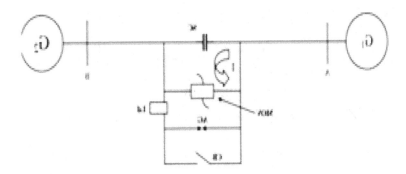

low recurrence tests. From that point forward, wavelet change's yields are utilized as contributions to ASVM to make a hybridwork to isolate shortcoming zones Sheshasaayee, D(2017).

2 CASE SURVEYS

2.1 Simulation of Case Surveys in PSCAD

For TCSC reenactment in PSCAD, in [13] is utilized. This method is appeared in Fig. 1. The ability of SC capacitor modifies in extent with terminating point. The possibility of suggested technique is assessed on 400 kV &50 Hz power framework. The two sides of this system are 2 sources that speak to 2 zones that are associated together. The source information at the two sides of TL is appeared in Table I. Recurrence subordinate (stage) model is utilized for itemized reenactment of TL. The TL information is accomplished from a genuine 400 kV single-circuit TL with 250 km length. thisinformation is introduced in Table II. To assess effect of TCSC area on issue of assurance, this component is put in 2 areas, firstin center of line 1 (125 kmof line 1) and second, in 75% of line 1 (187.5 km of line 1) that in extension of this manuscript for comfort of per users are alluded to as situation 1 and situation 2, separately. Three security zones are characterized for separation hand-off, 1st till 80% of principalline,2ndtill50% of subsequent line and 3rd till 25% of 3rd line.

In situation one which TCSC is put in the main line, four issue areas are accepted: in "50 km (1st zone, before TCSC), in 150 km (1st zone, after TCSC), in 250 km (2nd zone), in 325 km (3rd zone)". In situation 2, three shortcoming areas are accepted: in 100 km (1st zone), in 250 km (2nd zone), in 325 km (3rd zone). So as to assess practicality of suggested strategy under different framework condition, several types of flaw, issue obstruction, issue initiation edge(FIA), load edge (i), generator impedance, remuneration rate and deficiency area, which incorporate 28800 cases forsituation 1 and 21600 cases for situation 2, have been dissected using numerous spat PSCAD. Table III displays thought about cases. 25%, half, and 75% are picked TCSC pay rates for 250 km TL, their proportional incentive for SC capacitor is 120.45, 60.225 & 40.15 μfindinvidually. Haphazardly picked is cases for preparation procedure of SVMare 3600 cases for situation 1 and 2700 cases for situation 2. These cases are just 12.5% of all-out informational collection and appeared in Table IV. The Gaussian RBF part is utilized for preparing and testing procedure in SVM.

Table 1. Source level analysis

Parameters	Value
Source Voltage	400 kV
Frequency	50 Hz
Positive sequence impedance	$1.31+j15 \ \Omega$
Zero sequence impedance	$2.33+j26.6 \ \Omega$

2.2 OAA classification

The numerous paths isto utilize SVM to categorize information into higher than 2 classes.

1. Model 1 examination: Getting a suitabledestination from ASVM might be attained through met heuristic procedures or trial & error learning [13]. To use "trial& error learning "diverse values for gamma (g) & regularization factor (C)must be assessed on organization of training data to select better ones, then outcome will be utilized to categorize test data. For extent of these factors, outcomes in [13] areexcessive support. The better outcome is attained in21096 & 14.5 for C and g, correspondingly. With these values, 83.7% of achievement is attained in FZD.
2. Scenario 2 examination: In outcome examination of scenario 2, it might be observed that numerous correct FZD was reached with g=38.4 &C=1080000 and. The achievement rate is 69.7% that is lower than scenario 1. In section IV reasons of this less achievement rate is described.

2.3 OAO Classification

For this strategy, table ought to be recognized, and every SVM must make its choice on class among 2 classes. The class that accomplishes most elevated many votes are picked by the consolidated classifiers. Potential situations are appeared in Table V

Table 2.Tldata

Data type	value	Data type	value
Arrangement of wires	Horizontal	Vertical distance of conductors	15.45 m
Line length	250 km	SAG for all conductors	14 m
Number of bundle	2	Number of ground wires	2
Bundle spacing	45 cm	Height of ground wires above the lowest conductor	9.36 m
Conductor DC resistance	0.0553 ohm/km	Ground wire DC resistance	1.463 ohm/km
Conductor GMR	1.2161 cm	Ground wire radius	0.2445 cm
Height of conductors	41.46 m	Spacing between ground wires	18.70 m

Table 3. Deliberated Cases For System Examination

Case number	Number of fault cases for scenario 1	Number of fault cases for scenario 2	Parameters							
			$\%Z_{G1}$	$\%Z_{G2}$	$\%Xc$	R_f	FIA	δ	Fault types	Fault locations
1	5760	4320	100	100						4 and 3 location for scenario 1 and scenario 2 respectively
2	5760	4320	100	75	25,50 & 75	0,5 & 25,50	0,45 & 81,117	10,20 & 30	10	
3	5760	4320	100	125						
4	5760	4320	75	100						
5	5760	4320	125	100						
total number of cases in scenario 1= 5*5760=28800										
total number of cases in scenario 2= 5*4320=21600										

1) Scenario 1 Examination

The most extreme number of right location of deficiency zone utilizing OAO in situation 1 is 20629 cases that occurred in g=13.1&C=10000. This number is equal to 81.8% of achievement. Following up on OAO technique, with examination of outcomes, 1 might choose for cases that table can't decide in favor of their group. In the wake of examinationoutcomes, Table V is supplanted with Table VI. The numerous right location of deficiency zone utilizing Table VI is 21239 that occurred in g=10.4&C=10. True to form, achievement rate expanded to 84.2%.

2) Scenario 2 Examination

The many right identification of shortcoming zone utilizing OAO in situation 2 was 11802 cases that occurred in g=20&C=10. This number is equal to %62.2 of accomplishment. In the wake of supplanting Table V by Table VI many right cases expanded to 123399 in g=21.3&C=10 that is identical to 65.6% of progress.

2.4 Organization with OAA Technique in Changed Training Set

As might be observed, a generous level of accomplishment is not accomplished forsituation 1 and 2 with the two strategies. One of viable boundaries to accomplish a decent outcome in arrangement is having

Table 4. ASVM analysis

Case number	Number of fault cases for scenario 1	Number of fault cases for scenario 2	Parameters							
			$\%Z_{G1}$	$\%Z_{G2}$	$\%X_c$	R_f	FIA	δ	Fault types	Fault locations
1	720	540	100	100	50					4 and 3 location for scenario 1 and scenario 2 respectively
2	720	540	100	75	50	0,5 & 50	0,45 & 117	10 & 30	10	
3	720	540	100	125	50					
4	720	540	75	100	50					
5	720	540	125	100	50					
total number of cases in scenario 1= 5*720=3600										
total number of cases in scenario 2= 5*540=2700										

legitimate and perfect measure of preparing informational index. One can't anticipate that a calculation should foresee appropriately in inconspicuous zones. It is common to put 70% of informational collection into preparation set and 30% to test set. As of recently, just 12.5% of informational index is utilized in preparation set. Cases that their deficiency zone is distinguished mistakenly were dissected. A high extent of wrong forecasts occurred in information with load edges (ı̇) equivalent to 20, which was excluded from the preparation informational index. Hence, arbitrary informational collections with load points (ı̇ı̇) equivalent to 20 are included to preparation set. Table VII & Table VIII presents additional information for situation 1 and situation 2, individually. Deliberate that now 15.3% of all out information is utilized as preparation informational collection.

1) Scenario 1 Examination

The many correctdiscovery of FZ utilizing OAA in altered training set is 22993 cases that occurred in g=8.3&C=10. Altered training set expanded achievement rate from 83.7% to 94.2%.

2) Scenario 2 Examination

The numerous correct detection of FZ utilizing OAA in altered training set is 15880 cases that occurred in g=14.5&C=10. Changed training set improvedachievement rate from69.7% to86.7%.

2.5 ORGANIZATION WITH OAO TECHNIQUE IN CHANGED TRAINING SET

1) Scenario 1 examination

In this instance, better outcome in attained in g=12.1&C=10. With these values, achievement rate is enhanced to 86.7%. After examining outcomes, Table V is substituted with Table IX as "reference voting table". The outcomespresented91.6% achievement rate in g=12.1&C=10.

2) Scenario 2 Examination

The outcomes presented 76.1% achievement rate in g=15.3 &C=100000 that will be equal to 13926 correct FZD. After examining the outcomes, Table V is substituted with Table IX as "reference voting table". The outcomespresented 78.2% achievement rate in g=15.4&C=100000 that will be equal to 14318 correct FZD.

3 DISCUSSION ON THE RESULTS

As per the outcomes, the most elevated achievement rate is accomplished in situation 1 (introducing TCSC in the principal line). This achievement rate (94.2%) is reached with OAA strategy on adjusted preparing set. In scenario 2, most elevated achievement rate is 86.7% that is additionally reached with OAA technique on changed preparing set. Subsequently, the unmistakable part of choosing a legitimate preparing informational index had gotten apparent. To research why situation 2 had low achievement rate than situation 1, cases with some unacceptable location of shortcoming zone in the two situations are dissected in Table X & Table XI. They are likewise envisioned in Fig. 2 & Fig. 3 for improved agreement. By looking at these 2 tables, it very well may be inferred that expansion in wrong zone discoveries occurred in zone 2 & 3, while in zone 1 it nearly stayed steady. All in all, SVM committed more errors since pushing TCSC forward expanded its territory of impact. Moreover, in the two cases, zone 2 introduced the most elevated measure of wrong recognition and that is on the grounds that TCSC has its most noteworthy impact on this zone as it is set just before it. In this exploration ability and proficiency

Table 5. Numerous Wrong Predictions in Scenario 1

Predicted zone / Real zone	1	2	3	Total number of wrong predictions in each zone
1		99	211	310
2	249		512	761
3	193	143		336

of the proposed strategy was assessed on the example organization. It should be noticed that with an adjustment in the organization geography, gotten data from the separation transfer perspective would change. For this situation SVM boundary ought to be refreshed.

Table 6. Numerous Wrong Predictions in Scenario 2

Predicted zone / Real zone	1	2	3	Total number of wrong predictions in each zone
1		91	237	328
2	829		613	1442
3	381	283		644

4 CONCLUSIOIN

In this manuscript, another strategy for deficiency zone location of separation transfer in FACTS-based TLs is suggested.

For examining the impacts of TCSC area on shortcoming zone discovery of separation hand-off, two spots, one out of 50% of line length &other in 75% of line length are deliberated as two situations. Altogether, 4 cases have been considered for every situation. For situation 1, with OAA strategy, 83.7% of accomplishment, and with OAO technique, 84.2% of progress have been accomplished. In the wake of breaking down the outcome and expanding the preparation set, with OAA technique, 94.2% of accomplishment and with OAO strategy, 91.6% of progress are accomplished. For situation 2, with OAA

Figure 2. Numerous wrong predictions

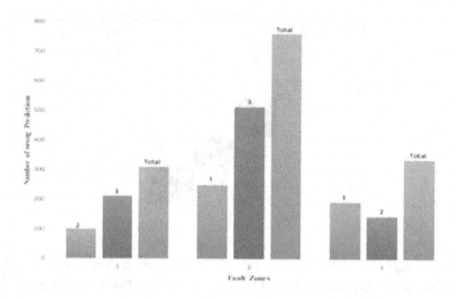

strategy, 69.7% of progress, and with OAO technique, 65.6% of accomplishment have been accomplished. In the wake of dissecting the outcome and expanding the preparation set, with OAA technique, 86.7% of accomplishment and with OAO strategy, 78.2% of progress are accomplished. The outcomes demonstrated that in tackling this sort of issue, eventually, OAA technique has similar bit of leeway over OAO strategy. With OAA technique, 94.2% and 86.7% of accomplishment are accomplished in situation 1 and 2, individually. It was indicated that introducing TCSC in the primary line prompts better outcomes

Figure 3. Numerous wrong predictions

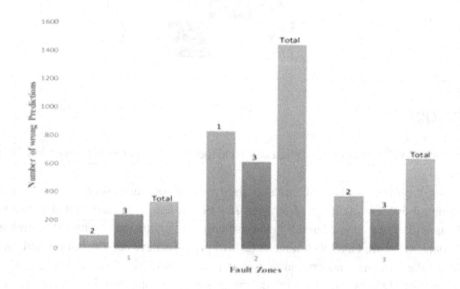

by diminishing its impact on zone 2 and 3 of the separation hand-off. TLs are crucial pieces of intensity framework. Improving their insurance is fundamental for unwavering quality of the system. This manuscript signifies new way to deal with increment the unwavering quality of framework by enhancing the assurance of TLs in issue occurrence and limiting the harm to other system hardware. Quick location and detachment of a defective TL, shield the system.

REFERENCES

Abinaya, S., & Arulkumaran, G. (2017). Detecting black hole attack using fuzzy trust approach in MA-NET. *Int. J. Innov. Sci. Eng. Res.*, *4*(3), 102–108.

Dash, P. K., Samantaray, S. R., & Panda, G. (2006). Fault Classification and Section Identification of an Advanced Series-Compensated Transmission Line Using Support Vector Machine. *IEEE Transactions on Power Delivery*, *22*(no. 1), 67–73. doi:10.1109/TPWRD.2006.876695

Dehkordi, Fereidunian, Dehkordi, & Lesani. (2014). HourǦahead demand forecasting in smart grid using support vector regression (SVR). *International Transactions on Electrical Energy Systems, 24*(12), 1650-1663.

Devaraju, Kumar, & Rao. (n.d.). A Real And Accurate Vegetable Seeds Classification Using Image Analysis And Fuzzy Technique. *Turkish Journal of Physiotherapy and Rehabilitation, 32*, 2.

Girgis, A. A. (1960). An adaptive protection scheme for advanced series compensated (ASC) transmission lines. *IEEE Transactions on Power Delivery*, *13*(2), 414–420.

Goldsworthy, D. L. (1987). A linearized model for MOV-protected series capacitors. *IEEE Transactions on Power Systems*, *2*(4), 953–957. doi:10.1109/TPWRS.1987.4335284

Imani. (2019). Effect of Changes in Incentives and Penalties on Interruptible/Curtailable Demand Response Program in Microgrid Operation. In *2019 IEEE Texas Power and Energy Conference (TPEC)*. IEEE.

Imani, Niknejad, & Barzegaran. (2019). Implementing Time-of-Use Demand Response Program in microgrid considering energy storage unit participation and different capacities of installed wind power. *Electric Power Systems Research, 175*, 105916.

Kalman, R. E. (n.d.). A new approach to linear filtering and prediction problems. *Journal of Basic Engineering, 82*(1), 35–45.

Kumar, D. S., Kumar, C. S., Ragamayi, S., Kumar, P. S., Saikumar, K., & Ahammad, S. H. (2020). A test architecture design for SoCs using atam method. *Iranian Journal of Electrical and Computer Engineering, 10*(1), 719.

Lepping, J. (2018). *Wiley Interdisciplinary Reviews. Data Mining and Knowledge Discovery*.

Megahed, A. I. (2006). Usage of wavelet transform in the protection of series-compensated transmission lines. *IEEE Transactions on Power Delivery*, *21*(3), 1213–1221.

Megahed, A. I., MonemMoussa, A., & Bayoumy, A. E. (2006). Usage of wavelet transform in the protection of series-compensated transmission lines. *IEEE Transactions on Power Delivery, 21*(3), 1213–1221. doi:10.1109/TPWRD.2006.876981

Obaid, A. J. (2020). *An efficient systematized approach for the detection of cancer in kidney.* Academic Press.

Parikh, U. B., Biswarup Das, & Prakash Maheshwari, R. P. (2008). Combined wavelet-SVM technique for fault zone detection in a series compensated transmission line. *IEEE Transactions on Power Delivery, 23*(4), 1789–1794. doi:10.1109/TPWRD.2008.919395

Provost, F., & Domingos, P. (2003). Tree induction for probabilitybased ranking. *Machine Learning, 52*(3), 199–215.

Sadeh, Ranjbar, Hadsaid, & Feuillet. (2000). Accurate fault location algorithm for series compensated transmission lines. *2000 IEEE Power Engineering Society Winter Meeting. Conference Proceedings (Cat. No. 00CH37077), 4,* 2527-2532.

Saha, M. M., Izykowski, J., Rosolowski, E., & Kasztenny, B. (1999). A new accurate fault locating algorithm forseries compensated lines. *IEEE Transactions on Power Delivery, 14*(3), 789–797. doi:10.1109/61.772316

Saikumar & Rajesh. (n.d.). A novel implementation heart diagnosis system based on random forest machine learning technique. *International Journal of Pharmaceutical Research, 12,* 3904–3916.

Saikumar, K., & Rajesh, V. (2020). Diagnosis of coronary blockage of artery using MRI/CTA images through adaptive random forest optimization. *Journal of Critical Reviews, 7*(14), 591–600.

Saikumar, K., Rajesh, V., Hasane Ahammad, S. K., Sai Krishna, M., Sai Pranitha, G., & Ajay Kumar Reddy, R. (2020). Cab for Heart Diagnosis with RFO Artificial Intelligence Algorithm. *International Journal of Research in Pharmaceutical Sciences, 11*(1).

Samantaray, S. R. (2013). A data-mining model for protection of FACTS-based transmission line. *IEEE Transactions on Power Delivery, 28*(2), 612–618. doi:10.1109/TPWRD.2013.2242205

Sheshasaayee, D., & Megala, R. (2017). A Conceptual Framework For Resource Utilization In Cloud Using Map Reduce Scheduler. *International Journal of Innovations in Scientific and Engineering Research, 4*(6), 188–190.

Song, Y. H., Johns, A. T., & Xuan, Q. Y. (1996). Artificial neural-networkbased protection scheme for controllable series- compensated EHV transmission lines. *IEE Proceedings. Generation, Transmission and Distribution, 143*(6), 535–540. doi:10.1049/ip-gtd:19960681

Song, Xuan, & Johns. (1997). Protection scheme for EHV transmission systems with thyristor controlled series compensation using radial basis function neural networks. *Electric Machines and Power Systems, 25*(5), 553-565.

Soumya, N., Kumar, K. S., Rao, K. R., Rooban, S., Kumar, P. S., & Kumar, G. N. S. (n.d.). 4-Bit Multiplier Design using CMOS Gates in Electric VLSI. *International Journal of Recent Technology and Engineering.*

Thomas, D. W. P., & Christopoulos, C. (1992). Ultra-high speed protection of series compensated lines. *IEEE Transactions on Power Delivery*, *7*(1), 139–145. doi:10.1109/61.108900

Xuan, Q. Y., Song, Y. H., Johns, A. T., Morgan, R., & Williams, D. (1996). Performance of an adaptive protection scheme for series compensated EHV transmission systems using neural networks. *Electric Power Systems Research*, *36*(1), 57–66. doi:10.1016/0378-7796(95)01014-9

Zhang, Nan, & Kezunovic. (2007). Transmission line boundary protection using wavelet transform and neural network. *IEEE Transactions on Power Delivery*, *22*(2), 859–869.

Chapter 29
K–Means Clustering Machine Learning Concept to Enable Social Distancing in Public Places

Abdullah Saleh Alqahtani

CFY Deanship King Saud University, Saudi Arabia

ABSTRACT

In endemic and pandemic situations, governments will implement social distancing to control the virus spread and control the affected and death rate of the country until the vaccine is introduced. For social distancing, people may forget in some public places for necessary needs. To avoid this situation, the authors develop the Android mobile application to notify the social distancing alert to people to avoid increased levels of the endemic and pandemic spread. This application works in both online and offline mode in the smart phones. This application helps the people to obey the government rule to overcome the endemic and pandemic situations. This application uses k-means clustering algorithms to cluster the data and form the more safety clusters for social distancing. It uses artificial intelligence to track the living location by the mobile camera without the internet facilities. It helps the user to follow social distancing even with no internet with user knowledge.

INTRODUCTION

Mobile phones are widely used electronic device in the world. This makes the technology revolution and moves the human life style to next level. The person can access the resource, makes phones calls, send text messages, taking photos, identify the current location of the people in the stand by location they can access any were at time. Virus spreads, endemic and pandemic situations the government will take some measures to control the spreads level. In some rare case of the virus spreads they implement the social distancing to avoid the spread in public places and mass gatherings. In recent times COVID 19 pandemic makes the big impact in the world. To avoid the virus spreads government implement the

DOI: 10.4018/978-1-7998-6870-5.ch029

social distancing act to control the virus spreads to next level of transmission. In general social distancing is important for human to maintain their hygiene level. Social distancing is also known as physical distancing to reduce the unfaithfulness of the people. Over flow of the country population will impact the health system and major issues in individual health. For those scenarios we follow the social distancing to overcome the virus spread and health related issues. Our proposed model works on the Bluetooth, wifi and internet technology for communication and data transfer purpose. This application model works on both offline and online modes. The measures are utilized in mix with great respiratory cleanliness, face veils and hand washing by a populace. During the COVID-19 pandemic, the World Health Organization (WHO) recommended preferring the expression physical separating rather than social distancing, with regards to the way that it is a physical separation which forestalls transmission; individuals can remain socially associated through innovation. To hinder the spread of irresistible illnesses and abstain from overburdening medicinal services frameworks, especially during a pandemic, a few social-separating measures are utilized, including the end of schools and work environments, segregation, isolate, confining the development of individuals and the wiping out of mass gathering.

Figure 1. Without and with social distancing

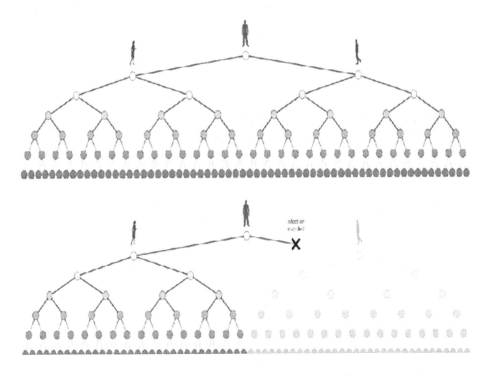

Android

Android is open source operating system which is owned by Google Corporation. This system is used operating for mobiles, tablets, and laptops. Android kernel was run on the basic of linux kernel. We choose android application why because it can widely use operating system. Android is most selling smart phones in the world since 2011. Android application can easily understand by the common people in the country. Its life cycle is much simpler than the other operating system. Android system provides security in the efficient way comparatively higher than other.

Around 70 percent of Android cell phones run Google's ecosystem, contending Android biological systems and forks incorporate Fire OS (created by Amazon) or Lineage OS. Anyway the "Android" name and logo are brand names of Google which force norms to limit "uncertified" gadgets outside their biological system to utilize Android branding. The source code has been utilized to create varia-

Figure 2. Android life cycle

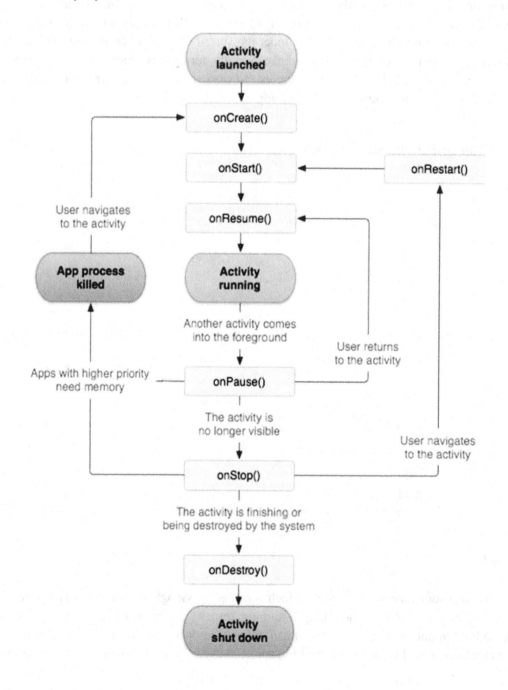

tions of Android on a scope of different hardware, for example, game consoles, advanced cameras, PCs and others, each with a particular UI. Some notable subordinates incorporate Android TV for TVs and Wear OS for wearables, both created by Google. Programming bundles on Android, which utilize the APK position, are commonly circulated through exclusive application stores like Google Play Store or Samsung Galaxy Store, or open source stages like Aptoide or F-Droid.

Bluetooth

Bluetooth is a widely used wireless technology for exchanging data between the mobiles in the short rage. This technology used minimum of 10 meters to 100 meters radius, this can use the UFH radio waves for transfer the data between two nodes. Its bandwidth is between 2.402 GHz to 2.480 GHz which is run the personnel area networks architecture. This technology works on the master slave model known as Pico-net, when master only sends the data during the data exchange.

Figure 3. Bluetooth architecture

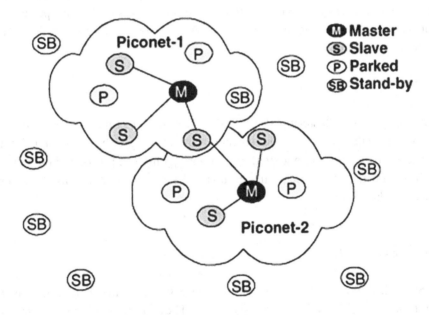

Wi-Fi

Wi-Fi is the family of wireless technology, which is used to connect the local area networks and internet. This use the 802.11 IEEE wireless communications stands. This technology is owned by non-profit wi-fi alliances. This will provides the privacy policy certification to the wi-fi technology for most successful way. Wi-Fi security is based on the encryption key, its based on WEP, WPS, WPS2 and latest security update is tested name called as WPS3. WPS3 is up gradation of the WPS2, WPS3 provide more security to user with encryption of 512 bit encryption which means it needs more time to break the password or may not possible to break the security.

Artificial intelligence (AI)

Artificial intelligence is a grown technology in the recent trends it's similar to human intelligence. The human mind solves the complex problems in efficient manner but, the problem count is increases it difficult to solve the problem. In case of AI it can compute the lacks of problem solution in same time. We use AI technology to the cameras to identify the distance between the user and objects. This technic help user when they are in offline mode.

GPS

Global Positioning System is widely used positioning system to monitor user location and also provide location advices to people. This technology was used across the world for location tracking, it's developed by USA. The GPS does not require the user to transmit any data, and it operates independently of any telephonic or internet reception, though these technologies can enhance the usefulness of the GPS positioning information. The GPS provides critical positioning capabilities to military, civil, and commercial users around the world. The United States government created the system, maintains it, and makes it freely accessible to anyone with a GPS receiver. In recent days Indian space service introduced the navigation system to the public use called as NAVIC system. This system will personally use for Indian local location tracking.

K-means Clustering

K-Means is an unsupervised clustering algorithm which is used to find groups within the data. it gives the observation in the data sets in the 3-dimensional way. This algorithm clustering the data by point of similar data in the cluster and differentiate the neighboring clusters. It is one of the iterative approaches to clustering the data with possible or similar groups in the clustering. K-means algorithm follows given steps to process the data.

LITERATURE SURVEY

Kamali T & Stashuk D W*et al.*2020 proposed the density based clustering structures to measure the neighbourhood distancing in the low density regions. They propose the neighbourhood distance entropy consistency algorithm to simulate the clustering model and use k-means algorithm to measure the neighbouring node distance. Jie Z et al. 2019 proposed the method to identify the social group in the local attribute. They use framework to eliminate the problem in the local networks. The method effectively communicates the networking nodes. Cristani, M *et al.* 2020 proposed the algorithm to measure the social distancing of the people by using visual mechanism. They use computer vision method to analysis the social distancing between two peoples. Tuan Le & Gerla M *et al.* 2016 proposed the social distancing based any cast routing method to identify the delay in the ad-hoc networks. They implement the proposed anycast social distancing metric to identify the social distancing of the group members connected in a network group. This method enables the social distancing of the member group and tolerant the network communication.

Lin Z *et al*. 2019 proposed algorithm to measure the GPS based location tracking by using the mobile GPS data. they use GPS trajectory data to simulate the model to measure the distance between the two users with help of mobility profile. Shahriar et al. 2014 proposed the method to identify the memory leak in the android applications. They develop the memory leak pattern to evaluate the application and memory leaks in efficient way. This model use API calls to identify the memory leaks in the application when the user connected to the network. Lam C. H & She J *et al*. 2019 proposed the distance estimation method to identify the moving object distance. The authors use the BLE beacon method to measure and identify the distance between the moving object. The analysing device is connected to the mobile phone and notifies the user to distance of the both speed and slow moving objects.

Han Yun *et al*. 2010 proposed the monitoring system for intelligent video monitoring. They use 3G technology to monitoring and analysis the video for long distance focusing. This notifies user by alarm to long distance video backup and security. Taniguchi, Y *et al*. 2015 proposed the sensor monitoring system for bicycle. This model placed in the bicycle front end to analysis the any obstacles in the system. This system uses the ultrasonic waves to analyse the obstacles and intimate to the user to bypass the obstacles and ensure save driving.Tatar & Bayar *et al*. 2018 proposed the Bluetooth based land vehicles. The authors use the Wi-Fi and Bluetooth technology to transfer the data between the nodes. This model controlled the land vehicle with the band width of the 5.4GHZ process.

Kajikawa *et al*. 2016 proposed the model to enable the fast connection method using the Bluetooth with energy consumption. The authors use the Bluetooth technology to monitor the data transfer in the classic Bluetooth mantes. This model ensures the fastest technology and passive scan. Sharma, A *et al*. 2018 proposed the model for preparing the self-organising map by using the Facebook data set. The authors discuss the kohonen SOM method to discover and analysis the Facebook data and generate the map for clustering environment. They use k-means clustering to organising the data set and map genera-tion. Yang & Ng *et al*. 2009 proposed the model to measure the distance in the clustering environment. The authors use the web opinions for the produce the distance based clustering. Alsayat& El-Sayed *et al*. 2016 proposed the k-means algorithm to clustering the data set and analyse the social media data. The authors use the OCD clustering for efficient data clustering. This model optimised the real time clustering of social data sets. Prabhu, J., et al.IoT role in prevention of COVID-19 and health care work-forces behavioural intention in India-an empirical examination. Jothikumar, R., et al exolained about Predicting life time of heart attack patient using improved c4. 5 classification algorithm. Jothikumar, R. (2016) desccbed about C4. 5 classification algorithm with back-track pruning for accurate prediction of heart disease.

EXISTING SYSTEM

In general COVID 19 pandemic spreads makes the various governments to force the people to obey the social distancing for controlling the virus spread in the country. They introduced various rules and some mobile application to monitor the people health via those applications. The American government introducing the mobile application to monitor the people's social distancing. This application works on the Bluetooth technology to monitor the social distancing of the common people. This application monitors the users distancing and monitors the location travel by the GPS technology. This will alert the user to aware the social distancing in the public places. The government of India also introducing the application called AarogyaSetu. This application is a type of chat bot to answer the doubts about the COVID 19 symptoms to people. This application also displays the map about the containment zone level and virus spread level in the country.

The above mentioned applications are developed only for the pandemic situations. This may use for people in only for the pandemic situations. All mentioned applications need internet to access the data and live location tracking. To overcome those issues the proposed methodology will works on both online and offline modes.

PROPOSED METHODOLOGY

The proposed model works on the Bluetooth and GPS technology for offline and online to monitor the social distancing of individuals. The architecture of the model is given below for better understanding. This model also uses NAVIC technology for better performance of the system in terms of Indian navigation.

Figure 4. Working model architecture

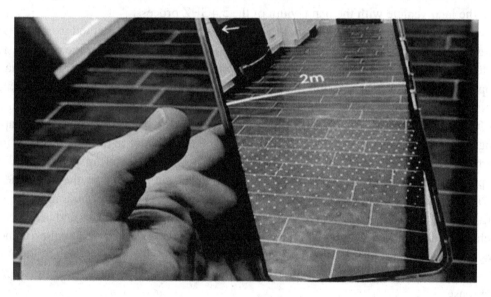

The proposed application installed to android device version 4.0 or more. This application runs on three modes one online by GPS and offline by Bluetooth, Wi-Fi and mobile camera with AI. Initially, it gets permission for accessing the Bluetooth and GPS in smartphones. If the user is online and goes to some ware it's tracking the current location status and sends the data to server. In server end there is model runs the k-means algorithm to prepare the clusters for that location with the help of other mobiles GPS data and sends to the user mobile. If the user goes mass gathering it will analysis each user distance by the living tracking mechanism. It makes an alarm notification to the user mobile even the phone in silent mode about the social distance violation the place. Its system measure distance the minimum of 2meter radius and the radius can be modified by the user wish.

The next method is offline mode or Bluetooth mode. Here the application automatically turns on the Bluetooth and monitors the user mobiles scanning the nearby Bluetooth device and measure the radius of the devices it the radius is less it will notify the user about to follows social distancing. It's also reduces the power consumption of the device and perform at low battery level. This application monitors the location by some build-in clustering algorithm to aware user about the social distance and also mentions the mass gathering area in your current location. This will also use Wi-Fi technology to analysis the local network group to monitor their social distancing. It helpful to the user in case the user use public Wi-Fi it generate the cluster for that connected devices and measures the distance between them. If the distance is reduces then notify the user by alarm. From this application user can also check the endemic or pandemic spreads of the country.

This application also works in camera mode to track the users social distancing by the implementation of AI with object detection method. It effectively view the distance between objects as well accurately follows the social distancing in user knowledge.

RESULT ANALYSIS

It uses AI technology to analysis the distance between user and other by live tracking it show in below figure5. This will detects the object and calculate the distance of the object and shows the distance with how fare the user and the respective object by live object detection technic.

Figure 5. AI based analysis for distancing

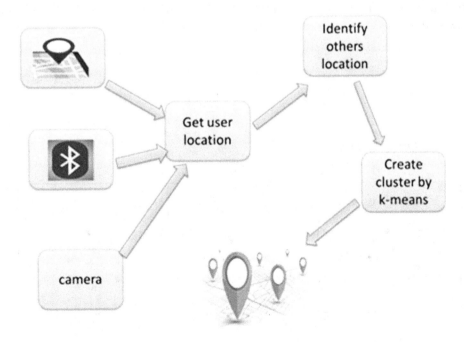

GPS Based Analysis

The GPS based calculation measured distance of the nearby objects by the internet. This method needs the network connection to analysis the data and sends the live tracking to user mobile with the help of the proposed application. This UI is shown is figure 6.

Figure 6. GPS based tracking

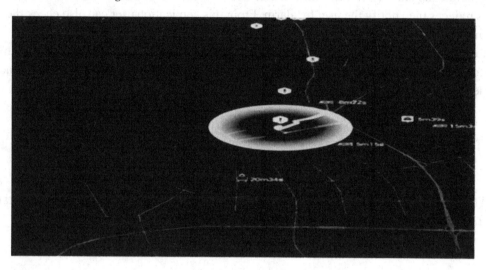

The k-means clustering with the application result is shown in figure 7. This will gets the data inputs form the user live location and calculate the distance by forming the clusters to identify the neighboring nodes distance. It ensures the system to form the accurate clusters to the application and calculate the distance between them.

Figure 7. k-means clustering formation

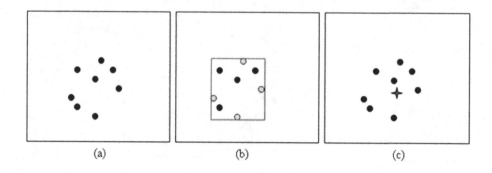

Bluetooth Tracking

Bluetooth based tracking needs Bluetooth connectivity and monitor the user location within rage to calculating the social distancing between Bluetooth connected devices. The result of this methodology is shown in figure 8.The user reaches the particular rage of distance coverage it will intimate the user to follow the social distancing by alarm message.

Figure 8. Bluetooth based tracking

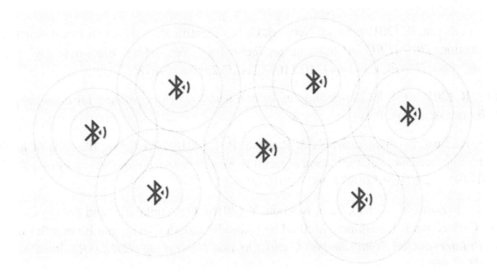

CONCLUSION

In this chapter, the analysis was performed based on the user data by the mobile application. The proposed model with most popular operating system is helps to achieve good results. The proposed mobile application with k-means clustering was the most efficient and more reliable model compared to previous mobile applications.

The results this application is useful to maintain the social distancing in any country with or without the internet and supports for any pandemic situations. The challenge in the application development is this data and live tracking of the user location. If the user is online it's easy to track but the user in offline it not easy to tracking the location. The combination of AI and application interface makes the offline tracking much easier and also effectively compares to Bluetooth or Wi-Fi technology. In future mobile application with AI is enough to maintain the social distancing in more accurate way.

REFERENCES

Alsayat & El-Sayed. (2016). Social media analysis using optimized K-Means clustering. *2016 IEEE 14th International Conference on Software Engineering Research, Management and Applications (SERA),* 61-66. 10.1109/SERA.2016.7516129

Android. (n.d.),. https://www.android.com/what-is-android/

Cristani, M., Bue, A. D., Murino, V., Setti, F., & Vinciarelli, A. (2020). The Visual Social Distancing Problem. *IEEE Access: Practical Innovations, Open Solutions, 8,* 126876–126886. doi:10.1109/AC-CESS.2020.3008370

Jie, Z., Li, Y., & Liu, R. (2019). Social Network Group Identification based on Local Attribute Community Detection. *2019 IEEE 3rd Information Technology, Networking, Electronic and Automation Control Conference (ITNEC),* 443-447. 10.1109/ITNEC.2019.8729078

Jothikumar, R. (2016). *C4. 5 classification algorithm with back-track pruning for accurate prediction of heart disease.* Academic Press.

Jothikumar, R., Susi, S., Sivakumar, N., & Ramesh, P. S. (2018). Predicting life time of heart attack patient using improved c4. 5 classification algorithm. *Research Journal of Pharmacy and Technology, 11*(5), 1951–1956. doi:10.5958/0974-360X.2018.00362.1

Kajikawa, N., Minami, Y., Kohno, E., & Kakuda, Y. (2016). On Availability and Energy Consumption of the Fast Connection Establishment Method by Using Bluetooth Classic and Bluetooth Low Energy. *2016 Fourth International Symposium on Computing and Networking (CANDAR),* 286-290. 10.1109/CANDAR.2016.0058

Kamali, T., & Stashuk, D. W. (2020, August). Discovering Density-Based Clustering Structures Using Neighborhood Distance Entropy Consistency. *IEEE Transactions on Computational Social Systems, 7*(4), 1069–1080. Advance online publication. doi:10.1109/TCSS.2020.3003538

Lam, H., & She, J. (2019). Distance Estimation on Moving Object using BLE Beacon. *2019 International Conference on Wireless and Mobile Computing, Networking and Communications (WiMob),* 1-6. 10.1109/WiMOB.2019.8923185

Le, T., & Gerla, M. (2016). Social-Distance based anycast routing in Delay Tolerant Networks. *2016 Mediterranean Ad Hoc Networking Workshop (Med-Hoc-Net),* 1-7. 10.1109/MedHocNet.2016.7528495

Lin, Z., Zeng, Q., Duan, H., Liu, C., & Lu, F. (2019). A Semantic User Distance Metric Using GPS Trajectory Data. *IEEE Access: Practical Innovations, Open Solutions, 7,* 30185–30196. doi:10.1109/ACCESS.2019.2896577

Prabhu, J., Kumar, P. J., Manivannan, S. S., Rajendran, S., Kumar, K. R., Susi, S., & Jothikumar, R. (2020). IoT role in prevention of COVID-19 and health care workforces behavioural intention in India-an empirical examination. *International Journal of Pervasive Computing and Communications.*

Shahriar, H., North, S., & Mawangi, E. (2014). Testing of Memory Leak in Android Applications. *2014 IEEE 15th International Symposium on High-Assurance Systems Engineering*, 176-183. 10.1109/HASE.2014.32

Sharma, M. K. S., & Dwivedi, R. K. (2018). Analyzing Facebook Data Set using Self-organizing Map. *2018 International Conference on System Modeling & Advancement in Research Trends (SMART)*, 109-112. 10.1109/SYSMART.2018.8746984

Taniguchi, Y., Nishii, K., & Hisamatsu, H. (2015). Evaluation of a Bicycle-Mounted Ultrasonic Distance Sensor for Monitoring Road Surface Condition. *2015 7th International Conference on Computational Intelligence, Communication Systems and Networks*, 31-34, 10.1109/CICSyN.2015.16

Tatar, G., & Bayar, S. (2018). FPGA Based Bluetooth Controlled Land Vehicle. *2018 International Symposium on Advanced Electrical and Communication Technologies (ISAECT)*, 1-6. 10.1109/ISAECT.2018.8618830

Yang, C., & Ng, T. D. (2009). Web opinions analysis with scalable distance-based clustering. *2009 IEEE International Conference on Intelligence and Security Informatics*, 65-70. 10.1109/ISI.2009.5137273

Yun, Hua, & Zhi. (2010). Intelligent video monitoring system based on 3G. *2010 International Conference on Educational and Network Technology*, 135-138. 10.1109/ICENT.2010.5532145

Chapter 30
An Efficient Selfishness Control Mechanism for Mobile Ad hoc Networks

D. Rajalakshmi
Sri Sairam Institute of Technology, India

Meena K.
Vel Tech Rangarajan Dr. Sagunthala R&D Institute of Science and Technology, India

ABSTRACT

A MANET (mobile ad hoc network) is a self-organized wireless network. This network is more vulnerable to security failure due to dynamic topology, infrastructure-less environment, and energy consumption. Based on this security issue, routing in MANET is very difficult in real time. In these kinds of networks, the mobility and resource constraints could lead to divide the networks and minimize the performance of the entire network. In real time it is not possible because some selfish nodes interacts with other nodes partially or may not share the data entirely. These kind of malicious or selfish nodes degrade the network performance. In this chapter, the authors proposed and implemented the effect of malicious activities in a MANETs using self-centered friendship tree routing. It's a novel replica model motivated by the social relationship. Using this technique, it detects the malicious nodes and prevents hacking issues in routing protocol in future routes.

1. INTRODUCTION

A Mobile ad hoc network (MANET) is a collection of devices; it creates the temporary connection without any infrastructure. It communicates the other nodes via multi-hop. No master slave relationship applicable in MANETs (Marchang et al., 2017).

Due to dynamic topology, the devices are simply entered the network and perform the communication then its moves the network in a very fashionable manner. The network size and topology is unpredictable because of this infrastructure less environment (Rajalakshmi et al., 2021). The network performance is

DOI: 10.4018/978-1-7998-6870-5.ch030

highly solicited on the entire nodes. It is mainly used in following applications: Military, Rescue operation, Business work, class rooms, conference and Personal Area Network (PAN).

MANET refers to a form of ad hoc wireless network or mesh networks, having computing mobile devices (nodes), which communicate through networks which are transmitted through any form of base station infrastructure or central administration systems within access points or cell phone networks situated within WLAN. Nodes move freely and randomly and organize arbitrarily and thus, they are unpredictable due to rapid changes in network topology changes rapidly. Such networks usually operate individually, often linked through the Internet.

MANETs, unlike conventional mobile wireless networks, don't rely on central coordinator but communicate self-organized. Mobile nodes communicate via in interrelated radio ranges, connected through wireless links, situated at a distance from other nodes and they transfer messages through a router. Ad hoc networks nodes, are capable of both sending/receiving messages through its respective hosts and routers can forward data for other nodes. The networks are additionally known as multi hop wireless ad- hoc networks.

MANET working groups in 1996 were set up by the Internet Engineering Task Force (IETF) in order to standardize IP routing conventions to suit routing systems which are wireless and function in dynamic and fixed topologies.

MANET, autonomous mobile nodal systems, functions in isolation or has multiple gateways regulated through fixed networks. Wireless transmitters have nodes and receivers which utilize broadcasting Omni dimensional antennas, bi-directional (point-to-point), or both of them unanimously. A system at any point of time is seen to be a random graph due to nodular movements, receiver/transmitter coverage patterns, interference in co-channels and transmission power levels. The topography of networks modifies itself with time through adjusting parameters of transmission and reception connections.

MANETs have these features:

- Unreliable links between wireless nodes causes inconsistencies in the links between ad hoc networks, due to restricted energy among wireless nodes.
- Due to rapid movements amongst nodes and their constantly changing topologies MANET topology changes: nodes constantly move in and out of other nodes radio ranges in ad hoc networks resulting in changing routing information.
- This wireless routing protocols lacks the corporation of multiple static security features meant for ad hoc environments. As the above topology changes continuously, it is important for adjacently paired nodes to include security routing characters to avoid attacks that attempt to use the statically configured routing protocol's vulnerabilities.

In MANET, the devices are simply move from one place to another place arbitrarily without any difficulties. Individual node may act as a router, it by pass the network traffic to other nodes in the network. MANETs can easily form and deform the networks without the need of any centralized arbiter.

MANETs are classified into four types, namely:

1. **VANET** - Vehicular Ad hoc Network, its an implementation of MANETs, Voluntary connection established between vehicles in wireless networks. It performs vehicle-to-vehicle and vehicle-to-roadside communication for safety purpose. This network leads to an Internet of autonomous vehicles.

2. **IMANET** - Internet Based Mobile Ad hoc Network is a wireless network, it supports the internet protocols like Transmission Control Protocol (TCP), Internet Protocol (IP), User Datagram protocol (UDP). These kinds of networks connect the mobile nodes and create the routes spontaneously.
3. **INVANET** - Intelligent Vehicular Ad hoc Network use WiFi and WiMAX providing better communications between vehicles in dynamic nature.
4. **FANET** - Flying Ad hoc Network, it's a group of small Unmanned Air Vehicles (UAV), used for military and commercial applications. It's a self-configurable network providing flexible communications between collaborative UAVs.

Routing protocols use several metrics to evaluate the best path for routing the packets to its destination. These metrics are a standard measurement that could be number of hops, which is utilized by the routing algorithm to establish the optimal path for the packet to its destination. The process of path determination is that, routing algorithms initialize and maintain routing tables, which contain the total route information for the packet. This route information varies from one routing algorithm to another. The routing algorithms generate the variety of information which is filled in the routing tables. The IP-address prefix and the next hop were the most common entries in the routing table. Routing tables Destination/next hop associations tell the router that a particular destination can be achieved optimally by sending the packet to a router representing the "next hop" on its way to the final destination and IP-address prefix specifies a set of destinations for which the routing entry is valid for the communication.

Challenges of a Routing Protocol

The various responsibilities of a routing protocol can be defined as exchanging the routing information, finding a feasible path to the destination based on hop length, minimum power required, and lifetime of the node. The other activities such as utilizing minimum bandwidth, gathering path break information, finding re-route to the destination are also considered. The various challenges are listed below. The mobility of a node may lead to the following problems such as packet collisions, frequent path breaks, transient loops and difficulty in resource reservation. All stated issues have to be solved efficiently for a good routing protocol. In the broadcast region, the channel is shared by all nodes. Depending upon the number of nodes and the network traffic handled, the bandwidth is made available for each node. The Bit Error Rate (BER) in a wireless channel is very high compared to that in its wired channel. Routing protocols designed for Ad hoc networks should consider BER, signal to noise ratio and path loss for improving the efficiency in routing. The load on the channel varies with the number of nodes present in a given geographical region. Hence when the number of node increases, the contention for the channel is high. This high contention for the channel may result in high number of collisions and wastage of bandwidth. A good routing protocol should consider this issue and a mechanism are needed for distributing the load uniformly across the network. The constraints such as battery power, computing power also limit the capability of a routing protocol.

Design Goals of Routing Protocols

The main goal of an ad-hoc network routing is to correctly and efficiently establish a route between a pair of nodes in the network so that a message can be delivered according to the expected quality of service parameters (Ilyas 2003). Dating back to the early 1980s, there have been a large number of routing protocols designed for multi hop ad-hoc networks. The following are typical design goals for ad-hoc network routing protocols:

Minimal control overhead: Control messages consume bandwidth and battery power for transmitting and receiving. Hence, reducing control messaging also helps to conserve battery power.

Multi hop routing capability: Because the wireless transmission range of mobile nodes is often limited, sources and destinations typically may not be within direct transmission range of each other. The routing protocol must be able to discover multi hop routes between the sources and destinations so that communication between those nodes is possible.

Dynamic topology maintenance: Once a route is established, it is likely that some of the links in the route will break due to node movement. Therefore, a viable routing path must be maintained, if there is node mobility.

Loop prevention: Routing loops occur when some node along a path selects a next hop to the destination which is also a node that occurred earlier in the path. When a routing loop exists, data and control packets may traverse the path multiple times until either the path is fixed and the loop is eliminated, or until the Time To Live (TTL) of the packet reaches zero. Because bandwidth is scarce and packet processing and forwarding are expensive, routing loops are extremely wasteful of resources.

In MANETs the routing protocols are used, to deliver the packet to intended recipient. It consists of set of rules and regulations that governs communication from source to destination. There are three types of routing Protocol (Mohammed Musthafa et al., 2020), namely:

1. Proactive Routing Protocol
2. Reactive Routing Protocol
3. Hybrid Routing Protocol

Proactive Routing Protocol

Each and every node should maintain a routing table periodically. Based on the network traffic the packets are delivered to the destination through some other neighbors. It's not a best approach for larger networks, because for all nodes maintaining a routing table is not an easy task. Examples: Destination Sequenced Distance Vector Routing (DSDV), Wireless Routing Protocol (WRP), Optimized Link State Routing Protocol (OLSR) etc.

Reactive Routing Protocol

Routing table is not updated periodically, that is whenever the network configuration has changed then only routing information is updated, and finds the optimal path to deliver the packets into corresponding destination. Examples: Ad-hoc On-demand Distance Vector (AODV), Dynamic Source Routing (DSR), Low Based Multipath Routing (LMR) etc.

Hybrid Routing Protocol

It combines best approaches of proactive and reactive routing protocols. Examples: Zone Routing Protocol (ZRP), Border Gateway Protocol (BGP) etc.

2. BACKGROUND

Due to the emerging era, MANET has populated in recent years. Without physical infrastructure, it performs the communication in very effective manner in the cases of dynamic technology configuration, energy consumption etc. The main characteristics of MANETs are fast installation, low bandwidth communication and narrow processing competency (Shakshuki et al., 2013). Based on the characteristics, it provides the routing information to the standard routing protocols. There are no rules and regulations for travelling the packets when inside the network region. So individual node they can choose the path accordingly to do the communication with one another (Rajalakshmi & Meena, 2019). Traditional Routing protocols are used in wireless communications and it gives the routing information to specified node. Some existing traditional routing protocols are AODV, OLSR, and DSR etc. Furthermore, each node has autonomously to detect and react on network topology changes (Rajalakshmi & Meena, 2019). In research point of view MANETs have several challenges in security perspectives. It has a limited battery power and limited communication area (Srinivasan & Murugaanandam, 2014).

Existing routing protocols, partially they detect the selfish nodes in terms of attacks (Active attacks, Passive attacks), criteria (Integrity, Security, Confidentiality, Authentication etc), performance metrics (Throughput, Propagation Delay etc), and network parameters (QOS, DoS etc).

Significance of Network

An Ad hoc network has the capacity to deploy in a wide range of application in emergency relief area such as disaster relief areas, temporary installations, vehicular networks, connecting organizations and hospitals. Figure 1 shows the significance of Mobile Ad hoc networks, that they can engage in any kind of technology and integrated with various kinds of environment. Ad hoc networks can be connected to pre-existing technologies like infrastructure networks. The integration of infrastructure network with Ad hoc is the easy way of integrating and utilizing the infrastructure network services in Ad hoc networks. Infrastructure networks are organized in a master slave technique and thus in a centralized infrastructure, which enables the network to have the base station fixed in one location. Since it is fixed the ultimate static network topology ensure the stable connectivity.

On the other hand, Ad hoc networks have less infrastructure requirements and there is no need of a base station since the network has the dynamic with frequent network topology. Ad hoc networks when integrated with Infrastructure a network which enables the interface with infrastructure network hardware such as access points. Basically Ad hoc network topologies with minimum number of nodes operate more efficiently, rather than having high number of nodes. The issues related to scalability are concerned with the network which provides connectivity and enables the user send urgent messages under any traffic condition.

Figure 1. Significance of MANET

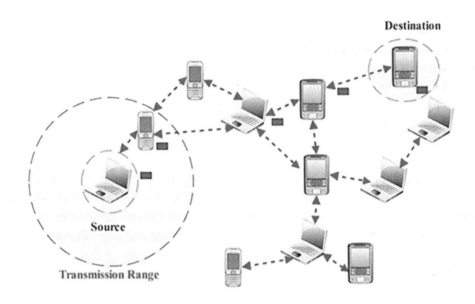

MANET NETWORK SECURITY CRITERIA

Passive Attacks

In passive attack, the intruder gain some knowledge and monitoring on confident connections to get valuable information of traffic without inject any fake information without affecting the system resources and normal functioning of the network. This type of attack serves the attacker to get information and makes the footprint of the invaded network in order to apply the attack successfully. Detecting this kind of attack is difficult because neither the system resources nor the critical network functions are physically affected to prove the intrusions.

Eavesdropping Attacks

The attacker simply listens to messages exchanged by two entities. For the attack to be useful, the traffic must not be encrypted. Any unencrypted information, such as a password sent in response to an HTTP request, may be retrieved by the attacker. This info is often later utilized by the malicious node. The key info like location, public key, non-public key, password etc. are often fetched by eavesdropper.

Traffic Analysis

The attacker looks at the metadata transmitted in traffic in order to deduce information relating to the exchange and the participating entities, e.g. the form of the exchanged traffic (rate, duration, etc.). In the cases where encrypted data are used, traffic analysis can also lead to attacks by cryptanalysis, whereby the attacker may obtain information or succeed in unencrypting the traffic.

Traffic analysis in ad hoc networks may reveal:

- The existence and location of nodes
- The communications network topology
- The roles played by nodes
- The current sources and destination of communications and
- The current location of specific individuals or functions

Active Attacks

These attack causes unauthorized state changes in the network such as denial of service, modification of packets, and the like. These attacks are generally launched by users or nodes with authorization to operate within the network. Classify active attacks into four groups: dropping, modification, fabrication, and timing attacks. It should be noted that an attack can be classified into more than one group.

Dropping Attacks

Malicious or selfish nodes deliberately drop all packets that are not destined for them. While malicious nodes aim to disrupt the network connection, selfish nodes aim to preserve their resources. Dropping attacks can prevent end-to-end communications between nodes, if the dropping node is at a critical point. It might also reduce the network performance by causing data packets to be retransmitted, new routes to the destination to be discovered, and the like. Unfortunately most routing protocols Dynamic Source Routing ((DSR) is an exception) have no mechanism to detect whether data packets have been forwarded or not. However, they can be detected by neighboring nodes through passive acknowledgement or hop-by-hop acknowledgement at the data link layer. An attacker can choose to drop only some packets to avoid being detected; this is called a selective dropping attack.

Besides data packets or route discovery packets, an attacker can also drop route error packets, causing the source node to be unaware of failed links (thus interfering with the discovery of alternative routes to the destination).

Modification Attacks

Insider attackers modify packets to disrupt the network. For example, in the sinkhole attack the attacker tries to attract nearly all traffic from a particular area through a compromised node by making the compromised node attractive to other nodes.

It is especially effective in routing protocols that use advertised information such as remaining energy and nearest node to the destination in the route discovery process. A sinkhole attack can be used as a basis for further attacks like dropping and selective forwarding attacks. A black hole attack is like a sinkhole attack that attracts traffic through itself and uses it as the basis for further attacks. The goal is to prevent packets being forwarded to the destination. If the black hole is a virtual node or a node outside the network, it is hard to detect.

Fabrication Attacks

Here the attacker forges network packets. In fabrication attacks are classified into "active forge" in which attackers send faked messages without receiving any related message and "forge reply" in which the attacker sends fake route reply messages in response to related legitimate route request messages. In the forge reply attack, the attacker forges a Route Reply message after receiving a Route Request message. The reply message contains falsified routing information showing that the node has a fresh route to the destination node on Ad hoc On demand Distance Vector (AODV) in order to suppress real routes to the destination.

It causes route disruption by causing messages to be sent to a non-existent node or putting the attacker itself into the route between two endpoints of a communication channel if the insider attacker has already have a route to the destination. Figure 2 shows an example of a forge reply attack. The best route (with minimum hop) from node S to node D is S-I1-I2-D. Malicious node M forges a Route Reply (RREP) message to the source node S through node I1.

Figure 2. A Forge Reply Attack

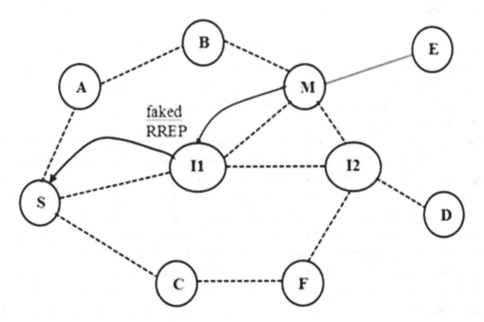

The message claims to come from the destination node D with higher destination sequence number to suppress the existing route. The faked message results in the updating of the route entry to the destination node in the routing tables of node S and I1. Node I1 forwards data packets to the malicious node instead of node I2 since node M seems to have a fresh route to node D, so the new route becomes S-I1-M-I2-D.

Attackers can initiate frequent packets to cause Denial of Service (DoS). Example DoS attacks that exploit MANETs' features are sleep deprivation torture attacks, routing table overflow attacks, ad hoc flooding attacks, rushing attacks, and the like.

The sleep deprivation torture attack consumes a node's battery power and so disables the node. It does so by persistently making service requests of one form or another. This attack was more powerful than better known DoS attacks such as Central Processing Unit(CPU) exhaustion, since most mobile nodes are run on battery power. The ad hoc flooding attack, introduced in new Routing Attack in Mobile Ad Hoc Networks, is another DoS attack against on-demand protocols, in which nodes send Route Request messages when they need a route.

The attacker exploits this property of Route Discovery by broadcasting many Route Request messages to a node that is not in the network. Another attack at the Route Discovery phase is the routing table overflow attack. Here the attacker sends a lot of route advertisements for nodes that do not exist. Since proactive protocols update routing information periodically before it is needed, this attack, this results in overflowing the victim nodes routing tables and preventing new routes from being created, is more effective in proactive protocols than in reactive protocols.

Another interesting fabrication attack on MANETs is the routing cache poisoning attack. A node can update its table with the routing information in the packets that it hears, even if it is not on the route of the packets. The attacker can make use of this property to poison the routes to a victim node by sending spoofed routing information packets, causing neighboring nodes to update their tables erroneously.

Timing Attacks

An attacker attracts other nodes by causing itself to appear closer to those nodes than it really is. DoS attacks, rushing attacks, and hello flood attacks use this technique. Rushing attacks occur during the Route Discovery phase. In all existing on-demand protocols, a node needing a route broadcasts Route Request messages and each node forwards only the first arriving Route Request in order to limit the overhead of message flooding.

So, if the Route Request forwarded by the attacker arrives first at the destination, routes including the attacker will be discovered instead of valid routes. Rushing attacks can be carried out in many ways: by ignoring delays at Media Access Control (MAC) or routing layers, by wormhole attacks, by keeping other nodes' transmission queues full, or by transmitting packets at a higher wireless transmission power. The hello flood attack is another attack that makes the adversary attractive for many routes. In some routing protocols, nodes broadcast Hello packets to detect neighboring nodes. These messages are received by all one-hop neighbor nodes, but are not forwarded to further nodes.

The attacker broadcasts many Hello packets with large enough transmission power that each node receiving Hello packets assumes the adversary node to be its neighbor. It can be highly effective in both proactive and reactive MANET protocols. A further significant attack on MANETs is the collaborative wormhole attack.

Here an attacker receives packets at one point in the network, tunnels them to an attacker at another point in the network, and then replays them into the network from this final point. Packets sent by tunneling forestall packets forwarded by multi-hop routes as shown in Figure 3 and it gives the attacker nodes an advantage for future attacks. Since the packets sent over tunneling are the same as the packets sent by normal nodes, it is generally difficult to detect wormhole attackers by software-only approaches such as Intrusion Detection System (IDS). That is why packet leashes (any information that is added to a packet designed to restrict the packet's maximum allowed transmission distance) have been introduced for preventing wormhole attacks.

Figure 3. Wormhole Attack

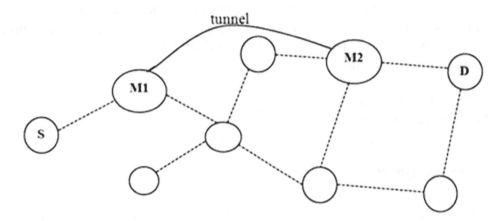

In large-scale network, the numbers of communication devices are low and the distance between one to another device becomes too high means the network falls into unreliable one and also some malfunctioning or misbehaviors are performed inside the network (Khan & Safi, 2018). This is unfavorable, because the communication demands are very high in the recent technologies.

3. RELATED WORK

In Mobile Ad hoc region, reactive routing protocol is more robust. Individual selfishness communication node achieves low cost-efficiency and truthfulness. Ad hoc-VCG works well for ad hoc networks, where communication between two nodes is generally long and the routing path does not change intensely during a communication (Rajalakshmi & Meena, 2020). If these conditions are not met, overhead of the route discovery phase stalls the network performance.

In this paper, a game-theoretic approach is used to produce better results with the existing system approaches. It analyzes the problem through theoretical perspectives. In selfish caching model, servers have two possible actions for each object. If the replica of a requested object is located at nearby node, the server accesses the remote replica. If the replica of a requested object is located at remote node, server caches the data by the object itself. Decisions about the caching and replicas are performed at locally, with their local costs. If one or more servers to do the same operation means the object is replicated in all servers. All servers specify the game theoretic approach and calculate the relevant cost for all servers (Guo et al., 2013).

In existing routing protocols, partially they detect the selfish nodes in terms of attacks (Active attacks, Passive attacks), criteria (Integrity, Security, Confidentiality, Authentication etc), performance metrics (Throughput, Propagation Delay etc), and network parameters (QOS, DoS etc). But the selfish nodes are handled by three categories (Anderegg & Eidenbenz, 2003; Balakrishnan et al., 2005). The Reputation-based Technique monitors the user behaviors and gets the routing information for delivering the packets from sender to receiver (Anderegg & Eidenbenz, 2003; Chun et al., 2004; Rajalakshmi & Meena, 2018). The Credit-payment technique gives credit information to others through this kind of transformation data is shared to others. The Game theory-based technique identifies the equilibrium point to improve the system performance. Above mentioned three categories concentrated on forwarding of packets. But in proposed system, give attention to forwarding of packets, Packet delivery ratio, routing overhead, and identifying the selfishness node behavior and activities.

Disadvantages of Existing System:

- It only focused on the packet forwarding.
- In DCG, network traffic is very worst.

4. PROBLEM DEFINITION

The MRA scheme is designed to resolve the weakness of Watchdog when it fails to detect misbehaving nodes with the presence of false misbehavior report. The false misbehavior report can be generated by malicious attackers to falsely report innocent nodes as malicious. This attack can be lethal to the entire network when the attackers break down sufficient nodes and thus cause a network division. The core of MRA scheme is to authenticate whether the destination node has received the reported missing packet through a different route. The source route broadcasts an RREQ message to all the neighbors within its communication range. Upon receiving this RREQ message, each neighbor appends their addresses to the message and broadcasts this new message to their neighbors. If any node receives the same RREQ message more than once, it ignores it. If a failed node is detected, which generally indicates a broken link in flat routing protocols like DSR, a RERR message is sent to the source node. When the RREQ message arrives to its final destination node, the destination node initiates an RREP message and sends this message back to the source node by reversing the route in the RREQ message.

It illustrates the issues of selfishness perspective, replica provision in MANET. It can easily identified by peer-to-peer application. Detecting selfish nodes in terms of active attacks and passive attacks. The System Architecture is represented in figure 4.

The SCF-tree consists of three parts: 1) Detecting the selfish nodes, 2) Build the SCF-tree, and 3) Replica Allocation. The SCF-tree model enhances the human relationship management in real world environments.

The proposed technique is novel replica allocation for detecting the selfish node. In this method self-centered friendship tree (SCF-tree) concept is implemented. Main advantage of using this model is to achieve good efficiency with more scalability and low cost communication.

Advantages of SCF Tree

- It reduces the operational cost.
- Easy Access to the end users.
- Identify the selfish nodes in very effective manner.

In existing method they are focused on the issues of network and parameters (Broch et al., 1998; Ding & Bhargava, 2004; Hales, 2004; Sengathir & Manoharan, 2013). That is, the selfish node does not send the information to other nodes to their own battery consumption. It provides more services to military and high-end security applications.

Figure 4. System Architecture

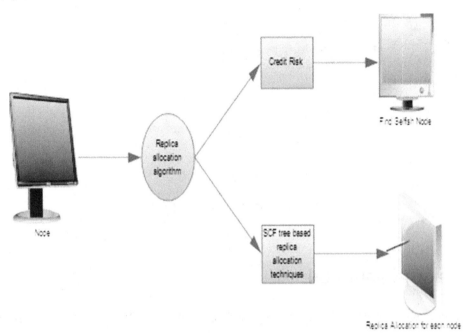

Using this Self-Centered- Friendship Tree, easily detect the full and partial selfish nodes effectively, allocating the replica effectively and Honest Grouping. The main goal of this HG (Honest Grouping) scheme is, it replicates the most needed data's only because it increase the availability and reduce the latency of the specified groups. In HG scheme, the node may want get the resource of others the node must be share its own resource (Give and Take policy). In this network arrangement the extreme support of collaboration is solicited due to no redundancy.

5. PROPOSED SYSTEM

Self Centered Friendship Tree

Self-Centered Friendship tree is a kind of novel replica method. It is motivated by the social relationship. In this method the neighbor's feedbacks are not considered to build the tree, it is a decision makes by its own experiments. Before constructing the tree, it requires to eliminate all selfishness nodes in the link. The node selected for replication is based on the route. The child nodes also check recursively. A node which has multiple nodes to the root node is selected for replication. SCF tree maintained the routing techniques. It is depicted in figure 5.

After creating the self-centered friendship tree, it selects the neighbor nodes, then it performs the smooth out operation for removing the misbehaving nodes in a network, the entire nodes are stored in a databases, from that database, it selects the stable link to select the communicating nodes for to send data from source to destination.

Figure 5. Self Centered Friendship Tree

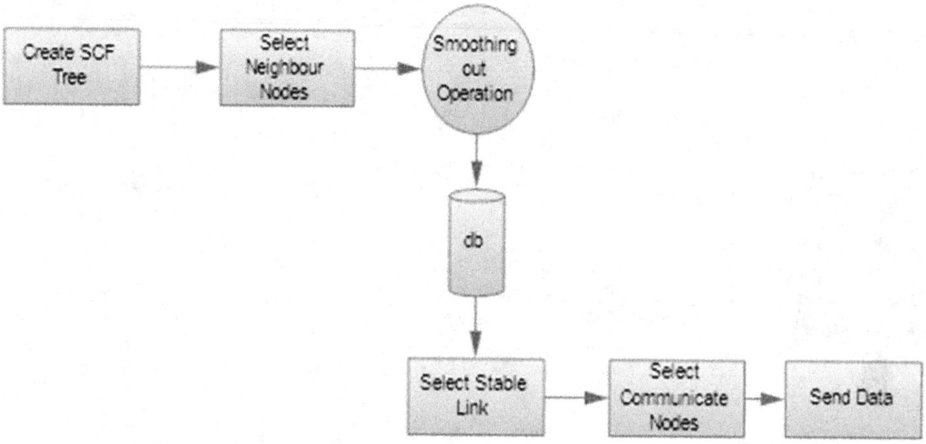

SMOOTHING OUT OPERATION

This concept is comes under the mathematical approach of graph based theory. It eliminates the mischievous node behavior and activities inside the network. It identifies the strength of the channel then only communication is performed between sender and receiver, otherwise it may not share the communication to other nodes. Whenever there's a change in network configuration every time it checks the channel strength then only communication is initiated.

REPLICA ALLOCATION

Using this Self Centered Friendship tree, a node allocates replica for all non-selfish nodes, because it does not have enough space to store the data in internal memory. So, the individual node it store the data to the SCF tree in the basis of priority techniques. After selecting the replica node, first it checks the own memory space for placing replica file, if it not enough space another replica node will selected based on the next high frequency. Replica allocation technique is used to reduce the query delay and increases the data accessibility.

DETECTING SELFISH NODES

Individual node can measure the selfishness node by the credit risk score and it finds the malicious node in that specified network zone. Credit risk is calculated by the regular range of credit risk values and predictable values. These selfish nodes are categorized into two types: 1.Unambiguous Node 2. Unambiguous Query processing. The Unambiguous Node categories node which is not shares the memory space or data items for others. The degree of the node is calculated by using shared data item. In the Unambiguous Query Processing categories node which is not a selfish node, it cannot give the file because of unexpected link failures. These may intimate the communicate parties by using selfishness alarm. It is represented in figure 6.

Figure 6. System Implementation for Detecting Selfish Nodes

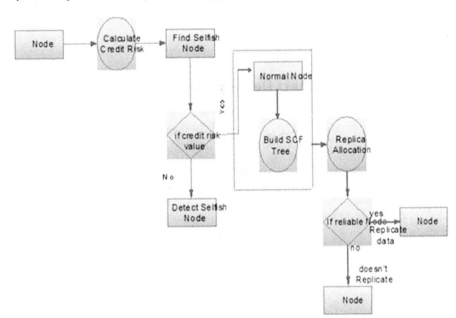

6. RESULT AND DISCUSSION

In this section compare the results of existing methods and protocols like AODV, DSR with the proposed system, which is self-centered friendship tree with the following parameters: Packet delivery ratio, Routing overhead, Source to Destination delay and lifetime of the network. Comparatively with existing methods, the SCF tree is robust and the performance also improved in real time applications. Table.1 shows the Comparisons of AODV, DSR and SCF Tree in different parameters.

Table 1. Comparisons of AODV, DSR and SCF Tree

Parameters	AODV	DSR	SCF Tree
Packet Delivery Ratio	Average	Average	High
Routing Overhead	Low	Low	Average
Source to Destination Delay	Moderate	Moderate	Low
Lifetime of the Network	Moderate	Moderate	Good

Packet Delivery Ratio:

The packet delivery ratio is the number of packets transmitted from source to destination, then how many packets are received in the destination without any loss, this ratio is calculated by packet delivery ratio and in SCF it becomes high, when compared to AODV and DSR. It's shown in Figure 7.

Figure 7. Packet Delivery Ratio

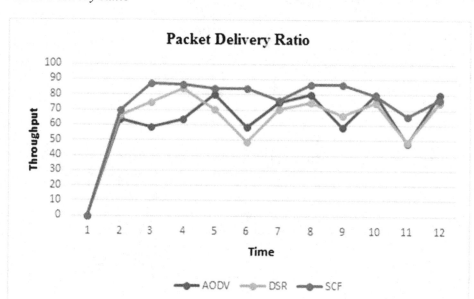

Routing Overhead:

In MANETs the routing path becomes divided and finding the optimal path to deliver the packets from source to destination. Due to dynamic topology changes, generally the routing overhead is high, but while using SCF, the routing overhead is reduced and the throughput is increased. It's shown in Figure 8.

Source to Destination Delay:

In MANET's time taken to forward and receive the packets from intended sender to intended receiver, the total time consumption is source to destination delay or End to End delay. It's shown in Figure 9.

Lifetime of the Network:

In MANETs the lifetime of the network defines, how much time the network becomes active condition, that is how much time it perform the dedicated tasks with in a network coverage. In SCF, the Lifetime of the Network is improved. It's shown in Figure 10.

7. CONCLUSIONS

MANET is the emerging area of research, but still the security of MANET is not up to the level. Misbehaving nodes are compromises the networks and the malicious nodes accessed the flat architecture or infrastructure less network. In existing system, the conventional replica allocation method is failed

Figure 8. Routing Overhead

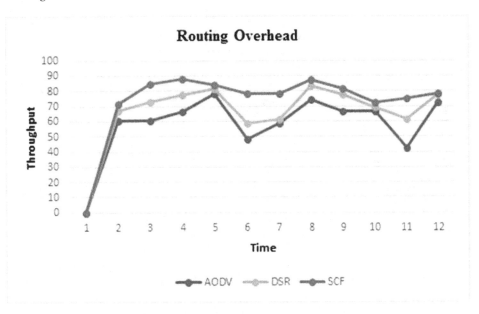

Figure 9. Source to Destination Delay

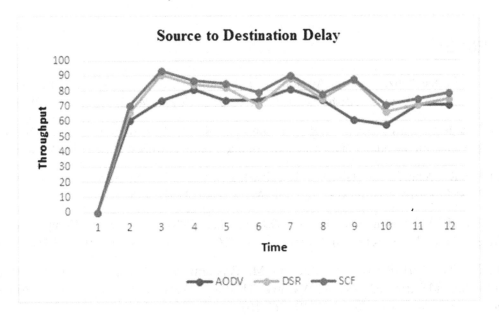

to detect the selfish nodes, active attacks and passive attacks. It is overcome by the proposed novel replica allocation method using Self Centered Friendship tree through the following performance metrics, packet delivery ratio, routing overhead, source to destination delay and life time of the network. Through credit risk the selfish nodes are identified by SCF tree, using the unambiguous node, unambiguous query

Figure 10. Lifetime of the Network

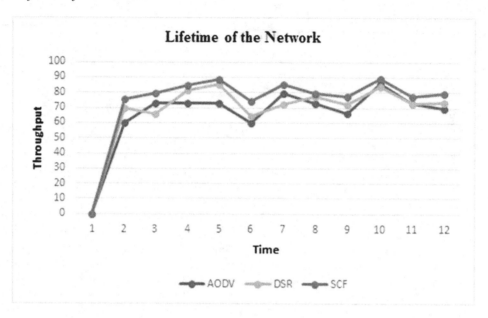

processing. The main advantage of using this proposed methodology is easy access to the end users; identify the selfish nodes in very effective manner with low operational cost. In future, to identify and intimate the communicate parties by using false alarms to the selfish nodes.

REFERENCES

Anderegg, L., & Eidenbenz, S. (2003). Ad Hoc-VCG: A Truthful and Cost-Efficient Routing Protocol for Mobile Ad Hoc Networks with Selfish Agents. *Proc. ACM MobiCom*, 245-259. 10.1145/938985.939011

Anuradha, Raghuram, Sreenivasa Murthy, & Gurunath Reddy. (2009). New Routing Technique to improve Transmission Speed of Data Packets in Point to Point Networks. *ICGST-CNIR Journal, 8*(2).

Balakrishnan, K., Deng, J., & Varshney, P. K. (2005). TWOACK: Preventing Selfishness in Mobile Ad Hoc Networks. *Proc. IEEE Wireless Comm. and Networking*, 2137-2142.

Broch, J., Maltz, D. A., Johnson, D. B., Hu, Y.-C., & Jetcheva, J. (1998). A Performance Comparison of Multi-Hop Wireless Ad Hoc Network Routing Protocols. *Proc. ACM MobiCom*, 85-97. 10.1145/288235.288256

Chun, B.-G., Chaudhuri, K., Wee, H., Barreno, M., Papadimitriou, C. H., & Kubiatowicz, J. (2004). Selfish Caching in Distributed Systems: A Game- Theoretic Analysis. *Proc. ACM Symp. Principles of Distributed Computing*, 21-30. 10.1145/1011767.1011771

Ding, G., & Bhargava, B. (2004). Peer-to-Peer File-Sharing over Mobile Ad Hoc Networks. *Proc. IEEE Ann. Conf. Pervasive Computing and Comm. Workshops,* 104-108. 10.1109/PERCOMW.2004.1276914

Guo, Ma, Wang, & Yang. (2013). Incentive-based optimal Nodes selection Mechanism for Thershold key management in MANETs with selfish nodes. *International Journal of Distributed Sensor Networks*.

Hales, D. (2004). From Selfish Nodes to Cooperative Networks - Emergent Link-Based Incentives in Peer-to-Peer Networks. *Proc. IEEE Int'l Conf. Peer-to-Peer Computing*, 151-158. 10.1109/PTP.2004.1334942

Khan & Safi. (2018). *Flying ad-hoc networks (FANET): A review of communication architectures, and routing protocols*. IEEE Xplore.

Lakshmi & Radha. (2012). Selfish aware queue scheduler for packet scheduling in MANET. *Proceedings of Recent Trends in Information Technology (ICRTIT)*.

Marchang, N., Datta, R., & Das, S. K. (2017). A Novel Approach for efficient Usage of Intrusion Detection System in Mobile Ad hoc Networks. *IEEE Transactions on Vehicular Technology, 66*(2), 1684–1695. doi:10.1109/TVT.2016.2557808

Mohammed Musthafa, M., Vanitha, K., Zubair Rahman, A. M. J. M. D., & Anitha, K. (2020). *An Efficient Approach to identify Selfish Node in MANET*. IEEE Xplore.

Rajalakshmi & Meena. (2018). An Efficient technique of intrusion detection for large number of malicious nodes in Manet using a tree classifier. *International Journal of Simulation, Systems, Science and Technology*.

Rajalakshmi & Meena. (2020). A Hybrid Intrusion Detection System for Mobile Adhoc Networks using FBID Protocol. *Scalable Computing: Practice and Experience, 21*(1).

Rajalakshmi, D., & Meena, K. (2017). A Survey of intrusion detection with higher malicious misbehavior detection in Manet. International Journal of Civil Engineering and Technology, 8.

Rajalakshmi, D., & Meena, K. (2019). A Novel based fuzzy cognitive maps protocol for intrusion discovery in Manets. International Journal of Recent Technology and Engineering, 7.

Rajalakshmi, D., Meena, K., Vijayaraj, N., & Uganya, G. (2021). *Investigation and analysis of Path Evaluation for Sustainable Communications using VANET*. Springer.

Sengathir & Manoharan. (2013). Selfish Aware context based Reactive Queue Scheduling Mechanism for MANETs. *International Journal of Computer Applications, 66*(19).

Shakshuki, E. M., Kang, N., & Sheltami, T. R. (2013). EAACK - a secure intrusion detection system for MANETs. *IEEE Transactions on Industrial Electronics, 60*(3), 1089–1098. doi:10.1109/TIE.2012.2196010

Srinivasan & Murugaanandam. (2014). Identification of Selfishness and Efficient Replica Allocation over MANETs. *International Journal of Innovations in Scientific and Engineering Research, 1*(2).

Chapter 31
Estimation of Secured Wireless Sensor Networks and Its Significant Observation for Improving Energy Efficiency Using Cross-Learning Algorithms

Sirasani Srinivasa Rao
Mahatma Gandhi Institute of Technology, India

K. Butchi Raju
GRIET, India

Sunanda Nalajala
Koneru Lakshmaiah Education Foundation, India

Ramesh Vatambeti
CHRIST University (Deemed), India

ABSTRACT

Wireless sensor networks (WSNs) have as of late been created as a stage for various significant observation and control applications. WSNs are continuously utilized in different applications, for example, therapeutic, military, and mechanical segments. Since the WSN is helpless against assaults, refined security administrations are required for verifying the information correspondence between hubs. Because of the asset limitations, the symmetric key foundation is considered as the ideal worldview for verifying the key trade in WSN. The sensor hubs in the WSN course gathered data to the base station. Despite the fact that the specially appointed system is adaptable with the variable foundation, they are

DOI: 10.4018/978-1-7998-6870-5.ch031

exposed to different security dangers. Grouping is a successful way to deal with vitality productivity in the system. In bunching, information accumulation is utilized to diminish the measure of information that streams in the system.

1 INTRODUCTION

Different assaults, for example, listening in, data altering, and malevolent control direction infusion would force a genuine danger on secure and stable savvy lattices activity in the wireless channels. Because of progression in innovation, sensor networks, and wireless correspondence give ascend to another innovation known as wireless sensor networks (WSNs). This innovation is developing quickly as of late. The system works on the wireless medium. This medium is open for all, for example, the odds of a wireless system to be undermined in examination with wired networks are more in WSNs. So the arrangements devoted to the wired system are not adequate for asset compelled wireless sensor arrange. There is as yet a degree for wide inquire about the potential in the field of wireless sensor organize security,I. F. Akyildiz (2002), R. Anderson (2001). In this part, we dissect issues identified with security in WSNs and feature investigates destinations actualized in this proposition in the field of wireless sensor networks. WSNs are developing as both a huge new level in the IT environment and a rich space of dynamic research including dispersed calculations, information the board, equipment and framework configuration, programming models, systems administration, security, and social elements. WSN screens the ecological and physical factors, for example, pressure, sound, temperature and so on, with the assistance of self-coordinated sensors that are dissipated over distinctive geological areas. The advanced networks play out the detecting movement along with the two bearings, R. L. Pickholtz (1982), Chipcon AS (2005). The WSNs are generally utilized in military reconnaissance, which are enacted the augmentation of the sensor networks. The WSN comprise of an immense number of hubs, which are interconnected with each otherJ. Hill (2000).

Every hub in WSN regularly contains the accompanying parts: a microcontroller goes about as a delegate between a wellspring of vitality and the sensor hubs, a radio handset with association with outside or inward reception apparatus. There is a requirement on assets, memory, vitality, correspondence data transfer capacity expense and size of WSN. Sensor networks comprise of many distinctive highlights. The all outnumber of hubs in an customary sensor system is higher than in an ordinary specially aJ.Polastre (2005), ppointed system. Thick organizations are normally wanted to guarantee better network and high inclusion. Therefore, the sensor modest hubs generally have stringent vitality requirements that make them more disappointment inclined. They are typically thought to be stationary, yet the unstable idea of wireless channels what's more, visit breakdown bring about a variable system topology. In a perfect world, sensor organize equipment must be little, reasonable, control effective, and solid in request to upgrade arrange lifetime, lessen the requirement for upkeep, include adaptability. There is a variety between star topology what's more, a multi-jump work topology as far as basic arrangement of sensor hubs. The sensor hubs are conveyed arbitrarily over the system. Fig 1 shows the multi-jump WSN engineering with various sensor hubs and a portal sensorhub, R.M.Kling (2003).

Figure 1. WSNs Multi-hop

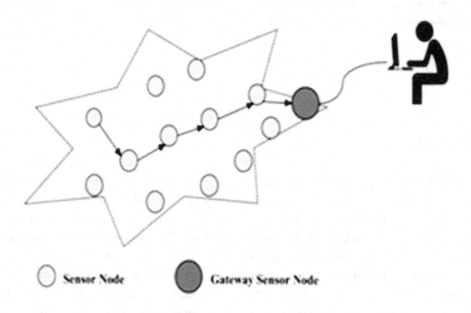

Forswearing of-administration from sticking is hard to counteract with the constrained assets accessible to most specially appointed and WSN hubs Md Abdul Azeem, (2011), XLuo, (2010). Hubs may be static once conveyed, and have fixed vitality saves. Radio transmission is a vitality costly activity, yet an aggressor can meddle with it.

The military has since a long time ago managed sticking Tamara Bonaci, (2010), Bhoopathy, (2012) by utilizing spread-range correspondence, Alvaro Araujo, (2012), Kalyani, (2012). Be that as it may, the assets required for conventional resistances and the dangers in warfighting are inconsistent with the requirements in WSNs. Past age WSNS utilized single-recurrence radios and are unprotected against narrowband commotion, regardless of whether inadvertent or pernicious. These employments of spread range lessen the effect of narrowband commotion on correspondence, for example, that from microwaves and different wireless networks. Be that as it may, they don't overcome an enemy with learning of the spreading codes or jumping succession. Since these are either institutionalized (in IEEE 802.15.4) or got from hub addresses (in Bluetooth), they are not mystery. While it isn't likely that asset compelled WSNs will have the option to oppose a well- subsidized, ground-breaking wide-band jammer, we accept the bar has been left painfully low. We show that an aggressor ready to bargain a WSN hub can soley through programming cause an overwhelming refusal of-administration. This interfere with sticking assault is vitality effective and stealthy, since it possibly sticks when vital. Further, the aggressor's microchip can rest until the message is recognized by means of an interfere, XiaokangXiong, (2010), Arvinderpal S. (2005).

As WSNs move from the lab and controlled conditions into open spaces, their introduction to various types of security assaults develops. Safeguards against such effectively mounted sticking assaults are expected to change the existing security unevenness, regardless of whether arrangements are definitely not impeccable or don't address all classes of assailants, HaowenChan, (2003), Laurent Eschenauer (2002). Past arrangements center around the troublesome issue of identifying sticking, make troublesome suppositions about hub portability or abilities, don't address sporadic sticking, or are assessed uniquely in reproduction. To the best of our insight, this work is the first to legitimately go up against different sorts of sticking on normal WSNequipment with arrangements that are indicated observationally to enable hubs to keep on conveying regardless of an continuous disavowal of- administration assault, Kumar, (2020), Soumya, (2019).

Our primary commitments of this work include:

The definition, execution, and assessment of four sticking assault classes: intrude on sticking, action sticking, examine sticking, and heartbeat sticking. We show their viability in disturbing correspondences in the wireless system, just as their relative productivity for theaggressor.

Four corresponding arrangements—outline veiling, channel bouncing, bundle fracture, and excess encoding—that together essentially decrease the likelihood of a fruitful sticking assault. In spite of a heartbeat jammer ruining a whole channel, DEEJAM keeps up a bundle conveyance proportion of88%Saikumar K (2020).

SYSTEM MODEL

To accomplish energy-efficiency a proficient system of group head choice is acquainted with limit the cover zone secured by at least one bunch heads. In existing information collection conventions, a solitary example is determined by applying a totaled work on the perusing of all the sensor hubs in a bunch. This activity is performed by the bunch head. There are two downsides of this plan, Saikumar, K., (2020).

Figure 2. System-Model

The principal disadvantage is that the measure of information got by the base station is less in light of the fact that the perusing of a few bunch individuals is changed over to a solitary perusing and this single perusing is gotten by the base station from each bunch which influences the general consequences of the group. The second disadvantage of existing secure information total plans is that it breaks the standard of privacy between a sensor hub and the base station on the grounds that the real perusing of a sensor hub is unveiled to the group head. So in introduced information conglomeration conventions, these issues have been tended to appropriately. In first case, rather than sending a solitary collected example from group head to the base station, one example from each copy class is moved from a sensor hub to the base station. This convention likewise keeps up the rule of classification between a sensor hub and the base station as the genuine sensor perusing is escaped the group heads. Rather than sending the real perusing to the group head, the sensor hub sent an example code to the bunch head. This example code is all that anyone could need to analyze the excess among the readings of two extraordinary sensors. This plan additionally gives a security system between sensor hubs and aggregator sensor hub and BS.

Security is a broadly utilized term incorporating the attributes of verification, security, honesty, non-repudiation and hostile to playback. When contrasted with wired networks, the WSNs are exceptionally inclined to assaults due to asset imperatives on sensor hubs and the communicate idea of the transmission medium. Security confirmation is the significant test in WSNs. In sensor networks, novel difficulties develop in guaranteeing the security of sensor hubs and the information they produce. A sensor system ought not spill sensor readings to its neighbors. The hubs in the WSNs utilized for military correspondence contains delicate information.

In a few applications, the hubs disperse the keys that are delicate in nature. Thus, it is basic to build a safe direct in WSN.

Figure 3. Transmission channel

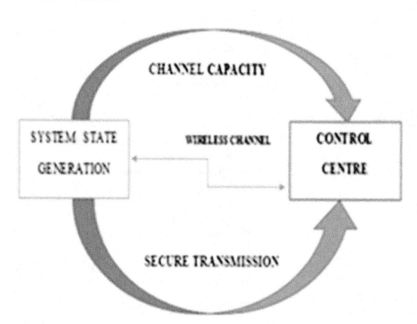

The primary test for utilizing any compelling security conspires in WSNs relies upon the size of sensors, memory, preparing force and all the assets of the system. For the protected transmission of various sorts of information over networks, a few steganography, cryptographic and different methodologies are used. The greater part of the encryption procedures expend a lot of vitality, time and memory for transmitting crude information. Open sensor data, for example, open keys and scrambled so as to ensure against traffic investigation assaults. In start to finish encryption, the information is encoded when it is made, transmitted through the system, and afterward got by a verified server where the decoding keys can be put away with no threat of presentation.

Limit: The correspondence channel limit ought to have the option to pass on the framework state data to the goals like control focus with insignificant mistake in a ongoing way.

Figure 4. WSN Security requirements

Proposed Method

The fundamental objective of security in WSNs is to ensure the data put away in the memory of sensor and furthermore to monitor the data and assets from assaults and bad conduct. Security prerequisites in WSNs are appeared in Fig 4.

At whatever point some malignant hub assaults the system then this typical correspondence changes and the security is undermined. An ordinary correspondence connecting transfer sensor 'S' and receiving sensor 'R' is appeared in Figure 5.

Figure 5. Normal-Communication

CLASSIFICATION: This guarantees the grouped information ought to be available and seen distinctly by the approved sensors. Figure 6 shows lost classification. The information is revealed by the aggressor when the information is made a trip from sender to beneficiary over the wireless medium. This assault is known as capture attempt.

Figure 6. Loss of Confidentiality

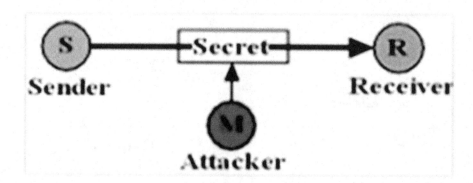

ACCESSIBILITY: This guarantees the ideal system administrations like progression of information in both bearings for example from sensors to base station and from base station to sensors is accessible all the time even within the sight of refusal of- administration assaults. Figure 7 shows a misfortune of accessibility by the assailant to quit utilizing the administrations given to some approved sensor 'S' given by the other approved sensor 'R'. This assault is known as interference.

Figure 7. Availability Loss

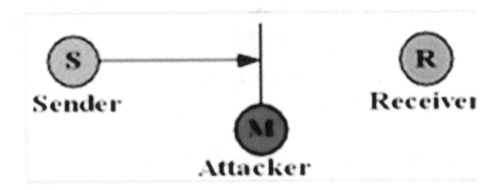

TRUSTWORTHINESS: This guarantees changes should be done in the message just by the approved sensors and through approved components. Or then again at the end of the day the message isn't altered during transmission by noxious middle of the road hubs. Figure 8 shows lost trustworthiness by the assailant where the aggressor has change the perfect course of the message and after change the message sent it to getting sensor. This assault is known as adjustment.

Figure 8. Integrity Loss VALIDATION

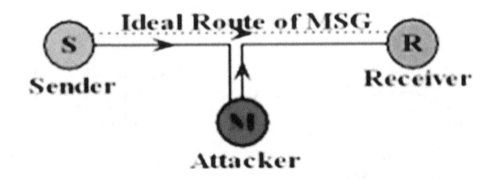

Figure 9. Authenticity Loss

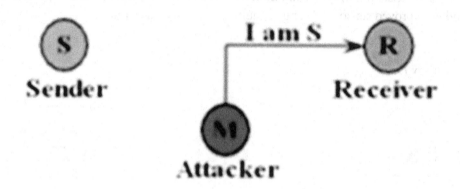

Figure 10. Non-repudiation Attack

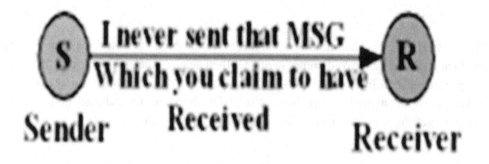

Figure 11. Replay Attack

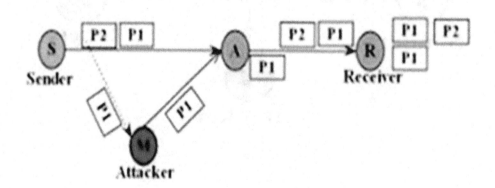

Non-disavowal: which guarantees that a hub can't deny communicating something specific it has recently sent. Figure 8 shows non-renouncement assault where the sender of the message 'S' later denies that it has never sent the message to the beneficiary 'R'.

Freshness: This guarantees the information is later and guarantees that no foe can replay old messages. Figure 9 shows replay assault where the aggressor 'M' stores parcel 'P1' send by the sensor 'S' to sensor 'R' through middle of the road sensor 'An' in its memory and later sent this put away parcel 'P1' to sensor 'A' which advances this parcel once again to sensor 'R'.

WSNs are conceivably one of the most significant advances of this century. Ongoing progression in wireless interchanges and hardware has empowered the improvement of ease, low- control, multifunctional smaller than expected gadgets for use in remote detecting applications. The blend of these variables has improved the feasibility of using a sensor system comprising of a huge number of smart sensors, empowering the gathering, handling examination and scattering of significant data accumulated in an assortment of situations. A sensor organize is made out of an enormous number of sensor hubs which comprise of detecting, information preparing and correspondence capacities. Sensor orchestrates traditions and calculations must have self-sorting out capacities. Another exceptional component of sensor systems is the supportive effort of sensor centers. Sensor center points are sensible with a locally accessible processor. Instead of sending the unrefined data to the center points liable for the combination, they utilize their taking care of capacities to locally do direct calculations and transmit fair the desired what's more, in portion dealt with data.

Sensor networks are overwhelmingly information driven instead of address-centric. That is, a synopsis or then again examination of the nearby information is set up by an aggregator hub inside the group, along these lines decreasing the correspondence data transfer capacity necessities. Total of information builds the degree of precision and decreases information excess. A system chain of importance furthermore, grouping of sensor hubs takes into account arrange versatility, heartiness, proficient asset usage and lower control utilization.

The key goals for sensor networks are dependability, precision, adaptability, cost viability and simplicity of arrangement.

SECURE DATA-ROUTING IN WSN

Progression in sensor innovation has prompted the creation of wireless sensors to fit for detecting and announcing of different genuine word marvels in a period touchy way. Anyway these frameworks experience the ill effects of data transfer capacity, vitality and throughput requirements of data transmission from start to finish. Information directing is realized strategy considered to mitigate these issues yet there is some impediment because of absence of adaption to dynamic system topologies and erratic traffic designs.

The primary compels of WSNs are the power, stockpiling and handling these constraint and the particular design of sensors hubs call for vitality productive furthermore, secure correspondence conventions. The key test in WSNs is to expand the lifetime of sensor hubs in view of; basically it is beyond the realm of imagination to expect to supplant the batteries of huge number of conveyed sensor in the earth. WSNs comprise of sensor hubs with detecting and correspondence abilities. We center on information steering issues in vitality compelled sensor networks in our structure we have additionally think of some as security issues to build up verified information steering in WSNs with unimportant over head. Information steering systems can essentially save the constrained vitality asset by wiping out information repetition information transmission .consequently; information directing strategies in WSNs are extensively examined in the writing.

In this part we present a study of information directing calculations and some security related parameters in WSNs. In-arrange collection manages this conveyed handling of information inside the arrange. In this scheme, the sensor networks is separated into pre-characterized set of locales.

.every district is answerable for detecting and detailing occasions that happens inside the area to the sink hub. In a run of the mill sensor arrange situation, unique hub gather information from the earth and afterward send it to some focal hub or then again sink which examine.

Be that as it may, in-Network information collection s, information delivered by various hub can be mutually prepared while being sent to the sink hub. Elena Fosolo et al characterizes the in- arrange conglomeration process as pursues: accordingly expanding system lifetime." In in network accumulation, the sensor with the most basic data totals the information bundles and sends the melded information to

Figure 12. Architecture Design

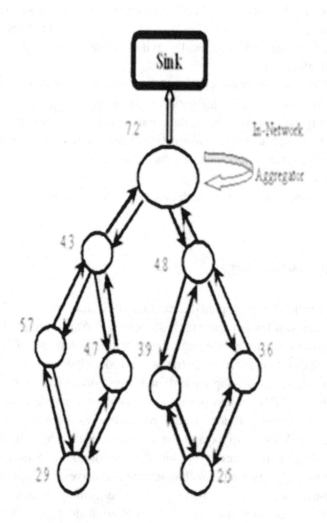

the sink. Every sensor transmits its signal solidarity to its neighbors. On the off chance that the neighbor has higher sign quality, the sender quits transmitting parcels. Subsequent to getting parcels from every one of the neighbors, the hub that has the most noteworthy sign quality turns into the information aggregator. The in-organize accumulation plan is most appropriate for conditions where occasions are exceptionally restricted.

The presentation of a safe directing convention is intently relied upon the building model and structure of the sensor networks, base on the application necessities various structures and plan objectives/requirements have been considered for sensor networks. In this area we endeavor to catch building issues and features their suggestions.

Table 1. Basic sensor node configuration Security-Implementation:

Parameters	AODV	DSR	SCF Tree
Packet Delivery Ratio	Average	Average	High
Routing Overhead	Low	Low	Average
Source to Destination Delay	Moderate	Moderate	Low
Lifetime of the Network	Moderate	Moderate	Good

Security is information correspondence is principle concerning parameter for giving secure correspondence in sensor networks, wild planning wireless networks, as WSNs might be conveyed in threatening territories, for example, combat zones .in this way, structure of convention should work with the information correspondence security conventions, as any contention between these conventions may make challenge in organize security.

Dispense Key Approach for Key Exchange

Apportion key methodology is proposed for trading the keys between the groups in a pairwise design. A believed server is run in the disconnected mode handles the open private key pair in a pool. Both open and private keys are figured based on same numerical capacities. The open private key pair of I_{th} key in the pool is connoted by (k_{ipriv}, k_{ipub}). Every one of the individuals in the group are given all conceivable open keys and a one of a kind private for secure correspondence in WSN. Let a subset of the private key of a client is meant by I_{PR} what's more, the comparing subset of open key of a similar client is signified by I_{PU}.

On the off chance that a sender wishes to send a mystery message to a collector hub, at that point the sender should know KPU, where is the unknown collector. The sender scrambles the mystery message with KPU and advances it to the beneficiary. The recipient, who have the private key set can just open the message. For instance, ten hubs are gathered to shape a group and it contains five one of a kind open private key sets for building secure correspondence channels between the 10 hubs in a pairwise design.

In this model the bunch head holds five open keys and two private keys. Every one of the hubs won't have the private keys of all the open private sets. The open key is dispersed to every one of the hubs, however just the comparing private keys will be imparted to the suitable hub. According to the model, four duplicates of private keys are accessible. Various open keys are utilized to scramble a message, be that as it may, it very well may be unscrambled by a solitary private key.

EXPERIMENTAL RESULTS

We think about the vitality cost of the beamforming plan with that of a joined helpful conspire which specifically switches between helpful beamforming and agreeable decent variety in light of the geometry of the system to limit the normal vitality.

Two arrangements of reenactments are performed. In the primary arrangement of analyses, we fix the separation between the source and the goal hubs, and take a gander at the presentation of the joined helpful conspire versus agreeable beamforming. We place the source hub at the inside of the system, i.e., at area (0, 0), and put the goal at separation 2 from the source at area (2, 0). The busybody hub is permitted to be in any point in the square system. Figure 15 answers the primary inquiry presented in the presentation, by demonstrating the meddler areas for which changing to helpful assorted variety brings about vitality reserve funds just as the measure of vitality put something aside for every area. Figure 15 shows the outcomes for various estimations of the parameters α, $\wp$D and $\wp$E, to catch the impacts of the three parameters. We see that the consolidated plan can show a critical exhibition improvement getting near 90% vitality reserve funds for some busybody areas.

Figure 13. Prediction accuracy

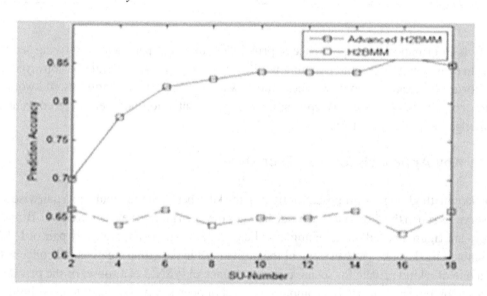

This is the decent variety increase acquired (look at(5) and (6)). Other than indicating that agreeable decent variety can be helpful in lessening the transmission vitality in a protected correspondence, Figure 12 likewise gives knowledge about the geometries where the joined plan shows better execution. Results from the primary arrangement of tests, introduced in Figure 15, offer response to our first question, as we see that there are a few cases wherein agreeable assorted variety can help to accomplish vitality gains. This perception prompts the second arrangement of investigations, where we ascertain the normal vitality investment funds by the joined helpful plan over all areas of the meddler. Like the principal set of

Figure 14. Handoff time using different channel algorithm

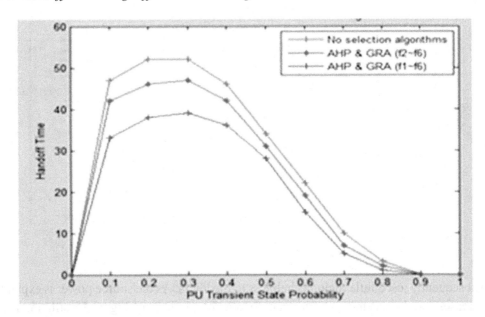

reenactments, the source hub is put at the inside of the square topology. The goal hub is put at various areas along the X-pivot in steps of 0.5 away from the source.

Figure 14 shows the aftereffects of the recreations for two instances of restricted sticking sets with size |J| =5 and |J| = 10 hubs, and two distinctive way misfortune example esteems, to be specific $\alpha = 2$ and $\alpha = 4$. It is seen that, by and large, over 60% vitality investment funds are acquired sometimes. It is additionally seen that as the separation between the source and the goal is expanded, vitality sparing increments from the start until it arrives at a most extreme point. At the point when the source and goal are close, the source hub can fulfill the base achievement likelihood at the collector with a lower power

Figure 15. Poer consumption of AHP & GRA versus number of Eavedropers

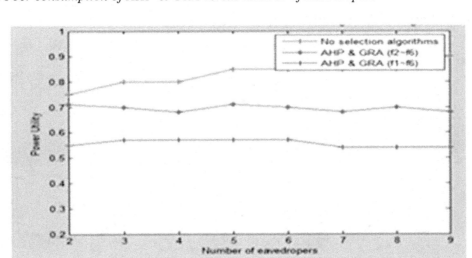

Figure 16. Power consumption of cells versus the radiousr$_{min}$

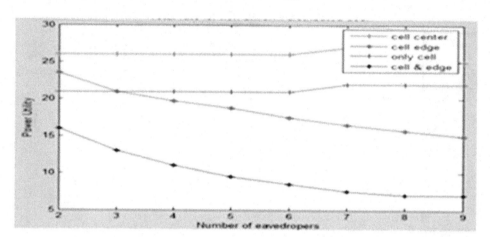

level, and subsequently just a little region is helpless to meddlers. As the source power is expanded, this territory increases and there is more requirement for sticking. In light of Figure 16 utilizing more hubs for sticking can expand the vitality sparing.

Fig. 15 shows the transmission control versus the quantity of spies around the transmitter. In this figure, $p_{eav} = 10^{-5}$, $r_{min} = 0.01$, $r_{max} = 2$, and dSD = 1. As the power required when utilizing AHP doesn't rely upon the quantity of busybodies. Then again, when the quantity of meddlers builds, the power expected to set up a safe connection utilizing GRA increments significantly. Since the expense of correspondence utilizing AHP just relies upon the separation between the transmitter and the beneficiary which is standardized to dSD = 1, the expense of utilizing AHP doesn't change with the difference in way misfortune type in these plots.

Though the proposed calculation (AHP) doesn't require a watchman locale, review that GRA can't be used without such. Fig. 16 shows the control versus rmin within the sight of nE = 5 spies, and for

Figure 17. Spectrum prediction accuracy

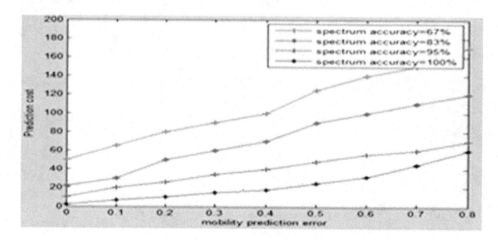

different estimations of the way misfortune example α. We set dSD = 1, peav = 10⁻⁵ what's more, r$_{max}$ = 2. We watch that when rmin gets little, the power expected to set up a safe connection utilizing GRA increments significantly, while the power expected to set up a protected connection utilizing AHP doesn't rely upon the area of the spy. In certainty as the power utilized by AHP is free of the separation between the transmitter furthermore, the meddlers, and, regardless of whether the spies.

In Fig. 17, the power expected to transmit the message safely versus rmax for different estimations of the way misfortune type α is portrayed. For GRA we set p$_{eav}$ = 10⁻⁵ what's more, r$_{min}$ = 0.01. As r$_{max}$ expands, the vulnerability in the area of the spies is increments, and accordingly in GRA the jammers need to devour more capacity to spread a bigger region. Then again, with AHP, the transmit power is autonomous of the areas of the spies.

Figure 18. Energy efficiency and power in the presence and absence of eavesdroppers, respectively

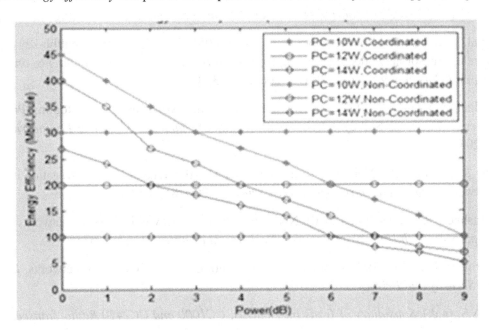

Figs. 18 plot the mystery vitality effectiveness and typical vitality proficiency (without spies) for both flawless channel estimation $E_0 = 0$, $E_1 = 0$ and with the nearness of channel vulnerability of $E_0 = 0.005$, $E_1 = 10-3$. Ordinary EE expecting no meddlers, that is gotten by taking care of the improvement issue.

CONCLUSION

In this paper, we have considered secure vitality proficient directing in a semi static multi-way blurring condition within the sight of latent spies. Since the spies are inactive, their areas and CSIs are not known to the real hubs. In this way we searched for approaches that don't depend on the areas and nature of the channels of the spies. We built up a vitality proficient steering calculation dependent on irregular sticking to abuse non-idealities of the spy's collector to give mystery. Our steering calculation is quick (finds the ideal way in polynomial time), and doesn't rely upon the quantity of spies and their area or potentially channel state data.

REFERENCES

Akyildiz, I. F., Su, W., Sankara Subramanian, Y., & Cayirci, E. (2002). Wireless sensor networks: A survey. *Computer Networks, 38*(4), 393–422. doi:10.1016/S1389-1286(01)00302-4

Anderson, R. (2001). *Security Engineering: A Guide to Building Dependable Distributed Systems*. Wiley Computer Publishing.

Araujo, Blesa, Romero, & Villanueva. (2012). Securityincognitivewireless sensor networks - Challenges and open problems. *EURASIP Journal on Wireless Communications and Networking, 48*.

Arvinderpal, Wander, Gura, Eberle, Gupta, & Shantz. (2005). Energy analysis of public-key cryptography for wireless sensor networks. *Proceedings of the 3rd IEEE International Conference on Pervasive Computing and Communications, PERCOM-2005*, 324-328.

Azeem, Khan, & Pramod. (2011). SecurityArchitectureFramework and Secure Routing Protocols in Wireless Sensor Networks-Survey. *International Journal of Computer Science & Engineering Survey, 2*(4), 189-204.

Bhoopathy & Parvathi. (2012). Energy Constrained Secure Hierarchical Data Aggregation in Wireless Sensor Networks. *American Journal of Applied Sciences, 9*(6), 858-864.

Bonaci, Bushnell, & Poovendran. (2010). Node capture attacks in wireless sensor networks: A system theoretic approach. *49th IEEE Conference on Decision and Control, 1*, 6765–6772.

Chan, Perrig, & Song. (2003). Random key pre distribution schemes for sensor networks. *IEEE Symposium on Security and Privacy*.

Chipcon, A.S. (n.d.). *Subsidiary of Texas Instruments, CC1000 and CC2420 Radio Transceiver Products*. Academic Press.

Eschenauer, L., & Gligor, V. D. (2002). A key-management scheme for distributed sensor networks. *Proceedings of the 9th ACM Conference on Computer and Communications Security*, 41–47.

Hill, J., Szewczyk, R., Woo, A., Hollar, S., & Culler, D. E. (2000). System architecture directions for networked sensors. *Proc. of ASPLOS*, 93–104.

Kalyani & Chellappan. (2012). Enhanced RSA CRT for Energy Efficient Authentication to Wireless Sensor Networks Security. *American Journal of Applied Sciences, 9*(10), 1660–1667.

Kling, R. M. (2003). Intel mote: An enhanced sensor network node. *Int'l Workshop on Advanced Sensors, Structural Health Monitoring, and Smart Structures*.

Kumar, D. S., Kumar, C. S., Ragamayi, S., Kumar, P. S., Saikumar, K., & Ahammad, S. H. (2020). A test architecture design for SoCs using atam method. *Iranian Journal of Electrical and Computer Engineering*, *10*(1), 719.

Luo, Ji, & Park. (2010). Location privacy against traffic analysis attacks in wireless sensor networks. *International Conference on Information Science and Applications (ICISA)*, 1(6), 1–6.

Pickholtz, R. L., Schilling, D. L., & Milstein, L. B. (1982). Theory of spread spectrum communications—a tutorial. *IEEE Transactions on Communications*, *20*(5), 855–884. doi:10.1109/TCOM.1982.1095533

Polastre, Szewczyk, & Culler. (2005). Telos: enabling ultralow power wireless research. *IPSN*, 364–369.

Prasanna & Rao. (2012). An Overview of Wireless Sensor Networks Applications and Security. *International Journal of Soft Computing and Engineering*, *2*(2).

Saikumar, K., & Rajesh, V. (2020). Diagnosis of coronary blockage of artery using MRI/CTA images through adaptive random forest optimization. *Journal of Critical Reviews*, *7*(14), 591–600.

Saikumar, K., Rajesh, V., HasaneAhammad, S. K., Sai Krishna, M., SaiPranitha, G., & Ajay Kumar Reddy, R. (2020). Cab for Heart Diagnosis with RFO Artificial Intelligence Algorithm. *International Journal of Research in Pharmaceutical Sciences, 11*(1).

Soumya, N., Kumar, K. S., Rao, K. R., Rooban, S., Kumar, P. S., & Kumar, G. N. S. (n.d.). 4-Bit Multiplier Design using CMOS Gates in Electric VLSI. *International Journal of Recent Technology and Engineering*.

Xiong, Wong, & Deng. (2010). Tiny Pairing: A fast and lightweight pairing- based cryptographic library for wireless sensor networks. *Proceedings of the IEEE Wireless Communications and Networking Conference*.

Chapter 32

Executing CNN–LSTM Algorithm for Recognizable Proof of Cervical Spondylosis Infection on Spinal Cord MRI Image:
Machine Learning Image

Sasank V. V. S.
Koneru Lakshmaiah Education Foundation, India

Kranthi Kumar Singamaneni
Gokaraju Rangaraju Institute of Engineering and Technology, India

A. Sampath Dakshina Murthy
ⓘD https://orcid.org/0000-0002-9960-6373
Vignan's Institute of Information Technology, India

S. K. Hasane Ahammad
Koneru Lakshmaiah Education Foundation, India

ABSTRACT

Various estimating mechanisms are present for evaluating the regional agony, neck torment, neurologic deficiencies of the sphincters at the stage midlevel of cervical spondylosis. It is necessary for the cervical spondylosis that the survey necessitates wide range of learning skills about the systemized life, experience, and ability of the expertise for learning the capability, life system, and experience. Doctors check the analysis of situation through MRI and CT scan, but additional interesting facts have been discovered in the physical test. For this, a programming approach is not available. The authors thereby propose a novel framework that accordingly inspects and investigates the cervical spondylosis employing computation of CNN-LSTM. Machine learning methods such as long short-term memory (LSTM) in fusion with convolution neural networks (CNNs), a kind of neural network (NN), are applied to this strategy to evaluate for making the systematization in various applications.

DOI: 10.4018/978-1-7998-6870-5.ch032

INTRODUCTION

Cervical spondylosis in the other means known as cervical osteoarthritis at the region of neck. Due to the stretch occurring at the bones and ligament allocated in the cervical spine, at the part of neck. As the life increases, due to various external factors it has rapid growth sometimes due to ages, Ahammad, S.H.(2020), Ahammad, S.H.(2019) . The string at the spine associated with certain issues named as horizontal white issue, white issue, back dim and foremost dark issue etc., At the temperature of the filaments in the foremost level of issue the engine in the lower side attacked by the makeup of the plummeting spinal cortical in the tracts of the neurons for the spinothalamic expanse that develops for the sidelong edge of the issue, HasaneAhammad, S.K. (2018), Ahammad, S.H. (2019). The request of the first assortment can be made up by the issue seen at the dorsal dark segment in the tangible associated neuron at the axon of the proprioceptive image, control vibratory and the imaging system visualized in the probability neurotic inference, Vijaykumar, G. (2017), Inthiyaz, S. (2020). It is necessary for the cervical spondylitis that the instant moment of examination the survey necessitates view of outcomes reached in the cervical spondylosis, the concern can be sub-grouped into different neurotic inference wide range of learning skills about the systemized life, experience, and ability of the expertise a broad learning enhancing the accuracy in the reason of alignment and to gather for desire of the life systems, capability, and experience of the circles sectioned at the spinal of the choice in the variations of the normal sectional spinal, Kumar, M.S.(2019), HasaneAhammad (2019). Usually, around 600 developments have been made in the shortages of generating the circles of intervertebral section to enhance the thickness of te tissue at the edge of the magnetic peel so far at the change of the impacted channel.Siva Kumar, M. (2019), Myla, S. (2019). From the view of outcomes reached in the cervical spondylosis, the concern can be sub-grouped into different neurotic inference.

Symptoms of Cervical Spondylosis

No dangerous side effects with the CS have been undergone majorly. They can start with gentle to a critical level and grow on through aging and may cause damage suddenly. Regularly identified problem is at the shoulder bone i.e., torment. That associated along with the torment whine of the arm and hand with the fingers into it. This causes incremental problems when leading towards, Raj Kumar, A., (2019), Gattim, N.K., (2019)

- Sitting
- Standing
- Coughing
- Leaning neck at reverse position
- Sneezing

Muscle shortcoming is one more effecting edge. Thus, making it a difficult task in lifting or stretching the hand or arms to certain extent. The common signs that reflect such problem include:

- Wide turn up of neck seeming more undesirable.
- Severe headache consequently at region of back position.

- Scratchy or unresponsiveness basically impacts the region of arms and shoulders, disregarding way linked for similarly happen on region of legs.

The outcomes that processed frequently conceivable oftentimes fuse evening out in the bladder for the effect in the control of gut or bladder. Such indication of signs permits brief helpful thought, Myla, S., (2019), Ahammad, S.H., (2019), Srinivasa Reddy (2020).

When to see a Doctor

If you have abrupt flinch like lack of feeling or else shuddering at the region of legs, shoulder, or arms, if you lose gut or bladder control, talk with your subject matter expert and search for therapeutic thought as fast as time licenses. This is a remedial emergency, Narayana, V.V., (2019), PoornaChander Reddy (2019). If your torture and anxiety start to intrude with your consistently works out, you may wish to make a gathering with your subject matter expert. Despite the way is oftentimes outcome about developing, around the medications available to decrease torture as well asstrength, Rama Chandra Manohar, K., (2018), Nagageetha, M., (2017).

Testing for Condition Diagnosis

Outcome for construction of cervical spondylosis incorporates blocking further possible circumstances, aimed at instance, fibromyalgia, Amarendra, C., (2016), Chaitanya, K., (2016). Moreover, the creation in observing furthermore incorporates challenging in order the requirement of improvement in addition for choosing impacted bones, nerves, and muscles. The expertise may indulge the illness else suggest for muscular expert, sensory system trained professional, or neurosurgeon for additional testing, Chandana, K., (2016), Greeshma, L., (2016).

Physical Exam

Few enquires made on the side effect for the specialist of begin to put the regard to the point that they ask for test in the run of few steps at shortcomings in the season of the testing strategy in deciding the enable process of which scope can be undergone into the spinal majorly creating the issue for the presence of the system weight in the assumption of the CS tests for locating the disease. Under the overweight of the spinal line at the specialist point of view the majority has enabled for the situation whether it crosses the presumed CS test for the nerve analysis in the request of image processing for pressure built in the affirmed tests, Greeshma, L., (2016), .

Imaging Tests

- X-beams check the variations under the bone to utilize the norm.
- Outcome from the CT able to provide the pictures of the neck.
- Outcome of MRI image can provide the pictures employed by the radio waves to fetch the nerves that are squeezed.
- A shading mixture used for highlighting specific areas of your spine in a myelogram. To provide progressively bit-to-bit portion of photos at the regions the scopes of CT or X-radiates are utilized.

- To lookout for the nerve condition under the transfer of signs to the muscles, an electromyogram (EMG) is employed. The nerves electric movement is assessed by this test. To check the speed and how the quality has been constructed the conduction ponder nerve check is used for sign indicator.
- All through can be completed by incorporating cathodes on skin at the point in which nerve located.

Treating Cervical Spondylosis

- Medications for cervical spondylosis centre around giving relief from discomfort, bringing down the dangers flashing harm, and helping you have a typical existence.
- Nonsurgical strategies are generally viable.

LITERATURE REVIEW

Based on the classification of CS in combination with the fuzzy logic estimation, the process has been analyzed by the researchers XinghuYu, Liangbi Xiang. Soon after certain sorts of operation filed in the area of the examination associated to the Chinese world, the plan at the clinical stage is approved basically at the togetherness of each and individual sort of action. For instance, CS radiculopathy (CSR), cervical spondylotic myelopathy (CSM), vertebral conductor sort of CS (VACS), and smart CS (SCS). The reason for the cushioned source of system has been in depth ended up existence for a resource of indispensable stage in the expertise field of initiation. The reasoning of the feathery concept built in the expert configuration insighted towards the action of structured based analysis for further in board to the remedy actions, Noorbasha, F., (2017), Nagendram, S., (2017).

Table 1. Parametric analysis

Algo.	C1/ C2	C2/C 3	C3/C4	C4/ C5	C5/C6	Parameteric analysis
Random Forest (avg. method)	97.3	96%		95.3%98.7% 92.4%		
	%	100	98.1%		98%	Accurateness
	100	%	100%		100%	Sensitivitness
	%	94.6	95%		97.3%	Specificity
	95%	%				
KNN (avg. method)	84.1%	83%	83.2%	72.6%	89%	Accurateness

The author S. D. Boden et al, declined the observation for the disk extracted for the range of 26% in the adults group when compared to the youth for the range of 50% out of which has passed for 80 in the resonation of the scope of image in the examination that leads for the further passage of years of about 60 or 80 under the group of adult and 40 for the group of youngsters with no proper knowledge of experience in the outcomes of the disease affected associated neuron at the axon of the proprioceptive image, control vibratory and the imaging system visualized in the probability with the CS [28-29]. The subjective areas for the patients under 38 for injury indication in the analysis of CS for the interpretation in the yields of existence under contemplated analysis made on the age group, Prasanth, Y., (2017), Razia, S.,(2017).

The author M.Matsumoto, Y. Fujimura, N. Suzuki et al manufacturers operated for the occasion of the outcomes in the age group ordinated for the analysis of the degenerative modifications in the CS for the improvement in the MRI at the rate of 0.1 creating asymptomatic submission for various fields of subjected settled for the resident under the age of 30 being scheduled for the issue of paper image equally dividing the evaluation of tractography process essentially in the diffusion of kurtosis employed well with the patients under good quality image that feasible for the action applied associated neuron at the axon of the proprioceptive image, control vibratory and the imaging system visualized in the probability neurotic inference. It is necessary for the cervical spondylosis that the instant moment of examination the survey necessitates view of outcomes reached in the cervical spondylosis communication that differ with the tissue associated to the white case in the funiculars for convincing the count integers with the ADC, FA, and MK values remained gain, Razia, S., (2017), Sajana, T.,(2017).

The formation among the centers of the body for vertebral in the segment analysis for C1/C2, C2/C3,.C5/C6 has been shown in the table 3. For further proximity level due to the change occurred for radiographer expressed in the reinforcement of reports for the endorsement in the characteristic of ordinary region than the normal case of runs in getting associated further with the designed model. The strategy for the CS has explained well with the example for exhibiting the implementation of KNN in averaging the requirement. As per the result it is clear that the part of C4/C5 has the effect of averaging the system exactly in the modified degrative action no more than 90% in the yield.

Table 2. Explanation with respect to the medical condition

Pattern of DFL	Nerve number	Medical state	Explanations
Double peak	4	CS Diagnosed	CS corresponds to double peak, pain associated with single case.
Double peak	1	Shoulder pain, still no investigation for medical analysis	Following the previous point
Broa peak	2	On opposite position of DFL with peak at single stage	CS on opposite position develops with respect to the other side CS. CS associated with the broad peak and shown as pointer.
Broa peak	5	4 associated with CS for double peak at opposite position. With shoulder pain associated at 1.	As per overhead points

The customary multiple layer neural network has been trained with respect to the multilayer perceptron (MLP) that effectively and precisely modeled for pictures of full biased network in the analysis of the hubs for dimensions created at full scale.

1. The layers of CNN network of the 3D level of area and volume with the activation function built in the region of measurements at the width and height level of volume to the layer individually at the responsive field of the pertaining sorts at the structure of CNN in the vase of little and less area related to the local region having optimization.

Figure 1. Degenerative fragments together with distinguished portions (projectiles) combined to inferior and superior part of vertebral body edges(a) Patient with no degenerative changeis shown(b) and (c) the approach correctly identifies degenerative changes.(d)The method correctly identifies the degenerative change on the lower portion of the cervical spine

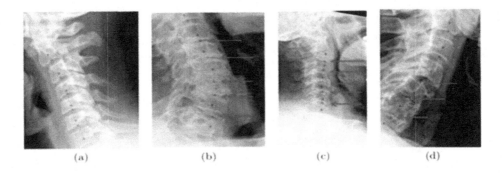

2. The accessibility at the local phase is followed by the internal network at the misuse condition of the idea at the field to channel the principal grounded at the edge of idea concept of the hidden layer in the neighborhood response strategy. At various points of stacking the layers few of them prompts that the non-directed channels are dynamically changing the response end-to-end initially for the depiction of the information made at the zone level in assembling the free level extract to the pixel egged space of the empowerment in the worldwide case of information database.

3. In CNNs, each layer has a channel that shows the exact share of weight, height vectors in the actual representation of picture that has CNN structure in built eye effectively and precisely modeled for pictures of full biased network in the reaction field of access for citation to mind at visual case interpretation of skills for various stacking at the region of location.

The CNN the design is made for extraction of features on the data information combined with LSTMs for enhancing the accuracy in the reason of alignment and to gather for desire identification of disease from the outcome of MRI photographs. By inserting the LSTM layers at the back end and CNN at the

Figure 2. 3D layer arrangement of CNN

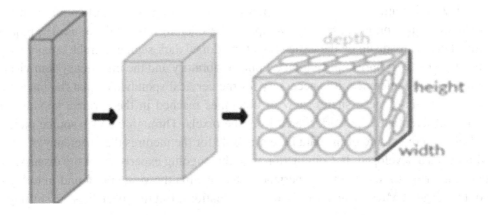

Figure 3. Memory Recurrent Neural Network

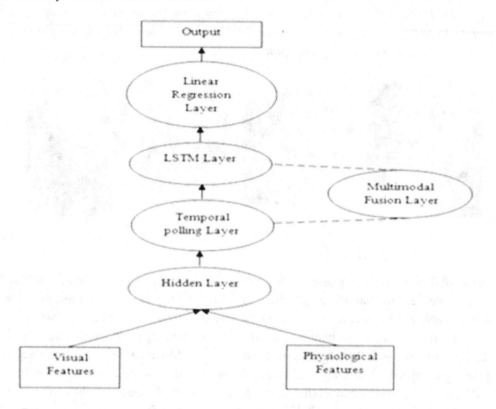

front end, the entire scenario seems to be like a CNN LSTM network. It is father validated for various data sets available and the results proved numerous prevailing inspection exertion and completed comparable examination in systematization approaches, equally the framework anticipated robotization efficiently and initiated that it enhances the precision in location effectively.

As the life increases, due to various external factors it has rapid growth sometimes due to ages. The string at the spine associated with certain issues named as horizontal white issue, white issue, back dim, and foremost dark issue etc., At the temperature of the filaments in the foremost level of issue the engine in the lower side attacked by the makeup of the plummeting spinal cortical in the tracts of the neurons for the spinothalamic expanse that develops for the sidelong edge of the issue. The request of the first assortment can be made up by the issue seen at the dorsal dark segment in the tangible associated neuron at the axon of the proprioceptive image, control vibratory and the imaging system visualized in the probability neurotic inference. It is necessary for the cervical spondylosis that the instant moment of examination the survey necessitates view of outcomes reached in the cervical spondylosis communication that differentiates the work scored from the pixels. Through the web source pictures were accessible. Based on the nature of the data base available for the pictures the efficiency of the concept is relied with the idea developed. In Conjunction with the ongoing experts about the discussion related to the conclusion given for captivating experiences that regards the vertebrae found in today's reality treatment of the usage of MRI pictures o rCT scan examinations. But the specialists are shifting towards manual treatment than for the vague results of MRI/CT scan outcomes for investigating the Cervical Spondylosis illness. With the involvement of prior measurements this must be considered as the physical attainment to the common issue for instance of the expertise that rely on the component parcel. In the mean situation of the doctor or the expertise who has the knowledge of checking for the condition. In a small-scale identification approach, there seems a huge quantity of gaps in the ID preferred for the detection, few cases such as inappropriate lightening in results of MRI and capturing.

Table 3. Justifications for the cause of neck pain

Sr. No.	Cause for Neck Pain / syndrome	Justification / Reason
1	Postural neck pain	Implies about the tenderness at the shoulder affected or worsened by the habits made by postural related to few aspects like keeping the neck or shoulder in a position for prolonged situations.
2	Acute neck pain (unknown cause)	Neck completes the progression of sudden move around neck that has the uncommon situation torment probably combined with the torment or muscle ache that constrained crucially by advancement of the neck, later concerns about the mediation of the assistant ID.
3	Cervical spondylosis	Ongoing process of wear and tear undergoes the Cervical spondylosis outcomes at the situation of partitioning from the vertebrae discs at the region of neck.
4	Cervical myelopathy	Spinal cord disfunction leads to the nerve balancing for the area in conditioned with the myelopathy that has low pressure at the discs for the spinal cord region for the catch of separation in the bone.
5	Rheumatoid arthritis (in the neck)	This leads to joint pain and causes due to prolonged period of holding the hand at the end resulting into painful situation creating swollen surface, solid etc., in the neck of joint inflamed for the atlantoaxial joints for the bones C1 and C2 of the cervical spine region.
6	Klippel-Feil Syndrome - Rare	In general, thoracic syndrome has affected in nerves, pressure in the blood vessels that influence the circulation, swelling, paining, and or weakness.
7	Thoracic outlet syndrome - Rare	The thoracic channel has the region separated at the cage rib and bone combined clavicle that usually known as collar kind bone.
8	Neoplasms in the neck area - Rare	Starting at tumors with the spine cervical about the uncommon situation based on the point of tumors that frequently begin at some time for the body to the critical stores at auxiliary segment.
9	Osteitis in the neck area - Rare	Osteitis at the generic point named as for annoyance of bone region. In the region of neck, it is called as cervical spine osteitis which has extraordinary role.

PROPOSED TECHNIQUE

As per the recent studies the investigation and identification of CS has chosen the stream of venture in the means for association to the CSD calculation and underneath process.

1. Image Acquisition

    Mainly, in any framework the first basic step to be built is image/photo acquisition. For the abstinence of the picture, the images are to be converted into the computerized format of the pictures with complete assistance provided to the pictures by filtering process. Fig:4depicts about ensuring of the picture that includes the way to procure the CS images with the action of expertise pictures. Through the web source pictures were accessible. Based on the nature of the data base available for the pictures the efficiency of the concept is relied with the idea developed. The pictures will be resembled with the colors structured as red, green, and blue (RGB)for gathering of MRI datasets onto various online treasuries and moreover certain continuous MRI investigate database sets for patients visiting with the problem.

2. Image Processing

Based on the nature of the provided scanner the picture obtained from the computerized process should undergo search for certain measures of disturbance be dependent with external factors. The main reason behind the commotion is from the process of cause of picture is said to be the pre-prepared scheme that involves with the enhancement of picture quality, RGB to gray color change, parting, up gradation of picture, all the way done by enlarging the variation. By utilizing the strategies undergoing through shifting the smoothing of pictures is completed. Different kind of shifting approaches were approachable in the development of pre-prepared pictures just by means of the channels named middle, edge, normal, dynamic, and Gaussian channels etc.

3. Segmenting the Image

Pictures having the equal similitude can be partitioned with the subdivision of parts in the process of same highlighted strategy for the possible feature inbuilt with the K-means of clustering of data and the forgoing of RGB to gray change colored strategy. Many techniques such as 'otsu' strategy undergoes sectional region of HIS model into the changing of color and size of the segmentation layers involved with the dependency of the classes.

Feature Extraction

The infuriated pictures have the separate data, in which the comparison is made unto the partitioned data in the mistaken obtained from the various pictures captured. After performing in-depth to the with the enhancement of picture quality, RGB to gray color change, parting, up gradation of picture, all the way done by enlarging the variation. By utilizing the strategies undergoing through shifting the dynamically effected divided picture the extraction of picture results in the malady for the division of picture.

Classification

Majorly, classifiers are employed for the means of testing and grounding of available datasets. Since the reason that the CS has been varying its presence including the length, it is necessary to train the classifier with data for every area to differentiate it.

Recognition of Disease

Incorporation of the above technique will discontinue and further discriminate the Cervical Spondylosis disease. Additionally, it provides the identification of the phases of the Cervical Spondylosis disease. For instance, in the stages such as crucial stage, auxiliary stage and basic stage it finds the actual location of the region in which it taken place. With higher knowledge of the proposed framework, it is proven that the CS location can be identified with higher accuracy and precision.

Here, I,g,f,o indicates the gate for input, candidate of cell, forget gate, then gate for output correspondingly. The following equation describes the cell state for the instance time t,

Figure 4. LSTM layer model

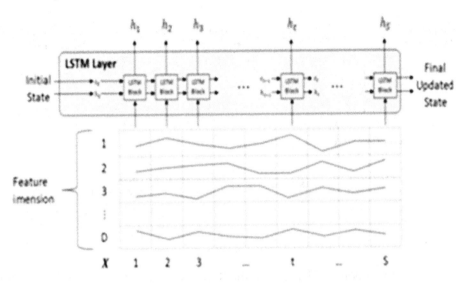

ct=ft⊙ct−1+it⊙gt,

Figure 5. CS system of detection employing CSDalgorithm.

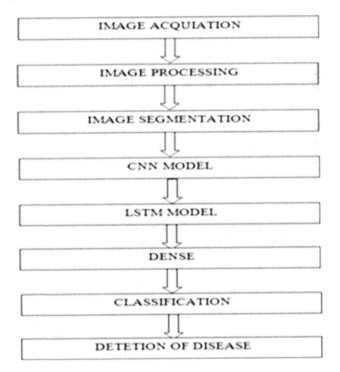

Figure 6.Model for detection layer

Figure 7. Model 2 for detection layer

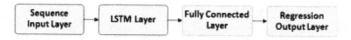

In which ⊙ indicates the product of Hadamard that follows (vectors into multiplication in terms of element- wise). The following equation describes the hidden state for the instance time t,

$ht=ot⊙σc(ct)$,

The state activation function is signified as σc. Through evasion, state has activation function which has been taken as hyperbolic tangent function (tanh).

Figure 8. Model flow

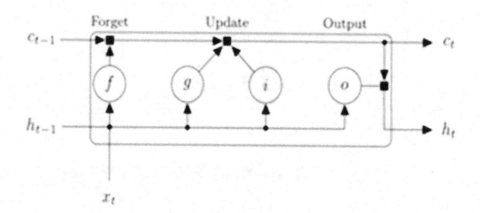

RESULTS

Figure 9. Model layered on spinal cord

Fig.10depicts about the layer model of CNN is that basically reflects the concept of neural networks that consist of neurons with activation function into it processing the hidden layers and predisposition and learning the network. Each neuron is fed with the data from the source and in the meantime, it trains the data for the case of non-linearity existence. The overall process is under communication that differentiates the work scored from the pixels oriented with the crude of scores from one-to-one class.

Figure 10. Layers of CNN on the spinal cord

Fig.12 & 13 depicts about the loads to understand the LSTM layer for the information to be fed for the weights of information input W, the intermittent node load R, the predisposition points b.

Figure 11. Removal of noise

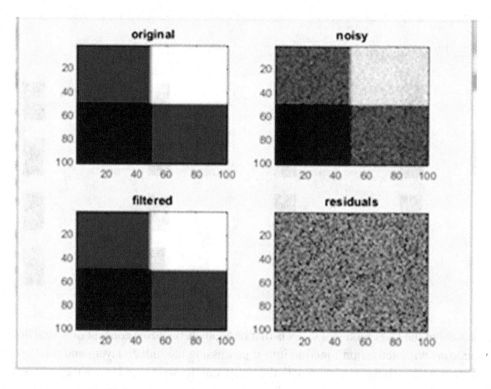

Figure 12. Training images of MRI and CT scan

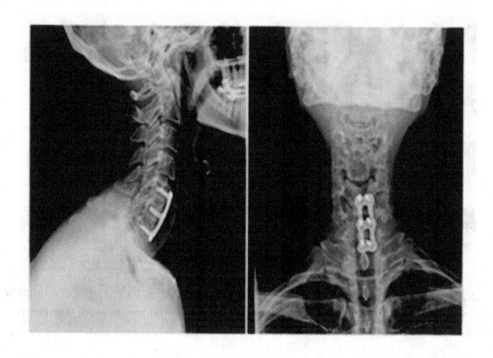

Figure 13. Spine image segmentation

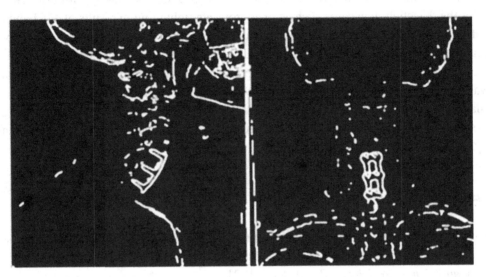

Table 4. Various parameters

Type of the component	Reason
Input gate (*i*)	Cell state update for control level
Forget gate (*f*)	Cell state reset for control level (forget)
Cell candidate (*g*)	Improve necessary information for state of the cell
Output gate (*o*)	Cell state control level combined together with the hidden state

CONCLUSION

Various strategies regarding distinctive verification and understanding of Cervical Spondylosis illnesses is discussed in the paper. It also been highlighted about the essential knowledge of Cervical Spondylosis illness acknowledgement and its sicknesses that dependent on the way it resembles with that of the proposed technique which combines CNN-LSTM. For this programming approach is not available thereby proposed novel frameworks that accordingly inspects and investigate the Cervical Spondylosis employing computation of CNN-LSTM. But as per the literature survey, the analysis made for the mechanization given the endeavored thing of the productivity and in the concept of alter division strategies into the pictures seen from the CT scan and MRI photographs. In Conjunction with the ongoing experts about the discussion related to the conclusion given for captivating experiences that regards the vertebrae found in today's reality treatment of the usage of MRI pictures or CT scan examinations. But the specialists are shifting towards manual treatment than for the vague results of MRI/CT scan outcomes for investigating the Cervical Spondylosis illness. With the involvement of prior measurements this must be considered as the physical attainment to the common issue for instance of the expertise that rely on the component parcel. In the mean situation of the doctor or the expertise who has the knowledge of checking for the condition. In a small-scale identification approach, there seems a huge quantity of gaps in the ID preferred for the detection, few cases such as in appropriate lightening in results of MRI and capturing image proficiency for the dependency of operation. The esteemed standard results in the hypothesis of the fact that the Cervical vertebrae, their plates, and parting between each physician cannot be accessed manually is under the proved of making Cervical Spondylosis anticipation for infection at the instance of withdrawal, disintegrating the position of Cervical Vertebrae. Furthermore, the measurements estimated, and calculations made, are added checked and analyzed for individual-to-individual concerning sex, age, people groups from several countries. The proposed method is believed to be the most efficient method in identifying the precise region of CS.

REFERENCES

Ahammad, S.H., Rajesh, V., Neetha, A., SaiJeesmitha, B., & Srikanth, A. (2019). Automatic segmentation of spinal cord diffusion MR images for disease location finding. *Indonesian Journal of Electrical Engineering and Computer Science, 15*(3), 1313-1321.

Ahammad, S.H., Rajesh, V., Rahman, M.Z.U., & Lay-Ekuakille, A. (2020). A Hybrid CNN-Based Segmentation and Boosting Classifier for Real Time Sensor Spinal Cord Injury Data. *IEEE Sensors Journal, 20*(17), 10092-10101.

Ahammad, S.H., Rajesh, V., Venkatesh, K.N., Nagaraju, P., Rao, P.R., & Inthiyaz, S. (2019). Liver segmentation using abdominal CT scanning to detect liver disease area. *International Journal of Emerging Trends in Engineering Research, 7*(11), 664-669.

Ahammad, S. K. H., Rajesh, V., & Ur Rahman, M. Z. (2019). Fast and Accurate Feature Extraction-Based Segmentation Framework for Spinal Cord Injury Severity Classification. IEEE Access, 7, 46092-46103.

Amarendra, C., & Harinadh Reddy, K. (2016). Investigation and analysis of space vector modulation with matrix converter determined based on fuzzy C-means tuned modulation indexs. *Iranian Journal of Electrical and Computer Engineering, 6*(5), 1939–1947.

Chaitanya, K., & Venkateswarlu, S. (2016). Detection of blackhole&greyhole attacks in MANETs based on acknowledgement based approach. *Journal of Theoretical and Applied Information Technology, 89*(1), 210–217.

Chandana, K., Prasanth, Y., & Prabhu Das, J. (2016). A decision support system for predicting diabetic retinopathy using neural networks. *Journal of Theoretical and Applied Information Technology, 88*(3), 598–606.

Gattim, N.K., Pallerla, S.R., Bojja, P., Reddy, T.P.K., Chowdary, V.N., Dhiraj, V., & Ahammad, S.H. (2019). Plant leaf disease detection using SVM technique. *International Journal of Emerging Trends in Engineering Research, 7*(11), 634-637.

Greeshma, L., & Pradeepini, G. (2016). Input split frequent pattern tree using mapreduce paradigm in hadoop. *Journal of Theoretical and Applied Information Technology, 84*(2), 260–271.

Greeshma, L., & Pradeepini, G. (2016). *Mining maximal efficient closed itemsets without any redundancy.* Academic Press.

Greeshma, L., & Pradeepini, G. (2016). Unique constraint frequent item set mining. *Proceedings - 6th International Advanced Computing Conference, IACC 2016,* 68-72.

HasaneAhammad, S.K., & Rajesh, V. (2018). Image processing based segmentation techniques for spinal cord in MRI. *Indian Journal of Public Health Research and Development, 9*(6), 317-323.

HasaneAhammad, S., Rajesh, V., Hanumatsai, N., Venumadhav, A., Sasank, N.S.S., Bhargav Gupta, K.K., & Inithiyaz. (2019). MRI image training and finding acute spine injury with the help of hemorrhagic and non hemorrhagic rope wounds method. *Indian Journal of Public Health Research and Development, 10*(7), 404-408.

Inthiyaz, S., Prasad, M. V. D., Usha Sri Lakshmi, R., Sri Sai, N. T. B., & Kumar, P. P. (2020). Agriculture based plant leaf health assessment tool: A deep learning perspective. *International Journal of Emerging Trends in Engineering Research, 7*(11), 690–694.

Kumar, M.S., Inthiyaz, S., Vamsi, C.K., Ahammad, S.H., Sai Lakshmi, K., VenuGopal, P., & BalaRaghavendra, A. (2019). Power optimization using dual sram circuit. *International Journal of Innovative Technology and Exploring Engineering, 8*(8), 1032-1036.

Myla, S., Marella, S.T., SwarnendraGoud, A., HasaneAhammad, S., Kumar, G.N.S., & Inthiyaz, S. (2019). Design decision taking system for student career selection for accurate academic system. *International Journal of Recent Technology and Engineering, 8*(9), 2199-2206.

Myla, S., Marella, S.T., Goud, A.S., Ahammad, S.H., Kumar, G.N.S., & Inthiyaz, S. (2019). Design decision taking system for student career selection for accurate academic system. *International Journal of Scientific and Technology Research, 8*(9), 2199-2206.

Nagageetha, M., Mamilla, S.K., & HasaneAhammad, S. (2017). Performance analysis of feedback based error control coding algorithm for video transmission on wireless multimedia networks. *Journal of Advanced Research in Dynamical and Control Systems, 9*(14), 626-660.

Nagendram, S., Rao, K. R. H., &Bojja, P. (2017). A review on recent advances of routing protocols for development of manet. *Journal of Advanced Research in Dynamical and Control Systems, 9*(2), 114-122.

Narayana, V.V., Ahammad, S.H., Chandu, B.V., Rupesh, G., Naidu, G.A., & Gopal, G.P. (2019). Estimation of quality and intelligibility of a speech signal with varying forms of additive noise. *International Journal of Emerging Trends in Engineering Research, 7*(11), 430-433.

Noorbasha, F., Manasa, M., Gouthami, R. T., Sruthi, S., Priya, D. H., Prashanth, N., & Rahman, M. Z. U. (2017). FPGA implementation of cryptographic systems for symmetric encryption. *Journal of Theoretical and Applied Information Technology, 95*(9), 2038–2045.

PoornaChander Reddy, A., Siva Kumar, M., Murali Krishna, B., Inthiyaz, S., & Ahammad, S.H. (2019). Physical unclonable function based design for customized digital logic circuit. *International Journal of Advanced Science and Technology, 28*(8), 206-221.

Prasanth, Y., Sreedevi, E., Gayathri, N., & Rahul, A. S. (2017). Analysis and implementation of ensemble feature selection to improve accuracy of software defect detection model. *Journal of Advanced Research in Dynamical and Control Systems, 9*(18), 601-613.

Raj Kumar, A., Kumar, G.N.S., Chithanoori, J.K., Mallik, K.S.K., Srinivas, P., & HasaneAhammad, S. (2019). Design and analysis of a heavy vehicle chassis by using E-glass epoxy & S-2 glass materials. *International Journal of Recent Technology and Engineering, 7*(6), 903-905.

Rama Chandra Manohar, K., Upendar, S., Durgesh, V., Sandeep, B., Mallik, K.S.K., Kumar, G.N.S., & Ahammad, S.H. (2018). Modeling and analysis of Kaplan Turbine blade using CFD. *International Journal of Engineering and Technology, 7*(3.12), 1086-1089.

Razia, S., & Narasingarao, M. R. (2017). A neuro computing frame work for thyroid disease diagnosis using machine learning techniques. *Journal of Theoretical and Applied Information Technology, 95*(9), 1996–2005.

Razia, S., Narasingarao, M. R., &Bojja, P. (2017). Development and analysis of support vector machine techniques for early prediction of breast cancer and thyroid. *Journal of Advanced Research in Dynamical and Control Systems, 9*(6), 869-878.

Sajana, T., & Narasingarao, M. R. (2017). Machine learning techniques for malaria disease diagnosis - A review. *Journal of Advanced Research in Dynamical and Control Systems, 9*(6), 349-369.

Siva Kumar, M., Inthiyaz, S., Venkata Krishna, P., JyothsnaRavali, C., Veenamadhuri, J., Hanuman Reddy, Y., & HasaneAhammad, S. (2019). Implementation of most appropriate leakage power techniques in vlsi circuits using nand and nor gate. *International Journal of Innovative Technology and Exploring Engineering, 8*(7), 797-801.

Srinivasa Reddy, K., Suneela, B., Inthiyaz, S., HasaneAhammad, S., Kumar, G.N.S., & Mallikarjuna Reddy, A. (2020). Texture filtration module under stabilization via random forest optimization methodology. *International Journal of Advanced Trends in Computer Science and Engineering, 8*(3), 458-469.

Vijaykumar, G., Gantala, A., Gade, M.S.L., Anjaneyulu, P., & Ahammad, S.H. (2017). Microcontroller based heartbeat monitoring and display on PC. *Journal of Advanced Research in Dynamical and Control Systems, 9*(4), 250-260.

Chapter 33
Soil Nutrients and pH Level Testing Using Multivariate Statistical Techniques for Crop Selection

Swapna B.

 https://orcid.org/0000-0002-7186-2842

Dr. M. G. R. Educational and Research Institute, India

S. Manivannan

Dr. M. G. R. Educational and Research Institute, India

M. Kamalahasan

Dr. M. G. R. Educational and Research Institute, India

ABSTRACT

The multivariate data analysis technique is used to determine the highly impacted data in soil and crop growth. The importance and relationship between soil variables were factored by using the regression analysis technique. The correlation matrix technique was used for comparing several variables to correlate positive and negative signs. From the soil testing procedure and understanding of results, it shows that soil nutrients and pH level have a powerful effect on variation in the usage of fertilizers, crop selection, and high crop yield. pH determination can be used to indicate whether the soil is suitable for the plant's growth or in need of adjustment to produce optimum plant growth. Based upon the predictive analysis results, nitrogen and potassium content are naturally high compared to other soil nutrients of this region and suggested fertilizers required for crop growth. To produce healthy crop yield, farmers should select the crops as per soil types, nutrients level, and pH level.

DOI: 10.4018/978-1-7998-6870-5.ch033

INTRODUCTION

Soil must have 45% of nutrients or minerals, 5% of organic matter, 25% of air and 25% of water to grow healthy and high-quality crops. Soil Nutrients play a vital role for crop selection, crop growth and high crop yield. It seems inadequate to say, the agricultural sector is in a state of distress and hollers with the need to be enhanced with productivity and crop yield. Agriculture is the science and art of cultivating crops and livestock. It has been there since ages, even before human beings started the early civilization. It is one such occupation that readily satisfies the person who is into it. Needless to mention, the agricultural sector contributes much to the nation's economy and progress but is often ignored as people are diverted towards the many advancements and growth in science and technology. Also due to some factors such as water scarcity and escalation of fertilizers, today agricultural work seems much not profitable.

This study takes into concern the growing apathy of the sector and aims at improving the crop yield with the help of sensor technology. The utilization of technology in the agricultural sector has a positive role to play since it contributes much to crop production and also sets an example for others to follow.

The main problem that lingers in the lack of high crop yield lies in the improper identification and selection of micronutrients and macronutrients of the soil. These macro and micro nutrients play a much bigger role in the growth of crops and influence their productivity to a considerable extent. Absence of any one of the micronutrients in the soil can limit and disrupt the growing nature of plants even when all other nutrients are present in adequate quantities. The proposed study aims at predicting the actual availability of the micro and macro nutrients with the help of traditional values from Agricultural department. Depending on the compatibility, a farmer communication system will be identified which will be useful for farmers in knowing the nature of their crops and thereby devise suitable improvements. Along with the micronutrients, electrical conductivity, humic matter and other physical parameters required for enhanced growth also validated. This being said, identifying the type of crop & proper fertilizers suitable for the soil to enhance the crop yield is quite a difficult task. Hence a user-friendly predictive analysis will be done to give timely information to farmers.

Plants need plenty of sun light, air and water. Measuring the different aspects will get the required data. One way to keep track of all these required nutrients is to measure the electrical conductivity of the soil. Electrical conductivity studies tell the need for the correct amount of nutrient requirements. The electrical conductivity of soil involves the measurement of the ions and their temperature-dependent behavior. Porosity and Caution exchange capacity (Soil texture) also will be measured, which are related to electrical conductivity. Testing of Electrical conductivity & pH gives an idea about the availability of nutrients to plants and their actual requirements.

In data analysis techniques, non-metric multi-dimensional scaling methods, canonical correspondence analysis, principal coordinate analysis, detrended correspondence analysis and co-inertia analysis methods in prediction were used for soil fertility (Kenkel, 2019). Soil types were analyzed using two way multiple analysis of variance. They concluded that clay soil is not a good soil for tiger nuts farming in Eastern Nigeria, fertilizer types were not important than soil types (Bartholomew, 2016). Soil physical and chemical properties were predicted by reflectance spectroscopy methods for samples from central region of Amazon, Brazil. Descriptive statistics, partial least squares regression and correlations were used to analyze soil parameters (Erika, 2017). Principal crops, average crop yield, average annual growth rates of major crops and gross area under irrigation by crops were predicted by using predictive analysis integration method (Purvagrover, 2016).

LITERATURE REVIEW

ANN, KNN, K means, Decision tree, support vector machine, fuzzy set methods were used for classification, clustering and regression of soil parameters (Geetha et al., 2015).

Soil fertility, forest nutrition and nutrient management areas were analyzed by using PCA, non-metric multidimensional scaling and cluster analysis methods for Teak (Tectonagrandis L.f.). Concluded cluster analysis is the best analysis for soil fertility (Jesus Fernandez 2014).

Knowledge Discovery in Database was used for crop yield prediction. Functional dependencies were used for agricultural soil parameters analysis (Abhishek et al., 2014).

PCA method was used to reduce and summarize the soil variables and analyze soil test data for fertilizer nutrients recommendation using Kathmandu soil test data for size 50. In Kathmandu, results showed that when pH increased then automatically nitrogen and organic matter gets decreased. Phosphorus and potassium are oppositely correlated (HariDahal 2007).

Normal distribution, linear discriminant functions and canonical discriminant analysis were used to analyze soil reaction and soil nutrients. It showed that crop rotation was the problem to get high crop yield (Vilamil 2008). Data mining techniques like K means, k nearest neighbor, artificial neural networks and support vector machine were used to estimate forest variables and discriminating good or bad crops (Mucherino et al., 2009). Agricultural data were analyzed by using several data mining techniques to predict the soil fertility index (Chinchulum 2010).

The Data Mining systems applied on Agricultural information incorporate k-implies, bi grouping, k closest neighbor, Neural Networks (NN) Support Vector Machine (SVM), Naive Bayes Classifier and Fuzzy methods. As can be seen the fittingness of information mining procedures is to a limited degree controlled by the various kinds of farming information or the issues being tended to (Sheela et al., 2015).

A soil quality index (SQI) that coordinates key soil property data would be valuable in limiting overflow impacts of aimless soil the board, for example, deficiencies in nourishment, water, vitality and decrease unfavorable repercussions of environmental change. Right now, new SQI that incorporates soil qualities is created utilizing halfway least squares relapse (PLSR), and contrasted and crop yields. The field information were procured in the year 2013 from 5 distinctive on-ranch destinations inside Ohio, USA that were under Natural Vegetation (NV), No-Till (NT), and Conventional Till (CT) the board (Paul et al., 2016)

Effective displaying procedures to assess the possibility to build rangeland SOC stocks are essentially critical to evaluate their job in the worldwide carbon cycle and quantum decrease. This investigation intended to assess boosted regression trees (BRT) and random forest (RF) models in anticipating SOC stocks from accessible persistent remotely detected factors utilizing two element choice systems. Prevailing factors that influence SOC stocks in the rangelands were likewise distinguished (Wang et al., 2018).

It furnishes end clients with down to earth rules of utilization for every calculation. Tried three groups of mainstream calculations, in particular direct models, choice trees and bolster vector machines; every way tried utilizing wind mean speed information and a few natural covariates as indicators. The outcomes exhibited that no single calculation could reliably be utilized to assess twist internationally, despite the fact that choice tree based strategies appeared to be frequently the best estimators (Veronesi et al., 2017).

Different scientific models like Decision Trees, Random Forests, Support Vector Machines, Bayesian Networks, and Artificial Neural Networks, etc, have been used for inciting the models and break down the outcomes. These strategies empower to dissect soil, atmosphere, and water system which are altogether engaged with crop development and exactness cultivating (Elavarasan et al., 2018)

The article built up the forecast of fruitfulness lists for soil natural carbon and four significant soil supplements (phosphorus pentoxide, iron, manganese and zinc) utilizing practically all the accessible relapse strategies, explicitly an assortment of 76 regressors which have a place with 20 families, including neural systems, profound learning, bolster vector relapse, irregular woodlands, packing and boosting, tether and edge relapse, Bayesian models and the sky is the limit from there (Sirsat et al., 2018)

The investigation was led for a multiclass soil compost suggestion framework for paddy fields. Besides, unique advancement techniques like Genetic Algorithm and Particle Swarm Optimization are utilized to tune the SVM parameters. At last, a near report on the presentation was additionally accomplished for the various decisions of the parameters, calling attention to their correctness's (Suchithra 2018)

Data mining algorithms were utilized to create models to estimate the result of leaf roller bother checking choices on 'Hayward' kiwifruit crops in New Zealand. Utilizing industry shower journal and vermin observing information assembled at a plantation square level for consistence purposes, 80 characteristics (free factors) were made in three classes from the splash journal information. (Hill et al., 2014)

Two stage usage of the REML approach for model based kriging, exemplified by anticipating soil natural carbon (SOC) focuses in mineral soils in Estonia from the enormous scope advanced soil map data and a formerly settled forecast model. The forecast model was a straight blended model in with soil type, physical mud content (molecule size < 0.01 mm) and A skyline thickness as fixed impacts and site, transect, plot, year, year transect arbitrary captures and site explicit irregular inclines for earth content (Ritz et al., 2015). The mid-infrared diffuse reflectance spectroscopy (4000–602 cm−1) was utilized to anticipate substance and textural properties for an internationally disseminated soil phantom library. They filtered 971 soil tests chose from the International Soil Reference and Information Center database. A high-throughput diffuse reflectance adornment was utilized with optics that prohibits specular reflectance as a potential wellspring of blunder (Urselmans et al., 2010).

The significant test of the horticulture is checking the soil health, determination of harvests, development of yields and water system process. The fundamental target of this examination paper is to structure a Wireless Sensor Networks (WSN) with Raspberry pi through IOT to develop the land with reasonable harvests (Swapna et al., 2018).

The research paper aimed to design and develop a control system using node sensors in the crop field with data management via smart phone and a web application. The three components are hardware, web application, and mobile application. The first component was designed and implemented in control box hardware connected to collect data on the crops. Soil moisture sensors are used to monitor the field, connecting to the control box. The second component is a web-based application that was designed and implemented to manipulate the details of crop data and field information (Zaminur et al., 2018).

The soil test report values are used to classify several significant soil features like village wise soil fertility indices of Available Phosphorus P, Available Potassium K, Organic Carbon OC and Boron B, as well as the parameter Soil Reaction pH. The classification and prediction of the village wise soil parameters aids in reducing wasteful expenditure on fertilizer inputs, increase profitability, save the time of chemical soil analysis experts, improves soil health and environmental quality. These five classification problems are solved using the fast learning classification technique known as Extreme Learning Machine ELM with different activation functions (Suchithra et al., 2020)

Machine Learning calculations for soil mapping in a tropical precipitous territory of an authority provincial settlement in the Zona da Mata locale in Brazil. Morphometric maps produced from an advanced height model, together with Landsat-8 satellite symbolism, and climatic maps, were among the arrangement of covariates to be chosen by the Recursive Feature Elimination calculation to anticipate soil types utilizing AI calculations (Souza et al., 2018)

A complete survey of research devoted to utilizations of machine learning in rural creation frameworks. The works examined were sorted in (a) crop the executives, remembering applications for yield forecast, malady location, weed discovery crop quality, and species acknowledgment; (b) domesticated animals the board, remembering applications for creature government assistance and animals creation; (c) water the board; and (d) soil the executives (Liakos et al., 2018).

Machine Learning with its prescient capacity to handle complex frameworks, may understand this obstruction in the advancement of privately based N suggestions. The goal of this examination was to investigate use of ML philosophies to anticipate financial ideal nitrogen rate (EONR) for corn utilizing information from 47 analyses over the US Corn Belt. Two highlights, a water balanced accessible water limit (AWCwt) and a proportion of in-season precipitation to AWCwt (RAWCwt), were made to catch the effect of soil hydrology on N elements (Zhisheng et al., 2018)

Anticipating durum wheat yield through various machine learning calculations, and contrast them with distinguish the one that best fits the model. So as to accomplish this objective, One-R, J48, Ibk and A priori calculations were run with information gathered by our exploration gathering of a RIL (recombinant inbreed lines) populace developing in six unique situations from the Province of Buenos Aires in Argentina (Romero et al., 2013).

Land Coverage changes prediction done by using satellite images for Vellore district of Tamilnadu to improve the crop production and crop selection (Prabhu et al., 2018). Crop Suitability for Vellore district was predicted by using neural network and fuzzy approximation techniques. Rough set data collected from Vellore district for prediction. Fuzzy proximity method used to increase the crop production in Vellore district (Anitha et al., 2017)

MATERIALS AND METHODS

Soil health data were used in multivariate data analysis employing correlation, regression, principal component analysis and Anova to analyze soil nutrients – fertilizer recommendation for soil and crop management. The soil test data for analysis were included from 94 villages of Vellore district for the year 2018-2019. The samples were analyzed for soil reaction (pH), organic carbon (OC), electrical conductivity (EC), macro nutrients (Nitrogen, Phosphorus, potassium) and micro nutrients (copper, iron, manganese, zinc, Sulphur, boron). There are 94 villages in Vellore district, Tamilnadu, India. Soil data of the above district collected from agricultural department. Data analyzed using multivariate data analysis techniques as shown in figure 1 with the help of SPSS software. Soil variables will predict using different data analysis techniques.

Multivariate Statistical Analysis Techniques are used to analysis the simultaneous relationship between one or more variables which will predict the important parameters among all variables. Descriptive analysis is the technique which is same as the mean, median and variance. It will give the average of the soil parameters for future use in the agriculture applications. Principle Component Analysis is one of the most common methods to determine the direction of the maximum variance of the data. It is the unsupervised dimensionality techniques

Soil data collected from department of agriculture through soil heath card data for Vellore district. Parameters like soil nutrient fertility and soil reaction (pH) level calculated with the help of soil variables

Figure 1. Workflow for prediction of soil variables using multivariate data analysis techniques

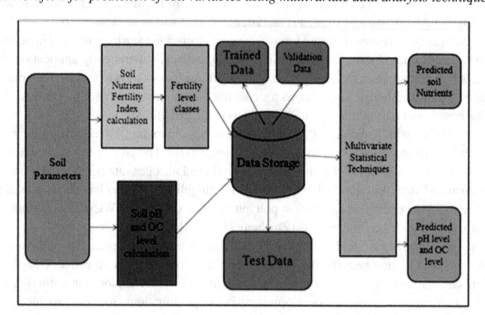

through data analysis techniques. Stored data can feed into the SPSS software as per the need. Data to be test with regression analysis, correlation analysis, principal component analysis, ANOVA method and descriptive analysis techniques. With the help of multivariate analysis techniques, soil variables should predict to increase the crop yield. We can predict the selective crop for specific land and type. Hence soil nutrients and soil reaction can predict the crop yield and crop production with high quality. Soil Nutrients and Soil parameters are very important to improve the crop production and soil fertility index in the vellore district of Tamilnadu. Different multivariate analysis techniques are used to predict the soil parameters like soil macro and micro nutrients.

Table 1. Descriptive statistics analysis using Vellore soil test data

	pH Highly acidic	Strongly Acidic	Moderately Acidic	Slightly acidic	Neutral	Moderately Alkaline	Critical for Germination	Harmful Crop Growth	Injurious to crops	Normal
N	94	94	94	94	94	94	94	94	94	94
Mean	.0000	.0319	2.5213	4.1596	.3723	49.1277	.0000	.0532	.4894	59.4681
Median	.0000	.0000	.0000	1.0000	.0000	39.0000	.0000	.0000	.0000	44.5000
Std. Deviation	.00000	.22964	7.64217	8.48567	1.18216	43.3453	.00000	.42419	3.38480	5.81623E1
Variance	.000	.053	58.403	72.007	1.398	1878.82	.000	.180	11.457	3.383E3
Minimum	.00	.00	.00	.00	.00	2.00	.00	.00	.00	2.00
Maximum	.00	2.00	55.00	53.00	8.00	225.00	.00	4.00	31.00	359.00

RESULTS AND DISCUSSIONS

Analysis for soil variables of 94 villages in Vellore district from a total sample size 470 was performed first to see the descriptive statistics and multivariate data analysis using principal component analysis, correlation analysis, Anova methods and regression analysis.

The descriptive statistics revealed the average values of soil attributes as moderate alkaline pH and normal growth of crop of Electrical conductivity as shown in **Table 1**. 94 villages in Vellore district soil samples are moderately alkaline in nature with normal range in electrical conductivity. To receive high crop yield, selected crops should be chosen for moderate alkaline soil reaction.

Figure 2. Control chart for pH level

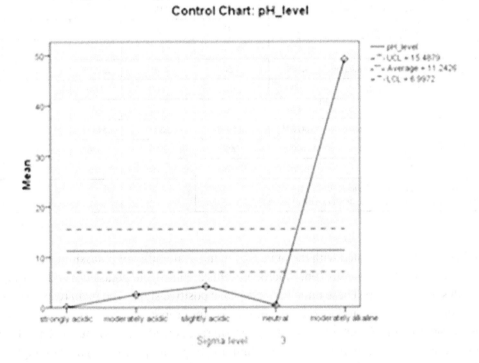

Figure 2 shows that the graph for different level of soil pH. There was no pH level of strongly acidic and neutral level in soils of the place. Mean of 3 percent soils have moderately acidic type of soils. The graph shows 5 percent mean values of slightly acidic pH in nature of Vellore district. Therefore approximately mean of 50 soils have moderate alkaline type of pH in nature. Hence prediction says that choosing of crops to increase the yield as per the moderate alkaline soil reaction (pH) level will be effective.

Table 2 displays the correlation matrix analysis for soil reaction, macronutrients, electrical conductivity and organic carbon for 470 samples. Nitrogen, organic carbon and potassium have positive signs and very high loadings. Thus, pH having positive sign move in same direction while nitrogen and organic carbon vary together. Nitrogen vary while pH having negative sign move in opposite direction. It means

Table 2. Correlation matrix for macro nutrients

		pH level	Electrical conductivity	Organic Carbon	nitrogen	Phosphorus	Potassium
pH level	Pearson Correlation	1					
	Sig. (2-tailed)						
	N	470					
Electrical conductivity	Pearson Correlation	-.025	1				
	Sig. (2-tailed)	.633					
	N	376	376				
Organic Carbon	Pearson Correlation	.001	.002	1			
	Sig. (2-tailed)	.985	.972				
	N	470	376	470			
Nitrogen	Pearson Correlation	-.106*	-.167**	.610**	1		
	Sig. (2-tailed)	.022	.001	.000			
	N	470	376	470	470		
Phosphorus	Pearson Correlation	-.108*	-.124*	.278**	.312**	1	
	Sig. (2-tailed)	.020	.016	.000	.000		
	N	470	376	470	470	470	
Potassium	Pearson Correlation	.416**	-.159**	.289**	.546**	.279**	1
	Sig. (2-tailed)	.000	.002	.000	.000	.000	
	N	376	376	376	376	376	376
*. Correlation is significant at the 0.05 level (2-tailed).							
**. Correlation is significant at the 0.01 level (2-tailed). .							

increased in soil pH is associated with the increased in the availability of potassium and organic carbon, decreased in the availability in nitrogen. Macro nutrients are nitrogen, potassium and phosphorous vary together with same direction. These three variables have positive signs. The main responsibility of these variables is soil fertility and product quality component for crop growth.

Table 3 shows the correlation matrix analysis for soil reaction, micronutrients, electrical conductivity and organic carbon. Copper and zinc have very high loadings and positive signs with high correlation. Thus, pH having negative sign move in opposite direction and copper vary together. Manganese, Sulphur and Iron have positive signs and very high loadings with high correlation. Manganese, Sulphur, zinc and Iron varies together while pH which have positive sign move in same direction.

This means that increased soil pH is associated with the increase of Manganese, Sulphur, zinc and Iron availability and decrease of copper availability. The Micro nutrients Copper, Manganese, Sulphur, Zinc, boron and Iron varies together with same direction. These three variables have positive signs. These variables are mainly responsible for immune system and in the development of healthy soil for crop growth. All micro nutrients vary together while the electrical conductivity having negative signs move in opposite direction.

From **Table 4** R value is 0.467 which implies that 46.7% of relationship exists between the dependent variables electrical conductivity, organic carbon and macro nutrients are nitrogen, phosphorus and

Table 3. Correlation matrix for micro nutrients

		pH level	Electrical conductivity	Copper	Iron	Manganese	Zinc	sulphur	Boron
pH level	Pearson Correlation	1							
	Sig. (2-tailed)								
	N	470							
Electrical Conductivity	Pearson Correlation	-.025	1						
	Sig. (2-tailed)	.633							
	N	376	376						
Copper	Pearson Correlation	-.094	-.049	1					
	Sig. (2-tailed)	.200	.507						
	N	188	188	188					
Iron	Pearson Correlation	.245**	-.003	.321**	1				
	Sig. (2-tailed)	.001	.963	.000					
	N	188	188	188	188				
Manganese	Pearson Correlation	.136	-.050	.691**	.805**	1			
	Sig. (2-tailed)	.063	.496	.000	.000				
	N	188	188	188	188	188			
Zinc	Pearson Correlation	.089	-.020	.902**	.543**	.818**	1		
	Sig. (2-tailed)	.225	.783	.000	.000	.000			
	N	188	188	188	188	188	188		
Sulphur	Pearson Correlation	.045	-.060	.739**	.624**	.845**	.753**	1	
	Sig. (2-tailed)	.539	.410	.000	.000	.000	.000		
	N	188	188	188	188	188	188	188	
Boron	Pearson Correlation	.148*	-.039	.281**	.617**	.598**	.366**	.477**	1
	Sig. (2-tailed)	.042	.592	.000	.000	.000	.000	.000	
	N	188	188	188	188	188	188	188	188

potassium and independent variable soil reaction (pH). The R square value is 0.218 implies that 21.8% of variation is explained by the independent and dependent variable.

Null hypothesis refers to non-linear relationship between the soil reaction and dependent variables. Alternative hypothesis refers to a linear relationship between the soil reaction (pH) and dependent variables.

Table 4. ANOVA with Regression analysis

Model	R	R Square	Adjusted R Square	Std. Error of the Estimate
1	.467[a]	.218	.207	5.30717
a. The dependent variables are Potassium, Electrical conductivity, Phosphorus, Organic Carbon, nitrogen				

Table 5. Regression Analysis

	Sum of Squares	df	Mean Square	F	Sig.
Regression	241.605	1	241.605	6.909	.092
Residual	13078.725	374	34.970		
Total	13320.330	375			
The independent variable is pH level.					

Table 6. ANOVA with Regression analysis for Nitrogen

R	R Square	Adjusted R Square	Std. Error of the Estimate
.135	.018	.016	5.914
The independent variable is nitrogen.			

In Anova the calculated F value is 6.909 and F significance is 0.092 as shown in **Table 5**. Since the value is more than the level of significance 0.05, reject the null hypothesis. Hence there is a linear relationship between the independent variable pH level and dependent soil variables like electrical conductivity, organic carbon and macro soil nutrients.

From **Table 6** R value is 0.135 which implies that 13.5% of relationship exists between the dependent variables electrical conductivity, and soil reaction (pH) and independent variable nitrogen. The R square value is 0.018 implies that 1.8% of variation is explained by the independent and dependent variable.

Null hypothesis refers there is no linear relationship between the nitrogen and dependent variables. Alternative hypothesis refers there is a linear relationship between the dependent soil variables and independent variable soil nutrient nitrogen.

Table 7. Regression Analysis for pH and Nutrients

	Sum of Squares	Df	Mean Square	F	Sig.
Regression	15805.803	1	15805.803	10.793	.001
Residual	547701.194	374	1464.442		
Total	563506.997	375			

The dependent variables are pH level and Electrical Conductivity.

Table 8. One sample test

Test Value = 11.42	T	Df	Sig. (2-tailed)	Mean Difference	95% Confidence Interval Difference	
					Lower	Upper
pH_level	-.139	469	.889	-.17745	-2.6771	2.3222

Table 9. Principal Component Analysis

Component	Initial Eigen values			Extraction Sums of Squared Loadings			
	Total	**% of Variance**	**Cumulative %**	**Total**	**% of Variance**	**Cumulative %**	
1	5.403	45.025	45.025	5.403	45.025	45.025	
2	2.523	21.021	66.046	2.523	21.021	66.046	
3	1.397	11.642	77.688	1.397	11.642	77.688	
4	.909	7.579	85.267				
5	.624	5.199	90.466				
6	.412	3.435	93.901				
7	.280	2.329	96.230				
8	.183	1.527	97.757				
9	.089	.739	98.496				
10	.078	.647	99.143				
11	.069	.575	99.719				
12	.034	.281	100.000				

Communalities	Initial	Extraction
Nitrogen	1.000	.761
Phosphorus	1.000	.841
Organic_Carbon	1.000	.868
pH_level	1.000	.678
Electrical_conductivity	1.000	.273
Potassium	1.000	.760
Copper	1.000	.949
Iron	1.000	.816
Manganese	1.000	.904
Zinc	1.000	.921
Sulphur	1.000	.797
Boron	1.000	.754

In Anova the calculated F value is 10.793 and F significance is 0.001 as shown in **Table 7**. Since the value is less than the level of significance 0.05, accept the null hypothesis. Hence there is no linear relationship between the dependent variables electrical conductivity, and soil reaction (pH) and independent variable nitrogen.

One sample test as shown in **Table 8** determines the characteristics of soil reaction (pH) level in the different sample. Sig value is 0.889 which is more than the level of significance 0.05, reject the null hypothesis. The soil variable pH shows the statistical significance on other soil variables. Hence there is a linear relationship between pH level and all other soil variables to predict the soil fertilizers usage and crop selection with respect to soil nutrients level. Therefore, analysis tells that pH level is the major variable to predict the soil type, crop selection, fertilizer dosage and yield of the product.

Table 9 shows the communalities which measure the amount of variance in all soil variables by using the principal component analysis techniques. It is also noted that the communality (h^2) can be defined as the sum of squared factor loadings. If a large amount of variance is extracted from the variable by the factor solution it is said to be as large communality. Soil variables with high values are well represented in the common factor space while the soil variables with low values are not well represented (Malhotra 2014, haridahal 2007).

By the application of variance components were retained explaining a total of about 77% variance as shown in the Table 9. A very high loading of Nitrogen, Potassium and organic carbon has been revealed as the major components responsible for 60% of the total variance by the correlation matrix. The other components are about 17% of the variance which has high loadings of micro nutrients. As of Vellore soils the major component may be known as "Plant growth component" and another component may be called as "Crop Nutrition Component".

CONCLUSION

The aim is to predict the soil nutrients and all other soil variables for selection of suitable crop for agricultural land. The outcome of the work greatly helps to make a data analytics decision system for Vellore district to manage the soil nutrient deficiency, fertilizer usage and crop selection problems. Results showed that soil test data analysis method helps to predict the chosen crops for suitable field. It is also used to recommend fertilizer dosage for the required crop. Hence the soil in Vellore district contains 60% of potassium, organic carbon and nitrogen naturally with the pH reaction of moderate alkaline. The most important nutrients were potassium and nitrogen if organic carbon is not acting as a nutrient. Even though phosphorus acts as an important nutrient, its loading is next to nitrogen which means that in order to maintain soil fertility, balancing of pH level is important than adding more fertilizer as its average levels. As per this level the farmers can choose their crops for more production and high yield.

REFERENCES

Anitha, A., & Acharjya, D. P. (2017). Crop suitability prediction in Vellore District using rough set on fuzzy approximation space and neural network. *The Natural Computing Applications Forum, 30*, 3633–3650.

Bartholomew. (2016). A simple two-way multiple analysis of variance-spss. *Research & Reviews: Journal of Statistics & Mathematical Sciences, 1*.

Chinchulum, A. (2010). Data mining techniques in agricultural & environmental sciences. *International Journal of Agricultural and Environmental Information Systems, 1*(1), 26–40. doi:10.4018/jaeis.2010101302

Dahal. (2007). *Factor analysis for soil test data: A methodological approach in environment friendly soil fertility management.* Academic Press.

de Paul Obade, V., & Lal, R. (2016). RTowards a standard technique for soil quality assessment. *Geoderma*, *265*, 96–102. doi:10.1016/j.geoderma.2015.11.023

Elavarasan, D., Vincent, D. R., Sharma, V., Zomaya, A. Y., & Srinivasan, K. (2018). Forecasting yield by integrating agrarian factors and machine learning models: A survey. *Computers and Electronics in Agriculture*, *155*, 257–282. doi:10.1016/j.compag.2018.10.024

Erika, F. M. (2017). Prediction of soil physical and chemical properties by visible & near infrared diffuse reflectance spectroscopy in the central amazon. *Remote Sensing*, *9*(4), 293. doi:10.3390/rs9040293

Fernández-Moya, J., Alvarado, A., Morales, M., San Miguel-Ayanz, A., & Marchamalo-Sacristán, M. (2014). Using multivariate analysis of soil fertility as a tool for forest fertilization planning. *Nutrient Cycling in Agroecosystems*, *98*(2), 155–167. doi:10.100710705-014-9603-3

Geetha, M. C. S. (2015). A survey on data mining techniques in agriculture. International Journal of Innovative Research in Computer & Communication Engineering, 3(2).

Hill, M. G., Connolly, P. G., Reutemann, P., & Fletcher, D. (2014). The use of data mining to assist crop protection decisions on kiwifruit in New Zealand. *Computers and Electronics in Agriculture*, *108*, 250–257. doi:10.1016/j.compag.2014.08.011

Indian Government. (2019). *Soil health card data*. Author.

Kenkel, N. C. (2019). On selecting an appropriate multivariate analysis. *Canadian Journal of Plant Science*.

Konstantinos, G. (2018). Machine Learning in Agriculture: A Review. *Sensors (Basel)*, *18*(8), 2674. doi:10.339018082674

Malhotra, N. K. (2004). *Marketing Research - An applied orientation*. Pearson Education Singapore and India.

Mankar & Burange. (2014). Data mining – An evolutionary view of Agriculture. *International Journal of Application or Innovation in Engineering and Management*.

Meier, M., Souza, E., & Francelino, M. R. (2018). Digital soil mapping using machine learning algorithms in a tropical mountains area. *Revista Brasileira de Ciência do Solo*, *42*, e0170421. doi:10.1590/18069657rbcs20170421

Mucherino, Petraqpapajorgji, & Pcardalos. (2009). A survey of data mining techniques applied to agriculture. *Oper Res Int J, 9*, 121-140.

Prabu & Margret Anouncia. (2018). Prediction of Land Cover Changes in Vellore District of Tamil Nadu by Using Satellite Image Processing. *Knowledge Computing and its Applications*, 87-100.

Purvagrover, R. J. (2016). PAID: Predictive agriculture analysis of data integration in India. *IEEE International conference on computing for sustainable global development*.

Qin, Z., Myers, D. B., Ransom, C. J., Kitchen, N. R., Liang, S.-Z., Camberato, J. J., Carter, P. R., Ferguson, R. B., Fernandez, F. G., Franzen, D. W., Laboski, C. A. M., Malone, B. D., Nafziger, E. D., Sawyer, J. E., & Shanahan, J. F. (2018). Application of machine learning methodologies for predicting corn economic optimal nitrogen rate. *Agronomy Journal*, *110*(6), 2596–2607. doi:10.2134/agronj2018.03.0222

Rahman, Mitra, & Islam. (2018). Soil Classification using Machine Learning methods and crop suggestion based on soil series. *IEEE 21st International conference of Computer and Information Technology (ICCIT)*.

Ritz, C., Putku, E., & Astover, A. (2015). A practical two-step approach for mixed model-based kriging, with an application to the prediction of soil organic carbon concentration. *European Journal of Soil Science*, *66*(3), 548–554. doi:10.1111/ejss.12238

Romero, J. R., Roncallo, P. F., Akkiraju, P. C., Ponzoni, I., Echenique, V. C., & Carballido, J. A. (2013). Using classification algorithms for predicting durum wheat yield in the province of Buenos Aires. *Computers and Electronics in Agriculture*, *96*, 173–179. doi:10.1016/j.compag.2013.05.006

Sheela, P. J., & Sivaranjani, K. (2015). A brief survey of classification techniques applied to soil fertility prediction. Int Conf Eng Trends Sci Hum, 80-83.

Sirsat, M. S., Cernadas, E., Fernández-Delgado, M., & Barro, S. (2018). Automatic prediction of village-wise soil fertility for several nutrients in India using a wide range of regression methods. *Computers and Electronics in Agriculture*, *154*, 120–133. doi:10.1016/j.compag.2018.08.003

Suchithra, M. S., & Maya, L. (2020). Improving the prediction accuracy of soil nutrient classification by optimizing extreme learning machine parameters. *Information Processing in Agriculture*, *7*(1), 72–82. doi:10.1016/j.inpa.2019.05.003

Suchithra, M. S., & Pai, M. L. (2018). Improving the performance of sigmoid kernels in multiclass SVM using optimization techniques for agricultural fertilizer recommendation system. *International Conference on Soft Computing Systems*, 857-868. 10.1007/978-981-13-1936-5_87

Swapna, B., & Manivannan, S. (2018). Analysis: Smart Agriculture and Landslides Monitoring System Using Internet of Things (IoT). *International Journal of Pure and Applied Mathematics*, *118*, 24.

Terhoeven-Urselmans, T., Vagen, T.-G., Spaargaren, O., & Shepherd, K. D. (2010). Prediction of soil fertility properties from a globally distributed soil mid-infrared spectral library. *Soil Science Society of America Journal*, *74*(5), 1792–1799. doi:10.2136ssaj2009.0218

Veronesi, F., Korfiati, A., Buffat, R., & Raubal, M. (2017). Assessing accuracy and geographical transferability of machine learning algorithms for wind speed modelling. In Information Science. Springer. doi:10.1007/978-3-319-56759-4_17

Villamil, M. B., Miguez, F. E., & Bollero, G. A. (2008). Multivariate analysis & visualization of soil quality data for no tillage systems. *Journal of Environmental Quality*, *37*(6), 2063–2069. doi:10.2134/jeq2007.0349 PMID:18948459

Wang, B., Waters, C., Orgill, S., Cowie, A., Clark, A., Li Liu, D., Simpson, M., McGowen, I., & Sides, T. (2018). Estimating soil organic carbon stocks using different modelling techniques in the semi-arid rangelands of eastern Australia. *EcolInd*, *88*, 425–438. doi:10.1016/j.ecolind.2018.01.049

Compilation of References

. Priya, K. B., Rajendran, P., Kumar, S., Prabhu, J., Rajendran, S., Kumar, P. J., & Jothikumar, R. (2020). Pediatric and geriatric immunity network mobile computational model for COVID-19. *International Journal of Pervasive Computing and Communications*.

Ahammad, S. K., & Rajesh, V. (2018). Image processing based segmentation techniques for spinal cord in MRI. *Indian Journal of Public Health Research & Development*, *9*(6), 317. doi:10.5958/0976-5506.2018.00571.5

Ana-Ramona, B. B., & Razvanand, F. A. (2010). Big Data and Specific Analysis Methods for Insurance Fraud Detection. *Database Systems Journal, 1*(1), 30-39. http://www.dbjournal.ro/archive/14/14_4.pdf

Anderegg, L., & Eidenbenz, S. (2003). Ad Hoc-VCG: A Truthful and Cost-Efficient Routing Protocol for Mobile Ad Hoc Networks with Selfish Agents. *Proc. ACM MobiCom*, 245-259. 10.1145/938985.939011

Bennett, C. H., Bessette, F., Brassard, G., Salvail, L., & Smolin, J. (1992). Experimental quantum cryptography. *Journal of Cryptology*, *5*(1), 3–28. doi:10.1007/BF00191318

Dahal. (2007). *Factor analysis for soil test data: A methodological approach in environment friendly soil fertility management*. Academic Press.

Devlin, C. A., Zhu, A., & Brazil, T. J. (2008). Peak to Average Power Ratio Reduction Technique for OFDM Using Pilot Tones and Unused Carriers. *Radio and Wireless Symposium*, 33 – 36. 10.1109/RWS.2008.4463421

Lemley, J., Bazrafkan, S., & Corcoran, P. (2017). Deep learning for consumer devices and services: Pushing the limits for machine learning, artificial intelligence, and computer vision. *IEEE Consumer Electronics Magazine*, *6*(2), 48–56. doi:10.1109/MCE.2016.2640698

Leroy, G., & Chen, H. (2001, December). Meeting medical terminology needs-the ontology-enhanced medical concept mapper. *IEEE Transactions on Information Technology in Biomedicine*, *5*(4), 261–270. doi:10.1109/4233.966101 PMID:11759832

Nesamani, Rajini, Josphine, & Salome. (2021). Deep Learning-Based Mammogram Classification for Breast Cancer Diagnosis Using Multi-level Support Vector Machine. In Advances in Automation, Signal Processing, Instrumentation, and Control. Springer.

Obrist, P. A., Gaebelein, C. J., Teller, E. S., Langer, A. W., Grignolo, A., Light, K. C., & McCubbin, J. A. (1978). The relationship among heart rate, carotid dP/dt, and blood pressure in humans as a function of the type of stress. *Psychophysiology*, *15*(2), 102–115. doi:10.1111/j.1469-8986.1978.tb01344.x PMID:652904

Ordanini, A., & Pasini, P. (2008). Service co-production and value co-creation: The case for a service-oriented architecture (SOA). *European Management Journal*, *26*(5), 289–297. doi:10.1016/j.emj.2008.04.005

Rao, P. K. VKumar, B. RSaiteja, ASrikar, NSreenivasulu, V. (2018). Experimental Investigation of Thermal stability of carbon nanotubes reinforced Aluminium matrix using TGA-DSC Analysis. *International Journal of Mechanical and Production Engineering Research and Development*, 8(3), 161–168. doi:10.24247/ijmperdjun201818

Sharma. (2016). A Real Time Speech to text conversion system using Bidirectional Kalman filter Matlab, India. *2016 Intl. Conference on Advances in Computing, Communications and Informatics (ICACCI)*.

Balakrishnan, K., Deng, J., & Varshney, P. K. (2005). TWOACK: Preventing Selfishness in Mobile Ad Hoc Networks. *Proc. IEEE Wireless Comm. and Networking*, 2137-2142.

Bavani, Rajini, Josephine, & Prasannakumari. (2019). Heart Disease Prediction System based on Decision Tree Classifier. Jour of Adv Research in Dynamical & Control Systems, 11(10).

Borangiu, T., Morariu, C., Morariu, O., Drăgoicea, M., Răileanu, S., Voinescu, I., ... Purcărea, A. A. (2015, February). A Service Oriented Architecture for total manufacturing enterprise integration. In *International Conference on Exploring Services Science* (pp. 95-108). Springer. 10.1007/978-3-319-14980-6_8

Buchman, T. G., Stein, P. K., & Goldstein, B. (2002). Heart rate variability in critical illness and critical care. *Current Opinion in Critical Care*, 8(4), 311–315. doi:10.1097/00075198-200208000-00007 PMID:12386491

Derrig, R. A., & Fraud, I. (2002). Insurance Fraud. *The Journal of Risk and Insurance*, 69(3), 271–287. doi:10.1111/1539-6975.00026

Hincapie & Sierra. (2012). *Advanced Transmission Techniques in WiMAX*. InTech.

Kührer, M., Hoffmann, J., & Holz, T. (2014, September). Cloudsylla: Detecting suspicious system calls in the cloud. In *Symposium on Self-Stabilizing Systems* (pp. 63-77). Springer. 10.1007/978-3-319-11764-5_5

Loepp, S., & Wootters, W. K. (2006). *Protecting Information*. Cambridge Univ. Press. doi:10.1017/CBO9780511813719

Siva, D., & Bojja, P. (2019). MLC based Classification of Satellite Images for Damage Assessment Index in Disaster Management. *Int. J. Adv. Trends Comput. Sci. Eng*, 8(3), 10–13. doi:10.30534/ijatcse/2019/24862019

Villamil, M. B., Miguez, F. E., & Bollero, G. A. (2008). Multivariate analysis & visualization of soil quality data for no tillage systems. *Journal of Environmental Quality*, 37(6), 2063–2069. doi:10.2134/jeq2007.0349 PMID:18948459

Vishnu, A. V., Kumar, P. J., & Ramana, M. V. (2018). Comparison among Dry, Flooded and MQL Conditions in Machining of EN 353 Steel Alloys-An Experimental Investigation. *Materials Today: Proceedings*, 5(11), 24954–24962. doi:10.1016/j.matpr.2018.10.296

Zadeh, L. A., Tadayon, S., & Tadayon, B. (2018). *U. S. Patent Application No. 15/919,170*. US Patent Office.

Ahammad, S. H., Rajesh, V., Neetha, A., Sai Jeesmitha, B., & Srikanth, A. (2019). Automatic segmentation of spinal cord diffusion MR images for disease location finding. *Indonesian Journal of Electrical Engineering and Computer Science*, 15(3), 1313–1321. doi:10.11591/ijeecs.v15.i3.pp1313-1321

Ansari, S., Aslam, T., Poncela, J., Otero, P., & Ansari, A. (2020). Internet of things-Based healthcare applications. In *IoT Architectures, Models, and Platforms for Smart City Applications* (pp. 1–28). IGI Global. doi:10.4018/978-1-7998-1253-1.ch001

Bhavana, D., Kumar, K. K., Kaushik, N., Lokesh, G., Harish, P., Mounisha, E., & Tej, D. R. (2020). Computer vision based classroom attendance management system-with speech output using LBPH algorithm. *International Journal of Speech Technology*, 23(4), 779–787. doi:10.100710772-020-09739-2

Corcoran, J., McGee, T. R., & Townsley, M. (2013). Malicious hoax calls and suspicious fires: An examination of their spatial and temporal dynamics. *Trends and Issues in Crime and Criminal Justice*, (459), 1.

Ishaq, Khan, & Gul. (2012). *Precoding in MIMO, OFDM to reduce PAPR (Peak to Average Power Ratio)* (M.A. thesis). Linnaeus University, Sweden.

Lewis, G. A., & Smith, D. B. (2008, September). *Service-oriented architecture and its implications for software maintenance and evolution. In 2008 Frontiers of Software Maintenance*. IEEE.

Li, G., Yuan, T., Li, C., Zhuo, J., Jiang, Z., Wu, J., Ji, D., & Zhang, H. (2020). Effective Breast Cancer Recognition Based on Fine-Grained Feature Selection. *IEEE Access: Practical Innovations, Open Solutions*, 8, 227538–227555. doi:10.1109/ACCESS.2020.3046309

Mucherino, Petraqpapajorgji, & Pcardalos. (2009). A survey of data mining techniques applied to agriculture. *Oper Res Int J, 9*, 121-140.

Nie, L., & Zhao, Y.-L. (2013, August). Bridging the Vocabulary Gap between Health Seekers and Healthcare Knowledge. *IEEE Transactions on Knowledge and Data Engineering*.

Nielsen, M. A., & Chuang, I. L. (2010). Quantum Computation and Quantum Information. Cambridge Univ. Press.

Parati, G., Di Rienzo, M., Bertinieri, G., Pomidossi, G., Casadei, R., Groppelli, A., Pedotti, A., Zanchetti, A., & Mancia, G. (1988). Evaluation of the baroreceptor-heart rate reflex by 24-hour intra-arterial blood pressure monitoring in humans. *Hypertension, 12*(2), 214–222. doi:10.1161/01.HYP.12.2.214 PMID:3410530

Rajalakshmi & Meena. (2018). An Efficient technique of intrusion detection for large number of malicious nodes in Manet using a tree classifier. *International Journal of Simulation, Systems, Science and Technology*.

Troester. (n.d.). *Big Data Meets Big Data Analytics*. SAS White Paper. https://eric.univ-lyon2.fr/~ricco/cours/slides/sources/big-data-meets-big-data-analytics-105777.pdf

Artis, M., Ayuso, M., & Guillén, M. (1999). Modelling Different Types of Automobile Insurance Fraud Behaviour in the Spanish Market. *Insurance, Mathematics & Economics, 24*(1-2), 67–81. doi:10.1016/S0167-6687(98)00038-9

Bai, L., Yang, D., Wang, X., Tong, L., Zhu, X., Bai, C., & Powell, C. A. (2020). *Chinese experts' consensus on the Internet of Things-aided diagnosis and treatment of coronavirus disease 2019. In Clinical eHealth* (Vol. 3). Elsevier.

Bhavana, D., Rajesh, V., & Kishore, P. V. V. (2016). A new pixel level image fusion method based on genetic algorithm. *Indian Journal of Science and Technology, 9*(45), 1–8. doi:10.17485/ijst/2016/v9i45/76691

Chinchulum, A. (2010). Data mining techniques in agricultural & environmental sciences. *International Journal of Agricultural and Environmental Information Systems, 1*(1), 26–40. doi:10.4018/jaeis.2010101302

Chun, B.-G., Chaudhuri, K., Wee, H., Barreno, M., Papadimitriou, C. H., & Kubiatowicz, J. (2004). Selfish Caching in Distributed Systems: A Game- Theoretic Analysis. *Proc. ACM Symp. Principles of Distributed Computing*, 21-30. 10.1145/1011767.1011771

Ji, H. (2019). Research on Feature Selection and Classification Algorithm of Medical Optical Tomography Images. *2019 IEEE/CIC International Conference on Communications in China (ICCC)*, 561-566. 10.1109/ICCChina.2019.8855804

Pavan, B., & Kumar, M. (2013, January-February). Comparitive Study Of PAPR Of DHT-Precoded OFDM System With DFT & WHT. *International Journal of Engineering Research and Applications, 3*, 527–532.

Salehi, Z., Sami, A., & Ghiasi, M. (2014). Using feature generation from API calls for malware detection. *Computer Fraud & Security, 2014*(9), 9–18. doi:10.1016/S1361-3723(14)70531-7

Seccareccia, F., Pannozzo, F., Dima, F., Minoprio, A., Menditto, A., Lo Noce, C., & Giampaoli, S. (2001). Heart rate as a predictor of mortality: The MATISS project. *American Journal of Public Health, 91*(8), 1258–1263. doi:10.2105/AJPH.91.8.1258 PMID:11499115

Sunitha, R., & Suman, MKishore, P. V. VEepuri, K. K. (2018). Sign language recognition with multi feature fusion and ANN classifier. *Turkish Journal of Electrical Engineering and Computer Sciences, 26*(6), 2871–2885.

Valipour, M. H. AmirZafari, B., Maleki, K. N., & Daneshpour, N. (2009, August). A brief survey of software architecture concepts and service oriented architecture. In *2009 2nd IEEE International Conference on Computer Science and Information Technology* (pp. 34-38). IEEE.

Abbott, Matkovsky, & Elder. (1998). An Evaluation of High-End Data Mining Tools for Fraud Detection. *Proc. of IEEESMC98.*

Bhadoria, R. S., Chaudhari, N. S., & Vidanagama, V. T. N. (2018). Analyzing the role of interfaces in enterprise service bus: A middleware epitome for service-oriented systems. *Computer Standards & Interfaces, 55,* 146–155. doi:10.1016/j.csi.2017.08.001

Bhavana, D., Kumar, K. K., Rajesh, V., Saketha, Y. S. S. S., & Bhargav, T. (n.d.). *Deep Learning for Pixel-Level Image Fusion using CNN.* Academic Press.

Bhavana, D., Kishore Kumar, K., & Bipin Chandra Medasani, S. K. B. (2021). Hand Sign Recognition using CNN. *International Journal of Performability Engineering, 17*(3), 314–321. doi:10.23940/ijpe.21.03.p7.314321

Bollegala, Matsuo, & Ishizuka. (2011). A Web Search Engine-Based Approach to Measure Semantic Similarity between Words. *IEEE Transactions on Knowledge and Data Engineering, 23*(7).

Camm, A. J. (1996, March 1). Heart rate variability: Standards of measurement, physiological interpretation, and clinical use. *Circulation, 93*(5), 1043–1065. doi:10.1161/01.CIR.93.5.1043 PMID:8598068

Kumari, B. S. (2014, June). PAPR and Fiber Nonlinearity Mitigation of DFT-Precoded OFDM System. *International Journal of Advanced Research in Electrical, Electronics and Instrumentation Engineering, 3,* 9801–9808.

Malhotra, N. K. (2004). *Marketing Research - An applied orientation.* Pearson Education Singapore and India.

Pham, H. T., Nguyen, M. A., & Sun, C. C. (2019). AIoT solution survey and comparison in machine learning on low-cost microcontroller. In *Proceedings of the 2019 International Symposium on Intelligent Signal Processing and Communication Systems,* pp. 1-2. IEEE. 10.1109/ISPACS48206.2019.8986357

Sudha, M. N., Selvarajan, S., & Suganthi, M. (2019). Feature selection using improved lion optimisation algorithm for breast cancer classification. *IJBIC, 14*(4), 237–246. doi:10.1504/IJBIC.2019.103963

Vidal, J. M., Monge, M. A. S., & Villalba, L. J. G. (2018). A novel pattern recognition system for detecting Android malware by analyzing suspicious boot sequences. *Knowledge-Based Systems, 150,* 198–217. doi:10.1016/j.knosys.2018.03.018

Vinoth Kumar, V., Karthikeyan, T., Praveen Sundar, P. V., Magesh, G., & Balajee, J. M. (2020). A Quantum Approach in LiFi Security using Quantum Key Distribution. *International Journal of Advanced Science and Technology, 29*(6s), 2345–2354.

Asada, H. H., Shaltis, P., Reisner, A., Sokwoo Rhee, & Hutchinson, R. C. (2003). Mobile monitoring with wearable photoplethysmographic biosensors. *IEEE Engineering in Medicine and Biology Magazine, 22*(3), 28–40. doi:10.1109/MEMB.2003.1213624 PMID:12845817

Burney, S.M.A., Arifeen, Q.U., Mahmood, N., & Bari, S.A.K. (n.d.). *Suspicious Call Detection Using Bayesian Network Approach*. Academic Press.

Ding, G., & Bhargava, B. (2004). Peer-to-Peer File-Sharing over Mobile Ad Hoc Networks. *Proc. IEEE Ann. Conf. Pervasive Computing and Comm. Workshops,* 104-108. 10.1109/PERCOMW.2004.1276914

Indian Government. (2019). *Soil health card data*. Author.

Jammes, F., Bony, B., Nappey, P., Colombo, A. W., Delsing, J., Eliasson, J., . . . Till, M. (2012, October). Technologies for SOA-based distributed large scale process monitoring and control systems. In *IECON 2012-38th Annual Conference on IEEE Industrial Electronics Society* (pp. 5799-5804). IEEE. 10.1109/IECON.2012.6389589

Kumar, S. U., & Inbarani, H. H. (2016). PSO-based feature selection and neighborhood rough set-based classification for BCI multiclass motor imagery task. *Neural Computing & Applications*, *28*(11), 3239–3258. doi:10.100700521-016-2236-5

Mailloux, L. O., Morris, J. D., Grimaila, M. R., Hodson, D. D., Jacques, D. R., Colombi, J. M., Mclaughlin, C. V., & Holes, J. A. (2015). A Modeling Framework for Studying Quantum Key Distribution System Implementation Nonidealities. *IEEE Access: Practical Innovations, Open Solutions*, *3*, 110–130. doi:10.1109/ACCESS.2015.2399101

Prakash, V. R., & Kumaraguru, P. (2012). Performance analysis of OFDM with QPSK using AWGN and Rayleigh Fading Channel. Conference: ICIESP, 10-15.

SchabenbergerO. (2018). https://iotpractitioner.com/what-is-the-aiot-exclusive-interview-with-oliver-schabenberger-at-iot-slam-live-2018/

Sepaskhah, A. R., & Ahmadi, S. H. (2012). A review on partial root-zone drying irrigation. *International Journal of Plant Production*, *4*(4), 241–258.

Sinha, Biswas, Raj, & Misra. (2020). Big Data Analytics On Matrimonial Data Set. *International Journal of Innovative Research in Applied Sciences and Engineering, 4*(4), 722-731. doi:10.29027/IJIRASE.v4.i4.2020.722-728

Tekin, Atan, & van der Schaar. (2015). *Discover the Expert: Context- Adaptive Expert Selection for Medical Diagnosis*. Academic Press.

Han, S. H., & Lee, J. H. (2005, April). An overview of peak-to-average power ratio reduction techniques for multicarrier transmission. *IEEE Wireless Communications*, *12*(2), 56–65. doi:10.1109/MWC.2005.1421929

Hori, M., Kawamura, T., & Okano, A. (1999, October). OpenMES: scalable manufacturing execution framework based on distributed object computing. In *IEEE SMC'99 Conference Proceedings. 1999 IEEE International Conference on Systems, Man, and Cybernetics (Cat. No. 99CH37028)* (Vol. 6, pp. 398-403). IEEE. 10.1109/ICSMC.1999.816585

Howell, T. A., Evett, S. R., O'Shaughnessy, S. A., Colaizzi, P. D., & Gowda, P. H. (2012). Advanced irrigation engineering: Precision and precise. *Journal of Agricultural Science and Technology A, 2*(1A), 1.

Kaggle. (n.d.). Available at: https://www.kaggle.com/c/c001hw1/data

Kanedera, N., Arai, T., Hermansky, H., & Pavel, M. (1999). On the relative importance of various components of the modulation spectrum for automatic speech recognition. *Speech Communication*, *28*(1), 43–55. doi:10.1016/S0167-6393(99)00002-3

Kumar, A., Fulham, M., Feng, D., & Kim, J. (2019). Co-learning feature fusion maps from PET-CT images of lung cancer. *IEEE Transactions on Medical Imaging*, *39*(1), 204–217. doi:10.1109/TMI.2019.2923601 PMID:31217099

Meier, M., Souza, E., & Francelino, M. R. (2018). Digital soil mapping using machine learning algorithms in a tropical mountains area. *Revista Brasileira de Ciência do Solo*, *42*, e0170421. doi:10.1590/18069657rbcs20170421

Monisha, M., & Sudheendra, G. (2017). Lifi-Light Fidelity Technology. *2017 International Conference on Current Trends in Computer, Electrical, Electronics and Communication (CTCEEC)*, 818-821. 10.1109/CTCEEC.2017.8455097

Nie, Li, & Akbari. (2014). *WenZher: Comprehensive Vertical search for Healthcare Domain.* Academic Press.

Pansiot, J. (2007). Ambient and wearable sensor fusion for activity recognition in healthcare monitoring systems. *IFMBE Proc.BSN*, 208-212. 10.1007/978-3-540-70994-7_36

Sengathir & Manoharan. (2013). Selfish Aware context based Reactive Queue Scheduling Mechanism for MANETs. *International Journal of Computer Applications, 66*(19).

Sharpe, R., van Lopik, K., Neal, A., Goodall, P., Conway, P. P., & West, A. A. (2019). An industrial evaluation of an Industry 4.0 reference architecture demonstrating the need for the inclusion of security and human components. *Computers in Industry, 108*, 37–44. doi:10.1016/j.compind.2019.02.007

Chance, B. (1992). *User-wearable Hemoglobinometer for measuring the metabolic condition of a subject.* Google Patents.

Crammer, Dredze, Ganchev, & Talukdar. (n.d.). *Automatic Code Assignment to Medical Text.* Department of Computer and Information Science, University of Pennsylvania, Philadelphia.

Data.world. (n.d.). *Marriage - dataset by fivethirtyeight.* Available at: https://data.world/fivethirtyeight/marriage

Gales, M. J. (1998). Maximum likelihood linear transformations for HMM-based speech recognition. *Computer Speech & Language, 12*(2), 75–98. doi:10.1006/csla.1998.0043

Hales, D. (2004). From Selfish Nodes to Cooperative Networks - Emergent Link-Based Incentives in Peer-to-Peer Networks. *Proc. IEEE Int'l Conf. Peer-to-Peer Computing*, 151-158. 10.1109/PTP.2004.1334942

Inoue, K. (2006, July-August). Quantum key distribution technologies. *IEEE Journal of Selected Topics in Quantum Electronics, 12*(4), 888–896. doi:10.1109/JSTQE.2006.876606

Jyothi & Sridevi. (2018). Low power, low area adaptive finite impulse response filter based on memory less distributed arithmetic. *Journal of Computational and Theoretical Nanoscience, 15*(6-7), 2003-2008.

Kadri, A., Yaacoub, E., Mushtaha, M., & Abu-Dayya, A. (2013, February). Wireless sensor network for real-time air pollution monitoring. In *2013 1st international conference on communications, signal processing, and their applications (ICCSPA)* (pp. 1-5). IEEE. 10.1109/ICCSPA.2013.6487323

Konstantinos, G. (2018). Machine Learning in Agriculture: A Review. *Sensors (Basel), 18*(8), 2674. doi:10.339018082674

Lin, Y. J., Chuang, C. W., Yen, C. Y., Huang, S. H., Chen, J. Y., & Lee, S. Y. (2019). An AIoT wearable ECG patch with decision tree for arrhythmia analysis. In *Proceedings of the 2019 IEEE Biomedical Circuits and Systems Conference (BioCAS)*, (pp. 1-4). IEEE. 10.1109/BIOCAS.2019.8919141

Nandurkar, S. R., Thool, V. R., & Thool, R. C. (2014, February). Design and development of precision agriculture system using wireless sensor network. In *2014 First International Conference on Automation, Control, Energy and Systems (ACES)* (pp. 1-6). IEEE. 10.1109/ACES.2014.6808017

Amitabh, V. D. (n.d.). *From Arranged to Online: A Study of Courtship Culture in India, Book: TOD16-Online-Courtship, Digital Publishing Toolkit, Indian marriage.* Github. Available at: https://github.com/DigitalPublishingToolkit/TOD16-Online-Courtship/blob/master/md/11_Dwivedi.md

Bahga & Madisetti. (2013). A Cloud-based Approach for Interoperable Electronic Health Records (EHRs). *IEEE Journal of Biomedical and Health Informatics, 17*(5).

Broch, J., Maltz, D. A., Johnson, D. B., Hu, Y.-C., & Jetcheva, J. (1998). A Performance Comparison of Multi-Hop Wireless Ad Hoc Network Routing Protocols. *Proc. ACM MobiCom*, 85-97. 10.1145/288235.288256

Jyothi, G. N., Debanjan, K., & Anusha, G. (2020). ASIC Implementation of Fixed-Point Iterative, Parallel, and Pipeline CORDIC Algorithm. In *Soft Computing for Problem Solving* (pp. 341–351). Springer. doi:10.1007/978-981-15-0035-0_27

Jyothi, G. N., & Sriadibhatla, S. (2019). Asic implementation of low power, area efficient adaptive fir filter using pipelined da. In *Microelectronics, electromagnetics and telecommunications* (pp. 385–394). Springer. doi:10.1007/978-981-13-1906-8_40

Lai, Y. H., Chen, S. Y., Lai, C. F., Chang, Y. C., & Su, Y. S. (2019). Study on enhancing AIoT computational thinking skills by plot image-based VR. *Interactive Learning Environments*, 1–14. doi:10.1080/10494820.2019.1580750

Poon, C. C., Yuan-Ting Zhang, & Shu-Di Bao. (2006). A novel biometrics method to secure wireless body area sensor networks for telemedicine and m-health. *IEEE Communications Magazine*, *44*(4), 73–81. doi:10.1109/MCOM.2006.1632652

Qin, Z., Myers, D. B., Ransom, C. J., Kitchen, N. R., Liang, S.-Z., Camberato, J. J., Carter, P. R., Ferguson, R. B., Fernandez, F. G., Franzen, D. W., Laboski, C. A. M., Malone, B. D., Nafziger, E. D., Sawyer, J. E., & Shanahan, J. F. (2018). Application of machine learning methodologies for predicting corn economic optimal nitrogen rate. *Agronomy Journal*, *110*(6), 2596–2607. doi:10.2134/agronj2018.03.0222

Zhong, R. Y., Dai, Q. Y., Qu, T., Hu, G. J., & Huang, G. Q. (2013). RFID-enabled real-time manufacturing execution system for mass-customization production. *Robotics and Computer-integrated Manufacturing*, *29*(2), 283–292. doi:10.1016/j.rcim.2012.08.001

Allen, J. (2007). Photoplethysmography and its application in clinical physiological measurement. *Physiological Measurement*, *28*(3), R1–R39. doi:10.1088/0967-3334/28/3/R01 PMID:17322588

Jyothi & Sridevi. (2018). Low power, low area adaptive finite impulse response filter based on memory less distributed arithmetic. *Journal of Computational and Theoretical Nanoscience*, *15*(6-7), 2003-2008.

Jyothi, Debanjan, & Anusha. (2020). *ASIC Implementation of Fixed-Point Iterative, Parallel, and Pipeline CORDIC Algorithm. In Soft Computing for Problem Solving* (pp. 341–351). Springer.

Lakshmi & Radha. (2012). Selfish aware queue scheduler for packet scheduling in MANET. *Proceedings of Recent Trends in Information Technology (ICRTIT)*.

Laleci, Yuksel, & Dogac. (2013). Providing Semantic Interoperability Between Clinical Care and Clinical Research Domains. *IEEE Journal of Biomedical and Health Informatics, 17*(2).

Rolón, M., & Martínez, E. (2012). Agent-based modeling and simulation of an autonomic manufacturing execution system. *Computers in Industry*, *63*(1), 53–78. doi:10.1016/j.compind.2011.10.005

Sheela, P. J., & Sivaranjani, K. (2015). A brief survey of classification techniques applied to soil fertility prediction. Int Conf Eng Trends Sci Hum, 80-83.

Yu, K. M., Chen, Y. C., Liu, C. H., Hsu, H. P., Lei, M. Y., & Tsai, N. (2019). IFPSS: intelligence fire point sensing systems in AIoT environments. In *Proceedings of the 2019 International Conference on Image and Video Processing, and Artificial Intelligence*. International Society for Optics and Photonics. 10.1117/12.2550315

Azhar, M. A. (2019). *Comparative Review of Feature Selection and Classification modeling, COMP-118-241-Ver-2.* IEEE.

Fritzi-Marie, T. (2013). Changing Patterns of Matchmaking: The Indian Online Matrimonial Market. *Asian Journal of Women's Studies*, *19*(4), 64–94. doi:10.1080/12259276.2013.11666166

Gisin, N., Ribordy, G., Tittel, W., & Zbinden, H. (2002, January). Quantum cryptography. *Reviews of Modern Physics*, *74*(1), 145–195. doi:10.1103/RevModPhys.74.145

Gutiérrez, J., Villa-Medina, J. F., Nieto-Garibay, A., & Porta-Gándara, M. Á. (2013). Automated irrigation system using a wireless sensor network and GPRS module. *IEEE Transactions on Instrumentation and Measurement*, *63*(1), 166–176. doi:10.1109/TIM.2013.2276487

Kannamal, E. (2020, January). Investigation of Speech recognition system and its performance. In *2020 International Conference on Computer Communication and Informatics (ICCCI)* (pp. 1-4). IEEE.

Marchang, N., Datta, R., & Das, S. K. (2017). A Novel Approach for efficient Usage of Intrusion Detection System in Mobile Ad hoc Networks. *IEEE Transactions on Vehicular Technology*, *66*(2), 1684–1695. doi:10.1109/TVT.2016.2557808

Monkaresi, H., Calvo, R. A., & Yan, H. (2014, July). A machine learning approach to improve contactless heart rate monitoring using a webcam. *IEEE Journal of Biomedical and Health Informatics*, *18*(4), 1153–1160. doi:10.1109/JBHI.2013.2291900 PMID:25014930

Shaffer, D. W. (2002). What is digital medicine? *Studies in Health Technology and Informatics*, 195–204. PMID:12026129

Shereen, M. A., Khan, S., Kazmi, A., Bashir, N., & Siddique, R. (2020). COVID-19 infection: Origin, transmission, and characteristics of human coronaviruses. *Journal of Advanced Research*, *24*, 91–98. doi:10.1016/j.jare.2020.03.005 PMID:32257431

Singh, Gupta, Siingh, Sharma, & Singh. (2013). Hybrid Clipping-Companding Schemes for Peak to Average Power Reduction of OFDM Signal. *International Journal of Advances in Engineering Science and Technology*, *2* (1), 114-119.

Suchithra, M. S., & Maya, L. (2020). Improving the prediction accuracy of soil nutrient classification by optimizing extreme learning machine parameters. *Information Processing in Agriculture*, *7*(1), 72–82. doi:10.1016/j.inpa.2019.05.003

Tao, F., & Qi, Q. (2017). New IT driven service-oriented smart manufacturing: Framework and characteristics. *IEEE Transactions on Systems, Man, and Cybernetics. Systems*, *49*(1), 81–91. doi:10.1109/TSMC.2017.2723764

Vinyals, O., Toshev, A., Bengio, S., & Erhan, D. (2015). Show and tell: A neural image caption generator. In *Proceedings of the IEEE conference on computer vision and pattern recognition* (pp. 3156-3164). 10.1109/CVPR.2015.7298935

Anuradha, Raghuram, Sreenivasa Murthy, & Gurunath Reddy. (2009). New Routing Technique to improve Transmission Speed of Data Packets in Point to Point Networks. *ICGST-CNIR Journal, 8*(2).

Chen, Z., Liu, S., & Wang, X. (2008, September). Application of context-aware computing in manufacturing execution system. In *2008 IEEE International Conference on Automation and Logistics* (pp. 1969-1973). IEEE. 10.1109/ICAL.2008.4636484

de Paul Obade, V., & Lal, R. (2016). RTowards a standard technique for soil quality assessment. *Geoderma*, *265*, 96–102. doi:10.1016/j.geoderma.2015.11.023

Jyothi, G. N., & Sridevi, S. (2018). Low power, low area adaptive finite impulse response filter based on memory less distributed arithmetic. *Journal of Computational and Theoretical Nanoscience*, *15*(6-7), 2003–2008. doi:10.1166/jctn.2018.7397

Konig, K. (2000, November). Multiphoton microscopy in life sciences. *Journal of Microscopy*, *200*(2), 83–104. doi:10.1046/j.1365-2818.2000.00738.x PMID:11106949

Lin, Y. J., Chuang, C. W., Yen, C. Y., Huang, S. H., Huang, P. W., Chen, J. Y., & Lee, S. Y. (2019). Artificial intelligence of things wearable system for cardiac disease detection. In *Proceedings of the 2019 IEEE International Conference on Artificial Intelligence Circuits and Systems (AICAS)*, (pp. 67-70). IEEE. 10.1109/AICAS.2019.8771630

Nahato. (2015). *Knowledge Mining from Clinical Datasets Using Rough Sets and Backpropagation Neural Network.* Comp. Math. Methods in Medicine.

Zhou, C., Bai, J., Song, J., Liu, X., Zhao, Z., Chen, X., & Gao, J. (2018). ATRank: An Attention-Based User Behavior Modeling Framework for Recommendation. *Association for the Advancement of Artificial Intelligence. Conference: AAAI 2017.*

Anderson, R. R., & Parrish, J. A. (1981). The optics of human skin. *The Journal of Investigative Dermatology*, *77*(1), 13–19. doi:10.1111/1523-1747.ep12479191 PMID:7252245

Rajalakshmi, D., & Meena, K. (2017). A Survey of intrusion detection with higher malicious misbehavior detection in Manet. International Journal of Civil Engineering and Technology, 8.

Romero, J. R., Roncallo, P. F., Akkiraju, P. C., Ponzoni, I., Echenique, V. C., & Carballido, J. A. (2013). Using classification algorithms for predicting durum wheat yield in the province of Buenos Aires. *Computers and Electronics in Agriculture*, *96*, 173–179. doi:10.1016/j.compag.2013.05.006

Sun, X., Gong, D., & Zhang, W. (2012). Interactive genetic algorithms with large population and semi-supervised learning. *Applied Soft Computing*, *12*(9), 3004–3013. doi:10.1016/j.asoc.2012.04.021

Vanus, J., Smolon, M., Martinek, R., Koziorek, J., Zidek, J., & Bilik, P. (2015). Testing of the voice communication in smart home care. Human-Centric Computing and Information Sciences, 5(15), 1-22. doi:10.118613673-015-0035-0

Wang, C., Bi, Z., & Da Xu, L. (2014). IoT and cloud computing in automation of assembly modeling systems. *IEEE Transactions on Industrial Informatics*, *10*(2), 1426–1434. doi:10.1109/TII.2014.2300346

AIoT Market. (2020). https://mindcommerce.com/artificial-intelligence-of-things/

Chen, Y., Sun, X., Gong, D., Zhang, Y., Choi, J., & Klasky, S. (2017). Personalized search inspired fast interactive estimation of distribution algorithm and its application. *IEEE Transactions on Evolutionary Computation*, *21*(4), 588–600. doi:10.1109/TEVC.2017.2657787

Deverajan, G. G., Muthukumaran, V., Hsu, C. H., Karuppiah, M., Chung, Y. C., & Chen, Y. H. (n.d.). Public key encryption with equality test for Industrial Internet of Things system in cloud computing. *Transactions on Emerging Telecommunications Technologies*, e4202.

Faber, D. J., Aalders, M. C. G., Mik, E. G., Hooper, B. A., van Gemert, M. J. C., & van Leeuwen, T. G. (2004). Oxygen saturation-dependent absorption and scattering of blood. *Physical Review Letters*, *93*(2), 028102. doi:10.1103/PhysRevLett.93.028102 PMID:15323954

Wang, B., Waters, C., Orgill, S., Cowie, A., Clark, A., Li Liu, D., Simpson, M., McGowen, I., & Sides, T. (2018). Estimating soil organic carbon stocks using different modelling techniques in the semi-arid rangelands of eastern Australia. *EcolInd*, *88*, 425–438. doi:10.1016/j.ecolind.2018.01.049

Zhang, L., Wang, X., & Dou, W. (2004, December). A K-connected energy-saving topology control algorithm for wireless sensor networks. In *International Workshop on Distributed Computing* (pp. 520-525). Springer. 10.1007/978-3-540-30536-1_57

Ni, J., Lin, X., & Shen, X. S. (2018). Efficient and secure service-oriented authentication supporting network slicing for 5G-enabled IoT. *IEEE Journal on Selected Areas in Communications*, *36*(3), 644–657. doi:10.1109/JSAC.2018.2815418

Roggan, A., Schädel, D., Netz, U., Ritz, J.-P., Germer, C.-T., & Müller, G. (1999). The effect of preparation technique on the optical parameters of biological tissue. *Applied Physics. B, Lasers and Optics, 69*(5-6), 445–453. doi:10.1007003400050833

Tian, J., Tan, Y., Zeng, J., Sun, C., & Jin, Y. (2018). Multiobjective infill criterion driven gaussian process-assisted particle swarm optimization of high-dimensional expensive problems. *IEEE Transactions on Evolutionary Computation, 23*(3), 459–472. doi:10.1109/TEVC.2018.2869247

Veronesi, F., Korfiati, A., Buffat, R., & Raubal, M. (2017). Assessing accuracy and geographical transferability of machine learning algorithms for wind speed modelling. In Information Science. Springer. doi:10.1007/978-3-319-56759-4_17

Zhu, T., Dhelim, S., Zhou, Z., Yang, S., & Ning, H. (2019). An architecture for aggregating information from distributed data nodes for industrial internet of things. In *Cyber-Enabled Intelligence* (pp. 17–35). Taylor & Francis. doi:10.1201/9780429196621-2

Elavarasan, D., Vincent, D. R., Sharma, V., Zomaya, A. Y., & Srinivasan, K. (2018). Forecasting yield by integrating agrarian factors and machine learning models: A survey. *Computers and Electronics in Agriculture, 155*, 257–282. doi:10.1016/j.compag.2018.10.024

Kim, B. S., & Yoo, S. K. (2006). Motion artifact reduction in photoplethysmography using independent component analysis. *IEEE Transactions on Biomedical Engineering, 53*(3), 566–568. doi:10.1109/TBME.2005.869784 PMID:16532785

Nagarajan, S. M., Muthukumaran, V., Murugesan, R., Joseph, R. B., & Munirathanam, M. (2021). Feature selection model for healthcare analysis and classification using classifier ensemble technique. *International Journal of System Assurance Engineering and Management*, 1-12.

Pan, L., He, C., Tian, Y., Wang, H., Zhang, X., & Jin, Y. (2018). A classification-based surrogate-assisted evolutionary algorithm for expensive many-objective optimization. *IEEE Transactions on Evolutionary Computation, 23*(1), 74–88. doi:10.1109/TEVC.2018.2802784

Park, K. T., Im, S. J., Kang, Y. S., Noh, S. D., Kang, Y. T., & Yang, S. G. (2019). Service-oriented platform for smart operation of dyeing and finishing industry. *International Journal of Computer Integrated Manufacturing, 32*(3), 307–326. doi:10.1080/0951192X.2019.1572225

Peeri, N. C., Shrestha, N., Rahman, M. S., Zaki, R., Tan, Z., Bibi, S., Baghbanzadeh, M., Aghamohammadi, N., Zhang, W., & Haque, U. (2020). The SARS, MERS and novel coronavirus (COVID-19) epidemics, the newest and biggest global health threats: What lessons have we learned? *International Journal of Epidemiology, 49*(3), 1–10. doi:10.1093/ije/dyaa033 PMID:32086938

Kushwah, R., Batra, P. K., & Jain, A. (2020). Internet of things- Architectural elements, challenges and future directions. In *6th International Conference on Signal Processing and Communication*, (pp. 1-5). IEEE. 10.1109/ICSC48311.2020.9182773

Ramollari, E., Dranidis, D., & Simons, A. J. (2007, June). A survey of service oriented development methodologies. In *The 2nd European Young Researchers Workshop on Service Oriented Computing* (Vol. 75). Academic Press.

Sirsat, M. S., Cernadas, E., Fernández-Delgado, M., & Barro, S. (2018). Automatic prediction of village-wise soil fertility for several nutrients in India using a wide range of regression methods. *Computers and Electronics in Agriculture, 154*, 120–133. doi:10.1016/j.compag.2018.08.003

Tian, Y., Yang, S., Zhang, L., Duan, F., & Zhang, X. (2018). A surrogate-assisted multiobjective evolutionary algorithm for large-scale task-oriented pattern mining. *IEEE Trans. Emerg. Top. Comput. Intell., 3*(2), 106–116. doi:10.1109/TETCI.2018.2872055

Wang, L., Lo, B. P. L., & Yang, G.-Z. (2007). Multichannel reflective PPG earpiece sensor with passive motion cancellation. *IEEE Transactions on Circuits and Systems*, *1*(4), 235–241. doi:10.1109/TBCAS.2007.910900 PMID:23852004

Akinsolu, M. O., Liu, B., Grout, V., Lazaridis, P. I., Mognaschi, M. E., & Di Barba, P. (2019). A parallel surrogate model assisted evolutionary algorithm for electromagnetic design optimization. *IEEE Trans. Emerg. Top. Comput. Intell.*, *3*(2), 93–105. doi:10.1109/TETCI.2018.2864747

Birse, R. M. (2004). rev. Patricia E. Knowlden. In Oxford Dictionary of National Biography. Academic Press.

Hossain, M. S., & Muhammad, G. (2016). Cloud-assisted industrial internet of things (iiot)–enabled framework for health monitoring. *Computer Networks*, *101*, 192–202. doi:10.1016/j.comnet.2016.01.009

Kushwah, R., Kulshreshtha, A., Singh, K., & Sharma, S. (2019). ECDSA for data origin authentication and vehicle security in VANET. In *Proceedings of the Twelfth International Conference on Contemporary Computing (IC3)*, (pp. 1-5). IEEE. 10.1109/IC3.2019.8844912

Suchithra, M. S., & Pai, M. L. (2018). Improving the performance of sigmoid kernels in multiclass SVM using optimization techniques for agricultural fertilizer recommendation system. *International Conference on Soft Computing Systems*, 857-868. 10.1007/978-981-13-1936-5_87

Komoda, N. (2006, August). Service oriented architecture (SOA) in industrial systems. In *2006 4th IEEE international conference on industrial informatics* (pp. 1-5). IEEE.

Song, Y., Jiang, J., Yang, D., & Bai, C. (2020). *Prospect and application of Internet of things technology for prevention of SARIs. In Clinical eHealth* (Vol. 3). Elsevier.

Swapna, B., & Manivannan, S. (2018). Analysis: Smart Agriculture and Landslides Monitoring System Using Internet of Things (IoT). *International Journal of Pure and Applied Mathematics*, *118*, 24.

Tian, J., Sun, C., Tan, Y., & Zeng, J. (2020). Granularity-based surrogate-assisted particle swarm optimization for high-dimensional expensive optimization. *Knowl.- Based Syst.*, *187*, 104815. doi:10.1016/j.knosys.2019.06.023

Waller, A. D. (1887). A demonstration on man of electromotive changes accompanying the heart's beat. *The Journal of Physiology*, *8*(5), 229–234. doi:10.1113/jphysiol.1887.sp000257 PMID:16991463

Bao, L., Sun, X., Chen, Y., Man, G., & Shao, H. (2018). Restricted boltzmann machine assisted estimation of distribution algorithm for complex problems. *Complexity*, *2018*, 2018. doi:10.1155/2018/2609014

Dai, H. N., Imran, M., & Haider, N. (2020). Blockchain-enabled internet of medical things to combat COVID-19. *IEEE Internet Things Magazine*, *3*(3), 52–57. doi:10.1109/IOTM.0001.2000087

Hill, M. G., Connolly, P. G., Reutemann, P., & Fletcher, D. (2014). The use of data mining to assist crop protection decisions on kiwifruit in New Zealand. *Computers and Electronics in Agriculture*, *108*, 250–257. doi:10.1016/j.compag.2014.08.011

Rivera-Ruiz, M. (2008). Einthoven's string galvanometer: The first electrocardiograph. *Texas Heart Institute Journal*, *35*, 174. PMID:18612490

Yaqoob, I., Ahmed, E., Hashem, I. A. T., Ahmed, A. I. A., Gani, A., Imran, M., & Guizani, M. (2017). Internet of things architecture: Recent advances, taxonomy, requirements, and open challenges. *IEEE Wireless Communications*, *24*(3), 10–16. doi:10.1109/MWC.2017.1600421

Hurst, J. W. (1998). Naming of the waves in the ECG, with a brief account of their genesis. *Circulation*, *98*(18), 1937–1942. doi:10.1161/01.CIR.98.18.1937 PMID:9799216

Lu, J., Wu, D., Mao, M., Wang, W., & Zhang, G. (2015). Recommender system application developments: A survey. *Decision Support Systems*, *74*, 12–32. doi:10.1016/j.dss.2015.03.008

McGovern, J., Ambler, S. W., Stevens, M. E., Linn, J., Jo, E. K., & Sharan, V. (2004). *A practical guide to enterprise architecture*. Prentice Hall Professional.

Muthu, B., Sivaparthipan, C. B., Manogaran, G., Sundarasekar, R., Kadry, S., Shanthini, A., & Dasel, A. (2020). IOT based wearable sensor for diseases prediction and symptom analysis in healthcare sector. *Peer-to-Peer Networking and Applications*, *13*(6), 2123–2134. doi:10.100712083-019-00823-2

Ritz, C., Putku, E., & Astover, A. (2015). A practical two-step approach for mixed model-based kriging, with an application to the prediction of soil organic carbon concentration. *European Journal of Soil Science*, *66*(3), 548–554. doi:10.1111/ejss.12238

Davis, J., Edgar, T., Porter, J., Bernaden, J., & Sarli, M. (2012). Smart manufacturing, manufacturing intelligence and demand-dynamic performance. *Computers & Chemical Engineering*, *47*, 145–156. doi:10.1016/j.compchemeng.2012.06.037

Gerber, R., & Nagel, N. H. (1996, September). Knowledge representation for the generation of quantified natural language descriptions of vehicle traffic in image sequences. In *Proceedings of 3rd IEEE international conference on image processing* (Vol. 2, pp. 805-808). IEEE. 10.1109/ICIP.1996.561027

Izbassarova, A., Duisembay, A., & James, A. P. (2020). Speech Recognition Application Using Deep Learning Neural Network. In *Deep Learning Classifiers with Memristive Networks* (pp. 69–79). Springer. doi:10.1007/978-3-030-14524-8_5

Jabbar, M. A., & Samreen, S. (2016). Heart disease prediction system based on hidden naïve bayes classifier. *2016 International Conference on Circuits, Controls, Communications and Computing (I4C)*.

Khan & Arora. (2019). Classification in Thermograms for Breast Cancer Detection using Texture Features with Feature Selection Method and Ensemble Classifier. *2nd International Conference on Issues and Challenges in Intelligent Computing Techniques (ICICT)*.

Lupia, T., Scabini, S., Pinna, S. M., Di Perri, G., De Rosa, F. G., & Corcione, S. (2020). 2019 novel coronavirus (2019-nCoV) outbreak: A new challenge. *Journal of Global Antimicrobial Resistance*, *21*, 22–27. doi:10.1016/j.jgar.2020.02.021 PMID:32156648

Mailloux, L. O., Grimaila, M. R., Hodson, D. D., Baumgartner, G., & McLaughlin, C. (2015, January). Performance evaluations of quantum key distribution system architectures. *IEEE Security and Privacy*, *13*(1), 30–40. doi:10.1109/MSP.2015.11

Mukhopadhyay, S. C. (2015). Wearable sensors for human activity monitoring: A review. *IEEE Sensors Journal*, *15*(3), 1321–1330. doi:10.1109/JSEN.2014.2370945

Muralibabu, & Sundhararajan. (2013). DCT Based PAPR and ICI Reduction Using Companding Technique in MIMO-OFDM System. *International Journal of Advanced Research in Computer Science and Software Engineering*, *3*, 49-53.

Rahman, Mitra, & Islam. (2018). Soil Classification using Machine Learning methods and crop suggestion based on soil series. *IEEE 21st International conference of Computer and Information Technology (ICCIT)*.

Rajalakshmi, D., Meena, K., Vijayaraj, N., & Uganya, G. (2021). *Investigation and analysis of Path Evaluation for Sustainable Communications using VANET*. Springer.

Ram Prasad, A. V. S., Ramji, K., Kolli, M., & Vamsi Krishna, G. (2019). Multi-Response Optimization of Machining Process Parameters for Wire Electrical Discharge Machining of Lead-Induced Ti-6Al-4V Alloy Using AHP–TOPSIS Method. *Journal of Advanced Manufacturing Systems*, *18*(02), 213–236. doi:10.1142/S0219686719500112

Sharma. (2018). Marriages made in heaven with a little help from Bayes. *Factor Daily*.

Burke, E. (1998). *Precision heart rate training*. Human Kinetics.

De Deugd, S., Carroll, R., Kelly, K., Millett, B., & Ricker, J. (2006). SODA: Service oriented device architecture. *IEEE Pervasive Computing*, *5*(3), 94–96. doi:10.1109/MPRV.2006.59

Koncar, J., Grubor, A., Maric, R., Vucenovic, S., & Vukmirovic, G. (2020). Setbacks to IoT implementation in the function of FMCG supply chain sustainability during COVID-19 Pandemic. *Sustainability*, *12*(18), 7391. doi:10.3390u12187391

Mao, M., Lu, J., Han, J., & Zhang, G. (2019). Multiobjective e-commerce recommendations based on hypergraph ranking. *Inform. Sci.*, *471*, 269–287. doi:10.1016/j.ins.2018.07.029

Terhoeven-Urselmans, T., Vagen, T.-G., Spaargaren, O., & Shepherd, K. D. (2010). Prediction of soil fertility properties from a globally distributed soil mid-infrared spectral library. *Soil Science Society of America Journal*, *74*(5), 1792–1799. doi:10.2136ssaj2009.0218

Al-Tudjman, F., & Deepak, B. D. (2020). Privacy-aware energy-efficient framework using the internet of medical things for COVID-19. *IEEE Internet Things Magazine*, *3*(3), 64–68. doi:10.1109/IOTM.0001.2000123

Eldridge, J. (2014). Recording a standard 12-lead electrocardiogram. An approved methodology by the Society of Cardiological Science and Technology (SCST). *Clinical Guidelines by Consensus*.

Gholami, M. F., Habibi, J., Shams, F., & Khoshnevis, S. (2010, March). Criteria-Based evaluation framework for service-oriented methodologies. In *2010 12th International Conference on Computer Modelling and Simulation* (pp. 122-130). IEEE. 10.1109/UKSIM.2010.30

Prabu & Margret Anouncia. (2018). Prediction of Land Cover Changes in Vellore District of Tamil Nadu by Using Satellite Image Processing. *Knowledge Computing and its Applications*, 87-100.

Rendle, S., Freudenthaler, C., Gantner, Z., & Schmidt-Thieme, L. (2009). Bpr: bayesian personalized ranking from implicit feedback. *Conference on Uncertainty in Artificial Intelligence*, 452–461.

Alvarez, R. A., Penín, A. J. M., & Sobrino, X. A. V. (2013). A comparison of three QRS detection algorithms over a public database. *Procedia Technology*, *9*, 1159–1165. doi:10.1016/j.protcy.2013.12.129

Anitha, A., & Acharjya, D. P. (2017). Crop suitability prediction in Vellore District using rough set on fuzzy approximation space and neural network. *The Natural Computing Applications Forum, 30*, 3633–3650.

Papazoglou, M. P., & Van Den Heuvel, W. J. (2006). Service-oriented design and development methodology. *International Journal of Web Engineering and Technology*, *2*(4), 412–442. doi:10.1504/IJWET.2006.010423

Singh, R. P., Javaid, M., Haleem, A., & Suman, R. (2020). Internet of things (IoT) applications to fight against COVID-19 pandemic. *Diabetes & Metabolic Syndrome*, *14*(4), 521–524. doi:10.1016/j.dsx.2020.04.041 PMID:32388333

Wang, X., Gao, C., Ding, J., Li, Y., & Jin, D. (2019). Cmbpr: Category-aided multichannel bayesian personalized ranking for short video recommendation. *IEEE Access: Practical Innovations, Open Solutions*, *7*, 48209–48223. doi:10.1109/ACCESS.2019.2907494

Chi, Y. M., Jung, T.-P., & Cauwenberghs, G. (2010). Dry-contact and noncontact biopotential electrodes: Methodological review. *IEEE Reviews in Biomedical Engineering*, *3*, 106–119. doi:10.1109/RBME.2010.2084078 PMID:22275204

Muthukumaran, V., & Manimozhi, I. (2021). Public Key Encryption With Equality Test for Industrial Internet of Things Based on Near-Ring. *International Journal of e-Collaboration*, *17*(3), 25–45. doi:10.4018/IJeC.2021070102

Rendle, S. (2012). Factorization machines with libfm. *ACM Transactions on Intelligent Systems and Technology, 3*(3), 1–22. doi:10.1145/2168752.2168771

Shaikh, R. A., & Nikilla, S. (2012). Real time health monitoring system of remote patient using ARM7. *International Journal of Instrumentation, Controle & Automação, 1*(3-4), 102–105.

Deverajan, G. G., Muthukumaran, V., Hsu, C. H., Karuppiah, M., Chung, Y. C., & Chen, Y. H. Public key encryption with equality test for Industrial Internet of Things system in cloud computing. *Transactions on Emerging Telecommunications Technologies*, e4202.

Lee, Y.-D., & Chung, W.-Y. (2009). Wireless sensor network based wearable smart shirt for ubiquitous health and activity monitoring. *Sensors and Actuators. B, Chemical, 140*(2), 390–395. doi:10.1016/j.snb.2009.04.040

Mdhaffar, A., Chaari, T., Larbi, K., Jmaiel, M., & Freisleben, B. (2017). IoT-based health monitoring via LoRaWAN. In *IEEE EUROCON 2017-17th International Conference on Smart Technologies*, (pp. 519-524). IEEE. 10.1109/EUROCON.2017.8011165

Xiao, J., Hao, Y., He, X., Zhang, H., & Chua, T. S. (2017). Attentional factorization machines: Learning the weight of feature interactions via attention networks. *Proceedings of the 26th International Joint Conference on Artificial Intelligence*, 3119–3125. 10.24963/ijcai.2017/435

Archana, R., & Indira, S. (2013). Soldier monitoring and health indication system. *International Journal of Science and Research*.

Cheng, H.-T., Koc, L., Harmsen, J., Shaked, T., Chandra, T., Aradhye, H., Anderson, G., Corrado, G., Chai, W., & Ispir, M. (2016). Wide & deep learning for recommender systems. In *Proceedings of the 1st Workshop on Deep Learning for Recommender Systems*. ACM. 10.1145/2988450.2988454

Muthukumaran, V., & Ezhilmaran, D. (2020). A Cloud-Assisted Proxy Re-Encryption Scheme for Efficient Data Sharing Across IoT Systems. *International Journal of Information Technology and Web Engineering, 15*(4), 18–36. doi:10.4018/IJITWE.2020100102

Parak & Korhonen. (2014). Evaluation of wearable consumer heart rate monitors based on photopletysmography. *IEEE EMBC*, 3670-3673.

Alfarhood, M., & Cheng, J. (2018). Deephcf: A deep learning based hybrid collaborative filtering approach for recommendation systems. In *2018 17th IEEE International Conference on Machine Learning and Applications (ICMLA)*. IEEE. 10.1109/ICMLA.2018.00021

Dias. (2009). Measuring physical activity with sensors: a qualitative study. *MIE*, 475-479.

Nagarajan, S. M., Muthukumaran, V., Murugesan, R., Joseph, R. B., & Munirathanam, M. (2021). Feature selection model for healthcare analysis and classification using classifier ensemble technique. *International Journal of System Assurance Engineering and Management*, 1-12.

Zhao, R., Yan, R., Chen, Z., Mao, K., Wang, P., & Gao, R. X. (2019). Deep learning and its applications to machine health monitoring. *Mechanical Systems and Signal Processing, 115*, 213–237. doi:10.1016/j.ymssp.2018.05.050

Muthukumaran, V. (2021). Efficient Digital Signature Scheme for Internet of Things. *Turkish Journal of Computer and Mathematics Education, 12*(5), 751–755.

Nweke, H. F., Teh, Y. W., Al-Garadi, M. A., & Alo, U. R. (2018). Deep learning algorithms for human activity recognition using mobile and wearable sensor networks: State of the art and research challenges. *Expert Systems with Applications, 105*, 233–261. doi:10.1016/j.eswa.2018.03.056

Salakhutdinov, R., Mnih, A., & Hinton, G. (2007). Restricted boltzmann machines for collaborative filtering. In *Proceedings of the 24th International Conference on Machine Learning, in: ICML '07*. ACM. 10.1145/1273496.1273596

Summers, R. L., Shoemaker, W. C., Peacock, W. F., Ander, D. S., & Coleman, T. G. (2003). Bench to bedside: Electrophysiologic and clinical principles of noninvasive hemodynamic monitoring using impedance cardiography. *Academic Emergency Medicine, 10*(6), 669–680. doi:10.1111/j.1553-2712.2003.tb00054.x PMID:12782531

Da Costa, C. A., Pasluosta, C. F., Eskofier, B., da Silva, D. B., & da Rosa, R. R. (2018). Internet of Health Things: Toward intelligent vital signs monitoring in hospital wards. *Artificial Intelligence in Medicine, 89*, 61–69. doi:10.1016/j.artmed.2018.05.005 PMID:29871778

Lee, S. (2013). A low-power and compact-sized wearable bioimpedance monitor with wireless connectivity. *Journal of Physics: Conference Series.*

Shibata, H., & Takama, Y. (2017). Behavior analysis of rbm for estimating latent factor vectors from rating matrix. In *2017 6th International Conference on Informatics, Electronics and Vision, 7th International Symposium in Computational Medical and Health Technology (ICIEV-ISCMHT)*. IEEE. 10.1109/ICIEV.2017.8338568

Alam, F., Mehmood, R., Katib, I., Albogami, N. N., & Albeshri, A. (2017). Data fusion and IoT for smart ubiquitous environments: A survey. *IEEE Access: Practical Innovations, Open Solutions, 5*, 9533–9554. doi:10.1109/ACCESS.2017.2697839

Ruiz, J. C. M. (2013). Textrode-enabled transthoracic electrical bioimpedance measurements-towards wearable applications of impedance cardiography. *J. Elect. Bioimp., 4*(1), 45–50. doi:10.5617/jeb.542

Suzuki, Y., & Ozaki, T. (2017). Stacked denoising auto encoder-based deep collaborative filtering using the change of similarity. *2017 31st International Conference on Advanced Information Networking and Applications Workshops (WAINA)*, 498–502.

Abouty, S., Renfa, L., Fanzi, Z., & Mangone, F. (2013, February). A Novel Iterative Clipping and Filtering Technique for PAPR Reduction of OFDM Signals: System Using DCT/IDCT Transform. *International Journal of Future Generation Communication and Networking, 6*(1), 1–8.

Elhoseny, H., Elhoseny, M., Abdelrazek, S., Bakry, H., & Riad, A. (2016). Utilizing service oriented architecture (SOA) in smart cities. *International Journal of Advancements in Computing Technology, 8*(3), 77–84.

Gayathri Devi, S., & Sabrigiriraj Feature Selection, M. (2018). Online Feature Selection Techniques for Big Data Classification: - A Review. *Proceeding of 2018 IEEE International Conference on Current Trends toward Converging Technologies*. 10.1109/ICCTCT.2018.8550928

Kenkel, N. C. (2019). On selecting an appropriate multivariate analysis. *Canadian Journal of Plant Science.*

Kononenko, I. (2001). Machine learning for medical diagnosis: History, state of the art and perspective. *Artificial Intelligence in Medicine, 23*(1), 89–109. doi:10.1016/S0933-3657(01)00077-X PMID:11470218

Lockdown. (2020). https://www.theguardian.com/world/2020/apr/04/lockdown-could-last-weeks-more-across-europe-officials-warn

Mailloux, L. O., Hodson, D. D., Grimaila, M. R., Engle, R. D., McLaughlin, C. V., & Baumgartner, G. B. (2016). Using modeling and simulation to study photon number splitting attacks. *IEEE Access: Practical Innovations, Open Solutions, 4*, 2188–2197. doi:10.1109/ACCESS.2016.2555759

Masumura, R., Ihori, M., Takashima, A., Moriya, T., Ando, A., & Shinohara, Y. (2020, May). Sequence-Level Consistency Training for Semi-Supervised End-toEnd Automatic Speech Recognition. In *ICASSP 2020-2020 IEEE International Conference on Acoustics, Speech and Signal Processing (ICASSP)* (pp. 7054- 7058). IEEE.

Minka, T., Cleven, R., & Zaykov, Y. (2018). *TrueSkill 2: An improved Bayesian skill rating system*. Available at: https://academic.microsoft.com/paper/2791828207/reference?showAllAuthors=1

Mohammed Musthafa, M., Vanitha, K., Zubair Rahman, A. M. J. M. D., & Anitha, K. (2020). *An Efficient Approach to identify Selfish Node in MANET*. IEEE Xplore.

Patel, S., Park, H., Bonato, P., Chan, L., & Rodgers, M. (2012). A review of wearable sensors and systems wit application in rehabilitation. *Journal of Neuroengineering and Rehabilitation, 9*(1), 21. doi:10.1186/1743-0003-9-21 PMID:22520559

Venkatarao, K., & Anup Kumar, T. (2019). An experimental parametric analysis on performance characteristics in wire electric discharge machining of Inconel 718. *Proceedings of the Institution of Mechanical Engineers. Part C, Journal of Mechanical Engineering Science, 233*(14), 4836–4849. doi:10.1177/0954406219840677

Yao, B. Z., Yang, X., Lin, L., Lee, M. W., & Zhu, S. C. (2010). I2t: Image parsing to text description. *Proceedings of the IEEE, 98*(8), 1485–1508. doi:10.1109/JPROC.2010.2050411

Islam, S. R., Kwak, D., Kabir, M. H., Hossain, M., & Kwak, K. S. (2015). The internet of things for health care: A comprehensive survey. *IEEE Access: Practical Innovations, Open Solutions, 3*, 678–708. doi:10.1109/ACCESS.2015.2437951

Liang, D., Krishnan, R. G., Hoffman, M. D., & Jebara, T. (2018). Variational autoencoders for collaborative filtering. In *The 2018 World Wide Web Conference*. ACM. 10.1145/3178876.3186150

Sherwood, A., McFetridge, J., & Hutcheson, J. S. (1998). Ambulatory impedance cardiography: A feasibility study. *Journal of Applied Physiology (Bethesda, Md.), 85*(6), 2365–2369. doi:10.1152/jappl.1998.85.6.2365 PMID:9843565

Chen, W. (2015). Wearable solutions using bioimpedance for cardiac monitoring. In Recent Advances in Ambient Assisted Living- Bridging Assistive Technologies, E-Health and Personalized Health Care. Academic Press.

Kim, D., Park, C., Oh, J., Lee, S., & Yu, H. (2016). Convolutional matrix factorization for document context-aware recommendation. *Proceedings of the 10th ACM Conference on Recommender Systems*, 233–240. 10.1145/2959100.2959165

Shu, Y., Li, C., Wang, Z., Mi, W., Li, Y., & Ren, T.-L. (2015). A pressure sensing system for heart rate monitoring with polymer-based pressure sensors and an anti-Interference post processing circuit. *Sensors (Basel), 15*(2), 3224–3235. doi:10.3390150203224 PMID:25648708

Xu, C., Zhao, P., Liu, Y., Xu, J., Sheng, V. S. S. S., Cui, Z., Zhou, X., & Xiong, H. (2019). Recurrent convolutional neural network for sequential recommendation. *The World Wide Web Conference*, 3398–3404. 10.1145/3308558.3313408

Gong, S. (2014). *A wearable and highly sensitive pressure sensor with ultrathin gold nanowires* (Vol. 5). Nat. Commun.

Jeremy & Jonathan. (n.d.). Comparison of frequentist and Bayesian inference. Class 20, 18.05. In *Reading 20: Comparison of frequentist and Bayesian Inference*. Mit.edu.

Cotes, J. E. (2009). Lung function: Physiology, measurement and application in medicine. John Wiley & Sons.

UCSD-AI4H 2020, UCSD-AI4H/COVID-CT. (2020). *GitHub*. Available at: github.com/UCSD-AI4H/COVID-CT

Goodman. (1978). Relationship of smoking history and pulmonary function tests to tracheal mucous velocity in non-smokers, young smokers, ex-smokers, and patients with chronic bronchitis 1–3. Am. Rev. Respir. Dis., 117, 205-214.

McBride. (2020). *COVID-19 Symptoms: Day-by-Day Chart of Coronavirus Signs*. Heavy.com.

Murphy, R. L. (1981). Auscultation of the lung: Past lessons, future possibilities. *Thorax, 36*(2), 99–107. doi:10.1136/thx.36.2.99 PMID:7268687

Science Daily. (2020). *Study Confirms 'classic' Symptoms of COVID-19.* www.sciencedaily.com/releases/2020/06/200624100047.htm

Harvard University. (n.d.). *Indian Online Matrimony Data Exploration.* Available at: https://projects.iq.harvard.edu/matrimony_data_exploration/home

Selley, W., Flack, F. C., Ellis, R. E., & Brooks, W. A. (1989). Respiratory patterns associated with swallowing: Part 1. The normal adult pattern and changes with age. *Age and Ageing, 18*(3), 168–172. doi:10.1093/ageing/18.3.168 PMID:2782213

Otis, A. B., Fenn, W. O., & Rahn, H. (1950). Mechanics of breathing in man. *Journal of Applied Physiology, 2*(11), 592–607. doi:10.1152/jappl.1950.2.11.592 PMID:15436363

Al-Khalidi, F. Q., Saatchi, R., Burke, D., Elphick, H., & Tan, S. (2011). Respiration rate monitoring methods: A review. *Pediatric Pulmonology, 46*(6), 523–529. doi:10.1002/ppul.21416 PMID:21560260

Aich, Al-Absi, Hui, & Sain. (2018). Prediction of Quality for Different Type of Wine based on Different Feature Sets Using Supervised Machine Learning Techniques. *ICACT Transactions on Advanced Communications Technology, 7*(3).

Bao, L., Sun, X., Chen, Y., Gong, D., & Zhang, Y. (2020). Restricted Boltzmann Machine-driven Interactive Estimation of Distribution Algorithm for personalized search, ScienceDirect. *Knowledge-Based Systems, 200,* 106030. doi:10.1016/j.knosys.2020.106030

Bartholomew. (2016). A simple two-way multiple analysis of variance-spss. *Research & Reviews: Journal of Statistics & Mathematical Sciences, 1.*

El Rifai, M. (2016). *Quantum secure communication using polarization hopping multi-stage protocols* (Ph.D. dissertation). School Elect. Comput. Eng., Univ. Oklahoma, Norman, OK.

Farhadi, A., Hejrati, M., Sadeghi, M. A., Young, P., Rashtchian, C., Hockenmaier, J., & Forsyth, D. (2010, September). Every picture tells a story: Generating sentences from images. In *European conference on computer vision* (pp. 15-29). Springer. 10.1007/978-3-642-15561-1_2

Joshi, H. D. (2012). *Performance augmentation of OFDM system* (Ph.D. dissertation). Jaypee Univ. of engineering and Technology.

Lavračʹc. (1998). Intelligent data analysis for medical diagnosis: Using machine learning and temporal abstraction. *Artif. Intell. Commun., 11*(3), 191–218.

Park, S., & Jayaraman, S. (2003). Enhancing the quality of life throug wearable technology. *IEEE Engineering in Medicine and Biology Magazine, 22*(3), 41–48. doi:10.1109/MEMB.2003.1213625 PMID:12845818

Prasad, P. I., Basha, S. G., & Janjanam, N. (2020). *Design, Fabrication and Evaluation of Thermal Performance of Parabolic Trough Collector with Elliptical Absorber Using Zn/H2O Nanofluid.* Academic Press.

Quarantine. (2020). https://ncdc.gov.in/WriteReadData/l892s/90542653311584546120.pdf

Reis, J. Z., & Gonçalves, R. F. (2018, August). The role of internet of services (ios) on industry 4.0 through the service oriented architecture (soa). In *IFIP International Conference on Advances in Production Management Systems* (pp. 20-26). Springer. 10.1007/978-3-319-99707-0_3

Shakshuki, E. M., Kang, N., & Sheltami, T. R. (2013). EAACK - a secure intrusion detection system for MANETs. *IEEE Transactions on Industrial Electronics, 60*(3), 1089–1098. doi:10.1109/TIE.2012.2196010

Yasir, M., Nababan, M. N., Laia, Y., Purba, W., & Gea, A. (2019, July). WebBased Automation Speech-to-Text Application using Audio Recording for Meeting Speech. *Journal of Physics: Conference Series, 1230*(1), 012081. doi:10.1088/1742-6596/1230/1/012081

Clark, F., & von Euler, C. (1972). On the regulation of depth and rate of breathing. *The Journal of Physiology, 222*(2), 267–295. doi:10.1113/jphysiol.1972.sp009797 PMID:5033464

Netzer, N., Eliasson, A. H., Netzer, C., & Kristo, D. A. (2001). Overnight pulse oximetry for sleep-disordered breathing in adults: A review. *Chest, 120*(2), 625–633. doi:10.1378/chest.120.2.625 PMID:11502669

Sackner, M. A., Watson, H., Belsito, A. S., Feinerman, D., Suarez, M., Gonzalez, G., Bizousky, F., & Krieger, B. (1989). Calibration of respiratory inductive plethysmograph during natural breathing. *Journal of Applied Physiology (Bethesda, Md.), 66*(1), 410–420. doi:10.1152/jappl.1989.66.1.410 PMID:2917945

Seymour, R. S. (2010). Scaling of heat production by thermogenic flowers: Limits to floral size and maximum rate of respiration. *Plant, Cell & Environment, 33*, 1474–1485. doi:10.1111/j.1365-3040.2010.02190.x PMID:20545882

Bianchi, W., Dugas, A. F., Hsieh, Y.-H., Saheed, M., Hill, P., Lindauer, C., Terzis, A., & Rothman, R. E. (2013). Revitalizing a vital sign: Improving detection of tachypnea at primary triage. *Annals of Emergency Medicine, 61*(1), 37–43. doi:10.1016/j.annemergmed.2012.05.030 PMID:22738682

Mathew, J., Semenova, Y., & Farrell, G. (2012). A miniature optical breathing sensor. *Biomedical Optics Express, 3*(12), 3325–3331. doi:10.1364/BOE.3.003325 PMID:23243581

Flisberg, P., Jakobsson, J., & Lundberg, J. (2002). Apnea and bradypnea in patients receiving epidural bupivacaine-morphine for postoperative pain relief as assessed by a new monitoring method. *Journal of Clinical Anesthesia, 14*(2), 129–134. doi:10.1016/S0952-8180(01)00369-5 PMID:11943527

Cabot, R. C. (2008). Case 12-2008: A newborn infant with intermittent apnea and seizures. *The New England Journal of Medicine, 358*(16), 1713–1723. doi:10.1056/NEJMcpc0801164 PMID:18420504

Stemp, L. I., & Ramsay, M. A. (2006). Oxygen may mask hypoventilation: Patient breathing must be ensured. *APSF Newsletter, 20*, 80.

Fohr, S. A. (1998). The double effect of pain medication: Separating myth from reality. *Journal of Palliative Medicine, 1*(4), 315–328. doi:10.1089/jpm.1998.1.315 PMID:15859849

Bandodkar, A. J., & Wang, J. (2014). Non-invasive wearable electrochemical sensors: A review. *Trends in Biotechnology, 32*(7), 363–371. doi:10.1016/j.tibtech.2014.04.005 PMID:24853270

Dhanya, Paul, Akula, Sivakumar, Jyothisha, & Nair. (2019). Comparative Study for Breast Cancer Prediction using Machine Learning and Feature Selection. *Proceedings of the International Conference on Intelligent Computing and Control Systems.*

Elmehdwi, Y., & Samanthula, B. K. (2014). Secure k-nearest neighbor query over encrypted data in outsourced environments. *Proc. IEEE 30th Int. Conf. Data Eng.*, 664–675.

Erika, F. M. (2017). Prediction of soil physical and chemical properties by visible & near infrared diffuse reflectance spectroscopy in the central amazon. *Remote Sensing, 9*(4), 293. doi:10.3390/rs9040293

Jayasuruthi, L., Shalini, A., & Vinoth Kumar, V. (2018). Application of rough set theory in data mining market analysis using rough sets data explorer. *Journal of Computational and Theoretical Nanoscience, 15*(6-7), 2126–2130. doi:10.1166/jctn.2018.7420

Jijina, N., & Pillai, S. S. (2014, April). Linear Precoding Schemes for PAPR Reduction in Mobile WiMAX OFDMA System. *International Journal of Advanced Research in Electrical, Electronics and Instrumentation Engineering, 3*, 8873–8884.

Kumar, G. N. S., & Srinath, A. (2020). Induction Motor for Pedestrian Transportation in Benz Circle Vijayawada. *Journal of Information Science and Engineering, 36*(2).

Liang, L., Lin, L., Jin, L., Xie, D., & Li, M. (2018). SCUT-FBP5500: A Diverse Benchmark Dataset for Multi-Paradigm Facial Beauty PredictionSCUT-FBP5500, Computer Science. *24th International Conference on Pattern Recognition (ICPR)*. 10.1109/ICPR.2018.8546038

Masumura, R., Sato, H., Tanaka, T., Moriya, T., Ijima, Y., & Oba, T. (2019). Endto-End Automatic Speech Recognition with a Reconstruction Criterion Using Speech-to-Text and Text-to-Speech Encoder-Decoders. *Proc. Interspeech, 2019*, 1606–1610. doi:10.21437/Interspeech.2019-2111

Mohammadi, M., & Mukhtar, M. (2018, September). Service-oriented architecture and process modeling. In *2018 International Conference on Information Technologies (InfoTech)* (pp. 1-4). IEEE.

Rajalakshmi, D., & Meena, K. (2019). A Novel based fuzzy cognitive maps protocol for intrusion discovery in Manets. International Journal of Recent Technology and Engineering, 7.

Social Distancing. (2020). https://www.health.gov.au/sites/default/files/documents/2020/03/coronavirus-covid-19-information-on-social-distancing.pdf

Yang, Y., Teo, C., Daumé, H., III, & Aloimonos, Y. (2011, July). Corpus-guided sentence generation of natural images. In *Proceedings of the 2011 Conference on Empirical Methods in Natural Language Processing* (pp. 444-454). Academic Press.

Hanly, P. J., Millar, T. W., Steljes, D. G., Boert, R., Frais, M. A., & Kryger, M. H. (1989). Respiration and abnormal sleep in patients with congestive heart failure. *Chest, 96*(3), 480–488. doi:10.1378/chest.96.3.480 PMID:2766808

Dolfin, T. (1983). Effects of a face mask and pneumotachograph on breathing in sleeping infants 1–3. *The American Review of Respiratory Disease, 128*, 977–979. PMID:6650989

Varady, P., Micsik, T., Benedek, S., & Benyo, Z. (2002). A novel method for the detection of apnea and hypopnea events in respiration signals. *IEEE Transactions on Biomedical Engineering, 49*(9), 936–942. doi:10.1109/TBME.2002.802009 PMID:12214883

Warren, R., Horan, S. M., & Robertson, P. K. (1997). Chest wall motion in preterm infants using respiratory inductive plethysmography. *The European Respiratory Journal, 10*(10), 2295–2300. doi:10.1183/09031936.97.10102295 PMID:9387956

Watanabe, K., Watanabe, T., Watanabe, H., Ando, H., Ishikawa, T., & Kobayashi, K. (2005). Noninvasive measurement of heartbeat, respiration, snoring and body movements of a subject in bed via a pneumatic method. *IEEE Transactions on Biomedical Engineering, 52*(12), 2100–2107. doi:10.1109/TBME.2005.857637 PMID:16366233

Wilhelm, F. H., Roth, W. T., & Sackner, M. A. (2003). The LifeShirt an advanced system for ambulatory measurement of respiratory and cardiac function. *Behavior Modification, 27*(5), 671–691. doi:10.1177/0145445503256321 PMID:14531161

Yumino, D., & Bradley, T. D. (2008). Central sleep apnea and Cheyne- Stokes respiration. *Proceedings of the American Thoracic Society, 5*(2), 226–236. doi:10.1513/pats.200708-129MG PMID:18250216

Toy, P., Popovsky, M. A., Abraham, E., Ambruso, D. R., Holness, L. G., Kopko, P. M., McFarland, J. G., Nathens, A. B., Silliman, C. C., & Stroncek, D. (2005). Transfusion-related acute lung injury: Definition andreview. *Critical Care Medicine*, *33*(4), 721–726. doi:10.1097/01.CCM.0000159849.94750.51 PMID:15818095

Wasserman, K. (1994). Coupling of external to cellular respiration during exercise: The wisdom of the body revisited. *American Journal of Physiology. Endocrinology and Metabolism*, *266*(4), E519–E539. doi:10.1152/ajpendo.1994.266.4.E519 PMID:8178973

Crapo, R. O. (1994). Pulmonary-function testing. *The New England Journal of Medicine*, *331*(1), 25–30. doi:10.1056/NEJM199407073310107 PMID:8202099

Anupkumar, T., Sharma, N., Srinath, A., & Dileep, G. S. (2019). Topology Optimization for Diffusive Heat Transfer in a Domain with Internal Heat Generation using Optimality Criteria. *International Journal of Vehicle Structures & Systems*, *11*(4), 447–451. doi:10.4273/ijvss.11.4.22

Elmehdwi, Y., Samanthula, B. K., & Jiang, W. (2015). k-Nearest neighbor classification over semantically secure encrypted relational data. *IEEE Trans. Knowledge Data Eng.* Available: https://ieeexplore.ieee.org/xpl/articleDetails.jsp?arnumber=6930802

El-Tarhuni, Hassan, & Sediq. (2010). A Joint Power Allocation and Adaptive Channel Coding Scheme for Image Transmission over Wireless Channels. *International Journal of Computer Networks & Communications, 12*(3).

Harun, N. Z., Zukarnain, Z. A., Hanapi, Z. M., & Ahmad, I. (2018). Evaluation of Parameters Effect in Multiphoton Quantum Key Distribution Over Fiber Optic. *IEEE Access: Practical Innovations, Open Solutions*, *6*, 47699–47706. doi:10.1109/ACCESS.2018.2866554

Jha, S. (2015). Matrimony.com develops recommendation engine to simplify matchmaking. *The Economic Times*. https://cio.economictimes.indiatimes.com/news/case-studies/matrimony-com-develops-recommendation-engine-to-simplify-matchmaking/46581684

McCall, B. (2020). COVID-19 and artificial intelligence: Protecting health-care workers and curbing the spread. *The Lancet. Digital Health*, *2*(4), e166–e167. doi:10.1016/S2589-7500(20)30054-6 PMID:32289116

Purvagrover, R. J. (2016). PAID: Predictive agriculture analysis of data integration in India. *IEEE International conference on computing for sustainable global development*.

Quang, Nguyen, Do, Wang, Heng, Chen, Ang, Philip, Singh, Pham, Nguyen, & Chua. (2019). Breast Cancer Prediction using Feature Selection and Ensemble Voting. *International Conference on System Science and Engineering (ICSSE)*.

Srinivasan & Murugaanandam. (2014). Identification of Selfishness and Efficient Replica Allocation over MANETs. *International Journal of Innovations in Scientific and Engineering Research, 1*(2).

Stoppa, M., & Chiolerio, A. (2014). Wearable electronics and smart textiles: A critical review. *Sensors (Basel)*, *14*(7), 11957–11992. doi:10.3390140711957 PMID:25004153

Toda, T., Chen, L. H., Saito, D., Villavicencio, F., Wester, M., Wu, Z., & Yamagishi, J. (2016, September). The Voice Conversion Challenge 2016. In Interspeech (pp. 1632-1636). Academic Press.

Vallabhaneni, R. B., & Rajesh, V. (2017). *On the performance characteristics of embedded techniques for medical image compression*. Academic Press.

Xu, L. D., Xu, E. L., & Li, L. (2018). Industry 4.0: State of the art and future trends. *International Journal of Production Research*, *56*(8), 2941–2962. doi:10.1080/00207543.2018.1444806

Reinhard, U., Müller, P. H., & Schmülling, R.-M. (1979). Determination of anaerobic threshold by the ventilation equivalent in normal individuals. *Respiration*, *38*(1), 36–42. doi:10.1159/000194056 PMID:493728

Hollmann, W., & Prinz, J. P. (1997). Ergospirometry and its history. *Sports Medicine (Auckland, N.Z.)*, *23*(2), 93–105. doi:10.2165/00007256-199723020-00003 PMID:9068094

Li, S., Lin, B.-S., Tsai, C.-H., Yang, C.-T., & Lin, B.-S. (2017). Design of wearable breathing sound monitoring system for real-time wheeze detection. *Sensors (Basel)*, *17*(12), 171. doi:10.339017010171 PMID:28106747

Jang. (2014). *Wearable respiration measurement apparatus*. US 8,636,671 B2.

Mogera, U., Sagade, A. A., George, S. J., & Kulkarni, G. U. (2014). Ultrafast response humidity sensor using supramolecular nanofibre and its application in monitoring breath humidity and flow. *Scientific Reports*, *4*(1), 4103. doi:10.1038rep04103 PMID:24531132

Kang, Y., Ruan, H., Wang, Y., Arregui, F. J., Matias, I. R., & Claus, R. O. (2006). Nanostructured optical fibre sensors for breathing airflow monitoring. *Measurement Science & Technology*, *17*(5), 1207–1210. doi:10.1088/0957-0233/17/5/S44

Bellazzi, R., & Zupan, B. (2008). Predictive data mining in clinical medicine: Current issues and guidelines. *International Journal of Medical Informatics*, *77*(2), 81–97. doi:10.1016/j.ijmedinf.2006.11.006 PMID:17188928

Geetha, M. C. S. (2015). A survey on data mining techniques in agriculture. International Journal of Innovative Research in Computer & Communication Engineering, 3(2).

Khan & Safi. (2018). *Flying ad-hoc networks (FANET): A review of communication architectures, and routing protocols*. IEEE Xplore.

Lei, L. (2018). Research on Logistic Regression Algorithm of Breast Diagnose Data by Machine Learning. *International Conference on Robots & Intelligent System*. 10.1109/ICRIS.2018.00049

Maithili, K., Vinothkumar, V., & Latha, P. (2018). Analyzing the security mechanisms to prevent unauthorized access in cloud and network security. *Journal of Computational and Theoretical Nanoscience*, *15*(6), 2059–2063. doi:10.1166/jctn.2018.7407

Murali, G., Nagavamsi, V., Srinath, A., & Prakash, M. A. (2020). *Battery Thermal Management System Using Phase Change Material on Trapezoidal Battery Pack with Liquid Cooling System*. Academic Press.

Prasad. (2004). *OFDM for Wireless Communications Systems*. Artech House Publishers.

Rai, S. (2019). Betterhalf: The millennials of India are turning to algorithms for love. *Hindustan Times*. Available at: https://tech.hindustantimes.com/tech/news/betterhalf-ai-the-millennials-of-india-are-turning-to-algorithms-for-love-story-TuytzxIJEItENBFJ9nCIzI.html

Rithika, H., & Santhoshi, B. N. (2016, December). Image text to speech conversion in the desired language by translating with Raspberry Pi. In *2016 IEEE International Conference on Computational Intelligence and Computing Research (ICCIC)* (pp. 1-4). IEEE. 10.1109/ICCIC.2016.7919526

Robinson, B. F., Epstein, S., Beiser, G. D., & Braunwald, E. (1966). Control of heart rate by the autonomic nervous system studies in man on the interrelation between baroreceptor mechanisms and exercise. *Circulation Research*, *19*(2), 400–411. doi:10.1161/01.RES.19.2.400 PMID:5914852

Theorin, A., Bengtsson, K., Provost, J., Lieder, M., Johnsson, C., Lundholm, T., & Lennartson, B. (2017). An event-driven manufacturing information system architecture for Industry 4.0. *International Journal of Production Research*, *55*(5), 1297–1311. doi:10.1080/00207543.2016.1201604

Zhang, Z., He, T., Zhu, M., Shi, Q., & Lee, C. (2020). Smart triboelectric socks for enabling artificial intelligence of things (AIoT) based smart home and healthcare. In *Proceedings of the 33rd International Conference on Micro Electro Mechanical Systems (MEMS)*, (pp. 80-83). Vancouver, Canada. IEEE. 10.1109/MEMS46641.2020.9056149

Armstrong, J. (2001). New OFDM Peak-to-Average Power Reduction Scheme. *IEEE VTS 53rd Vehicular Technology Conference, 1*, 756 – 760. 10.1109/VETECS.2001.944945

Chaitanya, P., Kotte, D., Srinath, A., & Kalyan, K. B. (2020). Development of smart pesticide spraying robot. *International Journal of Recent Technology and Engineering*.

Fernández-Moya, J., Alvarado, A., Morales, M., San Miguel-Ayanz, A., & Marchamalo-Sacristán, M. (2014). Using multivariate analysis of soil fertility as a tool for forest fertilization planning. *Nutrient Cycling in Agroecosystems, 98*(2), 155–167. doi:10.100710705-014-9603-3

Futoma, J., Morris, J., & Lucas, J. (2015, August). A comparison of models for predicting early hospital readmissions. *Journal of Biomedical Informatics, 56*, 229–238. doi:10.1016/j.jbi.2015.05.016 PMID:26044081

Gattim, N. K., Rajesh, V., Partheepan, R., Karunakaran, S., & Reddy, K. N. (2017). *Multimodal image fusion using Curvelet and Genetic Algorithm*. Academic Press.

Jose, J. (2021). *From Indian Matchmaking to ArrangedMarriages.co — a New World of Relationships Emerges Online*. NDTV, Gadgets. Available at: https://gadgets.ndtv.com/internet/features/arrangedmarriages-co-arranged-marriages-dating-apps-netflix-indian-matchmaking-tinder-okcupid-2346576

Maksuti, S., Tauber, M., & Delsing, J. (2019, October). Generic autonomic management as a service in a soa-based framework for industry 4.0. In *IECON 2019-45th Annual Conference of the IEEE Industrial Electronics Society* (Vol. 1, pp. 5480-5485). IEEE.

Matuszek, C. (2018). Grounded language learning: Where robotics and NLP meet. *Proceedings of the International Joint Conference on Artificial Intelligence*, 5687-5691. 10.24963/ijcai.2018/810

Muhammad, H. Z., Nasrun, M., Setianingsih, C., & Murti, M. A. (2018, May). Speech recognition for English to Indonesian translator using hidden Markov model. In *2018 International Conference on Signals and Systems (ICSigSys)* (pp. 255-260). IEEE.

Nesamani & Rajini. (2020). Evaluation of Ensemble Machines in Breast Cancer Prediction. Advances in Parallel Computing, 37, 391-395.

Rajalakshmi & Meena. (2020). A Hybrid Intrusion Detection System for Mobile Adhoc Networks using FBID Protocol. *Scalable Computing: Practice and Experience, 21*(1).

Snyder, F., Hobson, J. A., Morrison, D. F., & Goldfrank, F. (1964). Changes in respiration, heart rate, and systolic blood pressure in human sleep. *Journal of Applied Physiology, 19*(3), 417–422. doi:10.1152/jappl.1964.19.3.417 PMID:14174589

Wiesner, S. (1983). Conjugate coding. *ACM SIGACT News, 15*(1), 78–88. doi:10.1145/1008908.1008920

Abhijit. (2008). Phonetic Speech Analysis for Speech to Text Conversion. *2008 IEEE Region 10 Colloquium and the Third International Conference on Industrial and Information Systems*.

Bennett, C. H., & Brassard, G. (1984). Quantum cryptography: Public key distribution and coin tossing. *Proc. IEEE Int. Conf. Comput., Syst. Signal Process.*, 475–480.

Caplan, R. D., & Jones, K. W. (1975). Effects of work load, role ambiguity, and type A personality on anxiety, depression, and heart rate. *The Journal of Applied Psychology, 60*(6), 713–719. doi:10.1037/0021-9010.60.6.713 PMID:1194173

Guo, Ma, Wang, & Yang. (2013). Incentive-based optimal Nodes selection Mechanism for Thershold key management in MANETs with selfish nodes. *International Journal of Distributed Sensor Networks*.

Kekan, A. HKumar, B. R. (2019). Crack depth and crack location identification using artificial neural network. *Int. J. Mech. Product. Eng. Res. Develop*, *9*(2), 699–708. doi:10.24247/ijmperdapr201970

Leyh, C., Schäffer, T., Bley, K., & Forstenhäusler, S. (2016). Assessing the IT and software landscapes of Industry 4.0-Enterprises: the maturity model SIMMI 4.0. In *Information technology for management: New ideas and real solutions* (pp. 103–119). Springer.

Mankar & Burange. (2014). Data mining – An evolutionary view of Agriculture. *International Journal of Application or Innovation in Engineering and Management*.

Murdoch, C. (2017). Genius figures out how to use a fidget spinner to conquer Tinder. *Mashable India*. https://mashable.com/2017/05/24/fidget-spinner-tinder-trick/#sYjfwVF2H5qB

Nie, L. (2015). *Bridging the Vocabulary Gap between Health Seekers and Healthcare Knowledge*. Translational.

Schmidt, B., Ng, C., Yien, P., Harris, C., Saripalle, U., Price, A., Brewer, S., Scelsi, G., Slaviero, R., & Armstrong, J. (2006). Efficient Algorithms for PAPR Reduction in OFDM Transmitters Implemented using Fixed-Point DSPs. *IEEE 63rd Vehicular Technology Conference, 4*, 2023 – 2027. 10.1109/VETECS.2006.1683201

Suryanarayana, G., & Dhuli, R. (2017). Super-resolution image reconstruction using dual-mode complex diffusion-based shock filter and singular value decomposition. *Circuits, Systems, and Signal Processing*, *36*(8), 3409–3425. doi:10.100700034-016-0470-9

Thrall, J. H., Li, X., Li, Q., Cruz, C., Do, S., Dreyer, K., & Brink, J. (2018). Artificial intelligence and machine learning in radiology: Opportunities, challenges, pitfalls, and criteria for success. *Journal of the American College of Radiology*, *15*(3), 504–508. doi:10.1016/j.jacr.2017.12.026 PMID:29402533

Umadevi, S., Chandrakala, T., Bhavani, P., Leena Nesamani, S., & Ezhilmathi, T.M. (2020). Solutions to Critical Issues on Big Data using Machine Learning Techniques. *International Journal of Advanced Science and Technology, 29*(7S), 3054-3059.

Aadil, F., Ahsan, W., Rehman, Z. U., Shah, P. A., Rho, S., & Mehmood, I. (2018). Clustering algorithm for internet of vehicles (IoV) based on dragonfly optimizer (CAVDO). Springer. *The Journal of Supercomputing*, *74*(9), 4542–4567. doi:10.100711227-018-2305-x

Abbas, A. H., Habelalmateen, M. I., Jurdi, S., Audah, L., & Alduais, N. A. M. (2019, November). GPS based location monitoring system with geo-fencing capabilities. In. AIP Conference Proceedings: Vol. 2173. *No. 1* (p. 020014). AIP Publishing LLC. doi:10.1063/1.5133929

Abbasi, A. A., & Younis, M. (2007). A survey on clustering algorithms for wireless sensor networks. *Computer Communications*, *30*(14-15), 2826–2841. doi:10.1016/j.comcom.2007.05.024

Abdel Hafeez, K., Zhao, L., Shen, X., & Niu, N. (2013). Distributed Multichannel and Mobility-Aware Cluster-Based MAC Protocol for Vehicular Ad Hoc Networks. IEEE Transactions on Vehicular Technology, 62(8).

Abdelgadir & Saeed. (2020). Evaluation of Performance Enhancement of OFDM Based on Cross Layer Design (CLD) IEEE 802.11p Standard for Vehicular Ad-hoc Networks (VANETs), City Scenario. *International Journal of Signal Processing Systems, 8*(1).

Abdelgadir, M., Saeed, R. A., & AbuAgla, B. (2017). Mobility Routing Model for Vehicular Ad-hoc Networks (VANETs), Smart City Scenarios. *Vehicular Communications, 9*, 154–161. doi:10.1016/j.vehcom.2017.04.003

Abdelgadir, Saeed, & Babiker. (2018). Cross Layer Design Approach for Efficient Data Delivery Based on IEEE 802.11P in Vehicular Ad-Hoc Networks (VANETs) for City Scenarios. *International Journal on Ad Hoc Networking Systems, 8*(4).

Abinaya, S., & Arulkumaran, G. (2017). Detecting black hole attack using fuzzy trust approach in MANET. *Int. J. Innov. Sci. Eng. Res., 4*(3), 102–108.

Adi, Anwar, & Baig. (2020). Machine learning and data analytics for the IoT. *Neural Computing and Applications, 32*, 16205–16233.

Adler, R., Buonadonna, P., Chhabra, J., Flanigan, M., & Krishnamurthy, L. Kushalnagar, N., Nachman, L., & Yarvis, M. (2005). Design and deployment of industrial sensor networks: Experiences from the north sea and a semiconductor plant. *Proceedings of ACM SenSys*.

Afrati, F., Gionis, A., & Mannila, H. (2004, August). Approximating a collection of frequent sets. In *Proceedings of the tenth ACM SIGKDD international conference on Knowledge discovery and data mining* (pp. 12-19). ACM.

Ahammad, S. K. H., Rajesh, V., & Ur Rahman, M. Z. (2019). Fast and Accurate Feature Extraction-Based Segmentation Framework for Spinal Cord Injury Severity Classification. IEEE Access, 7, 46092-46103.

Ahammad, S.H., Rajesh, V., Neetha, A., Sai Jeesmitha, B., & Srikanth, A. (2019). Automatic segmentation of spinal cord diffusion MR images for disease location finding. *Indonesian Journal of Electrical Engineering and Computer Science, 15*(3), 1313-1321.

Ahammad, S.H., Rajesh, V., Neetha, A., SaiJeesmitha, B., & Srikanth, A. (2019). Automatic segmentation of spinal cord diffusion MR images for disease location finding. *Indonesian Journal of Electrical Engineering and Computer Science, 15*(3), 1313-1321.

Ahammad, S.H., Rajesh, V., Rahman, M.Z.U., & Lay-Ekuakille, A. (2020). A Hybrid CNN-Based Segmentation and Boosting Classifier for Real Time Sensor Spinal Cord Injury Data. *IEEE Sensors Journal, 20*(17), 10092-10101.

Ahammad, S.H., Rajesh, V., Venkatesh, K.N., Nagaraju, P., Rao, P.R., & Inthiyaz, S. (2019). Liver segmentation using abdominal CT scanning to detect liver disease area. *International Journal of Emerging Trends in Engineering Research, 7*(11), 664-669.

Ahammad, S. H., Rajesh, V., Rahman, M. Z. U., & Lay-Ekuakille, A. (2020). A Hybrid CNN-Based Segmentation and Boosting Classifier for Real Time Sensor Spinal Cord Injury Data. *IEEE Sensors Journal, 20*(17), 10092–10101. doi:10.1109/JSEN.2020.2992879

Ahammad, S. H., Rajesh, V., Venkatesh, K. N., Nagaraju, P., Rao, P. R., & Inthiyaz, S. (2019). Liver segmentation using abdominal CT scanning to detect liver disease area. *International Journal of Emerging Trends in Engineering Research, 7*(11), 664–669. doi:10.30534/ijeter/2019/417112019

Ahammad, S. K. H., Rajesh, V., & Ur Rahman, M. Z. (2019). Fast and Accurate Feature Extraction-Based Segmentation Framework for Spinal Cord Injury Severity Classification. *IEEE Access: Practical Innovations, Open Solutions, 7*, 46092–46103. doi:10.1109/ACCESS.2019.2909583

Ahmed, Z., Saeed, R. A., & Mukherjee, A. (2018). Vehicular Cloud Computing models, architectures, applications, challenges and opportunities. In Vehicular Cloud Computing for Traffic Management and Systems. IGI Global. Doi:10.4018/978-5225-3981-0

Ahmed, H. R. (2012). *Swarm Intelligence: Concepts, Models and Applications*. Queen's University.

Ahmed, Y., & Hashem, K. (2017). The role of big data analytics in Internet of Things. *Computer Networks, 129*, 459–471. doi:10.1016/j.comnet.2017.06.013

Akhbari, B., Mirmohseni, M., & Aref, M. (2009, June – July). Compress-and-forward strategy for the relay channel with non-causal state information. *Proc. IEEE Int. Symp. on Inf. Theory*, 1169 – 1173.

Akhilomen, J. (2013, July). Data mining application for cyber credit-card fraud detection system. In *Industrial Conference on Data Mining* (pp. 218–228). Springer. doi:10.1007/978-3-642-39736-3_17

Akkaya, K., Guvenc, I., Aygun, R., Pala, N., & Kadri, A. (2015). IoT-based occupancy monitoring techniques for energy-efficient smart buildings. *IEEE Wireless Communications and Networking Conference Workshops*, *1*, 58–73.

Akyildiz, I. F., Su, W., Sankara Subramanian, Y., & Cayirci, E. (2002). Wireless sensor networks: A survey. *Computer Networks*, *38*(4), 393–422. doi:10.1016/S1389-1286(01)00302-4

Alawi, Saeed, Hassan, & Alsaqour. (2014). Simplified gateway selection scheme for multi-hop relay vehicular ad hoc network. *International Journal of Communication Systems*, *27*(12), 3855–3873. doi:10.1002/dac.2581

Alawi, M., Saeed, R., Hassan, A., & Khalifa, O. (2012). Internet Access Challenges and Solutions for Vehicular Ad-Hoc Network Environment. *IEEE International Conference on Computer & Communication Engineering (ICCCE2012)*.

Alessa, A., & Faezipour, M. (2018, July). Tweet classification using sentiment analysis features and TF-IDF weighting for improved flu trend detection. In *International Conference on Machine Learning and Data Mining in Pattern Recognition* (pp. 174-186). Springer. 10.1007/978-3-319-96136-1_15

Ali, E. S., Hasan, M. K., Hassan, R., Saeed, R. A., Hassan, M. B., Islam, S., Nafi, N. S., & Bevinakoppa, S. (2021). Machine Learning Technologies for Secure Vehicular Communication in Internet of Vehicles: Recent Advances and Applications. *Journal of Security and Communication Networks*.

Ali, N., & Shah, V. & Bhuvanasundaram. (2016, April). Architecture for public safety network using D2D communication. In *2016 IEEE Wireless Communications and Networking Conference Workshops (WCNCW)* (pp. 206-211). IEEE. 10.1109/WCNCW.2016.7552700

Alsaqer, M., Hilton, B., Horan, T., & Aboulola, O. (2015). Performance assessment of geo-triggering in small geo-fences: Accuracy, reliability, and battery drain in different tracking profiles and trigger directions. *Procedia Engineering*, *107*, 337–348. doi:10.1016/j.proeng.2015.06.090

Alsayat & El-Sayed. (2016). Social media analysis using optimized K-Means clustering. *2016 IEEE 14th International Conference on Software Engineering Research, Management and Applications (SERA)*, 61-66. 10.1109/SERA.2016.7516129

Alves, H., & Souza, R. D. (2011). Selective decode-and-forward using fixed relays and packet accumulation. *IEEE Communications Letters*, *15*(7), 707–709.

Amarendra, C., & Harinadh Reddy, K. (2016). Investigation and analysis of space vector modulation with matrix converter determined based on fuzzy C-means tuned modulation indexs. *Iranian Journal of Electrical and Computer Engineering*, *6*(5), 1939–1947.

Amolins, K., Zhang, Y., & Dare, P. (2007). Wavelet based image fusion techniques—An introduction, review and comparison. *ISPRS Journal of Photogrammetry and Remote Sensing*, *62*(4), 249–263.

Anderson, R. (2001). *Security Engineering: A Guide to Building Dependable Distributed Systems*. Wiley Computer Publishing.

Android. (n.d.),. https://www.android.com/what-is-android/

Angehr. (2020). Artificial Intelligence and Machine Learning Applied at the Point of Care. *Front. Pharmacol.* doi:10.3389/fphar.2020.00759

Angelos. (2020). Tackling Faults in the Industry 4.0 Era—A Survey of Machine-Learning Solutions and Key Aspects. *Sensors (Basel)*, *20*(1), 109. doi:10.339020010109 PMID:31878065

Anuja, S. (2017). Cuckoo Search Optimization- A Review. *Materials Today: Proceedings*, *4*(8), 7262–7269. doi:10.1016/j.matpr.2017.07.055

Araujo, Blesa, Romero, & Villanueva. (2012). Securityincognitivewireless sensor networks - Challenges and open problems. *EURASIP Journal on Wireless Communications and Networking, 48.*

Arvinderpal, Wander, Gura, Eberle, Gupta, & Shantz. (2005). Energy analysis of public-key cryptography for wireless sensor networks. *Proceedings of the 3rd IEEE International Conference on Pervasive Computing and Communications, PERCOM-2005*, 324-328.

Arya, M., & Sastry G, H. (2020). DEAL– Deep ensemble algorithm framework for credit card fraud detection in real-time data stream with google tensorflow. *Smart Science, 8*(2), 71-83. doi:10.1080/23080477.2020.1783491

Asencio–Cortés, M.-E., Morales-Esteban, A., Shang, X., & Martínez-Álvarez, F. (2018). Earthquake prediction in California using regression algorithms and cloud-based big data infrastructure. *Computers & Geosciences*, *115*, 198–210. doi:10.1016/j.cageo.2017.10.011

Assis, Horita, & Freitas, Ueyama & De Albuquerque. (2018). A service-oriented middleware for integrated management of crowdsourced and sensor data streams in disaster management. *Sensors (Basel)*, *18*(6), 1689. doi:10.339018061689 PMID:29794979

Avvenuti, C., & Vigna, F. (2018). CrisMap: A big data crisis mapping system based on damage detection and geoparsing. *Information Systems Frontiers*, *20*(5), 993–1011. doi:10.100710796-018-9833-z

Azeem, Khan, & Pramod. (2011). SecurityArchitectureFramework and Secure Routing Protocols in Wireless Sensor Networks-Survey. *International Journal of Computer Science & Engineering Survey, 2*(4), 189-204.

Badamasi, Y. A. (2014). The working principle of an Arduino. Electronics. *Computer and Computation*, *1*, 1–4.

Bajic, B. (2018). Machine Learning Techniques for Smart Manufacturing: Applications and Challenges in Industry 4.0. *9th International Scientific and Expert Conference TEAM*

Balamurugan. (2019). Use Case of Artificial Intelligence in Machine Learning Manufacturing 4.0. *International Conference on Computational Intelligence and Knowledge Economy (ICCIKE)*

Ban, Y., Liu, H., Fei, T., Luo, H., Yu, L., & Yang, Q. (2020). Big data-based applied research on hazardous chemical safety regulation. In IOP Conference Series: Earth and Environmental Science (vol. 569, pp. 1-7). IOP Publishing Ltd. doi:10.1088/1755-1315/569/1/012056

Basanta-Val, A., & Wellings, G. (2016). Architecting time-critical big-data systems. *IEEE Transactions on Big Data*, *2*(4), 310–324. doi:10.1109/TBDATA.2016.2622719

Baweja, G., & Ouyang, B. (2002). Data acquisition approach for real-time equipment monitoring and control. *IEEE 2002 Advanced Semiconductor Manufacturing Conference (ASMC '02)*, 223–227. 10.1109/ASMC.2002.1001608

Beaulieu, N. C., & Hu, J. (2006). A noise reduction amplify-and-forward relay protocol for distributed spatial diversity. *IEEE Communications Letters*, *10*(11), 787–789. doi:10.1109/LCOMM.2006.060849

Behera, T. K., & Panigrahi, S. (2015, May). Credit card fraud detection: a hybrid approach using fuzzy clustering & neural network. In *2015 Second International Conference on Advances in Computing and Communication Engineering* (pp. 494-499). IEEE. 10.1109/ICACCE.2015.33

Bello-Orgaz, Gema, Jung, & Camacho. (2016). Social big data:Recent achievements and new challenges. *Information Fusion*, *28*, 45–59.

Ben Arbia, Alam, & Kadri, Hamida, & Attia. (2017). Enhanced IoT-based end-to-end emergency and disaster relief system. *Journal of Sensor and Actuator Networks*, *6*(3), 19.

BenMimoune, A., & Kadoch, M. (2017). *Relay technology for 5G networks and IoT applications. In Internet of Things: Novel Advances and Envisioned Applications*. Springer.

Bermejo, P., Gámez, J. A., & Puerta, J. M. (2011). Improving the performance of Naive Bayes multinomial in e-mail foldering by introducing distribution-based balance of datasets. *Expert Systems with Applications*, *38*(3), 2072–2080. doi:10.1016/j.eswa.2010.07.146

Bhagwat, P. (2001). Bluetooth: Technology for short-range wireless apps. *IEEE Internet Computing*, *5*, 96–103.

Bhatia, S., Kohli, P., & Singh, R. (2018). Neuro-symbolic Program Corrector for Introductory Programming assignments. Proceedings of the 40th International Conference on Software Engineering, 60 70. doi:10.1145/3180155.3180219

Bhatnagar, G. (2013). Human visual system inspired multi-modal medical image fusion framework. *Expert Systems with Applications*, *40*(5), 1708–1720.

Bhoopathy & Parvathi. (2012). Energy Constrained Secure Hierarchical Data Aggregation in Wireless Sensor Networks. *American Journal of Applied Sciences, 9*(6), 858-864.

Bhoye, T., More, S., Gatkawar, R., & Patil, R. (2018). Child Safety Wearable Device. *IJARIIE, 4*, 2395–4396.

Bhusari, V., & Patil, S. (2011). Detection of credit card fraud transaction using hidden markov model. *Research Journal of Engineering and Technology*, *2*(4), 203–206.

Bizer, B., Boncz, P., Brodie, M. L., & Erling, O. (2012). The meaningful use of big data: Four perspectives—four challenges. *SIGMOD Record*, *40*(4), 56–60. doi:10.1145/2094114.2094129

Blendon, R. J., Benson, J. M., Desroches, C. M., & Weldon, K. J. (2003). Using opinion surveys to track the public's response to a bioterrorist attack. *Journal of Health Communication*, *8*(sup1, S1), 83–92. doi:10.1080/713851964 PMID:14692573

Bonaci, Bushnell, & Poovendran. (2010). Node capture attacks in wireless sensor networks: A system theoretic approach. *49th IEEE Conference on Decision and Control*, 1, 6765–6772.

Boonsong, W., Senajit, N., & Prasongchan, P. (2021). Contactless Body Temperature Monitoring of In-Patient Department (IPD) Using 2.4 GHz Microwave Frequency via the Internet of Things (IoT) Network. *Wireless Personal Communications*, *1*, 1–7.

Braam, Huang, Chen, Montgomery, Vo, & Beausoleil. (2015). Wristband Vital: A wearable multi-sensor microsystem for real-time assistance via low-power Bluetooth link. *Internet of Things (WFIoT), 2015 IEEE 2nd World Forwn on, 1*, 87-91.

Brownlee, J. (2014). Classification accuracy is not enough: More performance measures you can use. *Machine Learning Mastery, 21*.

Brundashree, Kakhandaki, Kulkarni, Rod, & Matt. (2015). The PET-MRI Brain Image Fusion Using Wavelet Transforms. *International Journal of Engineering Trends and Technology, 23*(6), 304-307.

Buonaguidi, B., Mira, A., Bucheli, H., & Vitanis, V. (2021). Bayesian quickest detection of credit card fraud. *Bayesian Analysis*, *1*(1), 1–30.

Buribayeva, M., Miyachi, T., Yeshmukhametov, A., & Mikami, Y. (2015). An autonomous emergency warning system based on Cloud Servers and SNS. *Procedia Computer Science*, 60, 722–729. doi:10.1016/j.procs.2015.08.225

Cai, X., Xu, B., Jiang, L., & Vasilakos, A. V. (2016). IoT-based big data storage systems in cloud computing: Perspectives and challenges. *IEEE Internet of Things Journal*, 4(1), 75–87. doi:10.1109/JIOT.2016.2619369

Carcillo, F., Dal Pozzolo, A., Le Borgne, Y. A., Caelen, O., Mazzer, Y., & Bontempi, G. (2018). Scarff: A scalable framework for streaming credit card fraud detection with spark. *Information Fusion*, 41, 182–194. doi:10.1016/j.inffus.2017.09.005

Carley, T., Ba, M., Barua, R., & Stewart, D. (2003). Contention-free periodic message scheduler medium access control in wireless sensor/actuator networks. RTSS, 24th IEEE, 298–304.

Cascella, M., Rajnik, M., Aleem, A., Dulebohn, S., & Di Napoli, R. (2021). *Features, evaluation, and treatment of coronavirus (COVID-19)*. StatPearls.

Celikyilmaz, A., Hakkani-Tür, D., & Feng, J. (2010, December). *Probabilistic model-based sentiment analysis of twitter messages. In 2010 IEEE Spoken Language Technology Workshop*. IEEE.

Cen, J., Yu, T., Li, Z., Jin, S., & Liu, S. (2011). *Developing a disaster surveillance system based on wireless sensor network and cloud platform*. Academic Press.

Chaitanya, K., & Venkateswarlu, S. (2016). Detection of blackhole&greyhole attacks in MANETs based on acknowledgement based approach. *Journal of Theoretical and Applied Information Technology*, 89(1), 210–217.

Chamola, V., Hassija, V., Gupta, V., & Guizani, M. (2020). A comprehensive review of the COVID-19 pandemic and the role of IoT, drones, AI, blockchain, and 5G in managing its impact. *IEEE Access: Practical Innovations, Open Solutions*, 8, 90225–90265. doi:10.1109/ACCESS.2020.2992341

Chan, Perrig, & Song. (2003). Random key pre distribution schemes for sensor networks. *IEEE Symposium on Security and Privacy*.

Chandana, K., Prasanth, Y., & Prabhu Das, J. (2016). A decision support system for predicting diabetic retinopathy using neural networks. *Journal of Theoretical and Applied Information Technology*, 88(3), 598–606.

Chatterjee, M., & Das, S. (2001). Performance evaluation of a request-TDMA/CDMA protocol for wireless networks. *J. Interconn. Netw.*, 2(1), 49–67. doi:10.1142/S0219265901000257

Chen. (2019). Emerging Trends of ML-based Intelligent Services for Industrial Internet of Things (IIoT). *Computing, Communications and IoT Applications (ComComAp)*

Chen, L., Liu, Z., Wang, L., Dou, M., Chen, J., & Li, H. (2013). Natural disaster monitoring with wireless sensor networks: A case study of data-intensive applications upon low-cost scalable systems. *Mobile Networks and Applications*, 18(5), 651–663. doi:10.100711036-013-0456-9

Chen, M., Mao, S., & Liu, Y. (2014). Big data: A survey. *Mobile Networks and Applications*, 19(2), 171–209. doi:10.100711036-013-0489-0

Chen, R. C., Luo, S. T., Liang, X., & Lee, V. C. (2005, October). Personalized approach based on SVM and ANN for detecting credit card fraud. In *2005 International Conference on Neural Networks and Brain* (vol. 2, pp. 810-815). IEEE. 10.1109/ICNNB.2005.1614747

Chen, W., Letaief, K. B., & Cao, Z. (2012). Buffer-aware network coding for wireless networks. *IEEE/ACM Transactions on Networking*, 20(5), 1389–1401.

Chipcon, A.S. (n.d.). *Subsidiary of Texas Instruments, CC1000 and CC2420 Radio Transceiver Products*. Academic Press.

Chiu, C. C., & Tsai, C. Y. (2004, March). A web services-based collaborative scheme for credit card fraud detection. In *IEEE International Conference on e-Technology, e-Commerce and e-Service, 2004. EEE'04. 2004* (pp. 177-181). IEEE. 10.1109/EEE.2004.1287306

Choi & Bae. (2015). The real-time monitoring system of social big data for disaster management. In *Computer science and its applications* (pp. 809–815). Springer.

Çınar. (2020). *Machine Learning in Predictive Maintenance towards Sustainable Smart Manufacturing in Industry 4.0. In Sustainability.* MDPI. doi:10.3390u12198211

Collier, S. E. (2016). The emerging enernet: Convergence of the smart grid with the internet of things. *IEEE Industry Applications Magazine, 23*(2), 12–16. doi:10.1109/MIAS.2016.2600737

Correia, I., Fournier, F., & Skarbovsky, I. (2015, June). The uncertain case of credit card fraud detection. In *Proceedings of the 9th ACM International Conference on Distributed Event-Based Systems* (pp. 181-192). 10.1145/2675743.2771877

Cristani, M., Bue, A. D., Murino, V., Setti, F., & Vinciarelli, A. (2020). The Visual Social Distancing Problem. *IEEE Access: Practical Innovations, Open Solutions, 8,* 126876–126886. doi:10.1109/ACCESS.2020.3008370

Da Xu, L. (2018). Big data for cyber physical systems in industry 4.0: A survey. *Enterprise Information Systems, 13*(1), 1–22.

Da Xu, L., He, W., & Li, S. (2014). Internet of things in industries: A survey. *IEEE Transactions on Industrial Informatics, 10*(4), 2233–2243. doi:10.1109/TII.2014.2300753

Daneshvar, S., & Ghassemian, H. (2011). MRI and PET image fusion by combining IHS and retina-inspired models. *Information Fusion, 11*(2), 114–123.

Dash, P. K., Samantaray, S. R., & Panda, G. (2006). Fault Classification and Section Identification of an Advanced Series-Compensated Transmission Line Using Support Vector Machine. *IEEE Transactions on Power Delivery, 22*(no. 1), 67–73. doi:10.1109/TPWRD.2006.876695

De Gennaro, G., Dambruoso, P. R., Loiotile, A. D., Di Gilio, A., Giungato, P., Tutino, M., Marzocca, A., Mazzone, A., Palmisani, J., & Porcelli, F. (2014). Indoor air quality in schools. *Environmental Chemistry Letters, 12*(4), 467–482. doi:10.100710311-014-0470-6

De Montjoye, Y. A., Radaelli, L., Singh, V. K., & Pentland, A. S. (2015). Unique in the shopping mall: On the reidentifiability of credit card metadata. *Science, 347*(6221), 536–539. doi:10.1126cience.1256297 PMID:25635097

Deepak Kumar, B. (2011). Evaluation of the Medical Records System in an Upcoming Teaching Hospital—A Project for Improvisation. *Journal of Medical Systems.*

Dehkordi, Fereidunian, Dehkordi, & Lesani. (2014). HourĞahead demand forecasting in smart grid using support vector regression (SVR). *International Transactions on Electrical Energy Systems, 24*(12), 1650-1663.

Dehkordi, Sajjad, Fereidunian, Dehkordi, & Lesani. (2014). HourĞahead demand forecasting in smart grid using support vector regression (SVR). *International Transactions on Electrical Energy Systems, 24*(12), 1650-1663.

Delamaire, L., Abdou, H., & Pointon, J. (2009). Credit card fraud and detection techniques: A review. *Banks and Bank Systems, 4*(2), 57-68. http://usir.salford.ac.uk/id/eprint/2595/

Devaraju, Kumar, & Rao. (n.d.). A Real And Accurate Vegetable Seeds Classification Using Image Analysis And Fuzzy Technique. *Turkish Journal of Physiotherapy and Rehabilitation, 32,* 2.

Devaraju, Kumar, & Rao. (2020). A Real And Accurate Vegetable Seeds Classification Using Image Analysis And Fuzzy Technique. *Turkish Journal of Physiotherapy and Rehabilitation, 32*(2).

Developers, A. (2016). *Creating and monitoring geofences.* Academic Press.

Dheepa, V., & Dhanapal, R. (2012). Behavior based credit card fraud detection using support vector machines. *ICTACT Journal on Soft computing, 2*(7). doi:10.21917/ijsc.2012.0061

Dianle. (2020). *Edge Intelligence: Architectures, Challenges, and Applications.* ArXiv: 2003.12172v2

Dirar, Saeed, Hasan, & Mahmud. (2017). Persistent Overload Control for Backlogged Machine to Machine Communications in Long Term Evolution Advanced Networks. *Journal of Telecommunication, Electronic and Computer Engineering, 9*(3).

Dittrich & Quiané-Ruiz. (2014-2015). Efficient big data processing in Hadoop MapReduce. *Proceedings of the VLDB Endowment International Conference on Very Large Data Bases, 5*(12).

Dolui, K. (2017). Comparison of Edge Computing Implementations: Fog Computing, Cloudlet and Mobile Edge Computing. *IEEE Access: Practical Innovations, Open Solutions.*

Duarte-Melo, E. J., & Liu, M. (2003). Data-gathering wireless sensor networks: Organization and capacity. Comput. Netw. (COMNET). *Special Issue on Wireless Sensor Networks, 43*(4), 519–537.

Duman, E., & Ozcelik, M. H. (2011). Detecting credit card fraud by genetic algorithm and scatter search. *Expert Systems with Applications, 38*(10), 13057-13063. doi:10.1016/j.eswa.2011.04.110

Duman, E., & Elikucuk, I. (2013, June). Solving credit card fraud detection problem by the new metaheuristics migrating birds optimization. In *International Work-Conference on Artificial Neural Networks* (pp. 62–71). Springer. doi:10.1007/978-3-642-38682-4_8

Durai & Jebaseeli. (2020). A Robust Ecg Signal Processing And Classification Methodology For The Diagnosis Of Cardiac Health. *International Journal Of Innovations In Scientific And Engineering Research, 7*(8).

Dzhongova, E., Anderson, D., de Ruiter, J., Novakovic, V., & Oses, M. R. (2017). False alarm rates of liquid explosives detection systems. *Journal of Transportation Security, 10*(3-4), 145–169. doi:10.100712198-017-0184-7

Efeoglu, E., & Tuna, G. (2020). Detection of Hazardous Liquids Using Microwave Data and Well-Known Classification Algorithms. *Russian Journal of Nondestructive Testing, 56*(9), 742–751. doi:10.1134/S106183092009003X

Eltahir, A. A., & Saeed, R. A. (2018). V2V Communication Protocols in Cloud Assisted Vehicular Networks. In Vehicular Cloud Computing for Traffic Management and Systems. IGI Global. doi:10.4018/978-1-5225-3981-0.ch006

Eltahir, A. A., Saeed, R. A., Mukherjee, A., & Hasan, M. K. (2016). Evaluation and Analysis of an Enhanced Hybrid Wireless Mesh Protocol for Vehicular Ad-hoc Network. *EURASIP Journal on Wireless Communications and Networking, 2016*(1), 1–11. doi:10.118613638-016-0666-5

Eltahir, S., & Alawi. (2013). An enhanced hybrid wireless mesh protocol (E-HWMP) protocol for multihop vehicular communications. *2013 International Conference on Computing, Electrical and Electronics Engineering (ICCEEE),* 1 - 8.

Emmanouil & Nikolaos. (2015). Big data analytics in prevention, preparedness, response and recovery in crisis and disaster management. In *The 18th International Conference on Circuits, Systems, Communications and Computers (CSCC 2015), Recent Advances in Computer Engineering Series* (Vol. 32, pp. 476-482). Academic Press.

Erd, S., Schaeffer, F., Kostic, M., & Reindl, L. M. (2016). Event monitoring in emergency scenarios using energy efficient wireless sensor nodes for the disaster information management. *International Journal of Disaster Risk Reduction, 16,* 33–42. doi:10.1016/j.ijdrr.2016.01.001

Eschenauer, L., & Gligor, V. D. (2002). A key-management scheme for distributed sensor networks. *Proceedings of the 9th ACM Conference on Computer and Communications Security*, 41–47.

Fang, S., Xu, L., Zhu, Y., Liu, Y., Liu, Z., Pei, H., Yan, J., & Zhang, H. (2015). An integrated information system for snowmelt flood early-warning based on internet of things. *Information Systems Frontiers, 17*(2), 321–335. doi:10.100710796-013-9466-1

Fan, Y., Wang, C., Thompson, J., & Poor, H. V. (2007). Recovering multiplexing loss through successive relaying using repetition coding. *IEEE Transactions on Wireless Communications, 6*(12), 4484–4493.

Fatima. (2018). *Evaluation of Rule-Based Learning and Feature Selection Approaches for Classification.* Imperial College Computing Student Workshop.

Fertier, M., & Barthe-Delanoë, T., & Bénaben (2016, May). Adoption of big data in crisis management toward a better support in decision-making. *Proceedings of Conference on Information System for Crisis Response And Management (ISCRAM 16).*

Fisher, Hagon, Lattimer, Callaghan, Swithern, & Walmsley. (2018). *Executive summary world disasters report: Leaving no one behind.* Academic Press.

Garcia & Wang. (2013). Analysis of Big Data Technologies and Method - Query Large Web Public RDF Datasets on Amazon Cloud Using Hadoop and Open Source Parsers. *Semantic Computing (ICSC), IEEE Seventh International Conference.* http://architects.dzone.com/

García, J., Altimiras, F., Peña, A., Astorga, G., & Peredo, O. (2018). *A Binary Cuckoo Search Big Data Algorithm Applied to Large-Scale Crew Scheduling Problems.* Computational Intelligence in Modeling Complex Systems and Solving Complex Problems. doi:10.1155/2018/8395193

Gattim, N.K., Pallerla, S.R., Bojja, P., Reddy, T.P.K., Chowdary, V.N., Dhiraj, V., & Ahammad, S.H. (2019). Plant leaf disease detection using SVM technique. *International Journal of Emerging Trends in Engineering Research, 7*(11), 634-637.

Gattim, N. K., Pallerla, S. R., Bojja, P., Reddy, T. P. K., Chowdary, V. N., Dhiraj, V., & Ahammad, S. H. (2019). Plant leaf disease detection using SVM technique. *International Journal of Emerging Trends in Engineering Research, 7*(11), 634–637. doi:10.30534/ijeter/2019/367112019

Geetanjali. (2019). A Blockchain Framework for Securing Connected and Autonomous Vehicles. *Sensors (Basel), 2019*(19), 3165. doi:10.339019143165

Gianini, G., Fossi, L. G., Mio, C., Caelen, O., Brunie, L., & Damiani, E. (2020). Managing a pool of rules for credit card fraud detection by a game theory based approach. *Future Generation Computer Systems, 102*, 549–561. doi:10.1016/j.future.2019.08.028

Gibson, Andrews, Domdouzis, Hirsch, & Akhgar. (2014, December). Combining big social media data and FCA for crisis response. In *2014 IEEE/ACM 7th International Conference on Utility and Cloud Computing* (pp. 690-695). IEEE.

Girgis, A. A. (1960). An adaptive protection scheme for advanced series compensated (ASC) transmission lines. *IEEE Transactions on Power Delivery, 13*(2), 414–420.

Girotra, M., Nagpal, K., Minocha, S., & Sharma, N. (2013). Comparative survey on association rule mining algorithms. *International Journal of Computers and Applications, 84*(10).

Giusto, D. (2010). The internet of things 20th Tyrrhenian workshop on digital communications. Springer.

Goldsworthy, D. L. (1987). A linearized model for MOV-protected series capacitors. *IEEE Transactions on Power Systems*, *2*(4), 953–957. doi:10.1109/TPWRS.1987.4335284

Grant, M., & Boyd, S. (2014). *CVX: MATLAB software for disciplined convex programming, version 2.1*. http://cvxr.com/cvx

Greco, R., Ritrovato, P., Tiropanis, T., & Xhafa, F. (2018). IoT and semantic Web technologies for event detection in natural disasters. *Concurrency and Computation*, *30*(21), e4789. doi:10.1002/cpe.4789

Greeshma, L., & Pradeepini, G. (2016). *Mining maximal efficient closed itemsets without any redundancy*. Academic Press.

Greeshma, L., & Pradeepini, G. (2016). Unique constraint frequent item set mining. *Proceedings - 6th International Advanced Computing Conference, IACC 2016*, 68-72.

Greeshma, L., & Pradeepini, G. (2016). Input split frequent pattern tree using mapreduce paradigm in hadoop. *Journal of Theoretical and Applied Information Technology*, *84*(2), 260–271.

Grolinger, Capretz, Mezghani, & Exposito. (2013, June). Knowledge as a service framework for disaster data management. In *2013 Workshops on Enabling Technologies: Infrastructure for Collaborative Enterprises* (pp. 313-318). IEEE.

Grolinger, Higashino, Tiwari, & Capretz. (2013). Data management in cloud environments: NoSQL and NewSQL data stores. *Journal of Cloud Computing: Advances, Systems and Applications*, *2*(1), 22.

Gross, D., Shortle, J. F., Thompson, J. M., & Harris, C. M. (2008). *Fundamentals of Queueing Theory* (4th ed.). John Wiley & Sons, Inc.

Gubbi, B., Buyya, R., Marusic, S., & Palaniswami, M. (2013). Internet of Things (IoT): A vision, architectural elements, and future directions. *Future Generation Computer Systems*, *29*(7), 1645–1660. doi:10.1016/j.future.2013.01.010

Hadded, M., Zagrouba, R., Laouiti, A., Muhlethaler, P., & Saidane, L. A. (2015). A multi-objective genetic algorithm-based adaptive weighted clustering protocol in VANET. *Proc. of Evolutionary Computation (CEC), IEEE Congress on*, 994-1002.

Hafiz, H. R. S., Khan, Z. A., Iqbal, R., Rizwan, S., Imran, M. A., & Awan, K. (2019). *A Heterogeneous IoV Architecture for Data Forwarding in Vehicle to Infrastructure Communication*. Hindawi Mobile Information Systems.

Han, H., Le, G., & Du. (2011, October). Survey on NoSQL database. In *2011 6th international conference on pervasive computing and applications* (pp. 363-366). IEEE.

Han, L., Hipwell, J. H., Eiben, B., Barratt, D., Modat, M., Ourselin, S., & Hawkes, D. J. (2014). A Nonlinear Biomechanical Model based Registration Method for Aligning Prone and Supine MR Breast Images. IEEE Transactions on Medical Imaging, 33(3), 682-694.

Hao, C., & Ying, Q. (2011). Research of Cloud Computing Based on the Hadoop Platform. *Computational and Information Sciences (ICCIS), 2011 International Conference*.

Harris, T. (2010). *Cloud Computing- An Overview, Whitepaper*. Torry Harris Business Solutions.

Hasane Ahammad, S.K., & Rajesh, V. (2018). Image processing based segmentation techniques for spinal cord in MRI. *Indian Journal of Public Health Research and Development*, *9*(6), 317-323.

Hasane Ahammad, S., Rajesh, V., Hanumatsai, N., Venumadhav, A., Sasank, N. S. S., & Bhargav Gupta, K. K. (2019). MRI image training and finding acute spine injury with the help of hemorrhagic and non hemorrhagic rope wounds method. *Indian Journal of Public Health Research & Development*, *10*(7), 404–408. doi:10.5958/0976-5506.2019.01603.6

HasaneAhammad, S., Rajesh, V., Hanumatsai, N., Venumadhav, A., Sasank, N.S.S., Bhargav Gupta, K.K., & Inithiyaz. (2019). MRI image training and finding acute spine injury with the help of hemorrhagic and non hemorrhagic rope wounds method. *Indian Journal of Public Health Research and Development, 10*(7), 404-408.

HasaneAhammad, S.K., & Rajesh, V. (2018). Image processing based segmentation techniques for spinal cord in MRI. *Indian Journal of Public Health Research and Development, 9*(6), 317-323.

Hassan, M. B., Alsharif, S., Alhumyani, H., Ali, E. S., Mokhtar, R. A., & Saeed, R. A. (2021). An Enhanced Cooperative Communication Scheme for Physical Uplink Shared Channel in NB-IoT. *Wireless Personal Communications, 116*(2).

Havenith, G., Holmér, I., & Parsons, K. (2002). Personal factors in thermal comfort assessment: Clothing properties and metabolic heat production. *Energy and Building, 34*(6), 581–591. doi:10.1016/S0378-7788(02)00008-7

Hegazy, M., Madian, A., & Ragaie, M. (2016). Enhanced fraud miner: credit card fraud detection using clustering data mining techniques. *Egyptian Computer Science Journal, 40*(3).

Heinzelman, W., Chandrakasan, A., & Balakrishnan, H. (2000). An application-specific protocol architecture for wireless microsensor networks. *IEEE Transactions on Wireless Communications, 1*(4), 660–670. doi:10.1109/TWC.2002.804190

Hernández-Muñoz, J. M., Vercher, J. B., Muñoz, L., Galache, J. A., Presser, M., & Hernández Gómez, L. A. (2011). Smart Cities at the Forefront of the Future Internet . Lecture Notes in Computer Science, 6656.

Hill, J., Szewczyk, R., Woo, A., Hollar, S., & Culler, D. E. (2000). System architecture directions for networked sensors. *Proc. of ASPLOS*, 93–104.

Hormozi, E., Akbari, M. K., Hormozi, H., & Javan, M. S. (2013, May). Accuracy evaluation of a credit card fraud detection system on Hadoop MapReduce. In *The 5th conference on information and knowledge technology* (pp. 35-39). IEEE. 10.1109/IKT.2013.6620034

Hristidis, C., Chen, S.-C., Li, T., Luis, S., & Deng, Y. (2010). Survey of data management and analysis in disaster situations. *Journal of Systems and Software, 83*(10), 1701–1714. doi:10.1016/j.jss.2010.04.065

Huang, C. Jing & Chang (2015, November). DisasterMapper: A CyberGIS framework for disaster management using social media data. In *Proceedings of the 4th International ACM SIGSPATIAL Workshop on Analytics for Big Geospatial Data* (pp. 1-6). ACM.

Huang, C., Cervone, G., & Zhang, G. (2017). A cloud-enabled automatic disaster analysis system of multi-sourced data streams: An example synthesizing social media, remote sensing and Wikipedia data. *Computers, Environment and Urban Systems, 66*, 23–37. doi:10.1016/j.compenvurbsys.2017.06.004

Hu, C. Y., & Raymond, D. J. (2004). Lessons learned from hazardous chemical incidents—Louisiana hazardous substances emergency events surveillance (hsees) system. *Journal of Hazardous Materials, 115*(1), 33–38. doi:10.1016/j.jhazmat.2004.05.006 PMID:15518962

Huo, L., Chen, D., Kong, Q., Li, H., & Song, G. (2017). Smart washer—A piezoceramic-based transducer to monitor looseness of bolted connection. *Smart Materials and Structures, 26*(2), 025033. doi:10.1088/1361-665X/26/2/025033

Husnain & Anwar. (2020). An intelligent cluster optimization algorithm based on whale optimization algorithm for VANETs(WOACNET). *PLoSONE, 16*(4). doi:10.1371/journal.pone.0250271

Hussain, S. A. (2019). A Review of Quality-of-Service Issues in Internet of Vehicles (IoV). *IEEE Access: Practical Innovations, Open Solutions*.

Hyder, K. S., & Nah, F. (2019). Artificial Intelligence, Machine Learning, and Autonomous Technologies in Mining Industry. *Database Management*, *30*(2), 67–79. doi:10.4018/JDM.2019040104

Hyperledger Performance & the Scale Working Group. (2018). *Hyperledger blockchain performance metrics white paper.* https://www. hyperledger. org/resources/publications/blockchain-performance-metrics

Ibarra-Esquer, J. E., González-Navarro, F. F., Flores-Rios, B. L., Burtseva, L., & Astorga-Vargas, M. A. (2017). Tracking the Evolution of the Internet of Things Concept Across Different Application Domains. *Sensors (Basel)*, *17*(6), 1379. doi:10.339017061379 PMID:28613238

Ibrahim, S., Saeed, R. A., & Mukherjee, A. (2018). Resource management in Vehicular Cloud Computing. In Vehicular Cloud Computing for Traffic Management and Systems. IGI Global. doi:10.4018/978-1-5225-3981-0.ch004

Imani, Niknejad, & Barzegaran. (2019). Implementing Time-of-Use Demand Response Program in microgrid considering energy storage unit participation and different capacities of installed wind power. *Electric Power Systems Research*, *175*, 105916.

Imani. (2019). Effect of Changes in Incentives and Penalties on Interruptible/Curtailable Demand Response Program in Microgrid Operation. In *2019 IEEE Texas Power and Energy Conference (TPEC)*. IEEE.

Inthiyaz, S., Prasad, M. V. D., Usha Sri Lakshmi, R., Sri Sai, N. T. B., & Kumar, P. P. (2020). Agriculture based plant leaf health assessment tool: A deep learning perspective. *International Journal of Emerging Trends in Engineering Research*, *7*(11), 690–694.

Inthiyaz, S., Prasad, M. V. D., Usha Sri Lakshmi, R., Sri Sai, N. T. B., & Kumar, P. P. (2020). Ahammad, "Agriculture based plant leaf health assessment tool: A deep learning perspective", S.H. *International Journal of Emerging Trends in Engineering Research*, *7*(11), 690–694. doi:10.30534/ijeter/2019/457112019

Ishida & Ohyanagi. (2019). Implementation and evaluation of a visualization and analysis system for historical disaster records. *Journal of Ambient Intelligence and Humanized Computing*, 1–16.

Izadi, B., Ranjbarian, B., Ketabi, S., & Nassiri-Mofakham, F. (2013). Performance analysis of classification methods and alternative linear programming integrated with fuzzy delphi feature selection. *International Journal of Information Technology and Computer Science*, *5*(10), 9–20. doi:10.5815/ijitcs.2013.10.02

Jamali, V., Zlatanov, N., & Schober, R. (2015). Bidirectional buffer-aided relay networks with fixed rate transmission—Part I: Delay-unconstrained case. *IEEE Transactions on Wireless Communications*, *14*(3), 1323–1338.

Jamali, V., Zlatanov, N., & Schober, R. (2015). Bidirectional buffer-aided relay networks with fixed rate transmission—Part II: Delay-constrained case. *IEEE Transactions on Wireless Communications*, *14*(3), 1339–1355.

Jamali, V., Zlatanov, N., Shoukry, H., & Schober, R. (2015). Achievable rate of the half-duplex multi-hop buffer-aided relay channel with block fading. *IEEE Transactions on Wireless Communications*, *14*(11), 6240–6256.

Jameel, F., Javaid, U., Khan, W. U., Aman, M. N., Pervaiz, H., & Jäntti, R. (2020). Reinforcement Learning in Blockchain-Enabled IIoT Networks: A Survey of Recent Advances and Open Challenges. *Sustainability*, *2020*(12), 5161. doi:10.3390u12125161

James, A. P., & Dasarathy, B. V. (2014). Medical Image Fusion: A survey of the state of the art. *Information Fusion*.

Javed, Riaz, Ghafoor, Ali, & Cheema. (2014). MRI and PET Image Fusion Using Fuzzy Logic and Image Local Features. *The Scientific World Journal*, 1–8.

Jiang, X., Ma, Z., Yu, F. R., Song, T., & Boukerche, A. (2020). Edge Computing for Video Analytics in the Internet of Vehicles with Blockchain. In *Proceedings of the 10th ACM Symposium on Design and Analysis of Intelligent Vehicular Networks and Applications (DIVANet '20)*. Association for Computing Machinery. 10.1145/3416014.3424582

Jie, Z., Li, Y., & Liu, R. (2019). Social Network Group Identification based on Local Attribute Community Detection. *2019 IEEE 3rd Information Technology, Networking, Electronic and Automation Control Conference (ITNEC)*, 443-447. 10.1109/ITNEC.2019.8729078

Ji, H., Alfarraj, O., & Tolba, A. (2020). Artificial Intelligence-Empowered Edge of Vehicles: Architecture, Enabling Technologies, and Applications. *IEEE Access: Practical Innovations, Open Solutions*, 8, 61020–61034. doi:10.1109/ACCESS.2020.2983609

Ji, X., Chun, S. A., & Geller, J. (2016). Knowledge-based tweet classification for disease sentiment monitoring. In *Sentiment Analysis and Ontology Engineering* (pp. 425–454). Springer. doi:10.1007/978-3-319-30319-2_17

Jongman, W., Wagemaker, J., Romero, B., & de Perez, E. (2015). Early flood detection for rapid humanitarian response: Harnessing near real-time satellite and Twitter signals. *ISPRS International Journal of Geo-Information*, 4(4), 2246–2266. doi:10.3390/ijgi4042246

Josephine, M. S., Lakshmanan, L., Nair, R. R., Visu, P., Ganesan, R., & Jothikumar, R. (2020). Monitoring and sensing COVID-19 symptoms as a precaution using electronic wearable devices. *International Journal of Pervasive Computing and Communications*.

Joshi, R. (2016). *Accuracy, precision, recall & f1 score: Interpretation of performance measures*. Academic Press.

Jothikumar, R. (2016). *C4. 5 classification algorithm with back-track pruning for accurate prediction of heart disease*. Academic Press.

Jothikumar, R., Susi, S., Sivakumar, N., & Ramesh, P. S. (2018). Predicting life time of heart attack patient using improved c4. 5 classification algorithm. *Research Journal of Pharmacy and Technology*, 11(5), 1951–1956. doi:10.5958/0974-360X.2018.00362.1

Juan. (2020). Machine learning applied in production planning and control: a state-of-the-art in the era of industry 4.0. *Journal of Intelligent Manufacturing*.

Ju, H., Oh, E., & Hong, D. (2009). Catching resource-devouring worms in next-generation wireless relay systems: Two-way relay and full-duplex relay. *IEEE Communications Magazine*, 47(9), 58–65.

Kajikawa, N., Minami, Y., Kohno, E., & Kakuda, Y. (2016). On Availability and Energy Consumption of the Fast Connection Establishment Method by Using Bluetooth Classic and Bluetooth Low Energy. *2016 Fourth International Symposium on Computing and Networking (CANDAR)*, 286-290. 10.1109/CANDAR.2016.0058

Kalman, R. E. (n.d.). A new approach to linear filtering and prediction problems. *Journal of Basic Engineering*, 82(1), 35–45.

Kalyani & Chellappan. (2012). Enhanced RSA CRT for Energy Efficient Authentication to Wireless Sensor Networks Security. *American Journal of Applied Sciences*, 9(10), 1660–1667.

Kamali, T., & Stashuk, D. W. (2020, August). Discovering Density-Based Clustering Structures Using Neighborhood Distance Entropy Consistency. *IEEE Transactions on Computational Social Systems*, 7(4), 1069–1080. Advance online publication. doi:10.1109/TCSS.2020.3003538

Kamaruddin, S., & Ravi, V. (2016, August). Credit card fraud detection using big data analytics: use of PSOAANN based one-class classification. In *Proceedings of the International Conference on Informatics and Analytics* (pp. 1-8). 10.1145/2980258.2980319

Kanawaday & Sane. (2017). Machine learning for predictive maintenance of industrial machines using IoT sensor data. *2017 8th IEEE International Conference on Software Engineering and Service Science (ICSESS),* 87-90. 10.1109/ICSESS.2017.8342870

Kannan, R., Kalidindi, R., Iyengar, S. S., & Kumar, V. (2003). Energy and rate based MAC protocol for wireless sensor networks. *SIGMOD Record, 32*(4), 60–65. doi:10.1145/959060.959071

Karmakar, A. (2019). Industrial Internet of Things: A Review. *International Conference on Opto-Electronics and Applied Optics (Optronix)* 10.1109/OPTRONIX.2019.8862436

Kayarga, T., & Kumar, S.A. (2021). A Study on Various Technologies to Solve the Routing Problem in Internet of Vehicles (IoV). *Wireless Personal Communications.*

Khalil. (2020). *Deep Learning in Industrial Internet of Things: Potentials, Challenges, and Emerging Applications.* ArXiv.

Khan, A., & Al-Badi, A. (2020). Open Source Machine Learning Frameworks for Industrial Internet of Things. *Procedia Computer Science, 170,* 571–577. doi:10.1016/j.procs.2020.03.127

Khan, M. F. (2019). *Moth Flame Clustering Algorithm for Internet of Vehicle (MFCA-IoV).* IEEE., doi:10.1109/ACCESS.2018.2886420

Khatua, A., & Khatua, A. (2016, September). Immediate and long-term effects of 2016 Zika Outbreak: A Twitter-based study. In *2016 IEEE 18th International Conference on e-Health Networking, Applications and Services (Healthcom)* (pp. 1-6). IEEE.

Khayyam. (2019). *Artificial Intelligence and Internet of Things for Autonomous Vehicles. In Nonlinear Approaches in Engineering Applications.* Springer.

Kim, M. J., & Kim, T. S. (2002, August). A neural classifier with fraud density map for effective credit card fraud detection. In *International Conference on Intelligent Data Engineering and Automated Learning* (pp. 378-383). Springer. 10.1007/3-540-45675-9_56

Kim, T., Trimi, S., & Chung, J.-H. (2014). Big-data applications in the government sector. *Communications of the ACM, 57*(3), 78–85. doi:10.1145/2500873

Kling, R. M. (2003). Intel mote: An enhanced sensor network node. *Int'l Workshop on Advanced Sensors, Structural Health Monitoring, and Smart Structures.*

Kumar & Bawa. (2019). *A comparative review of meta-heuristic approaches to optimize the SLA violation costs for dynamic execution of cloud service.* Springer-Verlag GmbH Germany.

Kumar, M.S., Inthiyaz, S., Vamsi, C.K., Ahammad, S.H., Sai Lakshmi, K., VenuGopal, P., & BalaRaghavendra, A. (2019). Power optimization using dual sram circuit. *International Journal of Innovative Technology and Exploring Engineering, 8*(8), 1032-1036.

Kumar, R., & Hossain, A. (2019). Performance of random access Markov modelling for two-way buffer-aided relaying networks with wireless assisted links. Wireless Personal Comm., 108(3), 1995 – 2015.

Kumar. (2017). Ant Colony Optimization algorithm with internet of vehicles for intelligent Traffic Control System. *Computer Networks Journal.*

Kumar, D. S., Kumar, C. S., Ragamayi, S., Kumar, P. S., Saikumar, K., & Ahammad, S. H. (2020). A test architecture design for SoCs using atam method. *Iranian Journal of Electrical and Computer Engineering, 10*(1), 719.

Kumar, M. S., Inthiyaz, S., Vamsi, C. K., Ahammad, S. H., Sai Lakshmi, K., Venu Gopal, P., & Bala Raghavendra, A. (2019). Power optimization using dual sramcircuit. *International Journal of Innovative Technology and Exploring Engineering, 8*(8), 1032–1036.

Kumar, R., & Hossain, A. (2017). Optimisation of throughput of two-way buffer-aided relaying networks with wireless assisted links. *IET Communications, 11*(10), 1626–1632.

Kumar, R., & Hossain, A. (2019). Survey on half- and full- duplex relay based cooperative communications and its potential challenges and open issues using Markov chains. *IET Communications, 13*(11), 1537–1550.

Kumar, S., Maheshwari, V., Prabhu, J., Prasanna, M., & Jothikumar, R. (2021). Applying Blockchain in Agriculture: A Study on Blockchain Technology, Benefits, and Challenges. In *Deep Learning and Edge Computing Solutions for High Performance Computing* (pp. 167–181). Springer.

Küpper, A., Bareth, U., & Freese, B. (2011, October). Geofencing and background tracking–the next features in LBSs. *Proceedings of the 41th Annual Conference of the Gesellschaft für Informatik eV.*

Lai, Chang, Chao, Hossain, & Ghoneim. (2017). A Buffer-Aware QoS Streaming Approach for SDN-Enabled 5G Vehicular Networks. *IEEE Communications Magazine.*

Lam, H., & She, J. (2019). Distance Estimation on Moving Object using BLE Beacon. *2019 International Conference on Wireless and Mobile Computing, Networking and Communications (WiMob)*, 1-6. 10.1109/WiMOB.2019.8923185

Laneman, J., Tse, D., & Wornell, G. (2004). Cooperative diversity in wireless networks: Efficient protocols and outage behavior. *IEEE Transactions on Information Theory, 50*(12), 3062–3080. doi:10.1109/TIT.2004.838089

Langone, R. (2014). LS-SVM based spectral clustering and regression for predicting maintenance of industrial machines. *Engineering Applications of Artificial Intelligence.*

Langone, R., Alzate, C., Bey-Temsamani, A., & Suykens, J. A. K. (2014). Alarm prediction in industrial machines using autoregressive LS-SVM models. *2014 IEEE Symposium on Computational Intelligence and Data Mining (CIDM)*, 359-364. 10.1109/CIDM.2014.7008690

Le, T., & Gerla, M. (2016). Social-Distance based anycast routing in Delay Tolerant Networks. *2016 Mediterranean Ad Hoc Networking Workshop (Med-Hoc-Net)*, 1-7. 10.1109/MedHocNet.2016.7528495

LeDoux, F. S., Levine, A. S., & Kamp, R. N. (1983). New York State Department of Transportation bridge inspection and rehabilitation design program. *Transportation Research Record: Journal of the Transportation Research Board*, (899), 35–38.

Lepping, J. (2018). *Wiley Interdisciplinary Reviews. Data Mining and Knowledge Discovery.*

Liang, P. W., & Dai, B. R. (2013, June). Opinion mining on social media data. In *2013 IEEE 14th international conference on mobile data management* (Vol. 2, pp. 91-96). IEEE. 10.1109/MDM.2013.73

Liang. (2020). Toward Edge-Based Deep Learning in Industrial Internet of Things. *IEEE Internet of Things Journal, 7*(5).

Li, C., Chen, Z., Wang, Y., Yao, Y., & Xia, B. (2017). Outage analysis of the full-duplex decode-and-forward two-way relay system. *IEEE Transactions on Vehicular Technology, 66*(5), 4073–4086.

Li, J., & Liu, H. (2017). Challenges of feature selection for big data analytics. *IEEE Intelligent Systems, 32*(2), 9–15. doi:10.1109/MIS.2017.38

Li, L., Jin, X., Pan, S. J., & Sun, J. T. (2012, August). Multi-domain active learning for text classification. In *Proceedings of the 18th ACM SIGKDD international conference on Knowledge discovery and data mining* (pp. 1086-1094). 10.1145/2339530.2339701

Lin, F., & Wu, C. (2018). A spark-based high performance computational approach for simulating typhoon wind fields. *IEEE Access: Practical Innovations, Open Solutions, 6*, 39072–39085. doi:10.1109/ACCESS.2018.2850768

Lin, Z., Zeng, Q., Duan, H., Liu, C., & Lu, F. (2019). A Semantic User Distance Metric Using GPS Trajectory Data. *IEEE Access: Practical Innovations, Open Solutions, 7*, 30185–30196. doi:10.1109/ACCESS.2019.2896577

Li, S., Zhao, S., Yang, P., Andriotis, P., Xu, L., & Sun, Q. (2019). Distributed consensus algorithm for events detection in cyber-physical systems. *IEEE Internet of Things Journal, 6*(2), 2299–2308. doi:10.1109/JIOT.2019.2906157

Liu, H. (2016). *Rule Based Systems for Big Data: A Machine Learning Approach*. Springer. doi:10.1007/978-3-319-23696-4

Liu, X., & Ansari, N. (2017). Green relay assisted D2D communications with dual batteries in heterogeneous cellular networks for IoT. *IEEE Internet of Things Journal, 4*(5), 1707–1715. doi:10.1109/JIOT.2017.2717853

Liu, X., Li, X., Chen, X., Liu, Z., & Li, S. (2018). Location correction technique based on mobile communication base station for earthquake population heat map. *Geodesy and Geodynamics, 9*(5), 388–397. doi:10.1016/j.geog.2018.01.003

Li, Y., & Vucetic, B. (2008). On the performance of a simple adaptive relaying protocol for wireless relay networks. In *Proc. IEEE Semiannual Veh. Technol. Conf.* (pp. 2400–2405). Singapore: IEEE.

Lloret, T., Tomas, J., Canovas, A., & Parra, L. (2016). An integrated IoT architecture for smart metering. *IEEE Communications Magazine, 54*(12), 50–57. doi:10.1109/MCOM.2016.1600647CM

Lojka, T., Miškuf, M., & Zolotová, I. (2016). Industrial IoT Gateway with Machine Learning for Smart Manufacturing. In *Advances in Production Management Systems. Initiatives for a Sustainable World. APMS. IFIP Advances in Information and Communication Technology* (Vol. 488). Springer. doi:10.1007/978-3-319-51133-7_89

Luhn, H. P. (1960). Computer for verifying numbers. *US Patent, 2*(950), 48.

Luo, Ji, & Park. (2010). Location privacy against traffic analysis attacks in wireless sensor networks. *International Conference on Information Science and Applications (ICISA), 1*(6), 1–6.

Luo, S., & The, K. C. (2015). Buffer state based relay selection for buffer-aided cooperative relaying systems. *IEEE Transactions on Wireless Communications, 14*(10), 5430–5439.

Ma, B., Mansouri, H. S., & Wong, V. W. S. (2018). Full-duplex relaying for D2D communication in mm wave based 5G networks. *IEEE Transactions on Wireless Communications, 17*(7), 4417–4431.

Madureira, J., Paciência, I., Rufo, J., Ramos, E., Barros, H., Teixeira, J. P., & de Oliveira Fernandes, E. (2015). Indoor air quality in schools and its relationship with children's respiratory symptoms. *Atmospheric Environment, 118*, 145–156. doi:10.1016/j.atmosenv.2015.07.028

Maes, S., Tuyls, K., Vanschoenwinkel, B., & Manderick, B. (2002, January). Credit card fraud detection using bayesian and neural networks. In *Proceedings of the 1st international Naiso Congress on Neuro Fuzzy Technologies* (pp. 261-270). Academic Press.

Malini, N., & Pushpa, M. (2017, February). Analysis on credit card fraud identification techniques based on KNN and outlier detection. In *2017 Third International Conference on Advances in Electrical, Electronics, Information, Communication and Bio-Informatics (AEEICB)* (pp. 255-258). IEEE. 10.1109/AEEICB.2017.7972424

Mallik, R., Hazarika, A. P., Dastidar, S. G., Sing, D., & Bandyopadhyay, R. (2020). Development of an android application for viewing covid-19 containment zones and monitoring violators who are trespassing into it using firebase and geofencing. *Transactions of the Indian National Academy of Engineering, 5*(2), 163–179. doi:10.100741403-020-00137-3

Manoj, B. R., Mallik, R. K., & Bhatnagar, M. R. (2016). Buffer-aided multi-hop DF cooperative networks: A state-clustering based approach. *IEEE Transactions on Communications, 64*(12), 4997–5010.

Marques, G., Ferreira, C. R., & Pitarma, R. (2019). Indoor Air Quality Assessment Using a CO2 Monitoring System Based on Internet of Things. *Journal of Medical Systems, 43*(3), 1–12. doi:10.100710916-019-1184-x PMID:30729368

Marques, G., & Pitarma, R. (2017). Monitoring Health Factors in Indoor Living Environments Using Internet of Things. *Recent Advances in Information Systems and Technologies, 570*, 785–794. doi:10.1007/978-3-319-56538-5_79

Marques, G., & Pitarma, R. (2018). IAQ Evaluation Using an IoT CO2 Monitoring System for Enhanced Living Environments. *Trends and Advances in Information Systems and Technologies, 746*, 1169–1177. doi:10.1007/978-3-319-77712-2_112

Marques, G., & Pitarma, R. (2019). A Cost-Effective Air Quality Supervision Solution for Enhanced Living Environments through the Internet of Things. *Electronics (Basel), 8*(2), 170–182. doi:10.3390/electronics8020170

Marques, G., & Pitarma, R. (2019). Smartwatch-Based Application for Enhanced Healthy Lifestyle in Indoor Environments. *Computational Intelligence in Information Systems, 888*, 168–177. doi:10.1007/978-3-030-03302-6_15

Marques, G., & Pitarma, R. (2019). Using IoT and Social Networks for Enhanced Healthy Practices in Buildings. *Information Systems and Technologies to Support Learning, 111*, 424–432. doi:10.1007/978-3-030-03577-8_47

Marques, G., Roque Ferreira, C., & Pitarma, R. (2018). A System Based on the Internet of Things for Real-Time Particle Monitoring in Buildings. *International Journal of Environmental Research and Public Health, 15*(4), 821–834. doi:10.3390/ijerph15040821 PMID:29690534

Marrs, T. C., Maynard, R. L., & Sidell, F. R. (1996). *Chemical Warfare Agents: Toxicology and Treatment.* Wiley.

Massimo, Porcaro, & Iero. (2020). Edge Machine Learning for AI-Enabled IoT Devices: A Review. *Sensors (Basel), 20*(9), 2533. doi:10.339020092533

McClellan, M., Cervelló-Pastor, C., & Sallent, S. (2020). Deep Learning at the Mobile Edge: Opportunities for 5G Networks. *Applied Sciences (Basel, Switzerland), 10*(14), 4735. doi:10.3390/app10144735

Meador, C. A., & Moore, L. D. (2008). *U.S. Patent No. 7,431,202.* Washington, DC: U.S. Patent and Trademark Office.

Megahed, A. I. (2006). Usage of wavelet transform in the protection of series-compensated transmission lines. *IEEE Transactions on Power Delivery, 21*(3), 1213–1221.

Mei, Q., Ling, X., Wondra, M., Su, H., & Zhai, C. (2007, May). Topic sentiment mixture: modeling facets and opinions in weblogs. In *Proceedings of the 16th international conference on World Wide Web* (pp. 171-180). 10.1145/1242572.1242596

Mikalef, P., Pappas, I. O., Krogstie, J., & Giannakos, M. (2018). Big data analytics capabilities: A systematic literature review and research agenda. *Information Systems and e-Business Management, 16*(3), 547–578. doi:10.100710257-017-0362-y

Min, Q., Lu, Y., Liu, Z., Su, C., & Wang, B. (2019). Machine Learning based Digital Twin Framework for Production Optimization in Petrochemical Industry. *International Journal of Information Management, 49*, 502–519. doi:10.1016/j.ijinfomgt.2019.05.020

Mishra, A., & Ghorpade, C. (2018, February). Credit card fraud detection on the skewed data using various classification and ensemble techniques. In *2018 IEEE International Students' Conference on Electrical, Electronics and Computer Science (SCEECS)* (pp. 1-5). IEEE. 10.1109/SCEECS.2018.8546939

Mishra, M. K., & Dash, R. (2014, December). A comparative study of chebyshev functional link artificial neural network, multi-layer perceptron and decision tree for credit card fraud detection. In *2014 International Conference on Information Technology* (pp. 228-233). IEEE. 10.1109/ICIT.2014.25

Mo, L., Kim, Y. H. & Lee, J. K. (2015, June). Design of Disaster Collection and Analysis System Using Crowd Sensing and Beacon Based on Hadoop Framework. In *International Conference on Computational Science and Its Applications* (pp. 106-116). Springer. 10.1007/978-3-319-21410-8_8

Mohammad. (2014). Machine learning in wireless sensor networks: Algorithms, strategies, and applications. *IEEE Communications Surveys and Tutorials, 16*(4), 1996–2018. doi:10.1109/COMST.2014.2320099

Mohd, N. A. W. (2015). *A Comprehensive Review of Swarm Optimization Algorithms.* ResearchGate. doi:10.1371/journal.pone.0122827

Molina, Poyatos, Del Ser, García, Hussain, & Herrera. (2020). Comprehensive Taxonomies of Nature- and Bio-inspired Optimization: Inspiration versus Algorithmic Behavior, Critical Analysis and Recommendations. *arXiv journal.*

Moustafa, H., Kenn, H., Sayrafian, K., Scanlon, W., & Zhang, Y. (2015). Mobile wearable communications. *IEEE Wireless Communications, 22,* 1O–11.

Muenchaisri, P. (2019). Literature reviews on applying artificial intelligence/machine learning to software engineering research problems: preliminary. *Proceedings of the 2nd Software Engineering Education Workshop (SEED 2019).* http://ceur-ws.org

Mukherjee, A., Datta, J., Jorapur, R., Singhvi, R., Haloi, S., & Akram, W. (2012). Shared disk big data analytics with Apache Hadoop, High Performance Computing (HiPC). *19th International Conference.*

Müller, J. M., Kiel, D., & Voigt, K. I. (2018). What drives the implementation of Industry 4.0? The role of opportunities and challenges in the context of sustainability. *Sustainability, 10*(1), 247. doi:10.3390u10010247

Munson, J. P., & Gupta, V. K. (2002, September). Location-based notification as a general-purpose service. In *Proceedings of the 2nd international workshop on Mobile commerce* (pp. 40-44). 10.1145/570705.570713

Myla, S., Marella, S.T., Goud, A.S., Ahammad, S.H., Kumar, G.N.S., & Inthiyaz, S. (2019). Design decision taking system for student career selection for accurate academic system. *International Journal of Scientific and Technology Research, 8*(9), 2199-2206.

Myla, S., Marella, S.T., SwarnendraGoud, A., HasaneAhammad, S., Kumar, G.N.S., & Inthiyaz, S. (2019). Design decision taking system for student career selection for accurate academic system. *International Journal of Recent Technology and Engineering, 8*(9), 2199-2206.

Myla, S., Marella, S. T., Goud, A. S., Ahammad, S. H., Kumar, G. N. S., & Inthiyaz, S. (2019). Design decision taking system for student career selection for accurate academic system. *International Journal of Scientific and Technology Research, 8*(9), 2199–2206.

Myla, S., Marella, S. T., Swarnendra Goud, A., Hasane Ahammad, S., Kumar, G. N. S., & Inthiyaz, S. (2019). Design decision taking system for student career selection for accurate academic system. *International Journal of Recent Technology and Engineering, 8*(9), 2199–2206.

Nabar, R. U., Bölcskei, H., & Kneubühler, F. W. (2004). Fading relay channels: Performance limits and space-time signal design. *IEEE Journal on Selected Areas in Communications, 22*(6), 1099–1109.

Nagageetha, M., Mamilla, S.K., & Hasane Ahammad, S. (2017). Performance analysis of feedback based error control coding algorithm for video transmission on wireless multimedia networks. *Journal of Advanced Research in Dynamical and Control Systems, 9*(14), 626-660.

Nagageetha, M., Mamilla, S.K., & HasaneAhammad, S. (2017). Performance analysis of feedback based error control coding algorithm for video transmission on wireless multimedia networks. *Journal of Advanced Research in Dynamical and Control Systems, 9*(14), 626-660.

Nagendram, S., Rao, K. R. H., &Bojja, P. (2017). A review on recent advances of routing protocols for development of manet. *Journal of Advanced Research in Dynamical and Control Systems, 9*(2), 114-122.

Nakagawa, T., Yamada, W., Doi, C., Inamura, H., Ohta, K., Suzuki, M., & Morikawa, H. (2013, September). Variable interval positioning method for smartphone-based power-saving geofencing. In *2013 IEEE 24th Annual International Symposium on Personal, Indoor, and Mobile Radio Communications (PIMRC)* (pp. 3482-3486). IEEE. 10.1109/PIMRC.2013.6666751

Namuduri, S., Narayanan, B. N., Davuluru, V. S. P., Burton, L., & Bhansali, S. (2020). Review—Deep Learning Methods for Sensor Based Predictive Maintenance and Future Perspectives for Electrochemical Sensors. *Journal of the Electrochemical Society, 167*(3), 2020. doi:10.1149/1945-7111/ab67a8

Narayana, V.V., Ahammad, S.H., Chandu, B.V., Rupesh, G., Naidu, G.A., & Gopal, G.P. (2019). Estimation of quality and intelligibility of a speech signal with varying forms of additive noise. *International Journal of Emerging Trends in Engineering Research, 7*(11), 430-433.

Narayana, V. V., Ahammad, S. H., Chandu, B. V., Rupesh, G., Naidu, G. A., & Gopal, G. P. (2019). Estimation of quality and intelligibility of a speech signal with varying forms of additive noise. *International Journal of Emerging Trends in Engineering Research, 7*(11), 430–433. doi:10.30534/ijeter/2019/057112019

Nasrin, S., & Radcliffe, P. (2014). Novel protocol enables DIY home automation. *Telecommunication Networks and Applications, 1*, 212–216.

Nie, Y., & Zhang, W. (2018). Research on optimization design of hazardous chemicals logistics safety management system based on big data. *Chemical Engineering Transactions, 66*, 1477–1482.

Noji, E. K. (Ed.). (1997). *The public health consequences of disasters*. Oxford University Press.

Nomikos, N., Charalambous, T., Vouyioukas, D., Wichman, R., & Karagiannidis, G. K. (2018). Power adaptation in buffer-aided full-duplex relay networks with statistical CSI. *IEEE Transactions on Vehicular Technology, 67*(8), 7846–7850.

Noorbasha, F., Manasa, M., Gouthami, R. T., Sruthi, S., Priya, D. H., Prashanth, N., & Rahman, M. Z. U. (2017). FPGA implementation of cryptographic systems for symmetric encryption. *Journal of Theoretical and Applied Information Technology, 95*(9), 2038–2045.

Nurelmadina, Hasan, Mamon, Saeed, Akram, Ariffin, Ali, Mokhtar, Islam, Hossain, & Hassan. (2021). *A Systematic Review on Cognitive Radio in Low Power Wide Area Network for Industrial IoT Applications*. MDPI, Sustainability.

Obaid, A. J. (2020). *An efficient systematized approach for the detection of cancer in kidney*. Academic Press.

Orrù, P. F., Zoccheddu, A., Sassu, L., Mattia, C., Cozza, R., & Arena, S. (2020). Machine Learning Approach Using MLP and SVMAlgorithms for the Fault Prediction of a CentrifugalPump in the Oil and Gas Industry. *Sustainability, 2020*(12), 4776. doi:10.3390u12114776

Palanisamy, J. S. P. N., & Malmurugan, N. (2019). FPGA implementation of deep learning approach for efficient human gait action recognition system. International Journal Of Innovations In Scientific And Engineering Research, 6(11).

Parikh, U. B., Biswarup Das, & Prakash Maheshwari, R. P. (2008). Combined wavelet-SVM technique for fault zone detection in a series compensated transmission line. *IEEE Transactions on Power Delivery, 23*(4), 1789–1794. doi:10.1109/TPWRD.2008.919395

Park, J., Kim, T., & Kim, J. (2015, August). Image-based bolt-loosening detection technique of bolt joint in steel bridges. In *6th international conference on advances in experimental structural engineering.* University of Illinois, Urbana-Champaign.

Parvasi, S. M., Ho, S. C. M., Kong, Q., Mousavi, R., & Song, G. (2016). Real time bolt preload monitoring using piezoceramic transducers and time reversal technique—A numerical study with experimental verification. *Smart Materials and Structures, 25*(8), 085015. doi:10.1088/0964-1726/25/8/085015

Patel, R. D., & Singh, D. K. (2013). Credit card fraud detection & prevention of fraud using genetic algorithm. *International Journal of Soft Computing and Engineering, 2*(6), 292–294.

Patidar, R., & Sharma, L. (2011). Credit card fraud detection using neural network. *International Journal of Soft Computing and Engineering, 1*, 32–38.

Pawar, A. D., Kalavadekar, P. N., & Tambe, S. N. (2014). A survey on outlier detection techniques for credit card fraud detection. *IOSR Journal of Computer Engineering, 16*(2), 44–48. doi:10.9790/0661-16264448

Pereira, L. D., Cardoso, E., & da Silva, M. G. (2015). Indoor air quality audit and evaluation on thermal comfort in a school in Portugal. *Indoor and Built Environment, 24*(2), 256–268. doi:10.1177/1420326X13508966

Pham, Q.-V. (2020). *Swarm Intelligence for Next-Generation WirelessNetworks: Recent Advances and Applications.* arXiv:2007.15221v1 [cs.NI]

Philip Chen, C. L., & Zhang, C.-Y. (2014). Data-intensive applications, challenges, techniques and technologies: A survey on Big Data. *Information Sciences, 275*, 314–347. doi:10.1016/j.ins.2014.01.015

Pickholtz, R. L., Schilling, D. L., & Milstein, L. B. (1982). Theory of spread spectrum communications—a tutorial. *IEEE Transactions on Communications, 20*(5), 855–884. doi:10.1109/TCOM.1982.1095533

Pinjia, Wu, & Zhu. (2020). Open Ecosystem for Future Industrial Internet ofThings (IIoT): Architecture and Application. *CSEE Journal of Power and Energy Systems, 6*(1).

Piryani, R., Madhavi, D., & Singh, V. K. (2017). Analytical mapping of opinion mining and sentiment analysis research during 2000–2015. *Information Processing & Management, 53*(1), 122–150. doi:10.1016/j.ipm.2016.07.001

Pitarma, R., Marques, G., & Ferreira, B. R. (2017). Monitoring Indoor Air Quality for Enhanced Occupational Health. *Journal of Medical Systems, 41*(2), 1–10. doi:10.100710916-016-0667-2 PMID:28000117

Platt, J. (1998). *Sequential minimal optimization: A fast algorithm for training support vector machines.* Academic Press.

Polastre, Szewczyk, & Culler. (2005). Telos: enabling ultralow power wireless research. *IPSN*, 364–369.

Poonkuzhlai, P., Aarthi, R., Yaazhini, V. M., Yuvashri, S., & Vidhyalakshmi, G. (2021). Child Monitoring and Safety System Using Wsn and Iot Technology. *Annals of R.S.C.B., 25*, 10839–10847.

Poorna Chander Reddy, A., Siva Kumar, M., Murali Krishna, B., Inthiyaz, S., & Ahammad, S. H. (2020). Physical unclonable function based design for customized digital logic circuit. *International Journal of Advanced Science and Technology, 28*(8), 206–221.

PoornaChander Reddy, A., Siva Kumar, M., Murali Krishna, B., Inthiyaz, S., & Ahammad, S.H. (2019). Physical unclonable function based design for customized digital logic circuit. *International Journal of Advanced Science and Technology, 28*(8), 206-221.

Popovski, P., & Yomo, H. (2007). Wireless network coding by amplify-and-forward for bi-directional traffic flows. *IEEE Communications Letters, 11*(1), 16–18.

Prabhu, J., Kumar, P. J., Manivannan, S. S., Rajendran, S., Kumar, K. R., Susi, S., & Jothikumar, R. (2020). IoT role in prevention of COVID-19 and health care workforces behavioural intention in India-an empirical examination. *International Journal of Pervasive Computing and Communications.*

Prabu, S., & Kalaivani, M. (2019). An intelligent power adaptive model using machine learning techniques for wsn based smart health care devices. International Journal Of Innovations In Scientific And Engineering Research, 6(12).

Prasanna & Rao. (2012). An Overview of Wireless Sensor Networks Applications and Security. *International Journal of Soft Computing and Engineering, 2*(2).

Prasanth, Y., Sreedevi, E., Gayathri, N., & Rahul, A. S. (2017). Analysis and implementation of ensemble feature selection to improve accuracy of software defect detection model. *Journal of Advanced Research in Dynamical and Control Systems, 9*(18), 601-613.

Prince. (2020). IoT-Blockchain Enabled Optimized Provenance System for Food Industry 4.0 Using Advanced Deep Learning. *Sensors (Basel), 20*(10), 2990. doi:10.339020102990 PMID:32466209

Priya, K. B., Rajendran, P., Kumar, S., Prabhu, J., Rajendran, S., Kumar, P. J., ... Jothikumar, R. (2020). Pediatric and geriatric immunity network mobile computational model for COVID-19. *International Journal of Pervasive Computing and Communications.*

Priyan & Usha Devi. (2019). A survey on internet of vehicles: applications, technologies, challenges and opportunities. *Int. J. Advanced Intelligence Paradigms, 12.*

Provost, F., & Domingos, P. (2003). Tree induction for probabilitybased ranking. *Machine Learning, 52*(3), 199–215.

Pu & Kitsuregawa. (2013). *Big Data and disaster management: a report from the JST/NSF Joint Workshop.* Georgia Institute of Technology, CERCS.

Qadir, A., & Rasool, Z. (2016). Crisis analytics: Big data-driven crisis response. *Journal of International Humanitarian Action, 1*(1), 1–21. doi:10.118641018-016-0013-9

Qian, B. (2020). *Orchestrating the Development Lifecycle of Machine Learning-Based IoTApplications: A Taxonomy and Survey.* ArXiv: 1910.05433v5

Qiao, D. (2016). Effective capacity of buffer-aided full-duplex relay systems with selection relaying. *IEEE Transactions on Communications, 64*(1), 117–129.

Rahmath,, S., Nisha, Shyamala, C., Sheela, D., Abirami, M., Harshini, M., & Keerthana, R. (2021). Women and Children Safety System with IOT. *Turkish Journal of Computer and Mathematics Education, 12,* 768–772.

Raj Kumar, A., Kumar, G.N.S., Chithanoori, J.K., Mallik, K.S.K., Srinivas, P., & Hasane Ahammad, S. (2019). Design and analysis of a heavy vehicle chassis by using E-glass epoxy & S-2 glass materials. *International Journal of Recent Technology and Engineering, 7*(6), 903-905.

Raj Kumar, A., Kumar, G.N.S., Chithanoori, J.K., Mallik, K.S.K., Srinivas, P., & HasaneAhammad, S. (2019). Design and analysis of a heavy vehicle chassis by using E-glass epoxy & S-2 glass materials. *International Journal of Recent Technology and Engineering, 7*(6), 903-905.

Rakib, A., & Uddin, I. (2019). An Efficient Rule-Based Distributed Reasoning Framework for Resource-bounded Systems. *Springer. Mobile Networks and Applications, 24*(1), 82–99. doi:10.100711036-018-1140-x

Rama Chandra Manohar, K., Upendar, S., Durgesh, V., Sandeep, B., Mallik, K.S.K., Kumar, G.N.S., & Ahammad, S.H. (2018). Modeling and analysis of Kaplan Turbine blade using CFD. *International Journal of Engineering and Technology, 7*(3.12), 1086-1089.

Ram, A. (2017). Mobility adaptive density connected clustering approach in vehicular ad hoc networks. *International Journal of Communication Networks and Information Security.*

Ramaki, A. A., Asgari, R., & Atani, R. E. (2012). Credit card fraud detection based on ontology graph. *International Journal of Security, Privacy and Trust Management, 1*(5), 1–12. doi:10.5121/ijsptm.2012.1501

Ramirez-Cedeno, M. L., Ortiz-Rivera, W., Pacheco-Londono, L., & Hernandez-Rivera, S. P. (2010). Remote Detection of Hazardous Liquids Concealed in Glass and Plastic Containers. *IEEE Sensors Journal, 10*(3), 693–698. doi:10.1109/JSEN.2009.2036373

Rateb. (2020). Blockchain for the Internet of Vehicles: A Decentralized IoT Solution for Vehicles Communication Using Ethereum. *Sensors (Basel), 2020*(20), 3928. doi:10.339020143928

Rathore, A., Ahmad, A., Paul, A., & Rho, S. (2016). Urban planning and building smart cities based on the internet of things using big data analytics. *Computer Networks, 101*, 63–80. doi:10.1016/j.comnet.2015.12.023

Rathore, S., & Park, Y. P. J. H. (2019). BlockDeepNet: A Blockchain-Based Secure Deep Learning for IoT Network. *Sustainability, 2019*(11), 3974. doi:10.3390u11143974

Ray, M., Mukherjee, M., & Shu, L. (2017). Internet of things for disaster management: State-of-the-art and prospects. *IEEE Access: Practical Innovations, Open Solutions, 5*, 18818–18835. doi:10.1109/ACCESS.2017.2752174

Razia, S., Narasingarao, M. R., &Bojja, P. (2017). Development and analysis of support vector machine techniques for early prediction of breast cancer and thyroid. *Journal of Advanced Research in Dynamical and Control Systems, 9*(6), 869-878.

Razia, S., & Narasingarao, M. R. (2017). A neuro computing frame work for thyroid disease diagnosis using machine learning techniques. *Journal of Theoretical and Applied Information Technology, 95*(9), 1996–2005.

Richardson, P., Griffin, I., Tucker, C., Smith, D., Oechsle, O., Phelan, A., & Stebbing, J. (2020). Baricitinib as potential treatment for 2019-nCoV acute respiratory disease. *Lancet, 395*(10223), e30–e31. doi:10.1016/S0140-6736(20)30304-4 PMID:32032529

Riihonen, T., Werner, S., & Wichman, R. (2011). Hybrid full-duplex/half-duplex relaying with transmit power adaptation. *IEEE Transactions on Wireless Communications, 10*(9), 3074–3085.

Roglá, O. (2016). Social customer relationship management: Taking advantage of Web 2.0 and Big Data technologies. *SpringerPlus, 5*(1), 1462. doi:10.118640064-016-3128-y PMID:27652037

Ross, T. J. (2004). *Fuzzy logic with engineering applications* (Vol. 2). Wiley.

Russell, C. D., Millar, J. E., & Baillie, J. K. (2020). Clinical evidence does not support corticosteroid treatment for 2019-nCoV lung injury. *Lancet, 395*(10223), 473–475. doi:10.1016/S0140-6736(20)30317-2 PMID:32043983

Sadeh, Ranjbar, Hadsaid, & Feuillet. (2000). Accurate fault location algorithm for series compensated transmission lines. *2000 IEEE Power Engineering Society Winter Meeting. Conference Proceedings (Cat. No. 00CH37077), 4,* 2527-2532.

Sadeh, Ranjbar, Hadsaid, & Feuillet. (2000). Accurate fault location algorithm for series compensated transmission lines. In *2000 IEEE Power Engineering Society Winter Meeting. Conference Proceedings* (Cat. No. 00CH37077), (vol. 4, pp. 2527-2532). IEEE.

Saeed, R. A., Amran, B. H. N., & Aris, A. B. (2010). Design and evaluation of lightweight IEEE 802.11p-based TDMA MAC method for road side-to-vehicle communications. *IEEE, ICACT'10 Proceedings of the 12th international conference on Advanced communication technology,* 1483 – 1488.

Saha, M. M., Izykowski, J., Rosolowski, E., & Kasztenny, B. (1999). A new accurate fault locating algorithm forseries compensated lines. *IEEE Transactions on Power Delivery, 14*(3), 789–797. doi:10.1109/61.772316

Saikumar & Rajesh. (2019). A novel implementation heart diagnosis system based on random forest machine learning technique. *International Journal of Pharmaceutical Research, 12,* 3904–3916.

Saikumar & Rajesh. (n.d.). A novel implementation heart diagnosis system based on random forest machine learning technique. *International Journal of Pharmaceutical Research, 12,* 3904–3916.

Saikumar, K., Rajesh, V., Hasane Ahammad, S. K., Sai Krishna, M., Sai Pranitha, G., & Ajay Kumar Reddy, R. (2020). Cab for Heart Diagnosis with RFO Artificial Intelligence Algorithm. *International Journal of Research in Pharmaceutical Sciences, 11*(1).

Saikumar, K., Rajesh, V., HasaneAhammad, S. K., Sai Krishna, M., SaiPranitha, G., & Ajay Kumar Reddy, R. (2020). Cab for Heart Diagnosis with RFO Artificial Intelligence Algorithm. *International Journal of Research in Pharmaceutical Sciences, 11*(1).

Saikumar, K., & Rajesh, V. (2020). Diagnosis of coronary blockage of artery using MRI/CTA images through adaptive random forest optimization. *Journal of Critical Reviews, 7*(14), 591–600.

Saikumar, K., Rajesh, V., Hasane Ahammad, S. K., Sai Krishna, M., Sai Pranitha, G., & Ajay Kumar Reddy, R. (2020). Cab for Heart Diagnosis with RFO Artificial Intelligence Algorithm. *International Journal of Research in Pharmaceutical Sciences, 11*(1).

Sajana, T., & Narasingarao, M. R. (2017). Machine learning techniques for malaria disease diagnosis - A review. *Journal of Advanced Research in Dynamical and Control Systems, 9*(6), 349-369.

Sakharova, I. (2012, June). Payment card fraud: Challenges and solutions. In *2012 IEEE International Conference on Intelligence and Security Informatics* (pp. 227-234). IEEE. 10.1109/ISI.2012.6284315

Salamone, F., Belussi, L., Danza, L., Galanos, T., Ghellere, M., & Meroni, I. (2017). Design and Development of a Nearable Wireless System to Control Indoor Air Quality and Indoor Lighting Quality. *Sensors (Basel), 17,* 1021–1033.

Salazar, A., Safont, G., Soriano, A., & Vergara, L. (2012, October). Automatic credit card fraud detection based on non-linear signal processing. In *2012 IEEE International Carnahan Conference on Security Technology (ICCST)* (pp. 207-212). IEEE. 10.1109/CCST.2012.6393560

Samantaray, S. R. (2013). A data-mining model for protection of FACTS-based transmission line. *IEEE Transactions on Power Delivery, 28*(2), 612–618. doi:10.1109/TPWRD.2013.2242205

Sandhya, A., Nagendra, N. A., Srinivasa Rao, S., Patel, I., & Saikumar, K. (2021). Finding events of ecg signal using wavelet transform decomposition methods for sustainable application. *Journal of Green Engineering, 11*(2), 1321–1337.

Sang. (2020). Predictive Maintenance in Industry 4.0. *ICIST '2020.* 10.1145/1234567890

Saravanan, D., Nirmala Sumitra Rajini, S., & Dharmarajan, K. (2020). Efficient image Data Extraction using Image Clustering Technique. *Test Engineering and Management, 82,* 14574–14579.

Sardouk, M., & Merghem-Boulahia, G. (2013). Crisis management using MAS-based wireless sensor networks. *Computer Networks, 57*(1), 29–45. doi:10.1016/j.comnet.2012.08.010

Save, P., Tiwarekar, P., Jain, K. N., & Mahyavanshi, N. (2017). A novel idea for credit card fraud detection using decision tree. *International Journal of Computers and Applications, 161*(13), 6–9. Advance online publication. doi:10.5120/ijca2017913413

Serpanos, D., & Wolf, M. (2018). Industrial internet of things. In *Internet-of-Things (IoT) Systems* (pp. 37–54). Springer. doi:10.1007/978-3-319-69715-4_5

Shafie, A. E., Khafagy, M. G., & Sultan, A. (2014). Optimization of a relay-assisted link with buffer state information at the source. *IEEE Communications Letters, 18*(12), 2149–2152.

Shafie, A. E., Sultan, A., & Dhahir, N. A. (2016). Physical-layer security of a buffer-aided full-duplex relaying system. *IEEE Communications Letters, 20*(9), 1856–1859.

Shahid. (2020). A Novel Attack Detection Scheme for the Industrial Internet of Things Using a Lightweight Random Neural Network. *IEEE Access.*

Shahriar, H., North, S., & Mawangi, E. (2014). Testing of Memory Leak in Android Applications. *2014 IEEE 15th International Symposium on High-Assurance Systems Engineering,* 176-183. 10.1109/HASE.2014.32

Shah, Y. A. (2018). *CAMONET: Moth-Flame Optimization (MFO) Based Clustering Algorithm for VANETs.* IEEE Access. doi:10.1109/ACCESS.2018.2868118

Shajkofci, A. (2021). Correction of human forehead temperature variations measured by non-contact infrared thermometer. *IEEE Sensors Journal, 11,* 1–6. doi:10.1109/JSEN.2021.3058958

Shang, Jiang, Hemmati, Adams, Hassan, & Martin. (2013). Assisting developers of Big Data Analytics Applications when deploying on Hadoop clouds. *Software Engineering (ICSE), 35th International Conference.*

Shang, Jiang, Hemmati, Adams, Hassan, & Martin. (2013). Assisting developers of Big Data Analytics Applications when deploying on Hadoop clouds. *Software Engineering (ICSE), 35th International Conference.* hipi.cs.virginia.edu/ 2014/

Sharma, A., & Panigrahi, P. K. (2012). A review of financial accounting fraud detection based on data mining techniques. *International Journal of Computers and Applications, 39*(1), 37–47. Advance online publication. doi:10.5120/4787-7016

Sharma, M. K. S., & Dwivedi, R. K. (2018). Analyzing Facebook Data Set using Self-organizing Map. *2018 International Conference on System Modeling & Advancement in Research Trends (SMART),* 109-112. 10.1109/SYSMART.2018.8746984

Sharma, P. K., & Park, J. H. (2018). Blockchain based hybrid network architecture for the smart city. *Future Generation Computer Systems, 86,* 650–655. doi:10.1016/j.future.2018.04.060

Shearer, C. (2000). The CRISP-DM model: The new blueprint for data mining. *Journal of Data Warehousing, 5*(4), 13–22.

Sheshasaayee, D., & Megala, R. (2017). A Conceptual Framework For Resource Utilization In Cloud Using Map Reduce Scheduler. *International Journal of Innovations in Scientific and Engineering Research, 4*(6), 188–190.

Shim, Y., & Park, H. (2014). A closed-form expression of optimal time for two-way relay using DF MABC protocol. *IEEE Communications Letters, 18*(5), 721–724.

Shin, Cho, & Oh. (2018). SVM-Based Dynamic Reconfiguration CPS for Manufacturing System in Industry 4.0. Wireless Communications and Mobile Computing.

Shi, S., Li, S., & Tian, J. (2016). Markov modeling for two-way relay with finite buffer. *IEEE Communications Letters*, *20*(4), 768–771.

Shorabi, K., & Pottie, G. J. (1999). Performance of a novel self-Organization protocol for wireless ad hoc sensor networks. *Proceedings of IEEE VTC*, 1222–1226.

Sidorov, M., Nhut, P. V., Matsumoto, Y., & Ohmura, R. (2019). Lora-based precision wireless structural health monitoring system for bolted joints in a smart city environment. *IEEE Access: Practical Innovations, Open Solutions*, *7*, 179235–179251. doi:10.1109/ACCESS.2019.2958835

Siegel, J., & Perdue, J. (2012). Cloud Services Measures for Global Use: The Service Measurement Index (SMI). *SRII Global Conference (SRII), Annual*, 411-415. 10.1109/SRII.2012.51

Silva, F. A. (2014). Industrial Wireless Sensor Networks: Applications, Protocols, and Standards. *IEEE Industrial Electronics Magazine*, *8*, 67–68.

Singhal, T. (2020). A review of coronavirus disease-2019 (COVID-19). *Indian Journal of Pediatrics*, *87*(4), 281–286. doi:10.100712098-020-03263-6 PMID:32166607

Singh, S., & Raghavendra, C. S. (1998). PAMAS—Power aware multi-access protocol with signal- ing for ad hoc networks. *Proceedings of ACM SIGCOMM'98*, 5–26.

Sittón, I. (2018). Machine Learning Predictive Model for Industry 4.0. *International Conference on Knowledge Management in Organizations*, 501–510.

Siva Kumar, M., Inthiyaz, S., Venkata Krishna, P., JyothsnaRavali, C., Veenamadhuri, J., Hanuman Reddy, Y., & HasaneAhammad, S. (2019). Implementation of most appropriate leakage power techniques in vlsi circuits using nand and nor gate. *International Journal of Innovative Technology and Exploring Engineering*, *8*(7), 797-801.

Siva Kumar, M., Inthiyaz, S., & Venkata Krishna, P. (2019). Implementation of most appropriate leakage power techniques in vlsi circuits using nand and nor gate. *International Journal of Innovative Technology and Exploring Engineering*, *8*(7), 797–801.

Sivakumar, N., & Helenprabha, K. (2017). Hybrid medical image fusion using wavelet and curvelet transform with multi-resolution processing. *Biomedical Research*, *28*(6).

Sodhro, A. H., Luo, Z., Sodhro, G. H., Muzamal, M., Rodrigues, J. J. P. C., & Victor, H. C. (2019). Artificial Intelligence based QoS optimization for multimedia communication in IoV systems. *Future Generation Computer Systems*, *95*, 667–680. doi:10.1016/j.future.2018.12.008

Song, Xuan, & Johns. (1997). Protection scheme for EHV transmission systems with thyristor controlled series compensation using radial basis function neural networks. Electric Machines and Power Systems, 25(5), 553-565.

Song, Y., Alatorre, G., Mandagere, N., & Singh, A. (2013). Storage Mining: Where IT Management Meets Big Data Analytics, Big Data (BigData Congress). *IEEE International Congress*.

Song, Z., & Sekimoto, S. (2015, February). A simulator of human emergency mobility following disasters: Knowledge transfer from big disaster data. *Twenty-Ninth AAAI Conference on Artificial Intelligence*.

Song, Y. H., Johns, A. T., & Xuan, Q. Y. (1996). Artificial neural-networkbased protection scheme for controllable series- compensated EHV transmission lines. *IEE Proceedings. Generation, Transmission and Distribution, 143*(6), 535–540. doi:10.1049/ip-gtd:19960681

Soumya, N., Kumar, K. S., Rao, K. R., Rooban, S., Kumar, P. S., & Kumar, G. N. S. (2019). 4-Bit Multiplier Design using CMOS Gates in Electric VLSI. *International Journal of Recent Technology and Engineering.*

Soumya, N., Kumar, K. S., Rao, K. R., Rooban, S., Kumar, P. S., & Kumar, G. N. S. (n.d.). 4-Bit Multiplier Design using CMOS Gates in Electric VLSI. *International Journal of Recent Technology and Engineering.*

Spiegel-Ciobanu, V. E., Costa, L., & Zschiesche, W. (Eds.). (2020). *Hazardous Substances in Welding and Allied Processes.* Springer. doi:10.1007/978-3-030-36926-2

Srinivasa Reddy, K., Suneela, B., Inthiyaz, S., HasaneAhammad, S., Kumar, G.N.S., & Mallikarjuna Reddy, A. (2020). Texture filtration module under stabilization via random forest optimization methodology. *International Journal of Advanced Trends in Computer Science and Engineering, 8*(3), 458-469.

Srinivasa Reddy, K., Suneela, B., Inthiyaz, S., Hasane Ahammad, S., Kumar, G. N. S., & Mallikarjuna Reddy, A. (2020). Texture filtration module under stabilization via random forest optimization methodology. *International Journal of Advanced Trends in Computer Science and Engineering, 8*(3), 458–469.

Steffens, I. (2020). A hundred days into the coronavirus disease (COVID-19) pandemic. *Eurosurveillance, 25*(14), 2000550. doi:10.2807/1560-7917.ES.2020.25.14.2000550 PMID:32290905

Steinfeld, J. I., & Wormhoudt, J. (1998). Explosives detection: A challenge for physical chemistry. *Annual Review of Physical Chemistry, 49*(1), 203–232. doi:10.1146/annurev.physchem.49.1.203 PMID:15012428

Storey, V. C., & Song, I.-Y. (2017). Big data technologies and management: What conceptual modeling can do. *Data & Knowledge Engineering, 108*, 50–67. doi:10.1016/j.datak.2017.01.001

Sturm, B. L. (2013). Classification accuracy is not enough. *Journal of Intelligent Information Systems, 41*(3), 371–406. doi:10.100710844-013-0250-y

Sudac, D., Pavlović, M., Obhodas, J., Nad, K., Orlić, Ž., Uroić, M., Korolija, M., Rendić, D., Meric, I., Pettersen, H. E., & Valković, V. (2020). Detection of Chemical Warfare (CW) agents and the other hazardous substances by using fast 14 MeV neutrons. *Nuclear Instruments & Methods in Physics Research. Section A, Accelerators, Spectrometers, Detectors and Associated Equipment, 971*, 164066. doi:10.1016/j.nima.2020.164066

Sun, Q., Wu, Z., Duan, H., & Jia, D. (2019). Detection of Triacetone Triperoxide (TATP) Precursors with an Array of Sensors Based on MoS2/RGO Composites. *Sensors (Basel), 19*(6), 1281. doi:10.339019061281 PMID:30871286

Sun, W., Liu, J., & Yue, Y. (2019). AI-Enhanced Offloading in Edge Computing: When Machine Learning Meets Industrial IoT. *IEEE Network, 33*(5), 68–74. doi:10.1109/MNET.001.1800510

Suyama, A., & Inoue, U. (2016, June). Using geofencing for a disaster information system. In *2016 IEEE/ACIS 15th International Conference on Computer and Information Science (ICIS)* (pp. 1-5). IEEE. 10.1109/ICIS.2016.7550849

Swapna,, B., Gayathri,, S., Kamalahasan,, M., & Hemasundari,, H., SiraasGanth, M., & Ranjith, S. (2021). E-healthcare monitoring using internet of things. *IOP Conference Series. Materials Science and Engineering, 872*, 1–8.

Swapna, B., & Manivannan, S. (2018). Analysis: Smart Agriculture and Landslides Monitoring System Using Internet of Things (IOT). *International Journal of Pure and Applied Mathematics, 118*, 24–30.

Swapna, B., Manivannan, S., & Nandhinidevi, R. (2020). Prediction of soil reaction (pH) and soil nutrients using multivariate statistics techniques for agricultural crop and soil management. *International Journal of Advanced Science and Technology*, *29*, 1900–1912.

Swapna, Kamalahsan, & Sowmiya, Konda, & SaiZignasa. (2019). Design of smart garbage landfill monitoring system using Internet of Things. *IOP Conference Series. Materials Science and Engineering*, *561*, 012084–012091.

Swapna, & Shubhashree, George, & Hemasundari. (2020). IoT Based Intelligence Parking Management System using LOT Display. *International Journal of Advanced Science and Technology*, *29*, 2208–2213.

Szczytowski, P. (2014). Geo-fencing based disaster management service. In *Agent technology for intelligent mobile services and smart societies* (pp. 11–21). Springer.

Taboada, M., Brooke, J., Tofiloski, M., Voll, K., & Stede, M. (2011). Lexicon-based methods for sentiment analysis. *Computational Linguistics*, *37*(2), 267–307. doi:10.1162/COLI_a_00049

Tan, X., Huang, S., Zhong, Y., Yuan, H., Zhou, Y., Xiao, Q., Guo, L., Tang, S., Yang, Z., & Qi, C. (2017). *Detection and identification of flammable and explosive liquids using THz time-domain spectroscopy with principal component analysis algorithm.* Paper presented at the 2017 10th UK-Europe-China Workshop on Millimetre Waves and Terahertz Technologies (UCMMT). 10.1109/UCMMT.2017.8068488

Taniguchi, Y., Nishii, K., & Hisamatsu, H. (2015). Evaluation of a Bicycle-Mounted Ultrasonic Distance Sensor for Monitoring Road Surface Condition. *2015 7th International Conference on Computational Intelligence, Communication Systems and Networks*, 31-34, 10.1109/CICSyN.2015.16

Tannious, R., & Nosratinia, A. (2008). Spectrally-Efficient Relay Selection with Limited Feedback. *IEEE Journal on Selected Areas in Communications*, *26*(8), 1419–1428.

Tanwar, S., Bhatia, Q., Patel, P., Kumari, A., Singh, P. K., & Hong, W. (2020). Machine Learning Adoption in Blockchain-Based Smart Applications: The Challenges, and a Way Forward. *IEEE Access: Practical Innovations, Open Solutions*, *8*, 474–488. doi:10.1109/ACCESS.2019.2961372

Tatar, G., & Bayar, S. (2018). FPGA Based Bluetooth Controlled Land Vehicle. *2018 International Symposium on Advanced Electrical and Communication Technologies (ISAECT)*, 1-6. 10.1109/ISAECT.2018.8618830

Tene & Polonetsky. (2011). Privacy in the age of big data: A time for big decisions. *Stan. L. Rev. Online*, *64*, 63.

Thakur & Mann. (2014). Data mining for big data: A review. *International Journal of Advanced Research in Computer Science and Software Engineering*, *4*(5), 469–473.

Thelwall, M. (2010). Emotion homophily in social network site messages. *First Monday*, *15*(4). Advance online publication. doi:10.5210/fm.v15i4.2897

Thomas, D. W. P., & Christopoulos, C. (1992). Ultra-high speed protection of series compensated lines. *IEEE Transactions on Power Delivery*, *7*(1), 139–145. doi:10.1109/61.108900

Tong, Zhao, Yang, Cao, Chen, & Chen. (2021). A hybrid method for overcoming thermal shock of non-contact infrared thermometers. *Physics. Med-Ph*, *1*, 1-14.

Trakadas, P., Simoens, P., Gkonis, P., Sarakis, L., Angelopoulos, A., Ramallo-González, A. P., Skarmeta, A., Trochoutsos, C., Calvo, D., Pariente, T., Chintamani, K., Fernandez, I., Irigaray, A. A., Parreira, J. X., Petrali, P., Leligou, N., & Karkazis, P. (2020). An Artificial Intelligence-Based Collaboration Approach in Industrial IoT Manufacturing: Key Concepts, Architectural Extensions and Potential Applications. *Sensors (Basel)*, *2020*(20), 5480. doi:10.339020195480 PMID:32987911

Tralli, B., & Zlotnicki, D. (2005). Satellite remote sensing of earthquake, volcano,flood, landslide and coatal industrial hazards. *ISPRS Journal of Photogrammetry and Remote Sensing, 59*(4), 185–198. doi:10.1016/j.isprsjprs.2005.02.002

Trappey, C. V. (2019). Patent Value Analysis Using Deep Learning Models? The Case ofIoT Technology Mining for the Manufacturing Industry. *IEEE Transactions on Engineering Management.* Advance online publication. doi:10.1109/TEM.2019.2957842

Tuladhar, L. R., Shrestha, S., Ghimire, N., Acharya, N., & Tamrakar, E. T. (2021). Clinical Efficiency of Non-Contact Infrared Thermometer over Axillary Digital Thermometer and Mercury in Glass Thermometer with Paracetamol. *Nepal Medical College Journal, 23*(1), 31–36. doi:10.3126/nmcj.v23i1.36224

Van Vlasselaer, V., Bravo, C., Caelen, O., Eliassi-Rad, T., Akoglu, L., Snoeck, M., & Baesens, B. (2015). APATE: A novel approach for automated credit card transaction fraud detection using network-based extensions. *Decision Support Systems, 75,* 38–48. doi:10.1016/j.dss.2015.04.013

Van Western, Soeters, & Buchroithner. (1996). Potential and limitations of satellite remote sensing for geo-disaster reduction. *International Archieves of Photogrammetry and Remote Sensing, 31*(B6).

Varga, P., Peto, J., Franko, A., Balla, D., Haja, D., Janky, F., Soos, G., Ficzere, D., Maliosz, M., & Toka, L. (2020). 5G support for Industrial IoT Applications— Challenges, Solutions, and Research gaps. *Sensors (Basel), 20*(3), 828. doi:10.339020030828 PMID:32033076

Varmedja, D., Karanovic, M., Sladojevic, S., Arsenovic, M., & Anderla, A. (2019, March). Credit card fraud detection-machine learning methods. In *2019 18th International Symposium INFOTEH-JAHORINA (INFOTEH)* (pp. 1-5). IEEE. 10.1109/INFOTEH.2019.8717766

Venkatachalam. (2020). An automated and improved brain tumor detection in magnetic resonance images. *International Journal Of Innovations In Scientific And Engineering Research, 7*(8).

Vidhya, Patil, & Patil. (2019). An Efficient MRI Brain Image Registration and Wavelet Based Fusion. *International Journal of Recent Technology and Engineering, 8*(4), 10209-10218.

Vijaykumar, G., Gantala, A., Gade, M.S.L., Anjaneyulu, P., & Ahammad, S.H. (2017). Microcontroller based heartbeat monitoring and display on PC. *Journal of Advanced Research in Dynamical and Control Systems, 9*(4), 250-260.

Vijaykumar, G., Gantala, A., Gade, M. S. L., Anjaneyulu, P., & Ahammad, S. H. (2016). Microcontroller based heart-beat monitoring and display on PC. *Journal of Advanced Research in Dynamical and Control Systems, 9*(4), 250–260.

Wang, V., & Salehi, R. (2017, May). A large-scale spatio-temporal data analytics system for wildfire risk management. In *Proceedings of the Fourth International ACM Workshop on Managing and Mining Enriched Geo-Spatial Data* (pp. 1-6). ACM.

Wang, B. (2020). *Smart Manufacturing and Intelligent Manufacturing: A Comparative Review. Engineering Journal.*

Wang, W., Wu, Y., Yen, N., Guo, S., & Cheng, Z. (2016). Big data analytics for emergency communication networks: A survey. *IEEE Communications Surveys and Tutorials, 18*(3), 1758–1778. doi:10.1109/COMST.2016.2540004

White, T. (2010). *Hadoop: the Definitive Guide* (2nd ed.). O'Reilly Media.

Winning with the industrial Internet of things. (2015). *Accenture.* Available: https://www.accenture.com/t20160909T042713Z—w—/us- en/_acnmedia/Accenture/Conversion- Assets/DotCom/Documents/Global/PDF/Dual-pub_11/Accenture-Industrial-Internet-of-Things-Positioning-Paper-Report-2015.pdfla=en

Witten, I. H. (2017). Extending instance-based and linear models. In *Data Mining* (4th ed.). Elsevier. doi:10.1016/B978-0-12-804291-5.00007-6

Wu, Y., Chou, P. A., & Kung, S. Y. (2005, March). Information exchange in wireless network coding and physical-layer broadcast. In *Proc. 39th Annu. Conf. Sci. syst.*, (pp. 1 – 6). Academic Press.

Xia, B., Fan, Y., Thompson, J., & Poor, H. V. (2008). Buffering in a three-node relay network. *IEEE Transactions on Wireless Communications*, 7(11), 4492–4496.

Xiang, L. (2011). *Analysis on architecture of cloud computing based on Hadoop*. Computer Era.

Xie, Y., Liu, G., Cao, R., Li, Z., Yan, C., & Jiang, C. (2019, February). A feature extraction method for credit card fraud detection. In *2019 2nd International Conference on Intelligent Autonomous Systems (ICoIAS)* (pp. 70-75). IEEE. 10.1109/ICoIAS.2019.00019

Xiong, Wong, & Deng. (2010). Tiny Pairing: A fast and lightweight pairing- based cryptographic library for wireless sensor networks. *Proceedings of the IEEE Wireless Communications and Networking Conference*.

Xu & Dewei. (2012). Research on Clustering Algorithm for Massive Data Based on Hadoop Platform. *Computer Science & Service System (CSSS), 2012 International*.

Xu, Y. (2019). A Blockchain-Based Nonrepudiation Network Computing Service Scheme for Industrial IoT. IEEE Transactions on Industrial Informatics, 15(6).

Xuan, Q. Y., Song, Y. H., Johns, A. T., Morgan, R., & Williams, D. (1996). Performance of an adaptive protection scheme for series compensated EHV transmission systems using neural networks. *Electric Power Systems Research*, 36(1), 57–66. doi:10.1016/0378-7796(95)01014-9

Xu, H., Liu, X., Yu, W., Griffith, D. W., & Golmie, N. T. (2020). *Reinforcement Learning-Based Control and Networking Co-design for Industrial Internet of Things. IEEE Journal on Selected Areas in Communications*. doi:10.1109/JSAC.2020.2980909

Xu, Z., & Sotiriadis, A. (2018). CLOTHO: A large-scale Internet of Things-based crowd evacuation planning system for disaster management. *IEEE Internet of Things Journal*, 5(5), 3559–3568. doi:10.1109/JIOT.2018.2818885

Yang, F. (2014). An Overview of Internet of Vehicles. *China Communications Journal*. doi:10.1109/CC.2014.6969789

Yang, L., Guo, B. L., & Ni, W. (2008). Multimodality medical image fusion based on multiscale geometric analysis of contourlet transform. Neurocomputing, 72, 203–211.

Yang, S. H., & Plotnick. (2013). How the internet of things technology enhances emergency response operations. *Technological Forecasting and Social Change, 80*(9), 1854-1867.

Yang, C., & Ng, T. D. (2009). Web opinions analysis with scalable distance-based clustering. *2009 IEEE International Conference on Intelligence and Security Informatics*, 65-70. 10.1109/ISI.2009.5137273

Yang, L., Yan, H., & Lam, J. C. (2014). Thermal comfort and building energy consumption implications – A review. *Applied Energy, 115*, 164–173. doi:10.1016/j.apenergy.2013.10.062

Yang, T., Zhang, R., & Cheng, X. (2017). Graph coloring based resource sharing (GCRS) scheme for D2D communications underlaying full-duplex cellular networks. *IEEE Transactions on Vehicular Technology*, 66(8), 7506–7517.

Ye, W., Heidemann, J., & Estrin, D. (2002). An energy-efficient MAC protocol for wireless sensor networks. Proceedings of IEEE Infocom, 1567–1576.

Yelne, S., & Kapade, V. (2015). Human protection with the disaster management using an android application. *International Journal of Scientific Research in Science, Engineering and Technology, 1*(5), 15–19.

Yin, H., Wang, T., Yang, D., Liu, S., Shao, J., & Li, Y. (2016). A smart washer for bolt looseness monitoring based on piezoelectric active sensing method. *Applied Sciences (Basel, Switzerland), 6*(11), 320. doi:10.3390/app6110320

Yli-Ojanperä, M., Sierla, S., Papakonstantinou, N., & Vyatkin, V. (2019). Adapting an agile manufacturing concept to the reference architecture model industry 4.0: A survey and case study. *Journal of Industrial Information Integration, 15*, 147–160. doi:10.1016/j.jii.2018.12.002

Yu, B., Yang, L., Cheng, X., & Cao, R. (2015). Power and location optimization for full-duplex decode-and-forward relaying. *IEEE Transactions on Communications, 63*(12), 4743–4753.

Yu, D., Yang, F., Yang, C., Leng, C., Cao, J., Wang, Y., & Tian, J. (2016). Fast Rotation-Free Feature-Based Image Registration using Improved N-SIFT and GMM-Based Parallel Optimization. *IEEE Transactions on Biomedical Engineering, 63*(8), 1653–1664.

Yun, Hua, & Zhi. (2010). Intelligent video monitoring system based on 3G. *2010 International Conference on Educational and Network Technology*, 135-138. 10.1109/ICENT.2010.5532145

Yu, W. F., & Wang, N. (2009, April). Research on credit card fraud detection model based on distance sum. In *2009 International Joint Conference on Artificial Intelligence* (pp. 353-356). IEEE. 10.1109/JCAI.2009.146

Zareapoor, M., & Shamsolmoali, P. (2015). Application of credit card fraud detection: Based on bagging ensemble classifier. *Procedia Computer Science, 48*, 679–685. doi:10.1016/j.procs.2015.04.201

Zhang & Kezunovic. (2007). Transmission line boundary protection using wavelet transform and neural network. *IEEE Transactions on Power Delivery, 22*(2), 859–869.

Zhang, Nan, & Kezunovic. (2007). Transmission line boundary protection using wavelet transform and neural network. *IEEE Transactions on Power Delivery, 22*(2), 859–869.

Zhang, S. (2019). *Machine Learning and Deep Learning Algorithms for Bearing Fault Diagnostics? A Comprehensive Review.* arXiv preprint arXiv:1901.08247.

Zhang, D., Chan, C. C., & Zhou, G. Y. (2018). Enabling Industrial Internet of Things (IIoT) towards an emerging smart energy system. *Global Energy Interconnection, 1*(1), 39–47.

Zhang, J. (2017). Development of a Non-Contact Infrared Thermometer. *Proceedings of the International Conference Advanced Engineering and Technology Research, 1*, 10-22.

Zhang, Y., You, F., & Liu, H. (2009, August). Behavior-based credit card fraud detecting model. In *2009 Fifth International Joint conference on INC, IMS and IDC* (pp. 855-858). IEEE. 10.1109/NCM.2009.54

Zhao, F., Shin, J., & Reich, J. (2002). Information-driven dynamic sensor collaboration for target tracking. *IEEE Signal Processing Magazine, 19*(2), 61–72. doi:10.1109/79.985685

Zhao, X., Zhang, Y., & Wang, N. (2019). Bolt loosening angle detection technology using deep learning. *Structural Control and Health Monitoring, 26*(1), e2292. doi:10.1002tc.2292

Zheng, J., Simplot-Ryl, D., Bisdikian, C., & Mouftah, H.T. (2011). The internet of things. *Communications Magazine IEEE, 49*, 30-31.

Zhong, R. Y, (2017). *Intelligent Manufacturing in the Context of Industry 4.0: A Review. Engineering Journal.* doi:10.1016/J.ENG.2017.05.015

Zhou, B., Cui, Y., & Tao, M. (2015). Stochastic throughput optimization for two-hop systems with finite relay buffers. *IEEE Transactions on Signal Processing, 63*(20), 5546–5560.

Zhu, H., Liu, G., Zhou, M., Xie, Y., Abusorrah, A., & Kang, Q. (2020). Optimizing weighted extreme learning machines for imbalanced classification and application to credit card fraud detection. *Neurocomputing, 407*, 50-62. doi:10.1016/j.neucom.2020.04.078

Zhu, R., Hu, X., Li, X., Ye, H., & Jia, N. (2020). Modeling and Risk Analysis of Chemical Terrorist Attacks: A Bayesian Network Method. *International Journal of Environmental Research and Public Health, 17*(6), 2051. doi:10.3390/ijerph17062051 PMID:32204577

Zlatanov, N., & Schober, R. (2013). Buffer-aided relaying with adaptive link selection—Fixed and mixed rate transmission. *IEEE Transactions on Information Theory, 59*(5), 2816–2840.

Zolanvari, M., Teixeira, M. A., Gupta, L., Khan, K. M., & Jain, R. (2019, August). Machine Learning-Based Network Vulnerability Analysis of Industrial Internet of Things. *IEEE Internet of Things Journal, 6*(4), 6822–6834. doi:10.1109/JIOT.2019.2912022

Zolanvari, M., Teixeira, M. A., & Jain, R. (2018). Effect of Imbalanced Datasets on Security of Industrial IoT Using Machine Learning. *2018 IEEE International Conference on Intelligence and Security Informatics (ISI)*, 112-117. 10.1109/ISI.2018.8587389

Zulkernine, F., Martin, P., Ying, Z., Bauer, M., Gwadry-Sridhar, F., & Aboulnaga, A. (2013). Towards Cloud-Based Analytics-as-a-Service (CLAaaS) for Big Data Analytics in the Cloud, Big Data (BigData Congress). *IEEE International Congress.*

About the Contributors

Jingyuan Zhao is a research fellow at University of Toronto, Canada. She is a also professor at Beijing Union University (China). She obtained her PhD in Management Science and Engineering from University of Science and Technology of China (China) and completed a postdoctoral program in Management of Technology from University of Quebec at Montreal (Canada). Dr. Zhao's expertise is on management of technological innovation, technological strategy, regional innovation systems and global innovation networks, knowledge management, management information systems, and science and technology policy.

V. Vinoth Kumar is an Associate Professor in the Department of Computer Science and Engineering in MVJ College of Engineering, Bangalore, India. He is a highly qualified individual with around 8 years of rich expertise in teaching, entrepreneurship, and research and development with specialization in computer science engineering subjects. He has been a part of various seminars, paper presentations, research paper reviews, and conferences as a convener and a session chair, a guest editor in journals and has co-authored several books and papers in national, international journals and conferences. He is a professional society member for ISTE, IACIST and IAENG. He has published more than 15 articles in National and International journals, 10 articles in conference proceedings and one article in book chapter. He has filed Indian patent in IoT Applications. His Research interest includes Mobile Adhoc Networking and IoT.

* * *

Sampath Dakshina Murthy Achanta received his B.Tech in Electronics and Communication Engineering Degree in 2013 and M.Tech Degree in Digital Electronics and Communication Engineering in 2015. He is working as an Assistant Professor in the Department of Electronics and Communication Engineering at Vignan's Institute of Information Technology (A). Pursuing a Ph.D. in Electronics and Communication Engineering from Koneru Lakshmaiah Education Foundation Deemed University in Guntur, Andhra Pradesh. His research interest includes Gait analysis in Pattern Recognition, Image & Video Processing, Fuzzy Logic, and Neural Networks. He has published 36 papers in International Journals. He is having 2 patents; 1Book published and received various 2 International and 3 National awards. He is a Life member of I.S.R.S, IEI, IFERP, and MIET. He is a reviewer and editorial board for journals in Scopus and SCI. He has 5 years of teaching and research experience.

Wael Mohammad Alenazy is currently working as an Assistant Professor and Chair of Self-Development Skills at Common First Year Deanship, King Saud University, Riyadh, Saudi Arabia. He received his PhD and Master's degree from Faculty of Engineering and Information Technology, University of Technology, Sydney (UTS), Australia. His area of specialization includes Advanced ICT solutions on Smart Environments and Smart Classroom. His current area of research interests includes Smart Learning, Enhanced Smart Education, Image Processing and Augmented Reality.

Abdullah Saleh Alqahtani is currently working as an Associate Professor and Coordinator- IT in Computer Science at Common First Year Deanship, Department of Self-Development Skills, King Saud University, Riyadh, Saudi Arabia. He received his PhD and Master's degree from School of Computer Science, Engineering and Mathematics under the Faculty of Science and Engineering at Flinders University, Adelaide, Australia. His area of specializations includes Data Analytics on E-commerce, E-Business and Data Optimization by Structural Equation Modeling. His current area of research interests includes Optimization, Computational Algorithms, Data science, Deep Learning and Artificial intelligence.

Mona Bakri Hassan received her the M.Sc. degree in electronics engineering, Telecommunication in 2020, and B.Sc. (Honor) degree in electronics engineering, Telecommunication in 2013. She has completed CCNP Routing and Switching course and passed the CCNA Routing and Switching. She has gained experience in network operations center as network engineer in back office department in Sudan Telecom Company (SudaTel) and in vision valley as well. She has also worked as a teacher assistant in Sudan University of Science and Technology, which has given her a great opportunity to gain teaching experience. She published book chapters, and paper in wireless communications and networking in peer reviewed academic international journals. Her areas of research interest include Cellular LPWAN, IoT and wireless communication.

Bhavana D. works as an associate professor in the ECE department of Koneru Lakshmaiah Education Foundation Vaddeswaram, Guntur. She completed her PhD in 2017 on thermal image processing. Her area of interest is on image processing, IoT and AI.

Vijendra Babu D. received his B.E.from University of Madras, India, M.Tech. From SASTRA, Tanjore, India and Ph.D. from Jawaharlal Nehru Technological University, Hyderabad, India. He is currently designated as Vice Principal & Professor in the Department of Electronics and Communication Engineering. He has 21 Years of Experience in the field of Education, Research & Administration at various levels. He has served as HoD for 13 Years. He has obtained Grant for 2 Australian Patents & Published 5 Indian Patents. He has published 90+ papers in Referred International / National Journals & Conferences and also Reviewer in various leading Journals/Conferences. He has chaired in 19 International/National Conferences.Organised 100+ International & National Webinars/Workshops. He has delivered 30 Invited Lectures in the area of Wearable Devices, 5G Communication, Artificial Intelligence, Robotics, Image/Video Processing & Teaching Pedagogies in various Institutions. He has chaired several sessions in International / National Conferences. His research interests include IoT, Wearable Devices and Computer Vision applications. He is a recipient of Academic Leadership Award(2021),e-Innovation Award (2020),Most Engaging Educationist Award(2020),Digital Star of Education(2020),Best Paper Award in IEEE International Conference (2013), Outstanding Contribution to Teaching Award (2012)

& Active ISF Award (2011 & 2012). Apart from Academics, he is an active involvement in Professional societies as a Life Member in IEEE, CSI, IETE, BES (I), ISTE, ACEEE and IACSIT.

Sawsan Daoud is a Neurologist Assistant, Department of Neurology, Habib Bourguiba University Hospital, Sfax, Tunisia.

Amira Echtioui is a PhD student in engineering of computer systems from the National Engineering School of Sfax-Tunisia since 2017/2018. Received his degree of research master in Decisional Informatics from the Higher Institute of Applied Sciences and Technology of Gafsa-Tunisia in 2016, his degree in Applied License in Computer Networks: Information Technology and Telecommunications from the Higher Institute of Applied Sciences and Technology of Gafsa-Tunisia in 2012. She is a member of the Advanced Technologies for Medicine and Signal Research Group (ATMS) in Tunisia. Her research are satellite and medical image, remote sensing, biomedical and EEG signal processing, and brain-computer interface based on motor imagery.

Ebru Efeoglu received her B.S. degree in Geophysics Engineering from Kocaeli University and Management Information Systems from Anadolu University, Turkey. She received her Ph.D. degree in Computer Engineering from Trakya University, Turkey in 2021. She is currently an Assistant Professor at İstanbul Gedik University, Department of Management Information Systems. Her research interests include non-contact detection of hazardous liquids, machine learning and classification algorithms.

Mohamed Ghorbel (HDR, MC) was born in Sfax, Tunisia in 1967. He received the engineering diploma from the National School of Technical Education of Tunis in 1991, a Master degree in Electronics from the National School of Engineers of Sfax in 2000, a doctorate in electrical engineering in 2007 and the Habilitation degree (Post Doctorate degree) in 2014, from the University of Sfax, Tunisia. His research interests include design circuits and signal processing dedicated to medical devices and tools. Currently, he is Permanent Professor at National School of Electronics and Telecommunications of Sfax (ENETCom), Tunisia and head of the "Intelligent systems applied to medicine" research group in the "ATMS" Laboratory at ENIS Sfax, Tunisia.

Habib Hamam obtained the B.Eng. and M.Sc. degrees in information processing from the Technical University of Munich, Germany 1988 and 1992, and the PhD degree in Physics and applications in telecommunications from Université de Rennes I conjointly with France Telecom Graduate School, France 1995. He also obtained a postdoctoral diploma, "Accreditation to Supervise Research in Signal Processing and Telecommunications", from Université de Rennes I in 2004. He was for 10 years (2006-2016) a Canada Research Chair holder in "Optics in Information and Communication Technologies". He is currently a full Professor in the Department of Electrical Engineering at the University of Moncton. He is OSA senior member, IEEE senior member and a registered professional engineer in New-Brunswick. He is among others editor in chief in CIT-Review and associate editor of the IEEE Canadian Review. His research interests are in optical telecommunications, Wireless Communications, diffraction, fiber components, RFID, information processing, data protection and Deep learning.

Ashraf Hossain earned the Ph.D degree in Electronics & Electrical Communication Engg., from Indian Institute of Technology (IIT), Kharagpur, India in 2011. He received the M.Tech. and B.Tech.

degrees in Radio Physics & Electronics from the Institute of Radio Physics & Electronics, University of Calcutta, Kolkata, India in 2004 and 2002, respectively. He served Dept. of Electronics & Communication Engg., Aliah University, Kolkata and Haldia Institute of Technology as Asst. Professor. He is currently working as an Assistant Professor in the Dept. of Electronics & Communication Engg., National Institute of Technology (NIT), Silchar, Assam, India. His research interests include wireless sensor network, communication theory and systems. He is a Senior member of IEEE and IE (India).

Beschi I. S. is currently a faculty in Department of Computer Applications at St. Joseph's College (Arts & Science), Kovur, Chennai. He has completed his Bachelor of Science in Chemistry, Master of Computer Application, Master of Engineering in Computer Science and Master of Philosophy in Computer Science. He is aiming to perform a research on Bigdata and Machine learning. He has about 13 years of experience in teaching and 3 years of experience in the software industry. He has published more than six papers in various journals of repute globally. He has organized various seminars, webinars, workshops, and conferences. He is an active coordinator of Internal Quality Assurance Cell and an independent member of ISO committee in his college. His areas of interest are Artificial Intelligence, Bigdata, Internet of Things, Machine Learning and Deep Learning.

Balajee J. is currently pursuing a PhD in the School of Information Technology and Engineering, Vellore Institute of Technology University. He received his Bachelor of Computer Science degree in University of Madras, Chennai and Master of Computer Application degree from VIT University, Vellore and M.Phil degree Thiruvalluvar University, Vellore, respectively. I have worked as a Project Assistant for a project on Stability and aggregation of silver nanoparticles in natural aqueous matrices funded by the CSIR-Physical Sciences, Chennai, and Government of India. My current research interests include Big Data, Machine Learning, Deep learning, and IoT.

Ayesha Jahangir pursuing Bachelor of Technology in Computer Science and Engineering in Vel Tech Rangarajan Dr. Sagunthala R & D Institute of Science and Technology, Chennai, India. Her area of interest is Machine Learning and Natural Language Processing.

Arti Jain is working as Assistant Professor (Sr. Grade) in the Department of Computer Science & Engineering at Jaypee Institute of Information Technology, Noida, India. She is having academic experience of 18 years. She did her PhD in Computer Science & Engineering (2019) from JIIT Noida. She is member of IEEE, INSTICC, IAENG, IASSE, IFERP, and Life Member of TERA. She has more than 20 research papers in peer-reviewed International Journals, Book Chapters, and International Conferences. She has supervised one M.Tech Thesis and around 100 B.Tech projects. She is reviewer of reputed and peer-reviewed International Journals- Taylor & Francis, IGI Global, Wiley, TISA and Inderscience. She is TPC member and reviewer of several International Conferences- SCES, ICDSAA, UPCON 2020, CONFLUENNCE 2020, BigDML, ComITCon 2019. She is editorial board member of American Journal of Neural Networks and Applications.She is special session organizer in International Conference on Innovative Computing and Communication (ICICC 2020), India. Her research interest includes Natural Language Processing, Machine Learning, Data Science, Deep Learning, Social Media Analysis and Data Mining.

Rose Bindu Joseph is currently working as an Associate Professor in the Department of Mathematics at Christ Academy Institute for Advanced Studies, Bangalore. She received her Ph.D in Mathematics from VIT University, Vellore in the field of Interval Type-2 Fuzzy Theory. She has qualified NET for lectureship by CSIR-UGC. She holds a Master's degree and bachelor's degree in Mathematics from Mahatma Gandhi University, Kerala. She has more than 14 years of experience in academia and research. She has published more than 14 research papers in Scopus indexed journals and presented papers in many international conferences. Her research interests include fuzzy theory, machine learning, soft computing and artificial intelligence.

Amandeep Singh K. has completed B.E in Computer Science Engineering in Saveetha School of Engineering, M.Tech Computer Science Engineering in SRM University and pursuing research in the domain of IoT Security in the Department of Computer Science and Engineering of ABET accredited Dr. M.G.R. Educational and Research Institute, Deemed to be University with Graded Autonomous Status, Chennai, India. He also published six papers in International journals, applied for the funded proposal to government funding Agencies.

Meena K. is currently working as Professor in the Department of Computer Science & Engineering at Vel Tech Rangarajan Dr.Sagunthala R&D Institute of Science and Technology. She received her Ph.D Degree in Computer Science from the Department of Computer Science and Engineering, Manonmaniam Sundaranar University, India in 2014. She received her B.E- Electronics and Communication Engineering and M.E- Computer Science and Engineering Degree from the above mentioned University in 2002 and 2009 respectively. She has more than 16 years of research experience in Biometrics, computer vision, Machine Learning and pattern recognition. She has published more than 50 research articles in reputed International/ National Journals and conference. She has published 10 papers in SCI indexed journals including IET Computer Vision an IETE Journal of research. She is also the reviewer in reputed journals such as Institute of Engineering and Technology (IET), Inderscience and Institution of Electronics and Telecommunication Engineers (IETE). She is an Active Life member of IEEE, CSI, ISTE, ICST, IAEST, IAENG, SAISE and IACSIT. She has been invited as a Chief Guest for the Seminars, Resource Person for FDP and Guest Lectures in various Engineering colleges in Tamilnadu, Andrapradesh and Telangana. She is the academic council members for various autonomous institution in Tamil Nadu.

Pamidimukkala Kalpana is a student at Koneru Lakshmaiah Education Foundation. Her area of research is on machine learning, IOT and image processing.

Deepa Kumar is currently working as a Professor in Department of Computer Science and Engineering at Kalasalingam Academy of Research and Education (KARE), Virudhunagar, Tamilnadu, India. She is also serving as Dean, School of Computing. Her research interest includes Optimization Techniques, Network Routing, Distributed Computing, Network Security, Data Analytics, Machine Learning Techniques. She also takes care of KARE ACM student chapter as faculty mentor.

Jothi Kumar is currently working as a Professor in the Department of Computer Science and Engineering at Shadan College of Engineering and Technology, Peerancheru, Hyderabad. He has completed his Doctoral degree from the Noorul Islam Centre for Higher Education, Nagercoil, Tamilnadu in 2017 and M. Tech CSE from the Dr. M.G.R Educational and Research Institute, Chennai. Right now, he has

16 years of teaching experience in various foremost institutions. He has published around 20 Scopus indexed journals, 2 SCI indexed journals and one patent. He is also serving as the guest editor for many Scopus Indexed journals right now.

K. Kishore Kumar is currently working as an Associate Professor, in the department of Mechanical Engineering, Koneru Lakshmaiah Education Foundation. He completed his PhD in the year 2019.His area of interest is Robotics, Mechatronics, AI, ML and IOT.

R. Kumar completed his B.Tech in Electronics and Communication engineering from Madras university in 1991, M.Tech in Instrumentation Engineering from Anna university in 1994 and PhD from National Institute of Technology-Trichy in 2007.He worked as a Lecturer in the Department of Instrumentation and Control Engineering at National Institute of Technology-Trichy from 1996 to 2001 and as a Senior Lecturer in the same Department of National Institute of Technology-Trichy from 2001 to 2006. He subsequently joined Wipro Technologies Chennai and worked as a Consultant from 2006 to 2009 and served as a Senior Consultant from 2009 to 2016. He is currently working as Professor in the Department of Electronics and Instrumentation Engineering, Dean (Faculty Welfare) and Dean (Academic) in National Institute of Technology Nagaland, India since 2016. His areas of interest include Linux System Programming, MEMS, Sensors and Actuators, Wireless Sensor Networks (WSN), Internet of Things and Embedded Systems Design.

Rajeev Kumar received his Ph.D. and M. Tech. degrees in Dept. of Electronics & Communication Engineering (ECE) from National Institute of Technology (NIT), Silchar, Assam, India in 2019 and 2015, respectively. He obtained B.Tech. degree in Dept. of ECE from West Bengal University of Technology (WBUT), Kolkata, West Bengal, India in 2012. He held post as an Assistant Professor in the Dept. of ECE, School of Engineering, Central University of Karnataka (CUK), Kadaganchi, Kalaburagi, Karnataka, India. His current research interests include Cooperative Communications, Wireless sensor Network, Green Technology, Advanced Protocol design, IoT applications for next-generation radio networks, and Optimization Theory. He is Reviewer of IEEE Transactions on Communications, IEEE Access, IET Communications, Wireless Personal Communications. Dr. Kumar is also, selected as an Exemplary Reviewer of National Conference on Communication (NCC) at IIT Hyderabad for 2017. He is also, Reviewer of 26th (Virtual) Annual International Conference on ADCOM at NIT Silchar for 2020.

S. Kranthi Kumar pursued his PhD in Applied Cryptography and Network Security from the Department of CSE, GITAM Deemed to be University, A.P, India. He received his M.Tech, CSE in 2010 from Acharya Nagarjuna University, A.P, and M.Tech, IT in 2015 from JNTU Kakinada, A.P. 15 years of experience in teaching, Industry and Research helped Kranthi to bring in research oriented teaching-learning. His career as a Data Specialist in IBM provided an opportunity to explore Data Ware Housing and Big Data Analytics tools. He has done research on Computer Networks and Security. Currently, Kranthi is carrying out research on Applied Cryptography, Quantum Machine Learning for Large Cloud Data Security and Privacy. His 20 research papers in reputed journals indexed by SCI, WOS, SCOPUS and UGC and 5 patents that address solutions for real time problems in the area of Engineering speaks volumes about his penchant for research. He received a prestigious "Governors National Award for Excellence in Research and Development", GRA approved by Ministry of Science and Technology for the year 2017-18. He is a member of ISTE, CSI, IFERP, and IAENG.

Rashmi Kushwah is working as Assistant Professor (Sr. Grade) in the Department of Computer Science & Engineering at Jaypee Institute of Information Technology, Noida, India. She is having academic experience of 8 years. She did her PhD in Information Technology (2018) from Atal Bihari Vajpayee-Indian Institute of Information Technology and Management, Gwalior. She is member of IEEE, INSTICC, IAENG, and IFERP. She has more than 12 research papers in peer-reviewed International Journals, Book Chapters, and International Conferences. She has supervised three M.Tech Thesis and around 30 B.Tech projects. She is reviewer of reputed and peer-reviewed International Journals- IGI Global, Elsevier and Oxford University Press. She is editorial board member of International Journal of Science, Technology and Society (IJSTS). Her research interest includes IoT, Computer Netwrk, Wireless sensor Network, Mobile adhoc Network, soft computing and Optimization Techniques.

Abirami Manoharan, Ph.D., is currently working as an Assistant Professor in the Department of Electrical and Electronics Engineering, Government College of Engineering, Srirangam. Her research contributions include 10 SCI articles. She is a Senior Member in IEEE (SM 93333628) and also a Member in Institution of Engineers (India) (M 1558689).

Chokri Mhiri is head of the department of Neurology, Habib BOURGUIBA university hospital, CP 3029 - Sfax, TUNISIA. He is currently the president of the Tunisian Association of Neurology and past President of the Pan Arab Union of the Neurological Societies (PAUNS) from 2017 to 2019. After his medical studies in Sfax faculty of medicine (1979-85), he was resident at the national institute of neurology in Tunis between 1986 and 1988. He obtained a research master in neurosciences in 1990 from the university Pierre Marie Curie in Paris (France). He was assistant-professor than professor of neurology at Habib Bourguiba university hospital in Sfax. In 1996, he was nominated as head of the department of neurology in Sfax faculty of medicine. He assumed several responsibilities in his hospital (President of the medical board of Habib BOURGUIBA University hospital (2008-2013); Director of Clinical Investigation Center (CIC) of Habib BOURGUIBA University hospital since 2015…). He is member of scientific board, Sfax Faculty of Medicine since 2001 and President of Tunisian Association of Neurology between 2002 and 2006. His fields of interest are inflammatory diseases of CNS, Neurogenetics and Cerebrovasular disorders. He participated to the publication of about one hundred peer reviewed scientific papers.

Zahraa Tagelsir Mohammed, receive her B.Sc and M.Sc degrees in telecommunications engineering from Khartoum University , Sudan in 2014 and 2018 receptively. She is a lecturer in Electrical and Electronic Red Sea University. Her research interests in software defined radio (SDR), intelligent wireless communications, and IoT.

Sunanda N. is working as Assistant Professor in the Department of Computer Science Engineering from KLEF, Guntur, Andhra Pradesh. She has 17 years of teaching experience in reputed institutions in Telangana, A.P and Abroad. She is currently pursuing Ph.D from "Rayalaseema University", Kurnool. Her research areas of interests are cloud computing and security, network security.

Senthil Murugan Nagarajan is currently working as an Assistant Professor Senior at School of Computer Science and Engineering, VIT-AP University, Amaravati, Andhra Pradesh. He is expertise in various specializations of computer science and published papers in high reputed journals.

K. Nagi Reddy is currently heading the IT Department at Lords Institute of Engineering and Technology, Hyderabad. Prior to his appointment as a Professor at Lords college in 2018, he worked as an Assistant Professor and Associate Professor at Vasavi College of Engineering, Hyderabad and as an Associate Professor at RRS College of Engineering and Technology, Hyderabad. Dr.K.Nagi Reddy has completed his AMIE in ECE from the Institution of Engineers (India) in 1996 and M.Tech in Computer Science from JNTU Hyderabad in 2000. He also completed AMIE in Computer Science and Engineering from the Institution of Engineers (India) in 2003. He got his Ph.D degree from Rayalaseema University, Andhra Pradesh in 2017. Dr.K.Nagi Reddy has presented his work in many International and National Journals and Conferences. He is active member of professional bodies like AMIE, CSI, IAENG and ISTE. He organized several workshops and conferences for teachers of Engineering colleges.

Adada Neelothpala is a student in the ECE department of Koneru Lakshmaiah Education Foundation. Adada's area of interest is on IOT and AI.

V. R. Niveditha has completed B.E in Computer Science Engineering in PB college of Engineering, M.Tech Information Security and Cyber forensics and pursuing Research in the domain of Mobile Security in the Department of Computer Science and Engineering of ABET accredited Dr. M.G.R. Educational and Research Institute, Deemed to be University with Graded Autonomous Status, Chennai, India. She also published 12 papers in International journals, six papers in National conferences.

Dilip Kumar P. is a student of the Koneru Lakshmaiah Education Foundation. His area of research is on Robotics and artificial intelligence.

Janani P. received her Bachelor's degree in Electronics and Communication Engineering (B.E., E.C.E) from Peri Institute of Technology, Chennai, in the year 2015 and received her Master's in Technology (M.Tech, VLSI & Embedded system Design) in the year 2018 from Veltech Rangarajan Dr Sakunthala R and D Institute of Science and technology, Chennai. She has one-year industrial experience from CDAC Pune and 1 year of teaching experience. She has published scientific papers in international journals and Conferences, which also includes an IEEE conference. During hier stint as assistant professor, she has received various appreciations for motivating students to do industrial projects. She is an active member of professional body like IAENG . Her research area includes Bio-medical body area network, Internet of Things and Embedded System Design.

Visu P. is currently working as a professor in the department of Computer Science and Engineering in Velammal Engineering College, Chennai, Tamil Nadu, India. He obtained his B.Tech degree in Information Technology from Madras University, Chennai, Tamil Nadu, India, and M.E degree in Computer Science & Engineering from Sathyabama University, Chennai, Tamil Nadu, India, and Ph.D from Veltech Rangarajan Dr.Sakunthala R&D Institute of Science and Technology, Chennai, Tamilnadu, India. He is member of IEEE, ACM, CSI and life time member of ISTE. He published 6 articles in Wikipedia, 5 paper in SCI Indexed journals, 26 papers in International Journals, 10 Papers in national Journals and 58 papers in National and international conferences. He is also serving as the guest editor and reviewer for many SCI and Scopus Indexed journals. His research interests include Machine Learning, Computer Networks, Mobile Ad Hoc Networks and Swarm Intelligent Techniques.

Ibrahim Patel is working as Associate Professor in ECE department, B V. Raju Institute of Technology, Vishnupu., Narsapur, Medak (Dist), Andhra Pradesh, India. He has received B. Tech, M. Tech Degree in Biomedical Instrumentation and Ph.D from Andhra University. He is having 28 years of teaching and research experience and published 97 research papers in the International conference & journals, he has received 3 best paper award and his main research interest includes voice to sign language.

Archana Purwar has been working as an Assistant Professor in the Department of Computer Science Engineering and Information technology at Jaypee Institute of Information Technology, Noida, India. During her teaching career of more than 14 years, she has taught subjects such as database systems, software engineering, object-oriented programming, computer architecture and organization, data mining, and many more. Her area of interest lies in data mining, information retrieval, and soft computing. She has guided many graduate and undergraduate students.

Jayashree R., PhD, is working as an assistant professor at the College of Science and Humanities, SRM Institute of Science and Technology, Kattankulathur Campus, Chennai, India. She has served as coordinator for conducting various events in the National Level Symposium and Workshops organized by SRM IST. She published many papers in International and National Journals. She has participated and presented papers in International and National Conferences. She is a reviewer in Scopus and Web of Science Indexed journals: IEEE Access, IEEE Transactions on Computational Social Systems and The Journal of SuperComputers. She won a Gold Medal for a paper entitled "A splay tree-based approach for a recommending system in ranking a community-driven question answering website", which was presented under the Faculty category for Research Day 2019. She is an Editorial board member of the journal "American Journal of Education and Information Technology". Her area of interests includes Machine Learning, Deep Learning, and Educational Data Mining.

Kanniga Devi R. received her Bachelor of Engineering in Computer Science and Engineering from Madurai Kamaraj University, Madurai in the year 2001 and Master of Engineering from Anna University, Chennai in the year 2008. She pursued her doctoral program at Kalasalingam Academy of Research and Education, Tamilnadu, India in the year 2018. She is a member of CiE, CSI, IAEng, ISTE and IEEE. She has 17 years of teaching experience. She is the Principal Investigator of a DST project. She has published 30 research papers in International Journals and Conferences. She got Science Academy Summer Research Fellowship and Tamilnadu Young Scientist Fellowship. Her research interests include Cloud computing, Graph theory applications in Cloud Computing, and Machine Learning. She is currently working as an Associate Professor in the Department of Computer Science and Engineering at Kalasalingam Academy of Research and Education, Tamil Nadu, India.

D. Rajalakshmi is pursuing research in the Department of Computer Science and Engineering, Vel Tech Rangarajan Dr. Sagunthala R&D Institute of Science and Technology, Chennai, Tamil Nadu,India. She has completed her B.Tech degree in Information Technology from M.A.M College of Engineering, Anna University, Tamil Nadu, India and completed her M.E degree in Computer Science and Engineering from Sudharsan Engineering College, Anna University, Tamil Nadu, India. She having total of 14 years of teaching experience in engineering colleges. She worked in Shri Angalamman College of Engineering and Technology, Trichy, Tamil Nadu, around 9 years and presently she is working as an Assistant Professor in the Department of Computer Science and Engineering in Sri Sairam Institute of Technology,

Chennai, Tamil Nadu,India. She has published 15 papers in various National and International journals. She is a professional body member of ISTE, CSI and IAENG. She has attended 20 FDP/workshop/training programmes. Her research interests include MANETs and Network Security.

M. V. Ramana Rao is a Research Professional in the Computer Science & Engineering at Osmania University, Hyderabad, India. He obtained his Ph.D in Computer science & Engineering from Osmania University, Hyderabad, India. He received his B.Tech in Electrical Engineering from Regional Engineering College, Warangal and M.Tech in Computer Science & Engineering from JNTUH, Hyderabad, India. He has published his research contributions in peer reviewed International journals and conferences. His Research interests include Wireless Sensor Networks, Social networking and IOT.

A. Ratna Raju obtained his Bachelor's degree in Computer Science and Information Technology from Jawaharlal Nehru Technological University Hyderabad. Then he obtained his Master's degree in Computer Science and Engineering from Annamalai University Tamilnadu, India. His specializations include Computer Vision, Pattern Recognition, and Image Processing. His current research interests are Deep Learning, Network Security, Soft Computing.

B. Veera Sekhar Reddy obtained B.Tech (CSE) and M.Tech (CSE) from JNTU Anantapur. He is currently pursuing a Ph.D. in the Department of Computer Science and Engineering from MLR Institute of Technology (Research center Under JNTU). His research interests are Machine Learning and Natural Language processing.

S. Radha Rammohan is presently working as a Professor in the Department of Computer Applications, Dr. M. G. R. Educational and Research Institute, Deemed to be University with Graded Autonomy Status, Chennai, Tamil Nadu, India. He has 20 years of Experience in Teaching in India and Abroad, 7 Years in Industry. He has CCNA certification and was a Cisco Certified Authorized Instructor. Editorial Board Member for IJCA Journal. He has completed his PhD in 2010 and has published 26 research papers in International refereed Journals and three book chapters also published. He is a Member of the Board of Studies, Member of the Doctoral Committee. Also, an Expert Consultant for two funded Projects of the Ministry of Higher Education, Sultanate of Oman. He has 6 Patent published,5 Books Published . His area of research is Mobile Adhoc networks, IoT, Cloud Computing, Blockchain, and Deep Learning. He guided 1 Research Scholar on Mobile Adhoc Networks. He is currently guiding 5 Research Scholars in the domains, as mentioned earlier.

Magesh S. is a notable academician and a passionate entrepreneur. He commenced his academic career as Lecturer and after that elevated to the level of Assistant Professor, Associate Professor, and Head in the Department of Computer Science and Engineering with his distinguished career spanning in engineering institutions over a period of 15 years and 9 years of Corporate Experience. He has published 15 refereed International Indexed journals which include ESCI, SCOPUS, WoS, Compendex (Elsevier Engineering Index) to his credit, published 2 Patents and a Book with ISBN. He received the Distinguished Innovator & Edupreneur Award in the year 2017 and the Lifetime Achievement Award in the year 2019. Presently he serves as the Chairman & Director of Magestic Technology Solutions (P) Ltd, CEO of Maruthi Technocrat EServices and Chief Editor & Director, Jupiter Publications Consortium, Chennai, Tamil Nadu, India.

Saravanan S. received the B.E. degree in Electrical and Electronics Engineering from Arulmigu Kalasalingam College of Engineering, Virudhunagar District, affiliated to Madurai Kamaraj University, Madurai, Tamilnadu, India, in 1999 and M.E. degree in Power Systems Engineering from Thiagarajar College of Engineering, Madurai, affiliated to Anna University, Chennai, Tamil Nadu, India, in 2007. He is awarded doctoral degree from Kalasalingam University, Krishnankoil-626126, Tamilnadu, India, in 2015. He is presently working as a Professor in the Department of Electrical and Electronics Engineering, B V Raju Institute of Technology, Narsapur, Telangana India. Before to that, he worked as an Associate Professor in Department of Electrical and Electronics Engineering, Kalasalingam University, Tamil Nadu. He has published one text book titled Electrical Distribution System, NOVA publication, Delhi, India. He published more than 45 research articles in various renowned international journals/conferences and five patents. His research interests include the power system load forecasting, Generation Expansion Planning, Applications of soft computing technics in power systems and grid connected renewable energy resources.

Selva S. received her Bachelor of Engineering in Computer Science and Engineering from VPMM Engineering College for women, Srivilliputhur in the year 2010 and Master of Engineering from Kalasalingam University, Srivilliputhur in the year 2012. She had 5 years of teaching experience. She is currently pursuing her doctoral program at Kalasalingam Academy of Research and Education, Tamilnadu, India from 2019. Her research interests include Machine Learning, Deep Learning, and Wireless Sensor Networks.

Rashid A. Saeed received his PhD in Communications and Network Engineering, Universiti Putra Malaysia (UPM). Currently he is a professor in Computer Engineering Department, Taif University. He is also working in Electronics Department, Sudan University of Science and Technology (SUST). He was senior researcher in Telekom Malaysia™ Research and Development (TMRND) and MIMOS. Rashid published more than 150 research papers, books and book chapters on wireless communications and networking in peer-reviewed academic journals and conferences. His areas of research interest include computer network, cognitive computing, computer engineering, wireless broadband, WiMAX Femtocell. He is successfully awarded 3 U.S patents in these areas. He supervised more 50 MSc/PhD students. Rashid is a Senior member IEEE, Member in IEM (I.E.M), SigmaXi, and SEC.

Elmustafa Sayed Ali Ahmed received his M.Sc. in Electronics & Communication Engineering, Sudan University of Science & technology in 2012 and B.Sc. in 2008. Worked (former) as a senior engineer in Sudan Sea Port Corporation (5 years) as a team leader of new projects in wireless networks includes (Tetra system, Wi-Fi, Wi-Max, and CCTV). He is a senior lecturer in Electrical and Electronics Engineering Department in the Red Sea University. Currently he is a head of marine systems department in Sudan marine industries. Elmustafa published papers, and book chapters in wireless communications, computer and networking in peer reviewed academic international journals. His areas of research interest include, routing protocols, wireless networks, LPWAN and IoT. He is a member of IEEE Communication Society (ComSoc), International Association of Engineers (IAENG), Six Sigma Yellow Belt (SSYB), and Scrum Fundamentals certified (SFC).

Vaithyasubramanian Subramanian started his teaching carrier in the year 2002. His interested areas of research are formal languages, Petri net, Mathematical Modelling, Information security and CAPTCHAs. He has published various articles in his fields of research in reputed journals.

P. V. Praveen Sundar is currently working as an Assistant Professor in Computer Science Department, Adhiparasakthi College of Arts and Science. Previously Worked as a Director & HOD of PG Dept of Computer Applications (MCA) in Shanmuga Industries Arts and Science College, Tiruvannamalai. He was Published 13 Journals and presented a paper in 17 International Conferences. He contributed three Book chapters. His Area of Research includes e-Learning, Data Mining, Data Retrieval and Data Mining, Data Science, Databases, Distance Learning, e-Learning, e-Learning Organisational Issues, e-Learning Tools, Machine Learning, Supervised and Unsupervised Learning, Virtual Learning Environments, Web-based Learning Communities. He has been Acting as a Editorial Board Member in Various Journals. He has got Six International Academic Awards.

B. Swapna is currently working as Assistant Professor, ECE Department in Dr. M.G.R. Educational and Research Institute (Deemed to be University) Chennai, India. She is doing ph.D in Internet of things. Master's degree with honors in Applied Electronics secured Anna University 14th rank from SBC Engineering College, Arani. Bachelor's degree in Electronics and Communication Engineering from SBC Engineering College, Arani, India. She has 8 years of teaching experience. She has Published more than 10 articles in scopus indexed and web of science journals. She has attended and organized Seminar and workshops related to her area of research. Her areas of interest include VLSI, Embedded systems, IoT and Electronic circuits, Machine learning, Artificial Intelligence, etc.,

Abhishek Swaroop did his B.Tech. From G.B. Pant univ. Pantnagar, M.Tech. From Punjabi Univ. Patiala, he has completed his PhD from NIT Kurushetra. He has 20 years of teaching experience and 8 years Industrial experience. He has published more than 60 papers out of which 7 are SCI and 25 in Scopus. One Ph.D has been submitted under his supervision from NIT Kurukshetra and currently supervising one Ph.D. Scholar each from IIT Dhanbad and AKTU Lucknow.

Adilakshmi Thondepu is Professor & Head in the Department of Computer Science & Engineering at Vasavi College of Engineering, Hyderabad, India. She has vast academic and Research experience in various areas of Computer Science & Engineering. She obtained her Ph D in Computer science from Central University of Hyderabad, India. She received her BE in Electronics & Communication Engineering and M Tech Computer Science & Engineering. She has published her research contributions in peer reviewed international journals and conferences. Her Research interests include Data Mining, Wireless Sensor Networks and Artificial Intelligence.

Omar Trigui received his Engineer degree in Real-time computing from the Higher Institute of Applied Sciences and Technology of Sousse in 2010, his MS degree in Automatic Control and Industrial Computing from the National Engineering School of Sfax in 2012 and his PhD in Electrical Engineering from the National Engineering School of Sfax in 2017. He is a member of Advanced Technologies for Medicine and Signals (ATMS) Research Group, in Tunisia. His research interests include imaging, biomedical and EEG signal processing, and brain computer interface.

Gurkan Tuna is currently a Professor at the Department of Computer Programming of Trakya University, Turkey. He has authored several papers in international conference proceedings and refereed journals. He has been actively serving as a reviewer for a number of prestigious SCI-Expanded journals. His current research interests include wireless sensor networks, autonomous systems, and machine learning.

Akshay K. Uday has completed B.Tech in Civil Engineering in accredited Dr. M.G.R. Educational and Research Institute, M.Tech in Structural Engineering in PES University, Banashankari, Bengaluru, Karnataka and pursuing research in the domain of Basalt Retrofitting in Repair And Rehabilitation and Structural Health Monitoring in the Department of Civil Engineering accredited Dr. M.G.R. Educational and Research Institute, Deemed to be University with Graded Autonomous Status, Chennai, India. he also published papers in International reputed journals, and in National conferences and. he has published one patent and applied for the funded proposal to government funding Agencies.

D. Usha is currently working as Associate Professor in Department of Computer Science and Engineering, Dr. M.G.R. Educational and Research Institute, Chennai. She has 13 years of teaching experience. She completed her Doctorate of Philosophy in the year 2017 from Hindustan University. She has published more than 25 papers in various Conferences and Journals. Her area of research is Data Analytics and Data Mining. She is a member in IEEE and CSI. She is also playing the role of reviewer in International Journals.

Dhilip Kumar V. was awarded PhD at North Eastern Hill University (A Central University of INDIA) in 2018. Presently, he is working as an associate professor in the Department of computer science and engineering at Vel Tech Rangarajan Dr. Sagunthala R&D Institute of Science and Technology, Chennai. He has more than 9 years experience of teaching as well as research. He did his B.Tech Information Technology and M.E. Computer science engineering under Anna University, Chennai. He has published various international journals and international conferences in the field of wireless communication and vehicular communication. He is an editorial board member and reviewer of various international journals and conferences. His area of interest is wireless communication, Software defined network, Internet of things, Machine learning, etc.

Muthukumaran V. was born in Vellore, Tamilnadu, India, in 1988. He received the B.Sc. degree in Mathematics from the Thiruvalluvar University Serkkadu, Vellore, India, in 2009, and the M. Sc. degrees in Mathematics from the Thiruvalluvar University Serkkadu, Vellore, India, in 2012. The M. Phil. Mathematics from the Thiruvalluvar University Serkkadu, Vellore, India, in 2014 and Ph.D. degrees in Mathematics from the School of Advanced Sciences, Vellore Institute of Technology, Vellore in 2019. He has published more than 24 research articles in peer-reviewed international journals. He also presented more than 10 papers presented in national and international conferences. Best Researcher Award in every year in Vellore Institute of Technology, Vellore, India.

Vinothkumar V. is an Associate Professor at the Department of Computer Science and Engineering, MVJ College of Engineering, India. He is a highly qualified individual with around 8 years of rich expertise in teaching, entrepreneurship, and research and development with specialization in computer science engineering subjects. He has been professional society member of ISTE, IACIST and IAENG. He has co-chaired major Conferences Program Committees such as: ICACB'18, ICAIIS'19 etc. He has filed 3 patents and approved funding in CSIR. He has published as over than 25 papers in refereed journals, book chapter and conferences. He has demonstrable experience in leading large-scale research projects and has achieved many established research outcomes that have been published and highly cited in many significant Journals and Conferences. He is the Associate Editor of International Journal of e-Collaboration (IJeC) and Editorial member of International Journal of Intelligent Information Systems. His research interest includes Wireless networks, IoT and Machine learning. He is currently editing 2 books of IGI global. He is the guest editor of International Journal of e-Collaboration (IJeC)-IGI Global. International Journal of Pervasive Computing and Communications-Emerald Publishing, International Journal of System of Systems Engineering, International Journal Speech Technology (IJST)-Springer, International journal of Information Technology and Web Engineering (IJITWE), International Journal of Machine Learning and Computing (IJMLC), International Journal of Cloud Computing (IJCC), International Journal of Information Quality (IJIQ), Journal of Computational and Theoretical Nanoscience and International Journal of Intelligent Enterprise (IJIE).

Niveditha V. R. has completed B.E in Computer Science Engineering in PB college of Engineering, M.Tech Information Security and Cyber forensics and pursuing research in the domain of Mobile Security in the Department of Computer Science and Engineering of ABET accredited Dr. M.G.R. Educational and Research Institute, Deemed to be University with Graded Autonomous Status, Chennai, India. She also published fifteen papers in International reputed journals, six papers in National conferences and applied for the funded proposal to government funding Agencies. She has published four books and two patents.

Ramesh Vatambeti received his B.Tech from Sri Venkateswara University, Tirupati in Computer Science and Engineering and M.Tech in IT and PhD in CSE from Sathyabama University, Chennai. He has around 16 years of teaching experience from reputed Engineering Institutions. He works as Associate Professor in the Department of Computer Science and Engineering at CHRIST (Deemed to be University), Bangalore, India. He has published 3 books, 3 book chapters and more than 60 papers in refereed journals and conference proceedings. He is acting as Editorial Board Member for 6 International refereed journals. He is the reviewer for several refereed International Journals and acted as Session Chair and technical committee member for several International Conferences held in India and abroad. He is guiding 6 scholars for the award of Ph.D. and supervised more than 20 M.Tech and B.Tech projects. He received 7 awards from National and International organizations for teaching and research contributions. His research interests include Computer Networks, Mobile Ad-Hoc and Sensor Networks and Internet of Things.

M. Gokul Venkatesh is a Doctor by Profession in Hyderabad, India. He obtained his MBBS from Sidhartha Medical College, N.T.R University of Health Sciences, Vijayawada, A.P, India. He obtained COVID Warrior award from Govt. of A.P on his contributions to treat COVID patients. His Research interests include Community Medicine & Wearable Body Sensor Networks and Social networking.

Arun Yadav is working as assistant professor in National Institute of Technology, Hamirpur (H.P).

Divakar Yadav is currently working as Associate Professor in the Department of Computer Science & Engineering at National Institute of Technology, Hamirpur (HP), India. Prior to join this institute, he worked at Madan Mohan Malaviya University of Technology, Gorakhpur (UP) as Associate Professor and Jaypee Institute of Information Technology, Noida as Assistant as well as Associate Professor. He did his undergraduate (B. Tech.) in Computer Science & Engineering in 1999 from IET, Lucknow, Postgraduate (M.Tech.) in Information Technology in 2005 from Indian Institute of Information Technology, Allahabad and PhD in Computer Science & Engineering in 2010 from Jaypee Institute of Information Technology, Noida. He also worked as Post Doctoral Fellow at University of Carlos-III, Madrid (Spain) between 2011-2012. He supervised 4 PhD theses, 22 M.Tech. dissertations and many undergraduate projects. He published two books, 9 book chapters and more than 85 research articles in reputed International Journals and Conference Proceedings. His area of research includes information Retrieval and Machine Learning. He is senior member of IEEE.

Wassim Zouch was born in Sfax, Tunisia. He received his Engineering degree, his MS, and his PhD in Electrical Engineering from the National Engineering School of Sfax, Tunisia, in 2004, 2005, and 2010, respectively. He is a member of Advanced Technologies for Medicine and Signals (ATMS) Research Group, in Tunisia. Actually, he is with the Electrical and Computer Engineering Department, Faculty of Engineering, King Abdulaziz University, Jeddah, Saudi Arabia. His research interests include Microprocessor/Microcontrollers, Medical/Microwaves imaging, biomedical and EEG signal processing, and Brain Computer Interface.

Index

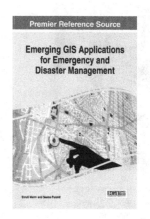
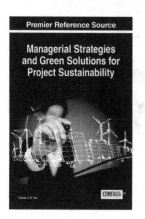

www.igi-global.com

Publisher of Peer-Reviewed, Timely, and
Innovative Academic Research Since 1988

IGI Global's Transformative Open Access (OA) Model:
How to Turn Your University Library's Database Acquisitions Into a Source of OA Funding

Well in advance of Plan S, IGI Global unveiled their OA Fee Waiver (Read & Publish) Initiative. Under this initiative, librarians who invest in IGI Global's InfoSci-Books and/or InfoSci-Journals databases will be able to subsidize their patrons' OA article processing charges (APCs) when their work is submitted and accepted (after the peer review process) into an IGI Global journal.

How Does it Work?

Step 1: **Library Invests in the InfoSci-Databases:** A library perpetually purchases or subscribes to the InfoSci-Books, InfoSci-Journals, or discipline/subject databases.

Step 2: **IGI Global Matches the Library Investment with OA Subsidies Fund:** IGI Global provides a fund to go towards subsidizing the OA APCs for the library's patrons.

Step 3: **Patron of the Library is Accepted into IGI Global Journal (After Peer Review):** When a patron's paper is accepted into an IGI Global journal, they option to have their paper published under a traditional publishing model or as OA.

Step 4: **IGI Global Will Deduct APC Cost from OA Subsidies Fund:** If the author decides to publish under OA, the OA APC fee will be deducted from the OA subsidies fund.

Step 5: **Author's Work Becomes Freely Available:** The patron's work will be freely available under CC BY copyright license, enabling them to share it freely with the academic community.

Note: This fund will be offered on an annual basis and will renew as the subscription is renewed for each year thereafter. IGI Global will manage the fund and award the APC waivers unless the librarian has a preference as to how the funds should be managed.

Hear From the Experts on This Initiative:

"I'm very happy to have been able to make one of my recent research contributions *freely available* along with having access to the *valuable resources* found within IGI Global's InfoSci-Journals database."

— **Prof. Stuart Palmer,**
Deakin University, Australia

"Receiving the support from IGI Global's OA Fee Waiver Initiative *encourages me to continue my research work without any hesitation*."

— **Prof. Wenlong Liu,** College of Economics and Management at Nanjing University of Aeronautics & Astronautics, China

For More Information, Scan the QR Code or Contact:
IGI Global's Digital Resources Team at eresources@igi-global.com.

Printed in the United States
by Baker & Taylor Publisher Services